POINT DEFECTS IN SOLIDS

Volume 1

General and Ionic Crystals

POINT DEFECTS IN SOLIDS

Volume 1: General and Ionic Crystals
Volume 2: Defects in Semiconductors
Volume 3: Defects in Metals

POINT DEFECTS IN SOLIDS

Edited by

James H. Crawford, Jr.

Chairman, Department of Physics
University of North Carolina
Chapel Hill, North Carolina

and

Lawrence M. Slifkin

Department of Physics
University of North Carolina
Chapel Hill, North Carolina

Volume 1
General and Ionic Crystals

PLENUM PRESS • NEW YORK–LONDON • 1972

Library of Congress Catalog Card Number 72-183562
ISBN 0-306-37511-7

A Division of Plenum Publishing Corporation
227 West 17th Street, New York, N.Y. 10011

United Kingdom edition published by Plenum Press, London
A Division of Plenum Publishing Company, Ltd.
Davis House (4th Floor), 8 Scrubs Lane, Harlesden, London, NW10 6SE, England

Printed in the United States of America

CONTRIBUTORS TO THIS VOLUME

Frederick C. Brown
Department of Physics
University of Illinois
Urbana, Illinois

A. D. Franklin
Institute for Materials Research
National Bureau of Standards
Washington, D.C.

Robert G. Fuller
Department of Physics
University of Nebraska
Lincoln, Nebraska

B. Henderson
Department of Physics
University of Keele
Staffordshire, England

A. E. Hughes
Materials Development Division
Atomic Energy Research Establishment-Harwell
Berkshire, England

M. N. Kabler
Solid State Division
Naval Research Laboratory
Washington, D.C.

Clifford C. Klick
Naval Research Laboratory
Washington, D.C.

A. S. Nowick
Henry Krumb School of Mines
Columbia University
New York, New York

W. A. Sibley
Department of Physics
Oklahoma State University
Stillwater, Oklahoma

E. Sonder
Solid State Division
Oak Ridge National Laboratory
Oak Ridge, Tennessee

PREFACE

Crystal defects can no longer be thought of as a scientific curiosity, but must be considered an important aspect of solid-state science. This is largely because many of the more interesting properties of crystalline solids are disproportionately dominated by effects due to a tiny concentration of imperfections in an otherwise perfect lattice. The physics of such lattice defects is not only of significance in a great variety of applications, but is also interesting in its own right. Thus, an extensive science of point defects and dislocations has been constructed during the past two and a half decades.

Stimulated by the technological and scientific interest in plasticity, there have appeared in recent years rather a large number of books dealing with dislocations; in the case of point defects, however, only very few broad and extensive treatments have been published. Thus, there are few comprehensive, tutorial sources for the scientist or engineer whose research activities are affected by point defect phenomena, or who might wish to enter the field. It is partially to fill this need that the present treatise aims.

Unlike the typical textbook, this treatise consists of chapters each of which has been written by a different author or set of authors. The reader pays for such a structure, of course, in that the approach, style, notation, and degree of sophistication are not as uniform as in the case of a single group of authors. On the other hand, the choice of different authors for each topic makes it possible to offer a range of expertise far more extensive than could likely be provided otherwise. Moreover, the greater diversity of points of view and the timeliness possible in such a cooperative venture further favor the multiauthor scheme. We hope that the reader will, therefore, willingly accept the inhomogeneity in style as a fair trade for the insights proffered by authors whose main research interests coincide with the fields about which they write. And if some lack of unanimity is detected with regard to yet-controversial questions, then perhaps this is the best way, after all, of presenting such ideas.

This first volume opens with a general thermodynamic and statistical description of point defects in crystals. The remainder of the book is concerned particularly with ionic crystals. Chapters 2 and 3 treat ion transport phenomena; the next four chapters discuss the extensive field of color centers; and the last chapter deals with electron transport, especially the polaron. Certain special topics pertaining more or less directly to ionic solids will be discussed in a later volume.

Some of the early chapters of this volume are quite general, while the later ones become rather detailed. Each is designed to be independently intelligible, so that the reader with some background knowledge and with specific interests need not work through prior chapters which he might consider to be peripheral.

The editors realize that not every topic of legitimate interest will be included in the complete treatise. Because of limitations on time and length, choices have had to be faced, and not everybody will agree with those that were made. While the choice of topics is inevitably conditioned in some measure by our professional preferences and prejudices, we have striven to present a reasonable approximation to a balanced treatment of the science of defects in crystals. We can only hope that we have been partially successful.

J. H. Crawford, Jr.
L. M. Slifkin

March, 1972

CONTENTS

Chapter 1. Statistical Thermodynamics of Point Defects in Crystals

A. D. Franklin

Chapter 4. Defect Creation by Radiation in Polar Crystals

E. Sonder and *W. A. Sibley*

Chapter 5. Properties of Electron Centers

Clifford C. Klick

Chapter 6. Hole Centers in Halide Lattices

M. N. Kabler

Chapter 7. Color Centers in Simple Oxides

A. E. Hughes and B. Henderson

Chapter 8. Conduction by Polarons in Ionic Crystals

Frederick C. Brown

Chapter 1

STATISTICAL THERMODYNAMICS OF POINT DEFECTS IN CRYSTALS

A. D. Franklin

Institute for Materials Research
National Bureau of Standards
Washington, D.C.

1. INTRODUCTION

Point defects play an important part in determining the physical properties of most crystalline substances, most notably those controlling the transport of matter and the properties that stem from it. They also strongly influence the resistivity of metals by scattering conduction electrons, the low-temperature thermal conductivity of all crystalline solids by scattering phonons, the electronic conduction and related properties of semiconductors by acting as donors or acceptors, and the optical properties of ionic solids by introducing electron states with optical transitions, to name a few examples. They are the primary principal product in irradiation of crystals by energetic particles, and their subsequent behavior very much determines the nature of the damage suffered by the crystal during irradiation and the recovery processes that occur during subsequent annealing.

Even a crystal of high purity under conditions of no irradiation contains point defects in thermal equilibrium. Some lattice sites are vacant, and some atoms are displaced from their normal lattice sites into interstitial positions or onto "wrong" lattice sites. For stoichiometric compounds of high purity, the concentrations of these point defects are very low, even at temperatures up to the melting point. A meaningful model, then, is to consider the crystal as a solvent containing a very dilute solution of simple, individual vacancies and interstitials. Long-range interactions among the defects and with impurity atoms, and short-range interactions that produce pairs or other clusters can be introduced in a first-order approximation.

This simple theory fails when concentrations become too high, or when interactions among the defects are too strong. Such conditions occur when, e.g., impurity levels are high, when the compound is nonstoichiometric, or under conditions of intense irradiation. However, a remarkable amount of defect-controlled behavior of most crystals can be understood with just this simple, first-order theory.

Since the theory is essentially a thermodynamic one, the partial molar thermodynamic properties of the defects themselves, treated as constituents, will be of paramount importance. In particular, the key quantities are the enthalpy and entropy required to form a given defect in an otherwise perfect crystal (the formation parameters), the enthalpy and entropy required to separate two paired defects and remove them infinitely far apart in an otherwise perfect crystal (the binding parameters), and the enthalpy and entropy required to move a defect from a lattice site (which includes the interstitial sites) to a position atop the potential energy barrier impeding its motion to the next site (the motion parameters).

In this chapter, we review the major aspects of this statistical thermodynamic theory of dilute solutions of point defects (Section 2), and then examine (Section 3) the experimental determination of the thermodynamic parameters. This latter discussion deals with noble gas solids (Section 3.1), metals (Section 3.2), semiconductors (Section 3.3), and ionic solids (Section 3.4). In this review, the author has hoped to provide a critical guide to recent attempts at determining the parameters, indicating, where possible, something about the reliance that might be placed on the results. The review should also serve as a survey of the experimental methods available for each of the listed classes of materials.

1.1. Symbols Used

The notation developed by Kroger and Vink[1] will be used to identify point defects. M and X represent metal and nonmetal; V and I represent vacancy and interstitial. Superscripts are used to describe the charge state of the defect; x denotes an effectively neutral defect, bearing the same charge as the lattice atom it replaces; a prime denotes one negative effective charge (i.e., the neutral defect plus one electron) and a dot denotes one positive effective charge. Thus in a homopolar semiconductor such as germanium, V^x is the bare vacancy formed by moving a germanium atom, without change of charge state, from a lattice site to a sink such as an edge dislocation or the surface. If this vacancy captures an electron it becomes V'; if it emits one, it becomes $V^{\bullet}$. In alkali halides, a bare cation vacancy is $V_M{}'$, with a single negative effective charge, while a cation interstitial would be $I_M{}^{\bullet}$. In monatomic metals and rare gas solids, where questions of charge state do not occur, the superscripts and subscripts are dropped. In alkali halides, where only bare vacancies of both signs are important, the notation has been simplified by using the subscript 1 to denote the cation vacancy and 2 the anion vacancy. In the silver halides, only silver vacancies and interstitials are important, and these are signified by V and I.

The defect thermodynamic parameters are given superscripts F, M, P, and A to indicate, respectively, formation, motion, pairing between thermally formed defects, and association of a thermal defect and an impurity ion. They are given subscripts to indicate the specific defect, or s for Schottky, f_M for cation Frenkel, and f_X for anion Frenkel defects. The M or X may be dropped where it is obvious in context which is meant.

With these conventions, the following symbols are used:

k	Boltzmann's constant
L	Avogadro's number
e	Electronic charge
T	Kelvin temperature
T_m	Melting temperature
T_q	Quench temperature
P	Pressure
C_P	Isobaric specific heat
C_V	Isochoric specific heat
S	Molar entropy
β	Volume thermal expansion coefficient
K	Isothermal compressibility
N	Number of formula units/unit volume
n	Total number of unit cells
v_0	Atomic volume
l	Specimen length
a	Lattice parameter, edge of cubic cell
d	Density
ϱ	Resistivity
χ_i	Mole fraction of ith defect
c	Mole fraction of impurity in excess
γ_i	Thermodynamic activity coefficient of ith species

g, h, s, v	Gibbs free energy, enthalpy, entropy, and volume associated with individual defect
Δg, etc.	Thermodynamic quantities released when isolated defects come together to form cluster
D^*	Self- or tracer-diffusion coefficient
Q	Activation energy for D^*
D_0	Preexponential factor for D^*
f	Correlation factor, as in diffusion
σ	Ionic conductivity
q_i	Effective charge on ith defect
Z	Number of nearest neighbors, or of orientations of defect or defect cluster
μ_i	Mobility of ith kind of defect
y	Mobility drag factor in Debye–Hückel theory
ν	Vibrational frequency
ω	Defect jump frequency

2. STATISTICAL THERMODYNAMIC THEORY

In principle, the methods of the canonical ensemble[2] may be used[3] to derive all the thermodynamic properties of a crystal containing specified defects. The thermodynamic functions can all be written in terms of the partition function

$$Z_p = \sum_i \exp(-E_i/kT) \tag{2.1}$$

where E_i is one of a complete set of energy eigenvalues for the crystal, which for simplicity will be taken to contain one mole. E_i depends on the composition, volume, and defect population, and is an eigenvalue of the equations of motion of the complete crystal.

The partition function, in particular, is related to the molar Helmholtz free energy A by

$$A = -kT \ln Z_p \tag{2.2}$$

This relation may be used to simplify the partition function. Consider a given configuration of the actual crystal (we will neglect impurities here) as derived from a perfect crystal by removing some atoms from the lattice sites to the surface, forming vacancies and new unit cells, and removing some from the surface, destroying unit cells, and putting them in interstitial positions. Allnatt and Cohen[3] write the Helmholtz free energy for the complete crystal as

$$A = A_0 + A_\chi + A_{\text{Mix}} + A_{\text{Int}} \tag{2.3}$$

where A_0 is the free energy of the perfect starting crystal, A_χ is the free energy derived from the introduction of the defects without interactions, as if each were going alone into an infinite perfect crystal, and A_{Mix} is the free energy of mixing of these noninteracting defects. A_{Int} is the free energy of interaction among the defects.

The equilibrium number of defects can be found by minimizing the appropriate free-energy function with respect to the numbers of the various types of defects that can be identified and treated as statistically independent entities. For most experimental situations, the pressure and temperature are held constant while equilibrium is established, and it is Gibbs free energy G that must be minimized:

$$\begin{aligned} G &= A + PV \\ &= G_0 + G_\chi + G_{\text{Mix}} + G_{\text{Int}} \end{aligned} \tag{2.4}$$

where P is the pressure and V the volume of the crystal.

If the interaction free energy A_{Int} is neglected, the rest of the terms in the combined equations (2.3) and (2.4) can be written readily. The term G_0 does not depend upon the defects, while

$$\begin{aligned} G_\chi &= \sum_i \chi_i g_i^F \\ G_{\text{Mix}} &= -kT \sum_i \ln \Omega_i(\chi_i) \end{aligned} \tag{2.5}$$

where g_i^F is the Gibbs free energy required to form a single defect of the ith kind in an otherwise perfect crystal, and $\Omega_i(\chi_i)$ is the number of ways to arrange the χ_i defects per mole of crystal on the available lattice sites. Minimization with respect to the χ_i of the sum $G_0 + G_\chi + G_{\text{Mix}}$ to give the equilibrium number of defects leads to the ideal-solution theory described in Section 2.1.

The more difficult parts of the theory involve evaluating A_{Int}. In the Debye–Hückel approach of Lidiard[4,5] (see Section 2.2), the interactions are written in terms of an averaged potential, the same for all ions of a given kind. Fong[6] assumes one-to-one pairing in such a way that the pairs become independent elements, each with the same energy states as all others. In either case, A_{Int} can be written as a sum (or a few sums) of identical terms; the corresponding partition function then becomes a product of identical terms, one for each defect or defect pair, and the problem can be solved for each defect or pair. Other, more complex solutions will be touched upon in the following sections.

2.1. Ideal Solution Theory

Equilibrium concentrations of point defects in crystals have traditionally been handled with a quasichemical mass-action formalism. For very low concentrations, where the defects can be considered as forming an ideal solution in the host crystal as solvent, the same results are obtained from the mass-action law approach as from a more detailed statistical mechanics treatment. The reader is referred particularly to the article by Howard and Lidiard[7] for a discussion of this correspondence, and to the book by Kroger[8] for a detailed examination of the results of the mass-action-law approach applied to a wide variety of situations.

Ideal solution behavior implies that the defects are noninteracting, statistically independent entities such that, except for exclusion effects, the probability of occupation of any one of the accessible sites by a given defect is not influenced by the distribution of the others among their accessible sites. "Exclusion effects" refers to the fact that a given lattice site can accept only one defect; if it is occupied by a given defect, then all others are excluded.

In real materials, the interactions between point defects are not negligible. However, there are several factors that help salvage ideal solution theory as a useful tool. For instance, in rare gas solids, the interatomic, and therefore interdefect, forces are very short-range. In metals, the long-range Coulombic part is screened by conduction electrons, and again the effective interdefect forces are short-range. This should also be true of degenerate semiconductors, and perhaps of semiconductors in general at sufficiently high temperatures. Where the interdefect forces are sufficiently short-range, pairs and other simple complexes among the defects can be introduced as separate entities into the chemical equilibrium equations and the mass-action laws applied to the whole system.

This treatment has also been used to describe trapping of electrons and holes (or the ionization of point defects) in semiconductors and insulators. It has been used as well for defect equilibria in ionic crystals and low-temperature semiconductors, where long-range Coulombic interactions, screened only by the dielectric polarization, are important. Here, the validity depends upon the very low concentrations of defects encountered in most stoichiometric compounds. However, even in the stoichiometric crystals, there is evidence[9–11] that intrinsic defect concentrations at high temperatures are high enough to require corrections to the ideal solution theory when one considers the more subtle aspects of this crystal behavior. In nonstoichiometric compounds, or in the presence of impurities, point defect concentrations are rarely so low that the interactions can be neglected.

Where the defect interaction energy decreases only slowly with increasing distance between the defects, a large number of distinguishable kinds of pairs are almost equally probable. The equations of equilibrium then become more complicated by virtue of this large number of unknowns. The details of the interactions between the defects have not been worked out well in any case as yet, but evidence is accumulating that important fractions of the defect pairs in alkali halides have greater separations than the nearest-neighbor distance (on the same sublattice).[12] It was, in fact, suggested some time ago,[13] on the basis of the Born model, that the next-nearest-neighbor pair composed of Sr^{2+} and a cation vacancy in KCl is more stable than the nearest-neighbor pair.

The thermodynamic treatment of point defects under equilibrium conditions can start with writing down a general Gibbs free-energy function G for the system in equilibrium (crystal plus other phases with which it may be in contact) in terms of all desired variables and then imposing as the condition of equilibrium that

$$\delta G = 0 = \sum_i (\partial G/\partial n_i)_{n_j,T,P}\, \delta n_i \tag{2.6}$$

where n_i is the number of entities of the ith kind. The derivative $(\partial G/\partial n_i)_{n_j,T,P}$ is the corresponding virtual chemical potential, given to within Stirling's approximation by

$$\xi_i = (\partial G/\partial n_i)_{n_j,T,P} = \xi_i^\circ + kT \ln(\chi_i/Z_i) \tag{2.7}$$

where ξ_i° is the free energy of formation of a single defect of the ith kind, excluding the entropy of mixing, χ_i is the fraction of available sites actually occupied by the defects, and Z_i is an internal partition function such that $-k \ln Z_i$ is the entropy associated with the internal degrees of freedom of the defect not specified by assigning it to a given site. In using this equation, care must be taken not to include the same entropy contribution in both ξ_i° and $-k \ln Z_i$. For instance, for a defect containing a single electron in a hydrogenlike ground state (e.g., the F-center in the alkali halides), spin degeneracy requires that $Z_i = 2$.

Various constraints must be taken into account, depending upon the system. Howard and Lidiard[7] introduce these by the method of Lagrangian multipliers, and Lidiard[14] has given a general treatment of defect equilibria on this basis. The constraints include: (1) in a compound, a fixed ratio of the number of cation sites to the number of anion sites must be maintained (Proust's "law of definite proportions"); (2) the number of impurity atoms must be preserved; and (3) the bulk of the crystal must be electrically neutral.

The ideal solution theory produces the following results. For single vacancies and for single interstitials in pure elemental metals (or disordered alloys) and in noble gas solids, and also for these same defects in the bare state in pure elemental semiconductors, the atom fractions of defects are given by

$$\chi = \exp(-g^F/kT) \tag{2.8}$$

where χ is the fraction of accessible sites occupied by vacancies, or interstitials, and g^F is the free energy required to produce a single defect in the perfect crystal by transport of the atom to or from the surface or equivalent sink or source. As used here, a bare defect is one produced by moving the atom to or from the surface without changing its charge. In a homopolar solid, bare vacancies and interstitials are uncharged, but in ionic crystals, they possess an effective charge and are "seen" by the rest of the crystals as charged entities.

Where identical vacancies tend to form divacancies, and identical interstitials to form diinterstitials, in these same systems, the ratio of the number of defect pairs to the total number of atom sites is

$$\chi_p = (Z/2)\exp[-(2g^F - \Delta g^P)/kT] \tag{2.9}$$

or

$$\chi_p/\chi^2 = (Z/2)\exp(\Delta g^P/kT)$$

where $Z_i = Z/2$ for the pair, with Z the number of distinguishable orientations of each pair, g^F is again the free energy necessary to form an isolated defect, and Δg^P is the free energy released when the two isolated defects form the pair. Notice that the equilibria involving isolated defects and pairs are independent of each other, and therefore Eqs. (2.8) and (2.9) hold simultaneously. Further, the equilibria involving interstitials and vacancies also hold simultaneously.

In metals and insulating homopolar solids such as the rare gas solids, the principal effect of impurities lies in their tendency to form pairs with vacancies and, at least in principle, with interstitials. Because of the short-range nature of the interactions involved, only nearest-neighbor pairs need be considered. Because impurity concentrations tend to be considerably larger than intrinsic defect concentrations, a correction for the exclusion of sites nearest-neighbor to the impurity atoms must also be made when considering the concentration of unassociated defects. For instance, in the limit of vanishingly small vacancy concentrations, Eq. (2.8) for the fraction of sites (normally available in the pure crystal) occupied by unassociated vacancies becomes

$$\chi_v = (1 - Z_n c_I) \exp(-g_v{}^F/kT) \tag{2.10}$$

where c_I is the fraction of sites occupied by impurity atoms and Z_n is the number of nearest neighbors. Unassociated vacancies and impurity atoms will form pairs, according to

$$\chi_A/\chi_v(c_I - \chi_A) = Z_n \exp(\Delta g^A/kT) \tag{2.11}$$

where $Z_i = Z_n$, χ_A is the fraction of sites occupied by impurity atoms associated with vacancies, and $-\Delta g^A$ is the additional free energy required to form a vacancy next to an impurity ion, above that required for an isolated vacancy.

For semiconductors, additional equilibria involving electrons and holes must also be considered. The treatment here will be the purely classical one applicable to nondegenerate semiconductors only. The equations above for bare defects are not affected, but bare vacancies tend to trap electrons, producing holes in the valence band, and bare interstitials tend to lose electrons to the conduction band. For low concentrations of electrons, an equilibrium equation can be written for the reaction

$$V^x \rightleftharpoons V' + h$$

where V^x represents a bare vacancy, V' the vacancy with a trapped electron and therefore with a single effective negative charge, and h a hole in the valence band. The equilibrium equation is

$$\chi_{V'}\chi_h/\chi_{V^x} = Z_{V'} \exp(-E_A/kT) \tag{2.12}$$

where $\chi_{V'}$ and χ_{V^x} are the fractions of the lattice sites occupied by charged and bare vacancies, respectively, and χ_h is the fraction of the available states at the top of the valence band occupied by holes. The factor $Z_{V'}$ is Z_i for the charged vacancy and takes into account the spin degeneracy of the trapped electron and Z_h takes into account the partition function of the hole in the valence band. E_A is the energy of the electron state in the charged vacancy, measured from the top of the valence band. The fraction of states occupied by holes is given by

$$\chi_h = \exp(-E_F/kT) \tag{2.13}$$

where E_F is the Fermi energy, measured from the same zero as E_A. Thus the ratio of charged to bare vacancies is

$$\chi_{V'}/\chi_{V^x} = Z_{V'} \exp[(E_F - E_A)/kT] \tag{2.14}$$

An exactly analogous equation holds for the bare interstitials acting as donors:

$$\chi_{M^\cdot}/\chi_{M^x} = Z_{M^\cdot} \exp[(E_D - E_F)/kT] \tag{2.15}$$

where M^x represents the bare interstitial (the occupied donor state) and $M^\cdot$ the interstitial having lost an electron and therefore bearing a single positive effective charge. The energy required to move an electron from the valence band onto $M^\cdot$ is E_D.

Impurities in homopolar semiconductors produce the effects noted above of pairing with vacancies and interstitials, perhaps even more strongly than in metals and rare gas solids since some of the defects are charged by virtue of the trapping of electrons and holes and there are consequently very strong Coulombic forces. In addition, however, the Fermi level is strongly influenced by the presence of donor or acceptor impurities, and this, acting through Eqs. (2.14) and (2.15), influences the numbers of charged vacancies and interstitials. Although the numbers of bare defects are not influenced in this way by the presence of impurities, the total numbers are.

In compounds, the formalism is similar but more complex. The several sublattices must be treated separately, but they are tied together by the constraints mentioned earlier. In all compounds, the law of definite proportions implies that aliovalent impurities (having valence different from the atoms they replace) will be accompanied either by vacancies or interstitials. There will be vacancies on the same sublattice or interstitials of the opposite kind if the impurity valence exceeds that of the atom it replaces, or vacancies on the opposite sublattice or interstitials of the same kind if the impurity valence is less than that of the host atom. In ionic compounds, the additional requirement of electroneutrality constrains the defects always to appear with equal numbers of positive and negative effective charges. While in a monatomic solid or homopolar compound, vacancies or interstitials on a given sublattice can appear independently of any other defect, in ionic compounds, defects that bear effective charges must be accompanied by other, charge-balancing defects bearing effective charges of the opposite sign.

For ionic crystals, the equations governing the equilibria involving bare vacancies and interstitials (which carry effective charges in ionic compounds) take the form of the Schottky and Frenkel equilibrium equations. For crystals with, e.g., chemical formula MX, these are:

$$(\chi_{V_M'})(\chi_{V_X^\cdot}) = \exp(-g_s{}^F/kT) \qquad \text{(Schottky)} \tag{2.16}$$

$$\left.\begin{aligned} \chi_{V_M'}\chi_{M_i^\cdot} &= \exp(-g_{fM}^F/kT) \\ \chi_{V_X^\cdot}\chi_{X_i'} &= \exp(-g_{fX}^F/kT) \end{aligned}\right\} \quad \text{(Frenkel)} \tag{2.17}$$

where $\chi_{V_M'}$ and $\chi_{V_X^\cdot}$ are the fractions of M and X sites occupied by bare vacancies, $\chi_{M_i^\cdot}$ and $\chi_{X_i'}$ are the fractions of interstitial sites occupied by bare M and X interstitials, $g_s{}^F$ is the free energy expended when an isolated bare M vacancy and an isolated bare X vacancy are formed by moving the appropriate ions, without change of charge, to the surface, but not including the configurational entropy of arrangement of the vacancies on their sublattices, and g_{fM}^F and g_{fX}^F are the similar free energies expended when the appropriate ions are moved from lattice to interstitial sites.

As in the case of the pure elemental semiconductors discussed above, there are in addition independent equilibria between the vacancies and interstitials on the one hand, and electrons and holes on the other. The equations are analogous to Eqs. (2.12), (2.14), and (2.15); Eqs. (2.16) and (2.17) also hold at all times.

In ionic compounds, because of the constraints tying the sublattices together, aliovalent impurities play a particularly important role. Thus, while Eqs. (2.16) and (2.17) govern the products of defect fractions, the absolute numerical values also involve relations equating, to within integral factors, the total number of sites on each sublattice, or the total charges of each sign. For instance, in a (monovalent) MX compound containing divalent cation impurities (C^{2+}), electroneutrality requires that

$$X_{C_M^\cdot} + \chi_{V_X^\cdot} + (N_{M_i}/N)\chi_{M_i^\cdot} = \chi_{V_M'} + (N_{X_i}/N)\chi_{X_i'} \tag{2.18}$$

where, per unit volume, N is the number of M sites and N_{M_i} and N_{X_i} are the numbers of sites available to the M and X interstitials, respectively. Since by Eq. (2.16), for instance, $\chi_{V_M'}$ and $\chi_{V_X^\cdot}$ are reciprocally related, the effect of adding C^{2+} to the system is to increase $\chi_{V_M'}$ and to decrease $\chi_{V_X^\cdot}$.

In addition, pair formation involving the various oppositely charged species must also be considered; each pair introduces a new mass action equation similar to Eq. (2.11). Simultaneous solution of Eqs. (2.16)–(2.18) plus the pair equilibria equations and an equation expressing conservation of impurity will then produce values for the various defect fractions.

2.2. Pair Formation with Long-Range Forces

The interdefect forces in ionic crystals are rather long-range, and the variation with their separation of the energy binding two defects into a pair is rather slow. It is therefore not necessarily a reasonable approximation to deal only with nearest-neighbor pairs. Indeed, as mentioned above, there is reason to believe that in systems like KCl containing divalent cations, the impurity–cation vacancy pair may have the lowest energy when the defects are in second-neighbor, rather than first-neighbor, positions (on the cation lattice).

In the framework of the theory of chemical equilibrium, this situation is handled by considering each possible separation of the defects ("excited state") as corresponding to a separate, identifiable chemical species. Equations of equilibrium, comparable to Eq. (2.9) or (2.11), can be written connecting the atom fractions of these species. In what follows, we shall consider only the simple case in which there are just two kinds of defects. Then,

$$\chi_{P_l} = Z_l[\exp(\Delta g_l^P/kT)]/q_P \tag{2.19}$$

where χ_{P_l} is the fraction of the defects of one sign that are in pairs in the lth state (with separation R_l), Δg_l^P is the binding free energy in this state, Z_l is the number of sites available in this state to the defect of the other sign, and q_P is what Fong[6] has called the molecular partition function for pair formation:

$$q_P = \sum_l Z_l \exp(\Delta g_l^P/kT) \tag{2.20}$$

As Fong points out, these equations are an approximation useful only at low temperatures and concentrations. In terms of the canonical ensemble used by Fong, this is because only when the interaction energy of a given defect with one other defect is much larger than its interaction energies with all others can the configurational partition function be factored into identical terms equal in number to the number of "pairs." As soon as the average separation between the partners in a pair begins to approach the distance between pairs, this simple treatment no longer holds.

The problem introduced by the long-range interactions is similar to that found in handling the thermodynamics of aqueous ionic solutions, and Lidiard[4,5] developed a treatment based upon a model composed of first-neighbor pairs and otherwise unassociated defects. In the zeroth-order approximation, the interactions among the unassociated defects and between these and the pairs are neglected. Then the equilibrium equation (2.9) becomes, for example, for an MX compound with Schottky defects,

$$\chi_P/\chi_{V'_M}\chi_{V^{\bullet}_X} = Z\exp(\Delta g^P/kT) \tag{2.21}$$

where χ_P is the fraction of lattice sites of one sign occupied by defects that are in pairs, Z is the number of adjacent sites available to the defect of the other sign in the pair, and $\chi_{V'_M}$ and $\chi_{V^{\bullet}_X}$ are the fractions of sites of either sign occupied by unpaired defects. The binding energy between the defects forming the pair is again Δg^P.

Lidiard pointed out that the most important long-range interaction in ionic crystals is the Coulombic interaction between effective charges. Since the pairs are electrically neutral, they are not affected by this interaction, but the unassociated defects are. Within the framework of the thermodynamic theory of dilute ionic solutions, this interaction can be taken into account by multiplying each fraction of sites occupied by unassociated defects in the denominator of Eq. (2.21) by the appropriate activity coefficient γ. Using the Debye–Hückel theory,[17] γ_i for the ith kind of defect is given by

$$\begin{aligned} \ln\gamma_i &= -q_i^2K/2\varepsilon kT(1+KR) \\ K^2 &= 4\pi\Big[\sum_i (q_i^2\chi_i/\Delta_i)\Big]\Big/\varepsilon kT \end{aligned} \tag{2.22}$$

where ε is the dielectric constant of the host crystal, R the "distance of closest approach" of the unassociated ions, χ_i the fraction of sites on the ith sublattice occupied by unassociated defects of charge q_i, and Δ_i is the volume per site of this ith sublattice.

The quantity R is an undetermined parameter. Since pairs at the distance of nearest neighbors (in terms of the possible separations of the partners in the pair) are excluded from the Debye–Hückel atmosphere, Lidiard set R equal to the next-nearest neighbor distance. This is an arbitrary decision since, as we have seen, the stability of next-nearest neighbor pairs is not very different from that of nearest-neighbor pairs in some systems of defects in ionic crystals, and even more extended pairs are probably still relatively stable. However, Lidiard was able to show that the results are not very sensitive to the choice of R, even when it is varied by a factor of five. By taking R as that distance beyond which the interaction is essentially Coulombic, and treating all pairs with smaller separations as "excited states," we again obtain Eq. (2.19), where P_l now refers to the fraction of all the pairs that are in the lth state, and the sum in Eq. (2.20) extends over the pair states only. Equation (2.21), with the Debye–Hückel correction, also holds for each pair state, so that χ_P, Z, and Δg^P become χ_{P_l}, Z_l, and Δg_l^P, respectively. In this case, Δg_l^P remains the binding energy of the pair in the lth state taking the reference state as the two isolated defects in an otherwise perfect crystal.

The Debye–Hückel correction must be made consistently wherever site fractions of unassociated defects bearing effective charges occur. For instance, the Schottky equilibrium, Eq. (2.16), will read

$$(\chi_{V_M'}\gamma_{V_M'})(\chi_{V_X^{\cdot}}\gamma_{V_X^{\cdot}}) = \exp(g_s^F/kT) \tag{2.23}$$

with g_s^F the free energy to form just the requisite number of the two defects isolated from each other in the otherwise perfect crystal. When this is inserted into the corrected version of Eq. (2.21) for the site fraction of cation vacancy–anion vacancy pairs in an MX crystal, the result is

$$\chi_P = Z\exp[-(g_s^F - \Delta g^P)/kT] \tag{2.24}$$

In general, the site fraction of any neutral complex formed from defects in thermal equilibrium will be independent of impurity concentration and unaffected by long-range interactions, as Eq. (2.24) shows for the vacancy pair.

The Debye–Hückel theory rests upon a form of approximation commonly used to simplify the statistical mechanics of interacting particles. Each particle is assumed to "feel" an average potential derived from all the other particles in their most probable configuration, so that the partition function for the canonical ensemble of the systems, each containing N particles, can be factored into N identical terms, or "molecular" partition functions, one for each particle. In the case in which the only interactions considered are those between paired particles, but at various separations [Eqs. (2.19) and (2.20)], the molecular partition function is for such a pair of interacting particles. In the Debye–Hückel theory as used by Lidiard, the defects are divided into two kinds, those associated into close pairs and those unassociated. The partition function for the canonical ensemble correspondingly contains two kinds of factors. For the unassociated ions, the energy is evaluated in the average potential due to the surrounding unassociated ions, while the associated pairs do not interact with the unassociated ions at all. The distribution of defects between the associated and unassociated states is then found by invoking the principle of minimum free energy.

Anderson[15] introduced an approximate method of handling the defect interactions that has also been widely used. In his method, only the nearest-neighbor defects are assumed to interact, and the partition function for the system is written containing factors depending upon these interactions. However, the number of such pairs is assumed to be that produced by a purely random distribution of the defects present. The total number of defects is calculated by way of the minimization of the free energy. The pair energy influences the defect population by modifying the energy of formation with an additive term proportional to the fraction of available sites occupied by the defects.

For instance, Eq. (2.8) for the atom fraction of the defects in a simple solid would be modified to

$$\chi = \exp[-(g^F - 2\chi \, \Delta g^P)/kT] \tag{2.25}$$

and the atom fraction of sites occupied by pairs would be

$$\chi_p = (Z/2)\chi^2 \tag{2.26}$$

where, as before, g^F is the free energy necessary to form the isolated defect, Δg^P is the free energy released when two isolated defects come together, and Z is the number of nearest neighbors to a given site.

Kroger[8] applies Anderson's treatment to an MX compound, under the conditions that the quantity of M remains fixed while that of X may be varied according to the pressure of gas X (assumed diatomic) in equilibrium with the solid, and assuming defects on the X sublattice only.* The atom fractions of X sites occupied by vacancies and interstitials are given by

$$\chi_i/(1 - \chi_i) = K_I P_{X_2}^{+1/2} \exp[-(g_i^F - 2\chi_i \, \Delta g_i^P)/kT] \tag{2.27}$$

$$x_v/(1 - \chi_v) = K_V P_{X_2}^{-1/2} \exp[-(g_v^F - 2\chi_v \, \Delta g_v^P)/kT] \tag{2.28}$$

where K_I and K_V are equilibrium constants containing the chemical potential of gaseous X_2 in the standard state, the free energy required to remove an X atom from the surface to the gas, and the vibrational contributions of the defects to the partition function; g_i^F and g_v^F are the energies expended when a surface atom is moved into an interstitial site or a lattice atom to the surface with the creation of a vacancy; and Δg_i^P and Δg_v^P are the free energies released when two isolated interstitials or vacancies

* There is some confusion at this point in Anderson's paper.[15] The text describes an AB compound with fixed amount of A and variable amount of B, but with the defects on the A sublattice. However, Eqs. (10) and (11) of the paper appear rather to apply to a fixed amount of B, with variable A and the defects on the A sublattice.

are brought together. The electrons and holes created by the change in composition are assumed to be nonlocalized.

Multiplication of Eqs. (2.27) and (2.28) produces

$$[\chi_i/(1-\chi_i)]\chi_v/(1-\chi_v) = K_F \exp[2(\chi_i\, \Delta g_i^P + \chi_v\, \Delta g_v^P)/kT]$$
$$K_F = K_I K_V \exp[-(g_i^F + g_v^F)/kT] \qquad (2.29)$$

a result which is independent of the pressure of X_2 in the gas phase, and expresses the Frenkel equilibrium condition for the X sublattice. K_F is the Frenkel equilibrium constant in the absence of the interaction between pairs of defects.

The weakest point in Anderson's treatment is the assumption that, in spite of the interaction between defects, their distribution is random and pairs are formed on the basis of chance alone. Several refinements of the theory intended to strengthen it on this point have been offered. Hagemark[16] introduced the idea of the exclusion of sites nearest-neighbor to the defects. The number of sites available to the defects, for the purpose of computing the configurational entropy, was allowed to decrease as the defect concentration increased. Atlas[18,19] has attempted a more ambitious improvement by admitting a distribution of interaction energies among the defects and attempting to evaluate the distribution of defects among these interaction states.

A completely general and, accordingly, very complex treatment of the contribution of interacting point defects to the partition functions of crystals has been given by Allnatt and Cohen.[3,20] The defect contribution is developed in a virial expansion in the defect concentration. This theory reproduces that of Lidiard,[4,5] based upon the Debye–Hückel theory of aqueous ionic solutions, as a first approximation. However, the convergence in the Allnatt–Cohen result can be rather slow, e.g., for divalent cations and cation vacancies in alkali halides at temperatures and concentrations generally of experimental interest. Thus, although rather clumsy to be of direct use in itself in its present form, the theory of Allnatt and Cohen does help set forth the limits of temperature and concentration within which the simpler theories may be used with confidence.

2.3. Surface and Dislocation Effects

The situations described so far pertain rigorously, for crystals containing charged defects, only to the regions of the crystal far from the defect sources (e.g., the surfaces and edge dislocations). Near surfaces and edge dislocations, the constraint of electroneutrality must be relaxed, since surface (or line) charges with compensating space charges can exist.[21,22]

For a qualitative discussion, we consider a monovalent *MX* crystal with Schottky defects predominating. In the intrinsic range, the Schottky equilibrium [Eq. (2.16)] together with the condition of electroneutrality requires that in the interior far from the defect sources,

$$\chi_0 = \chi_{V_M'} = \chi_{V_X^{\bullet}} = \exp(-g_s{}^F/2kT) \tag{2.30}$$

On the other hand, each type of vacancy must simultaneously be in equilibrium with its sources, so that Eq. (2.8) also holds everywhere, in the form

$$\chi_i = \exp[-(g_i{}^F + q_i\phi)/kT] \tag{2.31}$$

In this equation, ϕ is the electrostatic potential difference between the source and the position under examination in the crystal, q_i is the effective charge on the vacancy, and $g_i{}^F$ is the rest of the free energy required to form the single vacancy by removing the ion from an otherwise perfect crystal to a site at the source. Note that this quantity can vary with the type of source, and need not be the same for external surface as for a given type of edge dislocation, etc. Combination of Eqs. (2.30) and (2.31) for monovalent *MX* crystals then results in

$$(g_1{}^F - e\phi) + (g_2{}^F + e\phi) = g_1{}^F + g_2{}^F = g_s{}^F \tag{2.32}$$

where the subscript is 1 (2) for a cation (anion) vacancy. Thus, whatever the source, the sum of the formation free energies (neglecting the electrostatic energy) must be the Schottky energy.

The physical behavior of this model can be described as follows. Suppose $g_1{}^F < g_2{}^F$; then, in the absence of electrostatic effects, the source would produce more cation than anion vacancies, acquiring a positive charge in the process while the rest of the crystal became negative. In thermal equilibrium, including electrostatic effects, this process would proceed and the electrostatic potentials change until the free-energy expenditure, including the electrostatic energy, was the same for the generation of a cation vacancy as for an anion vacancy. The positive charge on the source would attract to itself negatively charged cation vacancies and repel the positively charged anion vacancies, so that in the neighborhood of the source, there would exist a negative space charge which would screen the charge on the source. In the interior, the crystal would remain electrically neutral.

Aliovalent impurities influence these effects markedly. As an example, consider the effect of substitional divalent cations, compensated by cation vacancies in a crystal where, as before, $g_1^F < g_2^F$. At temperatures high enough for the intrinsic defects to be much more numerous than the impurity ions, the presence of the impurities would produce little effect. At some lower temperature (the "isoelectric" temperature), the difference between cation and anion vacancy concentrations dictated by Eq. (2.31) with zero potential difference would just equal the impurity concentration. This implies that as the temperature is lowered from above, the potential difference and the source and space charges will be reduced and disappear altogether at this isoelectric temperature. At still lower temperatures, the potential difference will reverse in sign in order that enough excess cation vacancies be produced to compensate in the interior for the impurities. If $g_2^F < g_1^F$, no isoelectric point would exist in the presence of an excess of divalent cation impurities.

Lehovec[22] solved the problem for a pure crystal with plane surfaces. Eshelby *et al.*[23] introduced the effects of mobile impurities and noted the existence of the isoelectric temperature. They also pointed out that the space charge around the dislocations should contribute to the yield stress of the crystal, so that the yield stress should pass through a minimum at the isoelectric point. This effect provides the basis for most of the determinations of the isoelectric temperature T_0. Since

$$\begin{aligned} T_0 &= g_1^F/k \ln(\alpha^{-1}) \\ \alpha &= \tfrac{1}{2}c\{1 + [1 + (4\chi_0^2/c^2)]^{1/2}\} \end{aligned} \tag{2.33}$$

for a crystal for which $g_1^F < g_2^F$, where c is the concentration of divalent substitutional cations and χ_0 is given by Eq. (2.30), this determination of T_0 allows a separate determination of g_1^F. By varying the concentration c of impurity, the temperature variation of g_1^F and thus of the enthalpy h_1^F and entropy s_1^F of formation of the cation vacancies can be determined, at least in principle. Eshelby *et al.*[23] considered qualitatively the influences of simultaneous association between the divalent impurity ions and the cation vacancies, and also impurity precipitation. Under these circumstances, Eq. (2.33) remains valid, but for c, the actual concentration of free (unassociated) impurity ions must be used. Several minima can be produced in the plot of yield stress versus temperature.

Kliewer and Koehler[24] and Kliewer[25] introduced the influence of impurity–vacancy pairs and of pairs formed between cation and anion vacancies on a quantitative basis, and gave solutions for the potential distribution around line sources such as dislocations. For the monovalent MX crystal with $g_1^F < g_2^F$ and with divalent cation impurities, the impurities avoid the neighborhood of the sources above the isoelectric point, but collect around the sources below it. This latter tendency can be strong enough to saturate the crystal with respect to the impurity ions in the vicinity of the sources, even at very low impurity concentrations.

2.4. Pressure Effects

Assuming that the defect free energies, g^F for formation and g^M for motion, are in fact true thermodynamic free energies, it follows that

$$(\partial g/\partial P)_T = v \tag{2.34}$$

where v is the increase in volume of the system for each defect formed, so that a measurement of the variation of the free energy of formation or motion of a given species with applied hydrostatic pressure should provide an estimate of the volume change on formation of the species or on passing from the normal to the activated state during motion of the species.

For Schottky defects, the formation volume so determined will include the volume of a new unit cell created with the vacancies. The quantity of interest is the difference between the formation and unit-cell volumes in this case, and represents the sum of the volume changes (expansions or contractions) of the lattice around the vacancies. For Frenkel defects, no new unit cells are formed, and the formation volume is directly the sum of the volume changes around the interstitial and the vacancy.

3. EXPERIMENTAL DETERMINATION OF DEFECT THERMODYNAMIC PARAMETERS

3.1. Noble Gas Solids

3.1.1. *General References*

Several rather comprehensive reviews of the properties of noble gas solids have been published in the last few years. The reader's attention is drawn in particular to those of Boato,[26] Pollack,[27] and Horton,[28] although Horton does not discuss point defects.

3.1.2. *Experimental Methods for the Noble Gas Solids*

Theoretical calculations[29] suggest that the free energy of formation of interstitials is larger than that of vacancies, so that vacancies should predominate over interstitials in the noble gas solids. The vacancy formation free energy g_v^F is obtained from the temperature dependence of the atom fraction χ_v of vacant lattice sites using the appropriate form of Eq. (2.8):

$$\chi_v = \exp(-g_v^F/kT), \qquad g_v^F = h_v^F - Ts_v^F \tag{3.1}$$

The atom fraction of vacancies is determined in a variety of ways for these solids, all of which involve comparing some thermodynamic property of the crystal at high temperature to an estimate of the same property for the same crystal without vacancies.

The most direct way of determining χ_v is to compare the bulk density with the unit-cell density obtained from X-ray lattice parameter measurements.[30] That is,

$$\chi_v = (d_x - d_B)/d_x \tag{3.2}$$

where d_B is the bulk density of the sample and

$$d_x = n_c A_w / v_c \tag{3.3}$$

is the X-ray density, with n_c the number of atoms per unit cell (four for the cubic cell of the fcc lattice of the noble gas solids), v_c the unit-cell volume, and A_w the average atomic mass. However, the bulk densities are subject to large uncertainties due to accidental voids (e.g., vapor snakes) formed during solidification, and the vacancy concentrations from Eq. (3.2) are apt to be too high. Peterson *et al.*[31] have made a very precise determination of the X-ray density of solid argon up to the triple point. Combining these with the best available bulk density measurements,[32] values for χ_v are obtained that do not exhibit the temperature dependence expected on the basis of Eq. (3.1). At best, an upper limit of 1.3×10^{-3} for χ_v at the triple point can be deduced. A somewhat higher values, 3.7×10^{-3}, has been suggested by van Witzenburg.[33]

The difficulties that are encountered in the comparison of the bulk and X-ray densities can be avoided by comparing the changes that take place in each as the temperature is raised from a value sufficiently low for the vacancy concentration to be negligibly small. Simmons and Balluffi[30] have shown that the changes with temperature in bulk volume, $3\,\Delta l/l$, and unit cell volume, $3\,\Delta a/a$, differ only by expansions arising directly from the thermal creation of defects, provided the reference temperature is so low that there the number of thermal defects is negligible. Furthermore, as Eshelby[34] has shown, relaxations around point defects, if they are uniformly distributed, will equally affect $\Delta l/l$ and $\Delta a/a$. Therefore, only the appearance (vacancies) or disappearance (interstitials) of unit cells accompanying thermal defect generation will affect $\Delta l/l$ without affecting $\Delta a/a$. Since the relative change in the number of unit cells, $\Delta n/n$, is related to the defect population by

$$\Delta n/n = \chi_v - \chi_i$$

where χ_v and χ_i are, respectively, the vacancy and interstitial concentrations, for a cubic material, the approximation

$$\chi_v \approx 3[(\Delta l/l_0) - (\Delta a/a_0)] \tag{3.4}$$

can be used where $\chi_i \ll \chi_v$, with l_0 and a_0 the low-temperature values of overall specimen length and lattice parameter, and Δl and Δa the changes on going to some high temperature. Losee and Simmons[35] have made very careful measurements of both quantities on the same large-grained specimens of krypton and have found that χ_v determined in this way does exhibit the temperature dependence of Eq. (3.1). From this temperature dependence, the values for the vacancy formation enthalpy h_v^F and entropy s_v^F shown in Table I were obtained by Losee and Simmons.[35] These are the only reliable values as yet available for any noble gas solid. Figure 1 gives their values of $(\Delta l/l_0) - (\Delta a/a_0)$ as a function of temperature, and shows that the vacancy effect is really considerably larger than the scatter.

Table I. Vacancy Formation Parameters for Noble Gas Solids

Substance	Ref.	h_v^F, eV	s_v^F/k	Notes
Ar	38	0.056	3.4	C_p
	36	0.069	5	C_v
	41	0.069	5.1	S
	39	0.052	2.2–2.5	C_p; s_v^F/k estimated from Fig. 8, Ref. 39
	35	0.055	—	Corresponding states
Kr	38	0.077	3.4	C_p
	35	0.077	2.0	X-ray and bulk thermal expansion
Xe	35	0.108	—	Corresponding states

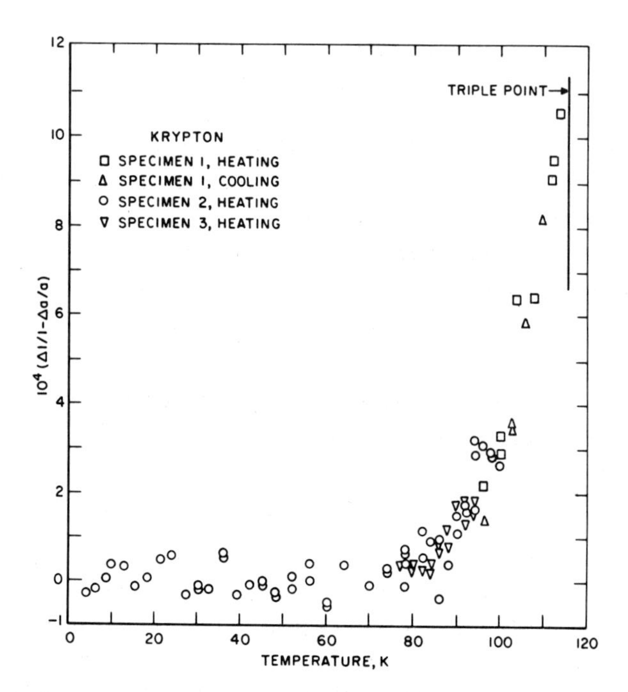

Fig. 1. Comparison of bulk length expansion measurements, $\Delta l/l$, with X-ray lattice expansion measurements, $\Delta a/a$, in solid krypton. The reference lengths are those values extrapolated to 0°K. From Losee and Simmons.[35]

Vacancies also make contributions to other thermodynamic properties,[14] and estimates of these contributions have been used to obtain values for the vacancy concentrations, for various temperatures, in a number of cases. The results are considerably less reliable than those obtained from the comparison of the bulk and X-ray thermal expansion.

We start with the assumption that the molar Gibbs free energy of the crystal may be written

$$G = G^\circ(T) + [\chi_v/(1 - \chi_v)]g_v{}^F L \approx G^\circ(T) + \chi_v g_v{}^F L \qquad (3.5)$$

where the superscript $^\circ$ denotes the properties of the hypothetical crystal without vacancies, L is Avogadro's number, and the vacancy formation enthalpies and entropies are assumed to be independent of temperature. Then, using

$$G^\circ = E^\circ + PV^\circ - TS^\circ$$

$$g_v{}^F = \varepsilon_v{}^F + Pv_v{}^F - Ts_v{}^F$$

and Eq. (3.1), the isobaric molar specific heat

$$C_P = (\partial H/\partial T)_P, \qquad H = H^\circ + \chi_v h_v{}^F L$$

becomes

$$C_P = C_P{}^\circ(T) + [\chi_v (h_v{}^F)^2 L/kT^2] \tag{3.6}$$

The isochoric molar specific heat is given by[36]

$$C_V = (\partial E/\partial T)_V, \qquad E = E^\circ + \chi_v \varepsilon_v{}^F L$$

$$C_V = C_V{}^0(T) + \frac{\chi_v L}{kT^2}\left(h_v{}^F - \frac{\beta^\circ v_v{}^F T}{K^\circ}\right)\left(h_v{}^F - \frac{\beta v_v{}^F T}{K}\right) \tag{3.7}$$

with β the coefficient of thermal expansion, K the isothermal compressibility, and $v_v{}^F$ the volume of formation per vacancy,

$$v_v{}^F = (\partial g_v{}^F/\partial P)_T$$

Use has been made of the relations

$$P = \int_0^T (\partial P/\partial T)_V \, dT = \beta^\circ T/K^\circ$$

(where T is in degrees Kelvin) to obtain the pressure at temperature T appropriate to the thermodynamic relation

$$\varepsilon_v{}^F = h_v{}^F - P v_v{}^F$$

The coefficient of thermal expansion referred to the volume V° of the vacancy-free crystal may be obtained by

$$\begin{aligned} \beta &= (1/V^\circ)(\partial V/\partial T)_P \\ V &= V^\circ(T) + \chi_v v_v{}^F L \\ \beta &= \beta^\circ(T) + (\chi_v h_v{}^F/kT^2)(v_v{}^F L/V^\circ) \end{aligned} \tag{3.8}$$

and the isothermal compressibility, also referred to V°, by

$$\begin{aligned} K &= K^\circ(T) + (\chi_v h_v{}^F/kT^2)(v_v{}^F L/V^\circ)(T v_v{}^F/h_v{}^F) \\ &= K^\circ(T) + (\beta - \beta^\circ)(T v_v{}^F/h_v{}^F) \end{aligned} \tag{3.9}$$

Finally, the entropy can be derived from Eq. (3.1-9) of Howard and Lidiard,[7] which is, after introducing Stirling's approximation,

$$S = S^\circ(T) + \chi_v s_v{}^F L + kL[(1 + \chi_v)\ln(1 + \chi_v) - \chi_v \ln \chi_v] \tag{3.10}$$

Utilizing the further approximation

$$\ln(1 + \chi_v) \approx \chi_v$$

and dropping terms in χ_v^2, this becomes[37]

$$S = S^\circ(T) + \chi_v L[k + (h_v^F/T)] \tag{3.11}$$

Equations (3.6)–(3.10), as well as (3.4), have been used to fit the corresponding experimental data and thereby to obtain estimates for h_v^F and s_v^F. Table I contains the results of some of the more recent of these attempts. In each case, the critical aspect of the determination is the estimation of the contribution to the particular physical property from the hypothetical crystal without vacancies. For instance, Beaumont *et al.*[38] used the isobaric specific heat, C_P to obtain h_v^F and s_v^F for solid argon, estimating $C_P^\circ(T)$ in Eq. (3.6) by a simple linear extrapolation of the low-temperature portion of a plot of C_P versus T^2 into the high-temperature region where vacancies are expected to contribute (Fig. 2). The plot of $\ln[T^2(C_P - C_P^\circ)]$ versus T^{-1} was linear, as suggested by Eqs. (3.6) and (3.1). It now appears that this linearity is not sufficient proof of the correctness of the determination of $C_P^\circ(T)$. Hillier and Walkley,[39] using the same data for argon,[40] estimated C_P° using a cell model with a 12–6 Mie–Lennard-Jones potential and obtained somewhat different values for the vacancy formation parameters. Foreman and Lidiard[36] converted the argon data to C_V and calculated C_V° [Eq. (3.7)] with an anharmonic Einstein model that produced good agreement with the experimental C_V below 60°*K*. Kuebler and Tosi[41] also used Flubacher *et al.*'s[40] argon C_P data to obtain the entropy. The Debye temperatures corresponding to these entropies for the temperature region from about 20 to about 60°K were fitted to expressions containing anharmonic contributions, and then extrapolated into the 60–80°K region to produce estimates of S° [Eq. (3.10)].

The values given by Losee and Simmons[35] for Kr were obtained by comparison of the bulk and X-ray thermal expansions and provide direct and reliable estimates of the vacancy concentration in krypton near the triple point, and hence reliable values for h_v^F and s_v^F. They then estimated values of h_v^F shown for the other noble gas solids using an empirical law of corresponding states[42] under which

$$h_v^{F*} = h_v^F T_{\text{tr}}^{-1}$$

is a constant. This includes all the noble gas solids where the triple point

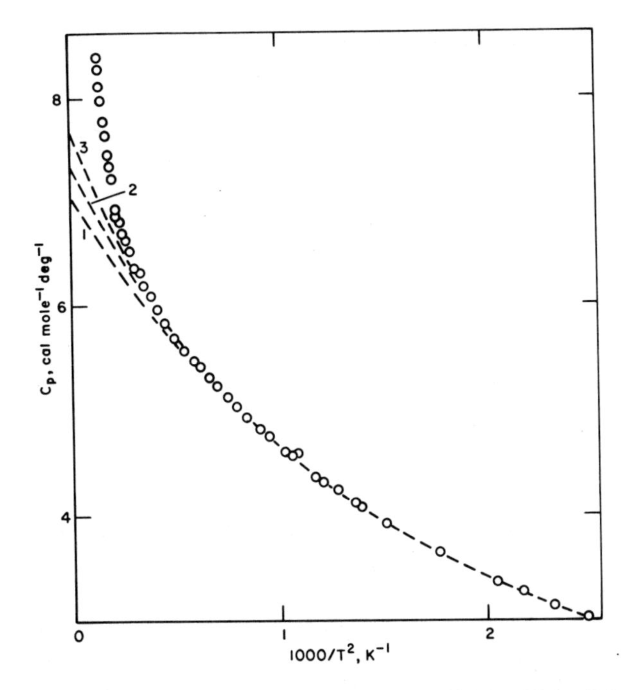

Fig. 2. Isobaric specific heat C_P as a function of T^{-2} for solid argon. The dashed curves 1, 2, and 3 represent different estimates of the ideal lattice specific heat. From Beaumont *et al.*[38] The author is indebted to The Institute of Physics and the Physical Society (London) for permission to reproduce this figure.

temperature T_{tr} is sufficiently high for quantum effects to be unimportant, but rules out He and probably Ne. They made the use of this law plausible by showing that the activation energy Q for self-diffusion could be in the same way converted to Q^*, and that this was the same in solid Ar and Xe to within a few per cent.

The extrapolations of the low-temperature properties into the high-temperature region to provide the estimates of the properties of the hypothetical vacancy-free crystal are quite uncertain. Anharmonic terms in the crystal potential make significant, but relatively unknown contributions to the crystal properties that become important in just the same temperature region as do the vacancy contributions. From their direct determination of the concentrations of vacancies in krypton, Losee and Simmons[43] were able to work the procedure backward [Eqs. (3.6)–(3.10)] and to obtain estimates of the properties of the hypothetical vacancy-free crystal.

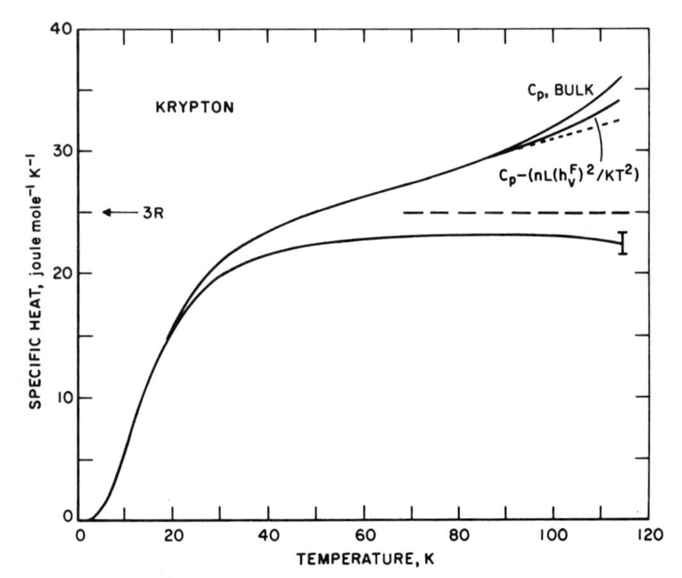

Fig. 3. Isobaric (C_P) and isochoric specific heats of solid krypton. The C_P curve marked $C_P - [nL(h_v^F)^2/kT^2]$ has been corrected for the vacancy contribution determined by Losee and Simmons[35] using their vacancy formation parameters. From Losee and Simmons.[35]

In Fig. 3, the corrected curve for the isobaric specific heat [labeled $C_P - nL(h_v^F)^2/kT^2$] rises significantly above the linear extrapolation that is often used. They could thus show that the extrapolation procedures in use do not in fact fit the observed behavior very well, and that vacancy parameters derived from these extrapolations cannot be considered reliable. For example, Manzhelii *et al.*[44] attributed a sharp rise in the Gruneisen constant for xenon at high temperatures to vacancies; more recent work (R. O. Simmons, private communication) has failed to confirm the existence of the rise.

3.2. Metals

3.2.1. *General References*

There are several recent books either devoted to point defects in metals or else containing at least major sections on them. These references include Damask and Dienes[45] and Girifalco and Welch.[46] The book by Thompson[47] contains an excellent elementary review of the configurations and energetics of point defects in metals and of the experimental techniques used to explore these subjects. Seeger and Schumacher[48] reviewed much of what was known of point defects in the noble metals up to 1965, and Seeger[49] has reviewed in greater detail both the equilibrium situation and diffusion via point defects. The Proceedings of the International Conference on Vacancies and Interstitials in Metals at Julich in 1968[49a] is particularly valuable.

3.2.2. "*Best*" *Values for Vacancy Formation Parameters in Metals*

Anticipating the results of the subsequent discussion of the parameters of individual metals, we collect the "best" values for the enthalpies of motion and formation, where reasonable estimates are possible, in Table II. The melting temperatures are listed in the second column. In the third and fourth columns are given the experimental values for the enthalpies of formation and motion, respectively, for single vacancies. Although two decimal places are given, in no case are these really known to better than about ± 0.1 eV. For the bcc metals, there is considerably less detailed knowledge of the defects present than there is for the fcc metals. The error is greater, and the numbers refer to the combined effects of single vacancies plus whatever clusters may be present. The numbers in parentheses were obtained by subtracting the other enthalpy from the activation energy of diffusion. The fifth column, labeled h^{calc}, is discussed below.

It is striking that, within present limits of error, the formation and motion enthalpies are about equal for all the metals listed except sodium. Furthermore, since the activation energy for diffusion is approximately

Table II. Collected Formation and Motion Enthalpies for Metals

Metal	T_m, °K	h_{1v}^F, eV	h_{1v}^M, eV	h^{calc}, eV	Structure
Au	1333	0.87	0.89	0.93	fcc
Ag	1234	0.99	0.86	0.86	fcc
Cu	1353	1.03	1.06	0.94	fcc
Al	983	0.73	0.65	0.65	fcc
Pt	2044	1.49	1.38	1.42	fcc
Pb	600	0.5	0.6	0.42	fcc
Nb	2688	2.0	(2.1)	1.9	bcc
Mo	2873	2.3	(1.7)	2.0	bcc
W	3663	3.3	3.3	2.6	bcc
Fe	1803	(1.5)	1.1	1.3	bcc
Na	370	~0.4	~(0.04)	0.3	bcc
Mg	924	~0.7	~0.7	0.65	~hcp

proportional to the melting temperature on the one hand, as discussed by Nachtrieb *et al.*,[50] Kidson and Ross,[51] and others, and is given roughly by the sum of the formation and motion enthalpies on the other, it follows from the equality of these enthalpies that their common value should also be approximately proportional to the melting temperature. The fifth column in the table lists the enthalpy calculated as $7 \times 10^{-4} T_m$(eV). These values reproduce the enthalpies of formation and motion within about ± 0.1 eV for the fcc metals, and ± 0.3 eV for the bcc metals except tungsten and sodium. In tungsten, the diffusion may well be of the "anomalous" variety as discussed below, while in sodium, the formation enthalpy is about equal to, but the motional enthalpy is much less than, the calculated value.

3.2.3. *Experimental Methods for Metals*

a. *Bulk and X-Ray Thermal Expansion.* This technique, described in detail in Section 3.1, gives a direct measure of the number of lattice points generated by thermal equilibrium. It is most useful where the excess concentration of defects of one sign exceeds $\sim 10^{-3}$ as the melting point is approached, and for metals with moderate ($<1500°$C) melting points. Under these circumstances—and only two metals (Au, Al) have as yet been examined in this detail—both the enthalpy and entropy of formation of the excess defects (vacancies) can be obtained from the defect concentration χ and its dependence upon temperature, using Eq. (2.8). In the other metals so far studied, the dependence of the (vacancy) concentration on temperature has been too small for this treatment. Instead, the magnitude of the vacancy concentration at the melting point, $\chi_v(T_M)$ was combined with a theoretically based estimate of s_{1v}^F to obtain h_{1v}^F. An error of $\pm 0.5k$ in s_{1v}^F produces an error in h_{1v}^F of less than 0.1 eV in most metals. The method does not distinguish between vacancies as singles and those in divacancies and larger clusters, but in all metals so far studied, the latter have made contributions to χ_v such that the effect on h_{1v}^F is well below the limit of error, and have been neglected.

b. *Self-Diffusion Data Combined with Motional Parameters.* For self-diffusion in metals by a single defect species (e.g., single vacancies) the activation energy Q is given by

$$Q = h^F + h^M \tag{3.12}$$

where h^F and h^M are, respectively, the enthalpies of formation and motion of the defects. If h^M is obtained from some other measurement, then h^F

can be calculated by difference. Implicit in the use of this equation is the assumption that the motional enthalpy is determined for migration of the defects in regions of perfect crystal, uninfluenced by dislocations, surfaces, grain boundaries, etc. In the next section, we will discuss how h^M may be obtained from experiments involving recovery of quenched, irradiated, or cold-worked specimens. In this section, we will consider the interpretation of self-diffusion measurements.

Divacancies are present in most metals at temperatures high enough to observe self-diffusion and complicate the interpretation of the self-diffusion experiments. The concentrations are governed by

$$\begin{aligned} &\text{Single vacancies:} \quad && \chi_{1v} = \exp(-g_v{}^F/kT) \\ &\text{Divacancies:} \quad && \chi_{2v} = Z\exp(-g_{2v}^F/kT) \end{aligned} \tag{3.13}$$

and with $g_{2v}^F = 2g_v{}^F - B$, the ratio of concentrations as vacancies in divacancy and in single-vacancy form is

$$\chi_{2v}/\chi_{1v} = Z\exp[-(g_v{}^F - B)/kT] \tag{3.14}$$

Since $g_v{}^F$ invariably is larger than the binding energy B, the relative importance of divacancies increases with increasing temperature. Thus at high temperatures, the influence of divacancies can become large.

This influence present something of a dilemma in the experimental determination of defect parameters. In order to obtain a good estimate of the diffusion activation energy, a wide temperature range is needed. But at the high end of a sufficiently wide temperature range, divacancies are apt to become important, producing curvature in the Arrhenius plot and spoiling the correspondence between the activation energy of diffusion and the sum of the enthalpies of formation and motion of single vacancies.

One possible resolution of this dilemma has been provided by Seeger and co-workers. Self-diffusion by single vacancies and divacancies simultaneously has been discussed by Seeger *et al.*[52] and Seeger and Schumacher.[53] The example of an fcc metal will be used for discussion. The self-diffusion coefficient of tracer atoms in a metal in which both single vacancies and divacancies contribute is

$$D^* = D_1 + D_2 \tag{3.15}$$

where

$$D_1 = a^2\nu_1[\exp(s_{1v}^M/k)][\exp(-h_{1v}^M/kT)]f_1\chi_{1v}$$

$$D_2 = \tfrac{2}{3}a^2\nu_2[\exp(s_{2v}^M/k)][\exp(-h_{2v}^M/kT)]f_2\chi_{2v}$$

In these equations, a is the lattice constant, the $\nu_i \exp(s_{iv}^M/k)$ are the frequency factors, the f_i are the correlation coefficients, and the atom fractions χ_{iv} are given by Eq. (3.13) with $Z = 6$ for an fcc metal. The correlation coefficient f_i takes into account the fact that not all jumps of the tracer atom contribute to the diffusion as expressed by the Fick's law diffusion coefficient D_i. When a tracer and a vacancy exchange places, the jump contributes to the Fick's law diffusion only to the extent that the jump is a random event with respect to the last jump. If this is the case, and one jump is completely independent of all others, f_i will be unity, but if each jump is, on the average, correlated to the last to some extent, then f_i will be less than unity. The correlation coefficient is a rather complicated quantity, depending upon the geometry of the lattice and also upon the variation of jump frequencies with position relative to the tracer.

Making the substitutions

$$\begin{aligned} D_{10} &= a^2\nu_1[\exp(s_{1v}^M/k)][\exp(s_{1v}^F/k)]f_1 \\ D_{20} &= a^2\nu_2[\exp(s_{2v}^M/k)][\exp(s_{2v}^F/k)]f_2 \\ Q_1 &= h_{1v}^F + h_{1v}^M \\ Q_2 &= h_{2v}^F + h_{2v}^M \end{aligned} \tag{3.16}$$

the self-diffusion coefficient can be written

$$D^* = D_{10}[\exp(-Q_1/kT)]\{1 + (D_{20}/D_{10})\exp[-(Q_2 - Q_1)/kT]\} \tag{3.17}$$

If only single vacancies matter, the term in curly brackets is unity and the Arrhenius plot, $\ln D^*$ versus T^{-1}, is linear. The term in curly brackets contains the correction for divacancies, and departs more from unity as the temperature rises. As long as the correction is small, the Arrhenius plot will have a slope approximated by

$$\frac{-Q_{\text{app}}}{k} = \frac{-1}{k}\left\{Q_1 + \frac{D_{20}}{D_{10}}(Q_2 - Q_1)\exp\left[-\frac{Q_2 - Q_1}{kT}\right]\right\} \tag{3.18}$$

provided also that D_{20}/D_{10} is temperature-independent. If the vacancy pair is not tightly bound, f_2 may exhibit a temperature dependence[53a] not contemplated in Seeger's treatment. Since $Q_2 > Q_1$ in most metals, the apparent activation energy Q_{app} will be larger than the sum Q_1 of the formation and motion enthalpies for single vacancies. This may be true even when the curvature represented by the temperature dependence of the right-hand side of Eq. (3.18) is too small to be detected. Seeger and co-workers have observed these effects in a series of detailed analyses of self-diffusion data for Cu,[54] Ag,[55] Au,[56] Ni,[52,53] and Pt.[57] They also introduced a temperature dependence of the enthalpies

$$h(T) = h_0 + \zeta(T - T_0) \tag{3.19}$$

which requires, since by thermodynamics, $(\partial h/\partial T)_P = T(\partial s/\partial T)_P$,

$$s(T) = s_0 + \zeta \ln(T/T_0) \tag{3.20}$$

T_0 here is a fixed reference temperature.

Out of a least-squares fit of this model to the self-diffusion data, guided by other information on defect properties, Seeger and his collaborators have been able to obtain values for Q_1, etc. These can then be combined with motional enthalpies and entropies obtained from kinetic data on the recovery of quenched, irradiated, or cold-worked specimens to produce estimates of the formation enthalpies and entropies. However, as detailed studies of ionic conductivity and diffusion data for alkali halides have shown (cf. Section 3.4.2), the fitting of slightly curved Arrhenius plots with sums of exponentials does not give well-defined values for the parameters, and the results must therefore be viewed with caution.

The measurement of the relative diffusion rates of isotopes, particularly two tracer isotopes into a specimen composed primarily of a stable third isotope, has been useful in some cases in sorting out the mechanism of diffusion. We follow here the treatment of LeClaire.[58] In most simple metals, the self-diffusion coefficient of isotope α is given by

$$D_\alpha^* = A\omega_\alpha f_\alpha \chi \tag{3.21}$$

where A is a geometric factor independent of which isotope is jumping, ω_α is the jump frequency for α into a neighboring vacancy-type defect, f_α is the correlation coefficient, and χ is the atom fraction of the defect. ω_α depends upon the tracer mass, through

$$\omega = \nu \exp(-g^M/kT) \tag{3.22}$$

If g^M, the free energy of motion, is independent of mass, then only $\nu \propto M^{-1/2}$ varies with the isotope, and

$$(\omega_\alpha - \omega_\beta)/\omega_\beta = (M_\alpha^{-1/2} - M_\beta^{-1/2})/M_\beta^{-1/2} = (M_\beta/M_\alpha)^{1/2} - 1 \tag{3.23}$$

However, in particular, the entropy of motion also depends upon the isotopic mass. This introduces a factor ΔK into the right-hand side of Eq.

(3.23), where ΔK is the fraction that resides in the migrating atom of the kinetic energy associated with the vibrational mode corresponding to passage over the energy barrier.

The correlation factor f also depends on the isotopic mass, through ω. In most metals, f is given by

$$f = u/(\omega + u) \tag{3.24}$$

where u is a function of the jump frequencies of atoms other than the tracer, and therefore is independent of the isotope. Using Eqs. (3.21), (3.23), and (3.34), an expression for the ratio of the isotopic diffusion coefficients results:

$$[(D_\alpha/D_\beta) - 1][(M_\beta/M_\alpha)^{1/2} - 1]^{-1} = f_\alpha \Delta K \tag{3.25}$$

If more than one (say n) atom migrates, as in an interstitialcy jump ($n = 2$), this equation is modified to

$$E_{\alpha\beta}(n) \equiv \left(\frac{D_\alpha}{D_\beta} - 1\right)\left\{\left[\frac{M_\beta + (n-1)M_0}{M_\alpha + (n-1)M_0}\right]^{1/2} - 1\right\}^{-1} = f_\alpha \Delta K \tag{3.26}$$

where M_0 is the mass of the stable isotope. The left-hand side of this equation is given the symbol $E_{\alpha\beta}(n)$ and can be obtained from the experiment plus an arbitrary choice of n. This quantity can be of assistance in determining the mechanism of diffusion as follows. Since both f_α and ΔK must be less than unity, n must be such that $E_{\alpha\beta}(n)$ is less than unity. For instance, Rothman and Peterson[59] found $E_{\alpha\beta}(1) = 0.684 \pm 0.014$ for Cu using Cu^{64} and Cu^{67}. Since $E_{\alpha\beta}(2)$ would be 1.3, a single-atom mechanism is the only one consistent with their data. Furthermore, the correlation factors are known for the various single-atom mechanisms that seem reasonable. Using these, they found ΔK values less than unity only with the interstitial ($\Delta K = 0.68$) and the single-vacancy ($\Delta K = 0.87$) mechanisms. Since interstitial mechanisms in the past have given ΔK close to unity,[60] the isotope-effect experiment supports either the single-vacancy mechanism, with $\Delta K = 0.87$, or a composite mechanism with ΔK closer to unity but the correlation factor f less than the single-vacancy value, 0.786, in the fcc lattice. The second mechanism might well be divacancy diffusion. However, in this case, the contribution of the divacancy mechanism to the overall diffusion rate is probably small, since the value observed for $E_{\alpha\beta}(1)$ was independent of temperature. In a mixed diffusion process, unless by chance the different mechanisms have the same temperature dependence, the relative contributions of the several mechanisms should change with temperature, and therefore so also should the effective value of $(D_\alpha/D_\beta) - 1$ in Eq. (3.26). In this connection, it is interesting to note that Rothman *et al.*[97] did find $E_{\alpha\beta}(1)$ to be temperature-dependent for Ag, for which there is other evidence (see Section 3.2.3, under "Silver") for a divacancy contribution. The temperature behavior of $E_{\alpha\beta}(1)$ may be one of the most reliable criteria for a mixed diffusion process.

c. *Quenching Experiments.* The special property of ductility possessed by many metal allows thin wires to be drawn, which may be very rapidly quenched, at maximum cooling rates of 10^5 deg/sec or higher. With such rapid quenches, one may hope to retain the equilibrium defects present at the high quench temperature T_q and to determine the concentration χ from an examination at lower temperatures from some property sensitive to the presence of the defects. The variation of the χ with T_q should yield the formation parameters via Eq. (2.8), while subsequent study of the rate of annealing-out of the quenched-in defects ("recovery") should provide a measure of the parameters governing the defect motion.

The resistivity is perhaps the most sensitive property of a metal to the presence of point defects. In general, the resistivity in a good metal can be rather well expressed by

$$\varrho = \varrho_0 + \varrho_L(T) \tag{3.27}$$

where $\varrho_L(T)$ represents the (temperature-dependent) scattering of the conduction electrons by the phonons, and ϱ_0 the resistivity due to fixed scatterers such as impurity atoms and point defects. By making the measurements at liquid helium temperatures, $\varrho_L(T)$ can be virtually eliminated. The residual resistivity ϱ_0 for low concentrations of defects is essentially linear in the concentration and the contributions from various sorts of defects do not interact:

$$\varrho_0 = \sum_i (\Delta\varrho_i)\chi_i \tag{3.28}$$

where $\Delta\varrho_i$ is the contribution per unit atom fraction from the ith kind of defect at concentration χ_i. In particular, while the defect concentration can approach 10^{-3} at the melting point, materials can be obtained with total impurity concentrations as low as 10^{-5} or better. Assuming the defects and impurities make about the same contribution to the resistivity, the defects can be arranged to produce a contribution to the resistivity at least two orders of magnitude larger than the background resistivity due to impurities. Thus, provided the quench really does retain all the defects, the precision with which they can be detected can be quite high. Where a single defect type dominates, the quenched-in resistivity $\Delta\varrho_q$ is given by

$$\Delta\varrho_q = \Delta\varrho_j[\exp(s_j{}^F/k)] \exp(-h_j{}^F/kT)$$

A number of studies of the influence of quench rate on the quenched-in resistivity have been made. Gregory[61] has shown that the possibility of trapping the vacancies requires that the enthalpy of motion be not too small relative to the enthalpy of formation of the defects. Since at equilibrium at high temperatures, the defect concentration is given by Eq. (2.8),

$$\chi = \exp(-g^F/kT) \tag{3.29}$$

the maximum rate at which defects must disappear in order to maintain equilibrium as the temperature is lowered is

$$\left(\frac{d\chi}{dt}\right)_1 = \frac{d\chi}{dT}\,\frac{dT}{dt} = \frac{h^F}{kT^2}\left[\exp\left(-\frac{g^F}{kT}\right)\right]\frac{dT}{dt} \tag{3.30}$$

On the other hand, in the presence of a fixed concentration J of sinks at which defects are annihilated, the maximum rate at which they can disappear if originally at thermal equilibrium is

$$\left(\frac{d\chi}{dt}\right)_2 \approx -J\nu[\exp(-g^M/kT)] \exp(-g^F/kT) \tag{3.31}$$

where ν is the frequency that appears in the expression for the diffusion coefficient for the defect [Eq. (3.15)]. The condition that the defects be retained in a quench from the melting temperature T_m is

$$|\,(d\chi/dt)_1\,| > |\,(d\chi/dt)_2\,|$$

or, ignoring entropy terms,

$$-dT/dt \gtrsim J\nu(kT_m{}^2/h^F) \exp(-h^M/kT_m) \tag{3.32}$$

Damask and Dienes[62] have suggested that the atom fraction J of sinks presented by dislocations can be calculated as $N_0 v_0^{2/3}$, where N_0 is the number of dislocation lines per square centimeter and v_0 is the volume per atom, $a^3/4$ for the fcc lattice with a the lattice parameter. With $a^3 = 6.4\times10^{-23}$ cm^3 and $N_0 = 10^7$, one finds $J \approx 10^{-8}$, about what Mori *et al.*[63] found from their studies of the influence of quench rate upon the quenched-in resistivity in gold. Using 10^{13} sec^{-1} for ν and $7\times10^{-4}T_m$ for both enthalpies, one finds from Eq. (3.32) for all the metals except sodium that quench rates in excess of 10^3–10^4 deg sec^{-1} would result in freezing-in of the vacancies present at the melting point. These, in fact, underestimate somewhat the quench rates required, as shown in Fig. 4. For sodium, and probably for all the alkali metals, a rate in excess of about 10^6 deg sec^{-1} would be required, a result that accounts for the fact that quenched-in resistivity due to vacancies has not been observed in those metals.

Figure 4 shows a typical case, drawn from the work of Mori *et al.*[63] on gold. Vacancies are the dominant defect, and while divacancies and larger clusters should be produced during the slower quenches, the scattering of electrons by vacancies is not much influenced by this clustering. Hence the departure from the linearity expected from Eq. (3 29), most pronounced for the slower quenches, is due to the annihilation of vacancies at sinks such as dislocations and surfaces and not their coalescence into divacancies.

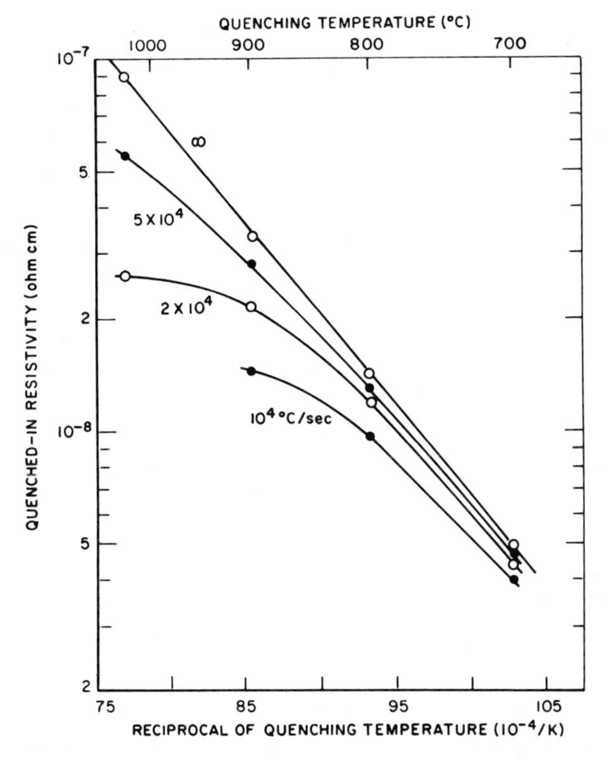

Fig. 4. Variation of quenched-in resistivity of gold with quenching temperature for the several quenching rates marked on the curves. The data on the line marked ∞ have been extrapolated to infinite rate of quench. From Mori *et al.*[63]

Quench rates can be increased by using smaller wires and by the choice of the quench medium. Extrapolation to infinite quench rate has been used to eliminate the loss of defects, but may also increase a strain-induced contribution to the resistivity that could wrongly be attributed to defects. As Fig. 4 shows, moderately rapid quenches from lower temperatures produce the right temperature dependence and, therefore, the correct derived formation enthalpy. Flynn *et al.*[64] have worked out a technique to account for the defect loss and so rectify data taken with less-than-optimum quench rates to produce correct formation enthalpies.

Quenches from low temperatures also reduce the relative concentration of divacancies formed during the quench, since both the divacancy concentration at equilibrium and the probability of formation of divacancies during the quench depend on the square of the concentration of single vacancies. Since the scattering of the conduction electrons is not much influenced by this formation of divacancies, there is little error introduced into the formation enthalpy of single vacancies determined from the quenched-in resistivity. However, since divacancies in most metals have a lower enthalpy of motion that single vacancies, and therefore move much more rapidly, they can profoundly alter the course of recovery during anneals after quenching.

Low-temperature quenches are necessary to obtain the single-vacancy enthalpy of motion.

d. *Measurement of High-Temperature Physical Properties under Equilibrium Conditions.* As discussed in Section 3.1 for noble gases, physical properties such as the specific heat, enthalpy, etc. at temperatures close to the melting point contain a contribution due to thermally generated defects. It is necessary to estimate the ideal-lattice value for the same crystal without defects in order to obtain the defect contribution by subtraction from the measured data. As with the noble gases, this is usually done for metals by using some extrapolation scheme to project data from low temperatures, where contributions from thermally generated defects are negligibly small, into the high-temperature region. This extrapolation involves considerable uncertainty.

In metals, as we have seen, the conduction of electricity by electrons provides a special property that can be used for this purpose. Table III, taken from Simmons,[65] compares the expected magnitudes of defect contributions to those of ideal-lattice contributions for various physical properties for silver just below the melting point, for which the thermal defects are predominantly vacancies.

Table III

	Volume expansion	Electrical resistivity	Energy content	Compressibility
Host	$3(\Delta l/l) = 0.0789$	$\varrho = 8.37\ \mu\Omega$	$\int_0^{T_m} C_p\, dT = 0.335$ eV	K
Defects	$\chi_v \beta_v = 1.0 \times 10^{-4}$	$\chi_v\, \Delta\varrho_v = 0.022\ \mu\Omega$	$\chi_v h_v^F = 1.85 \times 10^{-4}$ eV	$K_v = \chi_v \beta_v K$ (pure bulk effect)
Ratio	789	380	1810	10,000

It is clear that the resistivity offers a considerable advantage in sensitivity over the other properties. This approach is based upon Eqs. (3.27)–(3.29), with the assumption that the defect contributions are independent of temperature (Matthiessen's rule). The problem is to evaluate $\varrho_L(T)$, the lattice contribution, at high temperatures. Mott and Jones[66] point out that for an Einstein model, the resistivity is given by

$$\varrho_L \approx T\theta^{-2} \tag{3.33}$$

where θ is the characteristic temperature, usually taken as the Debye temperature. This is then used as an extrapolation formula by inserting, in some form, the temperature dependence of θ.

Formally, the transformation can be made to

$$d[\ln(\varrho_L/T)]/dT = 2\beta\gamma_G \tag{3.34}$$

Here, β is the volume thermal expansion coefficient, $V^{-1}(\partial V/\partial T)_P$, and γ_G is the Gruneisen constant, $-\partial(\ln\theta)/\partial(\ln V) = \beta V/C_V K_T$, where C_V is the constant-volume specific heat and K_T the isothermal compressibility. Provided β and γ_G are constants, this suggests that the plot of $\ln(\varrho/T)$ should be linear in T, which forms the basis for a simple linear extrapolation to higher temperatures that is sometimes used. However, due to anharmonic effects, β and γ_G are not expected to be constant. Direct evaluations of β and γ_G have been made from thermal data,[67] or their temperature dependence estimated assuming that the major influence of the temperature is through the volume expansion.[68,69] Figure 5, drawn from Misek and Polak,[69] illustrates well the difficulty in extrapolating the lattice resistivity. For curve (c), they estimated the temperature dependence of θ from the thermal expansion. Given the extrapolated lattice contribution to the resistivity, one writes

$$\varrho(T) - \varrho_L(T) = (\Delta\varrho)\chi$$

where $\Delta\varrho$ is the resistivity introduced per unit atom fraction of defects. It is usually assumed to be a constant (Matthiessen's rule), although there is evidence that this assumption may not be valid for aluminum,[70] platinum,[71] and perhaps gold.[72] The temperature dependence of $\varrho(T) - \varrho_L(T)$ thus makes possible an estimate of h_v^F, although an independent evaluation of $\Delta\varrho$ (and this quantity is not yet well known) is necessary to obtain s_v^F. It is also worth noting that the total vacancy concentration, including those in clusters, is measured. However, even divacancy concentrations are probably too low to have much effect on the derived values of h_v^F and s_v^F.

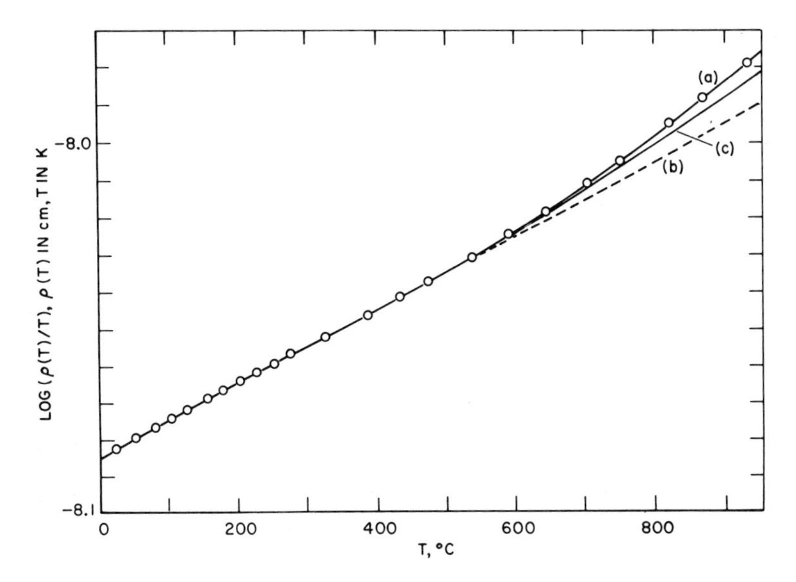

Fig. 5. Temperature dependence of $\ln[\varrho(T)/T]$ for gold. The experimental curve is marked (a), while the line marked (b) is a linear extrapolation from lower temperatures. Curve (c) is an estimate of the ideal lattice contribution. From Misek and Polák.[69]

e. *Recovery Experiments.* Nonequilibrium concentrations of defects can be introduced into metals in several ways. The methods used most are, in addition to the quenching from high temperature already discussed, irradiation with neutrons, electrons, and other energetic particles, and plastic deformation at low temperatures ("cold-work"). Upon subsequent anneal, the point defects migrate to various sinks and disappear as the concentration returns to the equilibrium value at the anneal temperature. If some property, such as the resistivity, is followed during the anneal, the rate of decay ("recovery") of the property to the equilibrium value can be used to obtain estimates of defect motion enthalpies.[49]

Let the property p be linear in the defect concentration χ. Then,

$$(dp/dt)_T = K'(d\chi/dt)_T \tag{3.35}$$

Depending upon the mechanism by which the defects disappear, $(d\chi/dt)_T$ can be a function of χ itself. For migration to fixed sinks, $(d\chi/dt)_T$ is expected to be proportional to χ. If the slow process is the bimolecular reaction to form divacancies, which then quickly migrate to permanent sinks, then $(d\chi/dt)_T$ should be proportional to χ^2. In many cases, the dependence is more complicated than in these simple examples.[73] In general, however,

$$\begin{aligned} (d\chi/dt)_T &= -F(\chi)\nu_0 \exp(-h^M/kT) \\ (dp/dt)_T &= -F(p)\nu_0 \exp(-h^M/kT) \end{aligned} \tag{3.36}$$

While it is not always possible to determine $F(p)$, there are several techniques by which h^M may be determined. In the change-of-slope method, the temperature is suddenly changed during the anneal. At this time, $F(p)$ is the same for the two temperatures, and may be eliminated by dividing one value of dp/dt by the other, giving

$$h^M = \frac{kT_1T_2}{T_2 - T_1} \ln \frac{(dp/dt)_2}{(dp/dt)_1}$$

Alternatively, two isotherms at different temperatures may be compared, taking the time in each to achieve the same degree of recovery. Integration of Eq. (3.36) shows that

$$h^M = [kT_1T_2/(T_2 - T_1)] \ln(t_1/t_2) \tag{3.37}$$

A third method, due to Meechan and Brinkman,[74] is more complex, and involves the annealing of two identical samples, one isothermally and the other "isochronally" (by a succession of equal-time anneals at successively higher temperatures).

The properties most commonly used to follow this recovery are the resistivity, internal friction and sound attenuation, elastic constants, and magnetic properties for ferromagnetic metals.

3.2.4. *Survey of Experimental Values for Vacancy Formation Parameters in Metals*

a. *Face-Centered Cubic Metals*

(1) *Noble Metals.* The noble metals Cu, Ag, and Au are perhaps the best studied of the metals. Extremely pure samples can be obtained; they are relatively inert and therefore not too difficult to handle; and the temperature range of interest, up to the melting point, is accessible without undue difficulty. Since very thin wires can be formed, specimens can be quenched from high temperatures with some hope of freezing-in the high-temperature equilibrium state for study at lower temperatures. In common with other face-centered cubic metals,[75] the noble metals are thought to possess much larger interstitial formation energies than vacancy formation

Table IV. Vacancy Parameters in Gold from X-Ray and Bulk Thermal Expansions

Ref.	h^F_{1v}, eV	s^F_{1v}/k	Δg^P_{2v}, eV	Notes
76	0.94	1.0	—	Original data
65	0.96	1.2	—	Ref. 76 data
56	0.86	0.5	<0.16	Ref. 76 data plus self-diffusion results

energies, so that only vacancies are present in significant numbers up to the melting point. The comparison of X-ray and bulk thermal expansion data supports the notion that vacancies are the predominant thermal defects at all temperatures.

(a) *Gold.* Of the noble metals, gold exhibits the largest vacancy concentrations $[\chi_v(T_m) = 7\times 10^{-4}]$ at the melting point,[76] and therefore offers the best chance for determining the various defect parameters with reasonable precision. Table IV displays the results of some of the more recent attempts to determine these parameters.

The data of Simmons and Balluffi[76] represent the most direct measurement of the concentration of vacancies in gold as a function of temperature. The scatter in their data is quite large, since each point is given by the rather small difference between two large quantities. Simmons[65] revised the parameters slightly in reviewing this work. Seeger and Mehrer[56] used the same data but represented them by rather different parameters. The discrepancies among these values for the enthalpy and entropy of formation of single vacancies in gold seem to arise wholly from differences in representation of the same data, and perhaps reflect the extent to which personal choice can enter into the presumably objective description of nature. The uncertainty in the enthalpy of formation of vacancies in gold as measured by the direct comparison of X-ray and bulk thermal expansions is at least as large as ± 0.05 eV and perhaps even ± 0.1 eV. This, it must be remembered, is the most direct method of determining this quantity and in one of the most favorable cases among metals.

The equilibrium measurements of Simmons and Balluffi do not distinguish between single vacancies and those existing as divacancies or in larger complexes. On the other hand, it now appears probable that the divacancy binding energy is less than 0.16 eV, e.g., Seeger and Mehrer,[56] which implies that less than 10% of the vacancies are present as divacancies at the melting point. Under these conditions, the temperature dependence of the concentration of all vacancies would be indistinguishable from that of the single vacancies alone. The Simmons–Balluffi value therefore probably applies to the single vacancies.

With this large an uncertainty in the single-vacancy formation parameters, it has been necessary to appeal to other data to try to reduce the uncertainty. Three other techniques have been rather widely used for gold. These are: (1) the use of self-diffusion data in conjunction with some independent estimate of the free energy of activation of the vacancy mobility; (2) the use of the dependence of the quenched-in resistivity (or, in one case, of the stored energy) on the temperature from which the specimen is quenched; and (3) the estimation of the contribution made by vacancies to such properties as specific heat or resistivity measured under high-temperature equilibrium conditions.

Self-diffusion data for gold have now been obtained over a wide range of temperatures: 600–1000°C,[77–81] and 286–412°C.[82] Seeger and Mehrer[56] have shown that over the whole temperature range, the data can almost be fitted by a single activated process,

$$D^* = D_0 \exp(-Q/kT) \tag{3.38}$$

with values for the parameters D_0 and Q very similar to those given by Makin *et al.*[78] ($D_0 = 0.091$ cm^2/sec and $Q = 1.81$ eV). Except at perhaps the highest temperatures, single vacancies are expected to be of paramount importance, so that Q should be very close to the sum of h_{1v}^F and h_{1v}^M. Estimates of h_{1v}^M have been obtained by many investigators, almost all using the recovery of quenched-in resistivity. It is now recognized that divacancies probably dominate this process in specimens quenched from temperatures above about 700–800°C. The observed activation energies for the recovery process tend to be low under these conditions, and to correspond to the energy of migration of the divacancies (0.6 eV,[83] 0.70 eV,[63] 0.56–0.67 eV,[84] 0.72 eV,[85] 0.71 eV,[85] 0.70 eV,[86] about 0.78 eV,[56] and 0.52 eV,[87]).

Estimates of the activation energy of migration of single vacancies, obtained from experiments in which the quench temperature was held below 700°C, are somewhat higher. Table V contains the results of some of these studies.

Impurities apparently play a role in determining the observed activation energies of the recovery process. Jeannotte and Machlin[84] found that the transition from divacancy (high temperature) to single-vacancy annealing occurred at a higher temperature when oxygen was present in the atmosphere during the prequench anneal than when it was not. Kino and Koehler[86]

Table V. Estimates of the Migration Energy of Single Vacancies in Gold

Ref.	h_{1v}^M, eV	Notes
84	0.80	Divacancies suppressed by oxygen
86	0.94	Divacancy contribution accounted for
89	0.78	—
88	0.85	From both quenched and irradiated specimens (3-MeV electrons, Stage III)

and Lee and Koehler[88] noted that the observed activation energies in specimens of very high-purity gold quenched from 700°C increased as the residual resistivity, a measure of the impurity concentration, decreased. Camanzi *et al.*[89] found a drop in the activation energy when 1% platinum was added to very pure gold.

There appears to be a tendency for the value of h_{1v}^M to rise as greater care is taken in its determination and more interfering effects are eliminated. The "best" value seems to be about 0.9 eV, with an uncertainty of perhaps ±0.05–0.1 eV. Combining this with the self-diffusion activation energy of about 1.8 eV produces a value of about 0.9 eV for the single-vacancy formation energy, with again an uncertainty not much less than ±0.1 eV. This estimate for h_{1v}^F is in essential agreement with the directly measured values given in Table IV.

The quenching experiments also allow a more direct estimate of the formation enthalpy h_{1v}^F, since for a perfect quench producing vacancies each with the same contribution to the resistivity, the quenched-in resistivity is expected to vary with the quench temperature T_q according to the equation

$$\Delta\varrho_q = K\exp(-h_{1v}^F/kT_q) \tag{3.39}$$

where K is a constant containing the contribution of each vacancy to the resistivity and the entropy contribution to the vacancy concentration. These estimates for h_{1v}^F range from about 0.91 eV[90] to 1.03 eV.[85] Careful attempts have been made to extrapolate to infinite quench rates by varying the quench time[63,64] or the specimen thickness,[91] and by quenching from temperatures so low that the results were insensitive to changes in the quench rates.[86] The estimates center around 0.95 eV, in essential agreement with the more directly determined values in Table IV. A similar value was also obtained by DeSorbo[83] from calorimetric measurements of the energy released ("stored energy") upon reannealing quenched specimens. The stored energy is expected to depend upon temperature according to an equation analogous to Eq. (3.29). The observed temperature dependence produced a value of 0.97 eV for h_{1v}^F.

There may be a balance between factors tending to produce quenched-in vacancy concentrations that are too small and those tending to make the concentrations too large. Loss of vacancies to sinks during quenching occurs, and various ways to account for this loss have been used.[63,64] Comparison specimens used to establish the resistivity of vacancy-free material can, in fact, contain some vacancies,[64] thus introducing an underestimation of the vacancy-produced resistivity in the quenched specimens. On the other hand, quenching produces strains, which in turn produce a contribution to the quenched-in resistivity,[86] sometimes included erroneously in the vacancy effect. No satisfactory way of removing or accounting for this effect is as yet available.

Equilibrium measurements have been made of several properties presumably sensitive to the presence of vacancies. The vacancy contributions have been estimated as the difference between the observed value and an "ideal lattice" value obtained by extrapolation from lower temperatures where vacancies are not expected to make a measurable contribution. Table VI displays the results of three such experiments. The uncertainties in estimating the vacancy-free ideal lattice values at the high temperatures make these results rather less reliable than those already discussed.

Activation volumes have been determined from pressure experiments, and are shown in Table VII. The atomic volume is denoted by v_0 and the activation volume for self-diffusion by v_{D^*}. As expected, the sum $(v_{1v}^F + v_{1v}^M)$ of the formation and motion volumes is close to that for self-diffusion.

Seeger and Mehrer[56] have produced a self-consistent set of vacancy parameters for gold from a combination of the self-diffusion data from various sources and the thermal expansion measurements of Simmons and Balluffi.[76] These parameters, with perhaps somewhat greater un-

Table VI. Vacancy Formation Parameters in Gold from Equilibrium Measurements

Ref.	h_{1v}^F, eV	s_{1v}^F/k	Notes
69	0.80	—	Resistivity; extrapolation via lattice parameter and Debye θ
222	0.56	—	Gruneisen constant; linear extrapolation
109	1.0	3.1	Heat capacity; linear extrapolation

Table VII. Volumes for Vacancy Formation, Motion, and Self-Diffusion in Gold

Ref.	v_{1v}^F/v_0	v_{1v}^M/v_0	v_{D^*}/v_0	Notes
223	—	0.15	—	Recovery of quenched-in resistivity
99	—	—	0.72	Tracer diffusion
90	0.53	—	—	Quenched-in resistivity
224	0.52	—	—	Magnitude of quenched-in resistivity
104	0.60	—	—	Pressure dependence of equilibrium resistivity

certainties than given by Seeger and Mehrer, represent the present state of knowledge of vacancy parameters in gold. They are given in Table VIII, with the formation volume from Table VII. Because of the uncertainty in both ν, the vibration frequency of the jumping atom, and s^M, its entropy of motion, Seeger and Mehrer estimate only the combined quantities, $\nu \exp(s^M/k)$, as they appear in the self-diffusion coefficient, Eq. (3.16).

(b) *Silver*. The situation for silver is somewhat more complicated than that for gold. Silver dissolves significant quantities of oxygen, and to achieve meaningful results, experiments involving high-temperature anneals need a protective atmosphere. The energy of formation of vacancies in silver is somewhat higher, and the melting point somewhat lower, than in gold. As a result, the number of vacancies at the melting point is lower, making the detection and measurement of vacancy-produced effects harder and less precise.

Nonetheless, reasonable consistency has been attained among vacancy parameters obtained by several different methods, as Table IX shows. Simmons and Balluffi,[65] from simultaneous measurements of the bulk

Table VIII. Vacancy Parameters for Gold According to Seeger and Mehrer[56]

Single vacancies	Divacancies
$h_{1v}^F = 0.87$ eV	$\Delta g_{2v}^P < 0.16$ eV
$h_{1v}^M = 0.89$ eV	$h_{2v}^M \leq 0.8$ eV
$s_{1v}^F = 0.5k$	—
$v_{1v}^F = 0.5v_0$ (from Table VII)	—
$\nu_1 \exp(s_{1v}^M/k) = 1.9 \times 10^{13}$ sec^{-1}	$\nu_2 \exp(s_{2v}^M/k) = 1.5 \times 10^{13}$ sec^{-1}

Table IX. Vacancy Parameters in Silver

Ref.	h_{1v}^F, eV	s_{1v}^F/k	h_{1v}^M, eV	h_{2v}^M, eV	Notes
225	1.10 ± 0.04	—	—	—	Quenched-in resistivity
226	1.06 ± 0.07	—	—	—	Quenched-in resistivity
67	1.09 ± 0.1	1.5	—	—	Bulk and X-ray thermal expansion
93	—	—	0.86 ± 0.06	0.58 ± 0.02	Quenched-in resistivity
227	1.04 ± 0.1	—	—	—	Quenched-in resistivity
95	—	—	0.88 ± 0.03	—	Quenched-in resistivity
228	1.10	—	—	0.57 ± 0.06	Quenched-in resistivity
94	1.10 ± 0.04	1.7	0.83 ± 0.05	0.57 ± 0.03	Quenched-in resistivity
110	1.06 ± 0.05	—	—	—	Equilibrium resistivity

and X-ray thermal expansion, found the vacancy concentration at the melting point $\chi_v(T_m)$ to be $1.7 \pm 0.5 \times 10^{-4}$, with single vacancies probably predominant. [Gertsriken and Novikov[92] investigated the thermopower of quenched and unquenched specimens, and from the comparison derived $\chi_v(T_m)$ to be 2.7×10^{-3}. They worked in air, without a protective atmosphere, and their experiments may have been influenced by dissolved oxygen. At any rate, their value for $\chi_v(T_m)$ seems unrealistically high in comparison to other data for fcc metals.] The precision of the experiments by Simmons and Balluffi did not allow a meaningful determination to be made of the temperature dependence of χ_v. To calculate the enthalpy of vacancy formation, they selected a value of $(1.5 \pm 0.5)k$ for the entropy of formation, and calculated h_{1v}^F using Eq. (3.1) and their measured $\chi_v(T)$. This procedure seems to be vindicated by the agreement found between their value for h_{1v}^F and those resulting from studies of the quenched-in resistivity.

While divacancies are not as numerous as single vacancies at equilibrium up to the melting point, they probably do form during quenching from high temperatures and anneal out rapidly at low temperatures (~0°C) upon subsequent annealing. Single-vacancy motion can be observed in specimens either quenched from temperatures low enough to avoid appreciable divacancy formation[93] or else after a short pulse anneal at temperatures of the order of 270°C to break up the divacancies already formed.[94] Ramsteiner *et al.*[95] observed single-vacancy annealing without these precautions, a result which may be related to the fact that their prequench heat treatment was carried out without an oxygen-excluding protective atmosphere.

Doyama and Koehler,[94] from a model of the dynamics of the formation and redissociation of divacancies, have estimated the divacancy binding energy to be 0.38 ± 0.05 eV. This is considerably greater than the value of about 0.1 eV which appears most reasonable for gold, and suggests that in silver, divacancies should play a rather greater role in mass transport processes than they do in gold.

The sum of the enthalpies of single-vacancy formation and motion agrees satisfactorily with the activation energy for self-diffusion in silver (1.91 eV,[96] 1.97 eV[97]). The self-diffusion process should be dominated by the single vacancies,[94] except perhaps at the highest temperatures. Rothman *et al.*[97] did detect a slight curvature of the self-diffusion Arrhenius plot ($\ln D^*$ versus T^{-1}), and found a temperature dependence in the isotope mass effect (see Section 3.2.2). Both observations could be attributed to divacancies. They estimate that the maximum divacancy contribution near the melting point is between 11 and 45% of the total diffusivity. Even more than with gold, the activation volume for self-diffusion is a large fraction of the atomic volume ($v_{D^*} \approx 0.9 v_0$).[98–100]

Mehrer and Seeger[55] have found a consistent set of vacancy parameters for silver which reconciles much of the experimental data. Their values are somewhat temperature-dependent. Those for 900°C are shown in Table X.

(c) *Copper*. In copper, as in silver, the concentration of vacancies, even at the melting point, is very low. Simmons and Balluffi[101] determined that $\chi_v(T_m) \approx (2 \pm 0.5) \times 10^{-4}$ from simultaneous measurements of the bulk and X-ray thermal expansions, with most of the vacancies present as single vacancies. However, the scatter in their data was too large for a meaningful determination to be made of the temperature dependence of χ_v. In order to obtain an estimate of the enthalpy of formation, they chose $1.5k$ as the most probable value for the entropy and used Eq. (3.1) to obtain 1.17 eV for h_{1v}^F. An uncertainty of $\pm 0.5k$ in the adopted value of s_{1v}^F, plus

Table X. Vacancy Parameters for Silver at 900°C According to Mehrer and Seeger[55]

Single vacancies	Divacancies
$h_{1v}^F = 0.99$ eV	$\Delta h_{2v}^P = 0.24$ eV
$s_{1v}^F \approx 0.5k$	$\Delta s_{2v}^P = 2.6k$
$h_{1v}^M = 0.86$ eV	$h_{2v}^M = 0.58$ eV

the experimental uncertainty in χ_v, produced an uncertainty of about ± 0.11 eV in h_{1v}^F.

Other methods of determining the vacancy parameters for copper are also rather uncertain. For instance, the additional resistivity induced by quenching from high temperature is very small by virtue of the small vacancy concentrations, and is quite sensitive as well to the presence of oxygen, which is very hard to remove.

The most ambitious attempt to date to determine the vacancy parameters for copper is that of Mehrer and Seeger.[54] They fitted the very careful self-diffusion data of Rothman and Peterson,[59] including isotope effect measurements, with a model in which both single vacancies and divacancies contribute and in which a linear temperature dependence for both the formation and motion energies is allowed. It is interesting to note that although divacancies contribute as much as 30% of the diffusion current at the melting point (by virtue of their relatively high mobility), the Arrhenius plot for diffusion exhibits only a very slight curvature, easily missed in the experimental uncertainty. Even so, although the overall activation energy for self-diffusion derived by Rothman and Peterson from their data was 2.19 ± 0.01 eV, Mehrer and Seeger find the activation energy of the single-vacancy contribution to be 2.09 eV. The absence of curvature in the Arrhenius plot for self-diffusion does not guarantee that the observed activation energy is accurately the sum of the enthalpies of formation and motion of single vacancies.

In their analysis, Mehrer and Seeger use the data of Simmons and Baluffi[101] for χ_v together with an assumed value of $0.3k$ for the formation entropy at 400°C. They deduce a substantially lower value for h_{1v}^F than Simmons and Balluffi do from the same data. This analysis also produces

Table XI. Vacancy Formation Parameters in Copper

Ref.	h_{1v}^F, eV	s_{1v}^F/k	Δh_{2v}^P, eV	$\Delta s_{2v}^P/k$	Notes
101	1.17	1.5	—	—	Bulk and X-ray thermal expansion
229	1.06	—	—	—	Quenched-in resistivity, $T_q = 590$–765°C
230	1.05	3.7	—	—	Equilibrium heat capacity, linear extrapolation
55	1.03	0.3	0.12	~2	Ref. 101 data plus self-diffusion

Table XII. Vacancy Formation Parameters for Aluminum

Ref.	$\chi_v(T_m)$	h^F_{1v}, eV	s^F_{1v}/k	Notes
231, 232	—	0.79 ± 0.04	2.4	Quenched-in resistivity, $T_q = 260$–$320°C$
30	9.4×10^{-4}	0.74 ± 0.76	2.0–2.4	Bulk and X-ray thermal expansions
103	—	0.77	—	Equilibrium resistivity
233	—	0.75 ± 0.02	2.3	Quenched-in resistivity, $T_q = 280$–$500°C$
234	—	0.76 ± 0.03	—	Quenched-in resistivity, $T_q = 300$–$400°C$
91	—	0.76 ± 0.02	—	Quenched-in resistivity, extrapolated to zero thickness
235	8.3×10^{-4}	0.71 ± 0.04	1.76 ± 0.56	Bulk and X-ray thermal expansions
236	8.7×10^{-4}	—	—	Bulk and X-ray thermal expansions
102	—	0.71 ± 0.03	—	Quenched-in resistivity, corrected for vacancy loss
104	—	0.74	—	Equilibrium thermopower

a rather low binding enthalpy for the divacancy, $\Delta h \approx 0.12$ eV. The results of some of the most recent attempts to determine the vacancy formation parameters in copper are contained in Table XI.

(2) *Other fcc Metals.* (a) *Aluminum.* The vacancy concentration at the melting point $\chi_v(T_m)$ is about 10^{-3} (atom fraction) in aluminum, sufficiently large that fairly firm estimates of the vacancy formation parameters may be obtained. Table XII lists the results of some of the more recent determinations of these parameters for aluminum. The three determinations of $\chi_v(T_m)$ made by comparing the bulk and X-ray thermal expansions agree among themselves impressively well.

The values for h^F_{1v} and s^F_{1v} determined in the various ways also agree rather well. Bass[102] suggests that the "best" value for h^F_{1v} is about 0.73 eV, which seems reasonable in view of the data in Table XII. The uncertainty appears to be of the order of ± 0.05 eV. Federighi (1965) has pointed out that the high value for h^F_{1v} (0.79 eV) given by DeSorbo and Turnbull relies rather heavily upon one experimental point; the rest of their data are consistent with a considerably lower value. The values derived from high-temperature equilibrium measurements (Simmons and Balluffi[103] measured the resistivity and Bourassa *et al.*[104] the thermopower) depend upon extrapolations from lower temperatures of hypothetical vacancy-free properties, and must therefore be considered less reliable than the others.

Table XIII. Enthalpies of Vacancy Motion in Aluminum

Ref.	h_{1v}^{M}, eV	h_{2v}^{M}, eV	Notes
232	0.65	—	Quenched-in resistivity, $T_q = 260$–$320°C$
108	—	0.50 ± 0.04	Quenched-in resistivity, $T_q = 395°C$
237	0.62	≤ 0.46	Quenched-in resistivity, $T_q = 400$–$600°C$
238	0.62 ± 0.04	—	Stage III recovery, electron irradiation-induced resistivity

Recovery of the quenched-in resistivity has been studied as a means of deriving activation energies of motion in aluminum. Some of the more recent results are shown in Table XIII. As with gold, the activation energy associated with the recovery process depends upon the temperature and rate of quench. Quenching from high temperatures probably results in the production of divacancies during the quench itself, while low-temperature quenches probably result in the predominance of single vacancies.

The sum of h_{1v}^{F} and h_{1v}^{M}, from Tables XII and XIII, is 1.4 eV or perhaps less, while tracer self-diffusion measurements by Lundy and Murdoch[105] produced an activation energy Q of 1.48 ± 0.05 eV, which is slightly larger. Spokas and Slichter[106] found a value of 1.4 ± 0.1 eV for Q from NMR measurements, and Butcher *et al.*[107] interpreted their high-temperature steady-state creep data in terms of a diffusion-controlled process with $Q = 1.45 \pm 0.04$ eV. The small discrepancy between the sum of the enthalpies of motion and formation of single vacancies on the one hand, and Q on the other, if indeed it is real, may well be the result of a small contribution to the self-diffusion process by divacancies. As shown by Mehrer and Seeger[54] in their analysis of copper, a small concentration of divacancies can produce an appreciable increase in Q over the single-vacancy value without introducing an observable curvature in the self-diffusion Arrhenius plot. Such a divacancy contribution is consistent with the estimate by Doyama and Koehler[108] that the binding energy of divacancies in aluminum is of the order of 0.17 eV. It does not seem necessary to invoke, as does Bass,[102] a value for h_{1v}^{M} as high as 0.75 eV in order to obtain agreement between the observed self-diffusion activation energy and the sum of the single-vacancy enthalpies of motion and formation.

A few measurements involving pressure as a variable have been performed on aluminum, from which activation volumes have been derived. Bourassa *et al.*[104] examined at high temperature the pressure dependence of the excess thermoelectric power attributable to vacancies, and estimated that $v_{1v}^F/v_0 = 0.52$, which is very similar to the value for gold. On the other hand, Butcher *et al.*[107] found from the pressure dependence of the high-temperature creep a value for the activation volume for self-diffusion of $1.3v_0$, much larger than for gold. Such a large value, if confirmed, means either a very large activation volume of motion for the single vacancy ($\sim 0.8v_0$) or a much larger role for divacancies ($v_{2v}^F \approx 0.9v_0$, according to Bourassa *et al.*[104]) in self-diffusion than is at present expected.

(b) *Platinum*. Sufficient work has been done on pure platinum to produce apparently reliable values for the vacancy parameters. Table XIV gives the results of several of the most recent attempts to determine the vacancy formation parameters. The value given by Kraftmakher and Lanina[239] depends upon the extrapolation of the properties of the vacancy-free crystal from lower temperatures, and is therefore uncertain to an unknown degree. The quenching experiments seem to illustrate the sensitivity of the apparent formation energy to the experimental details. The work of Ascoli *et al.*[110] and of Bacchella *et al.*[111] was done with high-purity (99.999%) platinum, and care was taken to examine the influence of the quenching rate. However, air was present during the high-temperature prequench anneals, the temperature intervals were not large, and most data were obtained from quenches from above 1000°C. The data of Gertsricken and Novikov[92] were taken with much less pure material and included quench temperatures well above 1000°C and spanning only a narrow range. On the other hand, Jackson[112] was careful to include only data from quenches starting at 1000°C or below (to minimize vacancy loss during quenching), used high-purity metal, and minimized plastic deformation. His value for h_{1v}^F appears to be the most reliable published value presently available. With an assumed value of unity for s_{1v}^F/k, this gives 6×10^{-4} for $\chi_v(T_m)$, at the melting point.

Table XIV. Vacancy Formation Parameters for Platinum

Ref.	h_{1v}^F, eV	s_{1v}^F/k	Notes
110	1.23	—	Quenched-in resistivity, $T_q = 850$–1600°C
111	1.20 ± 0.04	—	Quenched-in resistivity, $T_q = 1050$–1630°C
92	1.41 ± 0.04	—	Quenched-in thermopower, $T_q = 1300$–1660°C
112	1.51 ± 0.04	—	Quenched-in resistivity, $T_q < 1000$°C
239	1.6	4.5	Equilibrium resistivity and heat capacity

Table XV. Activation Energy of Motion of Single Vacancies in Platinum

Ref.	h_{1v}^M, eV	Notes
110	1.42	Quenched-in resistivity, $T_q = 1150$–$1400°C$
111	1.48 ± 0.08	Quenched-in resistivity, $T_q \approx 1100$–$1200°C$
240	1.43 ± 0.1	Recovery of resistivity, neutron-irradiated at 50°C and also cold-worked at liquid nitrogen temperature
112	1.38 ± 0.05	Quenched-in resistivity, $T_q = 1000°C$, also cold-worked at low temperature
241	1.36 ± 0.08	—
115	1.3 ± 0.1	Quenched-in resistivity, $T_q = 1000°C$
57	1.33 ± 0.05	Quenched-in resistivity, $T_q = 1000°C$

A number of estimates summarized in Table XV have been obtained for the activation energy of motion of the single vacancy. Only those quenching experiments are included for which the quench temperatures were relatively low, so that significant divacancy concentrations were avoided. When the quench temperatures are of the order of 1600°C, the apparent motional energy is near 1.1 eV, much lower than the values in the Table XV, and probably reflects divacancy motion.

Schumacher *et al.*[57] chose 1.38 eV as the "best" value for h_{1v}^M. Combining this with Jackson's value[112] for h_{1v}^F, 1.51 eV, produces an expected value of 2.89 eV for the activation energy of self-diffusion by single vacancies. The experimental activation energies, obtained from tracer diffusion measurements in the temperature range 1250–1725°C, are 2.96 ± 0.06 eV[51] and 2.89 ± 0.04 eV,[113] in excellent agreement.

Divacancy binding energies have been estimated from the resistivity recovery of wires quenched from higher temperatures. Kopan[114] and Polák[115,116] arrived at a value of about 0.2 eV, significantly less than Jackson's value of 0.4 eV.

Schumacher *et al.*[57] have analyzed a variety of data, including those from experiments on quenched-in resistivity and its recovery, and self-diffusion. They have arrived at the following self-consistent set of vacancy parameters for platinum that produced the best overall fit:

$$h_{1v}^F = 1.49 \text{ eV} \qquad \Delta h_{2v}^B = 0.11 \text{ eV}$$

$$s_{1v}^F = 1.3k \qquad \Delta s_{2v}^B = 2.5k$$

$$h_{1v}^M = 1.38 \text{ eV} \qquad h_{2v}^M = 1.11 \text{ eV}$$

$$\nu_1 \exp(s_{1v}^M/k) = 3.5 \times 10^{13} \text{ sec}^{-1} \qquad \nu_2 \exp(s_{2v}^M/k) = 1.1 \times 10^{13} \text{ sec}^{-1}$$

(c) *Lead.* Quench experiments have not been possible with lead because sufficiently thin wires cannot be drawn. Feder and Nowick,[116a] using X-ray data from d'Heurle *et al.*,[117] compared the bulk and X-ray thermal expansions in 99.999% pure lead up to the melting point. The vacancy concentration at the melting point $\chi_v(T_m)$ was only 2×10^{-4}, but an estimate of the temperature dependence of χ_v was established from the experiment. It was found that h_{1v}^F was 0.5 ± 0.1 eV and s_{1v}^F was $(0.7 \pm 2.0)k$.

With this low vacancy concentration, it seems likely that only single vacancies are important in self-diffusion. Hudson and Hoffman[118] found the self-diffusion activation energy to be 1.08 ± 0.03 eV. The activation energy of motion of single vacancies is therefore about 0.6 eV.

The activation volume for single-vacancy motion was also determined by Hudson and Hoffman,[118] $v_{1v}^M/v_0 = 0.64$, and Nachtrieb *et al.*,[119] $v_{1v}^M/v_0 = 0.77$. These values are similar to those found for the noble metals.

(d) *Nickel.* No reliable direct measurements of the formation parameters for vacancies in Ni appear to exist. Mehrer *et al.*[120] studied the recovery of the coercive field and the initial magnetic susceptibility after neutron irradiation at 80°C. They attributed a recovery stage occurring at 100–300°C to divacancies at the lower-temperature end and to single vacancies at the higher temperatures, and derived the values $h_{1v}^M = 1.46 \pm 0.07$ eV and $h_{2v}^M - \Delta h_{2v}^p = 0.5 \pm 0.06$ eV. An estimate of the divacancy motional parameters was obtained (F. Walz and H. Mehrer, as communicated privately to Seeger and Schumaker[53]) from magnetic relaxation, yielding $h_{2v}^M = 0.82 \pm 0.03$ eV and $\nu_2 \exp(s_{2v}^M/k) = 2\times10^{13}\ \text{sec}^{-1}$.

Seeger *et al.*[52] and Seeger and Schumaker[53] have combined these results with self-diffusion data of Hoffman *et al.*[120a] in an analysis in terms of both single-vacancy and divacancy contributions. Allowing for a temperature dependence of the defect parameters, they find the following room-temperature values:

$$h_{1v}^F = 1.35\ \text{eV} \qquad \Delta h_{2v}^p = 0.28\ \text{eV}$$
$$s_{1v}^F = 1.2k \qquad \Delta s_{2v}^p = 1.4k$$
$$h_{1v}^M = 1.48\ \text{eV} \qquad h_{2v}^M = 0.82\ \text{eV}$$
$$\nu_1 \exp(s_{1v}^M/k) = 1.6\times10^{14}\ \text{sec}^{-1} \qquad \nu_2 \exp(s_{2v}^M/k) = 2\times10^{13}\ \text{sec}^{-1}$$

The calculated atom fraction of all vacancies at the melting point is $\chi_v(T_m) = 4.7\times10^{-4}$.

b. *Body-Centered Cubic Metals*

(1) *Alkali Metals.* Vacancies appear to be the dominant defects in the alkali metals, although in lithium, significant numbers of interstitials

may also be present in thermal equilibrium, and recently, Brown *et al.*[120a] have suggested that diffusion in sodium involves interstitials. Attempts to quench-in high-temperature vacancy populations in sodium have been unsuccessful,[68] apparently because the enthalpy of vacancy motion is considerably less than the vacancy formation enthalpy.[61] As a consequence, direct determination of the enthalpy of motion for vacancies has not been possible, and therefore diffusion data cannot be used to obtain vacancy formation enthalpies. Equilibrium measurements of the high-temperature resistivity and heat capacity have provided estimates of the formation enthalpies of vacancies. The isotopic mass effect in self-diffusion has been used to a considerable extent, particularly with sodium, to help sort out the mechanism of diffusion, and thus to gain some idea of the defect populations.

Formation enthalpies attributable to single vacancies are listed in Table XVI for the alkali metals. The only values available for K, Rb, and Cs are derived from equilibrium measurements of resistivity and heat capacity. In addition to the usual uncertainties associated with this type of experiment, it has often been observed that with the alkali metals, some sort of premelting phenomena occurs, presumably due to impurities. Martin[121] has pointed out that in the case of Rb and Cs, a high-temperature

Table XVI. Vacancy Formation Parameters in the Alkali Metals

Metal	Ref.	$\chi_v(T_m)$	h_v^F, eV	s_v^F/k	Notes
Li	126	—	0.40 ± 0.01	—	Equilibrium resistivity
	125	—	0.34 ± 0.04	0.9 ± 0.8	Bulk and X-ray thermal expansions
Na	126	—	0.395 ± 0.004	—	Equilibrium resistivity
	68	7×10^{-4}	0.2	—	Equilibrium resistivity
	242	8×10^{-4}	0.14	-3 ± 2	Bulk and X-ray thermal expansions
	121	10×10^{-4}	0.46	7.3	Equilibrium heat capacity
	124	7.5×10^{-4}	0.42 ± 0.03	5.8 ± 1.1	Bulk and X-ray thermal expansions
K	126	—	0.395 ± 0.004	—	Equilibrium resistivity
	121	14×10^{-4}	0.42	7.8	Equilibrium heat capacity
Rb	121	25×10^{-4}	0.31	5.6	Equilibrium heat capacity
Cs	121	32×10^{-4}	0.35	7.5	Equilibrium heat capacity
	243	—	0.28	4.9	Equilibrium heat capacity

eutectic occurs with oxygen, making the interpretation of high-temperature heat capacity data in terms of the equilibrium thermal defects additionally hazardous. Diffusion coefficients for these metals always seem to be anomalously high a few degrees below the melting point,[122] a fact attributed by Mundy *et al.*[123] in the case of Na to K as an impurity. By analogy to Na, which has been more extensively studied, the dominant defects in K, Rb, and Cs are assumed to be vacancies. The values for them in Table XVI, $\chi_v(T_m) = 1\text{–}3\times 10^{-3}$, $h_v^F = 0.3\text{–}0.4$ eV, and $s_v^F/k = 5\text{–}8$, are generally reasonable for predominantly single vacancies in the light of the results for Na.

The sodium data exhibit remarkably good agreement among the different techniques, not only for the concentration of vacancies at the melting point, $\chi_v(T_m) = 7\text{–}10\times 10^{-4}$, but also for h_v^F and s_v^F/k if the data of Sullivan and Weymouth are not included in the comparison. There is a large scatter in these latter data, plus hysteresis in $\Delta l/l$, probably due to the formation of oxide on the surface of the specimen.[124] The value of $\chi_v(T_m)$ found by comparing the bulk and X-ray thermal expansions is reasonable, but the precision is not good enough to allow the temperature dependence to be very accurately determined. The formation parameters in Table XVI probably apply to single vacancies. Feder and Charbnau[124] have shown that reasonable concentrations of divacancies would not change the single-vacancy parameters much. Interstitials could also be present, but might be expected to dominate the diffusion process if present in significant concentrations. There is evidence, discussed below, that diffusion is primarily by way of vacancies, so that in sodium interstitials are probably not important. We can therefore take the parameters of Feder and Charbnau as the "best" values for single vacancies in sodium. The rather large value for the formation entropy suggests a very relaxed vacancy. No direct measurements have yet been made on the volume of formation, but it is expected, because of the relaxed nature of the vacancy in sodium, to be smaller relative to the atomic volume than was the case for the fcc metals.

There less extensive data for Li. Feder[125] suggested that vacancies predominate over interstitials, on the basis of a comparison of the bulk and X-ray thermal expansions. His value for h_v^F is in reasonable agreement with MacDonald's,[126] drawn from equilibrium resistivity measurements. Moreover, it is in general accord with the values from the other alkali metals. There is, however, the unexplained tendency for h_v^F to maximize at Na. Feder's value for the formation entropy for Li is definitely different from those for other alkali metals. It suggests a much less relaxed vacancy in Li.

Table XVII. Self-Diffusion Parameters in the Alkali Metals

Metal	Ref.	D_0, cm²/sec	Q, eV	Notes
Li	244	0.24	0.57	NMR
	245	0.39	0.59	Tracer, temperature range 60–170°C
	127	—	0.52	NMR
	246	—	0.51	NMR
	247	0.12	0.53	Tracer, temperature range 60–170°C
	122	0.12	0.548	Tracer, $^6D_7$,[a] 35–178°C
		—	0.557	Tracer, $^7D_6$, 35–178°C
Na	248	0.242	0.45	Tracer, polycrystal, 0–95°C
	244	0.20	0.43	NMR
	127	—	0.41	NMR, −50 to −42°C
	123	0.145	0.438	Tracer, 0–97°C
K	249	0.31	0.423	Tracer, 0–60°C
Rb	244	0.23	0.41	NMR

[a] Note: $^7D_6$ denotes diffusion of ^{6}Li as the tracer in ^{7}Li as the host.

Diffusion measurements, particularly involving the comparison of the rates of diffusion of several isotopes, have been of recent importance for the alkali metals. Table XVII contains the activation energies Q and preexponential factors D_0 for the alkali metals. The activation volumes for diffusion of Li ($0.28v_0$),[127] Na ($0.51v_0$, Nachtrieb *et al.*[50]; $0.41v_0$, Hultsch and Barnes[127]), and K ($0.55v_0$)[128] are positive, suggesting strongly that a vacancy mechanism is most likely. The rather small value compared with about, e.g., $0.7v_0$ for Au, is consistent with a relaxed vacancy.

The comparison of the formation and diffusion parameters in Na suggest an anomalously low value for the enthalpy of motion of the vacancy. For single vacancies on a bcc lattice, D_0 is related to the entropy of motion by

$$D_0 = fa^2\nu \exp(s^{D^*}/k)$$

where $f = 0.727$ is the correlation coefficient, a is the unit cube edge, and ν is a vibrational frequency often identified with the Debye frequency. The D_0 values in Table XVII for Li and Na produce values for s^{D^*}/k in the range 3.5–4.5. Taking $Q = h_v^F + h_v^M$ and $s^{D^*} = s_v^F + s_v^M$, the diffusion parameters for Na in Table XVII combined with Feder and Charbnau's[124] parameters for vacancy formation from Table XVI imply a very low enthalpy and a negative entropy of motion for the vacancies. While the low enthalpy of motion is perhaps consistent with a very relaxed vacancy, there is as yet no explanation of the negative entropy. In Li, the entropy of motion is positive and of normal magnitude ($s_v^M/k = 2$–3). In this regard, it is interesting to note that computer simulation of the vacancies in alkali metals, including Na, do not produce extensive relaxations,[129] although they do reproduce reasonably well the experimental enthalpies of formation and diffusion.

Sodium is a particularly favorable case in which to study the isotope mass effect, since two radioactive tracers (^{22}Na and ^{24}Na) exist in addition to the inactive ^{23}Na, with a relatively large mass difference and with half-lives that are within the range useful for diffusion studies and yet sufficiently different to allow separation of the activities of the two isotopes. For instance, it was found[60] that in NaCl, the isotope mass effect conforms to the single-cation-vacancy mechanism of sodium diffusion that has been thoroughly established through other experiments. In metallic sodium, the results are not so clear. Mundy *et al.*[123] found $E_{\alpha\beta}(1) = 0.36$ [see Eq. 3.26)]. While an interstitialcy mechanism cannot be ruled out on this evidence alone, it favors a relaxed vacancy mechanism, with the correlation coefficient $f = 0.727$ appropriate to the bcc lattice and a single vacancy, and a low value for ΔK, about 0.5, arising from the sharing of the kinetic energy at the saddle point among the several ions involved in the relaxed vacancy. Brown *et al.*,[129a] on the other hand, suggest that diffusion probably involves a crowdion interstitial with a formation energy very similar to that of the vacancy. Their model could account for the low migration energy and the low value observed for $E_{\alpha\beta}(1)$.

For Li, the computer calculations of Torrens and Gerl[129] suggest that diffusion by way of a split interstitial mechanism should be about as probable as by a vacancy mechanism. While the values for D_0 and Q in Table XVII are not particularly surprising, the isotope studies of Lodding *et al.*[122] are hard to reconcile with a simple vacancy mechanism. Since there are only two Li isotopes available (mass numbers 6 and 7, both nonradioactive) the usual isotopic mass effect experiments cannot be performed. Lodding *et al.* instead measured the tracer diffusion of each in a host of the other. The D_0 values were identical, but the Q's differed just at the limit of precision. The ratio of the two diffusion coefficients, ${}^7D_6/{}^6D_7$ (where ${}^7D_6$ denotes the tracer diffusion coefficient of ^6Li in ^7Li), was too large (about 1.3) for simple vacancy or interstitial mechanisms on a classical basis. However, they found that the experimental ratio could be approximately reproduced by a model in which the unit jump is the tunneling of a split interstitial. According to this model, the activation energy of motion is only about 0.05 eV, leaving 0.5 eV for the formation energy of the interstitial. It has

Table XVIII. Formation, Motion, and Diffusion Parameters for BCC Transition Metals

A. Formation

Metal	Ref.	h_v^F, eV	s_v^F/k	Notes
Nb	250	2.0	4.2	Equilibrium heat capacity
Mo	251	2.24	5.7	Equilibrium heat capacity
Mo	252	2.4	—	Quenched, vacancy loop count
Mo	253	1.86	3.9	Equilibrium enthalpy
W	132	3.3	1.4	Quenched, resistivity

B. Motion

Metal	Ref.	h_v^M, eV	Notes
W	132	1.9 ± 0.3	Recovery of quenched-in resistivity
W	131	3.3	Recovery after irradiation, field-ion microscope
Fe	254	1.1	Recovery of quench-in resistivity
Zr	255	0.8	Recovery of quench-in resistivity

C. Diffusion

Metal	Ref.	D_0, cm²/sec	Q, eV
Nb	255	1.3	4.1
Mo	256	0.1	4.0
W	133	42.8	6.6
Fe	256	0.5	2.6

also been suggested (e.g., Ott *et al.*[130] and the references therein) that a number of impurities diffuse in Li as interstitials. As Lodding *et al.*[122] note, however, the simultaneous operation of a vacancy mechanism cannot be ruled out.

(2) *Transition Metals.* Very little as yet is known about the intrinsic defects in the bcc transition metals. Because of the very high melting points, equilibrium properties would have to be measured at temperatures above the range presently accessible for high-precision measurements. Some thermodynamic measurements have been made, but no comparison of bulk and X-ray thermal expansions has yet been possible. Some quenching

experiments have been done, but detailed examination of the influence of quench rates has not yet been included.

Table XVIII lists the observed formation, motion, and diffusion parameters. There is reasonable agreement among the several determinations for Mo, obtained by equilibrium heat content measurements, but also including an estimate from the vacancy loops formed in materials quenched from high temperatures.

The motional enthalpies were determined by following the recovery of quenched or, in one case, irradiated material. Using the relation $Q = h^F + h^M$, estimates of the formation and motion enthalpies can be made for Nb, Mo, W, and Fe. Contrary to earlier suggestions (see Gregory[61]), the formation and motional enthalpies are about equal. The two values for W do not agree, reflecting uncertainty as to which stage in the recovery curve of W represents vacancy motion. The motional enthalpy of Jeannotte and Galligan,[131] 3.3 eV, fits better than does that of Schultz,[132] 1.9 eV, with the formation enthalpy of Schultz and the activation energy of diffusion of Andelin *et al.*,[133] 6.6 eV.

At the present time, the whole defect picture among several bcc transition metals (Zr, Ti, U, V, Cr, Hf) is clouded by the fact that the self-diffusion coefficient does not obey the simple Arrhenius behavior. The plot of $\ln D^*$ versus T^{-1} is curved or broken into segments (with the possible exception of W, the diffusion data in Table XVIII are all normal). Nowick[134] has suggested that in those metals (Zr, Ti, U, Hf) exhibiting a transformation at high temperatures into the bcc (β) phase, the anomalous diffusion is in some way related to the existence of the transformation. The anomalous behavior in V and Cr may reflect a strong divacancy contribution at high temperatures.

c. *Hexagonal Metals—Magnesium.* Several attempts, summarized in Table XIX, have been made to determine the formation parameters of the thermal equilibrium defects in magnesium. The comparison of bulk and

Table XIX. Vacancy Formation Parameters in Magnesium

Ref.	h_v^F, eV	s_v^F, eV	$\chi_v(T_m)$	Notes
135	0.89 ± 0.06	—	1×10^{-4}	Resistivity, quenched
136	0.81	—	—	Resistivity, equilibrium
257	0.58 ± 0.01	0 ± 0.3	7×10^{-4}	Bulk and X-ray thermal expansions

X-ray thermal expansions clearly shows these to be vacancies. The formation enthalpies as determined from resistivity measurements on quenched specimens[135] agree well with those determined from the equilibrium resistivity at high temperatures,[136] but this agreement may well be fortuitous. In his study of quenched specimens, Beevers[135] did not investigate the influence of quench rate, which, as seen in the discussion of the fcc metals above, can markedly effect the apparent formation enthalpy. The equilibrium resistivity measurements[136] depended upon a linear extrapolation into the high-temperature region of the low-temperature resistivity for an estimate of the ideal-lattice resistivity. The value for h_v^F derived from the comparison of bulk and X-ray thermal expansions[137] is considerably lower. However, there is considerable curvature in the plot of $\ln \chi_v$ versus T^{-1} from the data of Janot *et al.*[137] and the true value may well lie somewhat higher than the one they report.

The correlation between the enthalpies of vacancy formation and motion on the one hand and the melting temperature on the other, which, as noted in Section 3.2.1, appears to hold at least approximately for all but a few metals, suggests that h_v^F and h_v^M should be equal and given by $7\times10^{-4}T_m$ eV. For Mg, this gives about 0.7 eV, which is not unreasonable h_v^F from Table XIX. Shewmon[138] has measured the tracer self-diffusion in Mg, which he finds to be almost isotropic, with

$$D_0 = 1.0\text{–}1.5\ \text{cm}^2/\text{sec}, \qquad Q = 1.40\text{–}1.41\ \text{eV}$$

These values are consistent with a vacancy mechanism, and with both h_v^F and h_v^M of about 0.7 eV.

3.3. Semiconductors

The nature of the intrinsic defects in semiconductors, as well as the energetics of their formation and motion, is not yet well understood. Compound semiconductors are also subject to wide departure from stoichiometry, and the defects present are determined by this rather than by thermal equilibrium. The simple concepts used in this chapter, applicable only to the very low concentrations in which thermal defects occur, do not apply to the compound semiconductors.

3.3.1. *Germanium and Silicon*

Seeger and Chik[139] have published an extensive review of the present knowledge of point defects and mechanisms of diffusion in Ge and Si, to which the reader is particularly directed. Other recent reviews are by Seeger[140] and Seeger and Swanson.[141]

In the elemental semiconductors Ge and Si, the classical methods for studying the thermal equilibrium point defects have failed because the concentrations, even at the melting point, are apparently too low. Janot *et al.*[137] have found that the bulk and X-ray thermal expansions of Ge agree with each other to the melting point to within almost 10^{-5}, indicating this figure as an upper limit for the difference between vacancy and interstitial concentrations.

The self-diffusion coefficients, discussed below, are very small. At the melting point, for both Ge and Si, D^* attains values only 10^{-3} to 10^{-4} times those of most metals. These small values, provided defect mechanisms are responsible, indicate that the defect concentrations and/or the defect mobilities are very low. Other evidence suggests that both vacancy and interstitial mobilities are in fact rather high. In electron-irradiated silicon, Watkins showed by EPR experiments that the neutral vacancy migrated[142,143] with the low value for h_v^M, the enthalpy of motion, of 0.33 eV, and the doubly negatively charged vacancy exhibited[144] a value of 0.18 eV. No free interstitials have been positively identified in Si, which may indicate[143] appreciable mobility even at 4°K. In Ge also, indirect evidence[145,146] suggests that vacancies migrate at temperatures below 80°K, with an enthalpy of motion as low as 0.2 eV. Thus the low diffusion coefficients appear to imply very low defect concentrations. Seeger and Chik[139] estimate that for both Ge and Si, an upper limit for $\chi_v(T_m)$ is about 10^{-7}, three or four orders of magnitude less than found in most metals. This value is of the order of or less than the impurity concentration in the purest material, so that even at the melting point, effects due to thermal equilibrium defects may be masked by impurities.

The low concentrations and high mobilities mean also that quenching experiments will be of little use in studying thermal equilibrium defects. Since impurity concentrations equal or exceed the defect concentrations even at the melting point, precipitated-impurity re-solution effects will occur and tend to mask the presence of the quenched-in defects. In *n*-type Ge, for instance, when Cu is present, a quench from 800°C produces *p*-type material. The explanation involves the introduction of acceptors by the re-solution of the Cu which has precipitated at dislocations and/or other vacancy sources.

As Gregory[61] has pointed out, the combination of low defect concentration and high mobility implies that the defect will reach sinks during the quench and not be retained, an effect already noted above for sodium. In silicon, in which there is evidence for both vacancies and interstitials in comparable numbers (Frenkel pairs) at high temperatures, the self-annihilation of these pairs should be very rapid and make retention of the high-temperature defect population particularly difficult. The formation of divacancies is also very probable because of the high vacancy mobility and because the vacancy binding energy may be appreciable,[147] so that what vacancies are retained are very apt to be as divacancies or higher clusters.

Radiotracer self-diffusion experiments have been done in both Ge and Si, although the usable temperature range is rather narrow because of the very small diffusion coefficients. Table XX exhibits two representative sets of results. The values of D_0 are considerably larger than expected on the basis of a simple vacancy mechanism, e.g., ~0.1–1.0 as for most metals. Seeger and Chik[139] have suggested that diffusion in Ge is by way of extended vacancies (such as might be formed by replacing 11 atoms of crystalline Ge with 10 atoms in configuration of the amorphous state), while in Si, extended interstitials (10 atoms of crystal replaced by 11 atoms in the liquid configuration) also play a role. The formation entropy of these extended defects (8–10k for the vacancy, 13–15k for the interstitial) could account for the large D_0 values. Again, the situation is similar to that found with Na.

In semiconductors, as discussed in Section 2.1, the defects may trap electrons or holes, and thus exist in several charge states. The distribution of defects among the possible charge states then depends, via the Fermi level, upon the presence of n-type or p-type dopants (impurities). In electron-irradiated Si, the EPR and infrared absorption spectroscopy work of Watkins and others[144] has identified monovacancies existing in the positively charged (acceptor, symbol $V^{\bullet}$), neutral or bare (symbol V^{x}), singly negatively charged (donor, symbol V'), and doubly negatively charged (donor, symbol V'') states. The latter occurs only in n-type material, the other three in p-type. Similar experiments have not been possible in Ge. Figure 6 illustrates the vacancy states in Si.

Table XX. Self-Diffusion Parameters for Germanium and Silicon

	Ref.	D_0, cm²/sec	Q, eV	Temperature range, °C
Ge	258	7.8±3.4	2.95±0.04	766–928
Si	259	1800	4.86	1220–1400

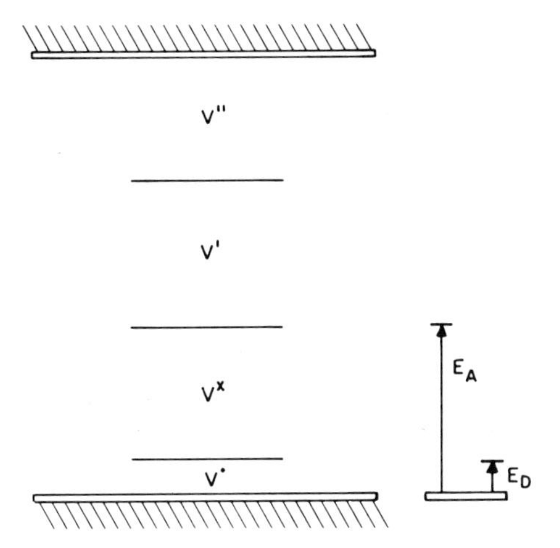

Fig. 6. Charge states of the single vacancy in silicon. The heavy bars in the band gap show the values of the Fermi energy at which a particular electron level fills. Thus for Fermi energies below the lowest bar (in the region marked $V^{\bullet}$), the vacancy has lost an electron and is present as $V^{\bullet}$. When the Fermi energy rises past this bar, the vacancy traps an electron and becomes V^{x}, etc. E_A and E_D refer to the discussion of Eq. (3.42) in the text. After Watkins.[144]

The influence of impurities in Ge on the self-diffusion[139,148] can be interpreted in terms of neutral and singly negatively charged vacancies. Suppose in general not only the neutral vacancy, but also a negatively and a positively charged vacancy contribute to the diffusion current:

$$D^* = D^*_{V^x} + D^*_{V'} + D^*_{V^\bullet} \tag{3.40}$$

with

$$D^*_{V^x} = D_{V^x} \chi_{V^x} f_{1V}, \quad \text{etc.} \tag{3.41}$$

Here, $D_i{}^*$ denotes the diffusion of tracer atoms by way of the ith type of vacancy, D_i is the diffusion coefficient of the ith type of vacancy itself, and f_{1V} is the monovacancy correlation coefficient on the diamond lattice, value 0.5. Inserting Eqs. (2.14) and (2.15), we obtain

$$\begin{aligned} D^* = D^*_{V^x}\{1 &+ [(Z_{V'} D_{V'}/D_{V^x}) \exp(-E_A/kT)][\exp(E_F/kT)] \\ &+ [(Z_{V^\bullet} D_{V^\bullet}/D_{V^x}) \exp(E_D/kT)] \exp(-E_F/kT)\} \end{aligned} \tag{3.42}$$

where $Z_{V'}$ and $Z_{V^\bullet}$ are, respectively, the spin degeneracies of the positively and negatively charged defects, E_F is the Fermi level measured from the top of the valence band, E_A is the energy required to move an electron from the valence band onto the bare vacancy, producing V', and E_D is the energy required to move an electron from the valence band onto $V^{\bullet}$, producing the bare vacancy (Fig. 6).

According to Eq. 3.42, a minimum in D^* should occur as E_F is increased from p-type to n-type by doping. However, for Ge, it is found that even with heavy doping, no minimum occurs. D^* increases monotonically from p-type to n-type. Valenta and Ramasastry[148] interpreted this to mean that no stable positively charged vacancy exists. In p-type material, only the bare vacancies are present. Since the concentration of the bare vacancies is not affected by the position of the Fermi level, they are also present in the n-type material, but augmented by the negatively charged vacancies.

In Si, self-diffusion is considerably more complex than it is in Ge, and Seeger and Chik[139] have suggested that above about 900°C, an extended interstitial mechanism dominates, while below this temperature, the main mechanism is by way of the extended vacancy.

Taking the activation energies for diffusion (Table XX) as the sum of enthalpies of formation and motion, and estimating the latter as 0.2–0.3 eV, Seeger and Chick[139] have suggested that the enthalpy of formation of vacancies in Ge is about 2.4 eV, and in Si, about 4 eV.

3.3.2. *Graphite*

Thrower[149] has recently reviewed the current knowledge of defects, including point defects, in graphite. The reader is referred to his article for a detailed discussion of this topic.

The energies of formation of vacancies and interstitials in graphite appear to be quite high, 7 eV or greater, while the energies of motion, at least of the interstitials in the basal plane, appear to be relatively low. This combination of circumstances means that the equilibrium concentrations of defects are very low, as it also does for germanium and silicon, and are very difficult to retain by quenching for study at low temperatures. As a result, very little reliable information exists on the nature and energetics of the thermal defects in graphite, and most conjectures are based upon rather indirect evidence.

Graphite is a layered structure (Fig. 7), each layer composed of a hexagonal net of carbon atoms. The interatomic distance between nearest neighbors in the layer (1.42 Å) is much less than the closest distance (3.55 Å) between atoms of adjacent layers. The network patterns of alternate

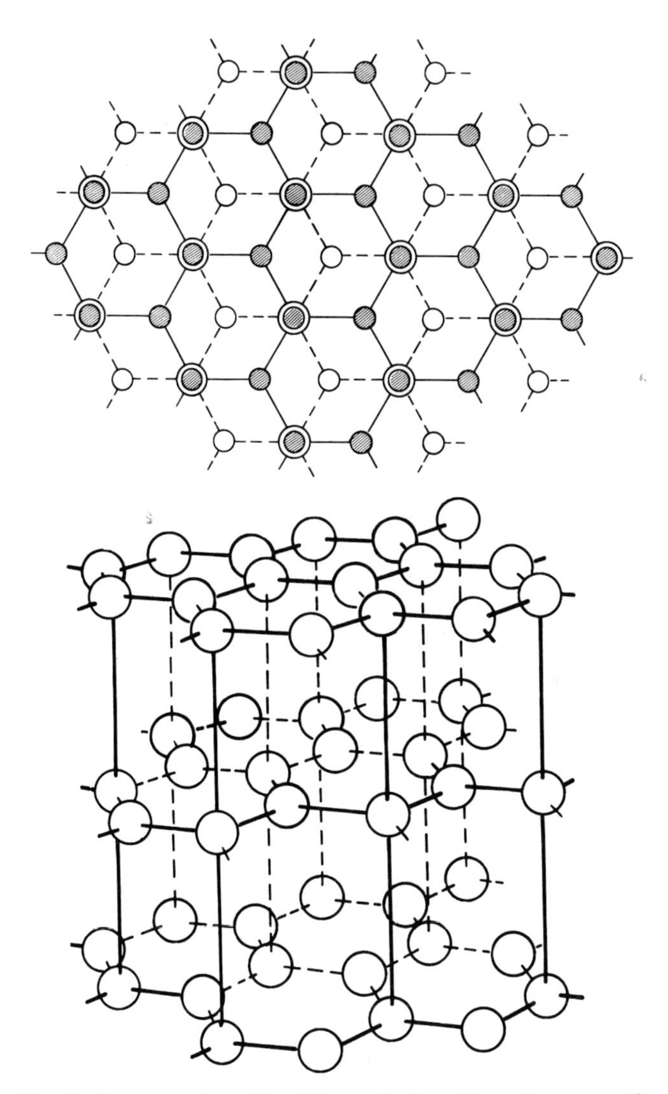

Fig. 7. Structure of graphite: (a) a projection of two planes upon the basal plane. ◍, A planes; ○, B planes; ◎, superimposed A and B planes. (b) a perspective 3-dimensional view looking into the lattice.

layers are superimposable one upon the other, forming a hexagonal structure with an *ABAB* stacking, the layers forming the basal planes, the *c* axis lying perpendicular to them. This layered structure is reflected in the considerable anisotropy of the physical properties, and in the strong tendency to cleave along the basal planes.

Hennig[150] utilized this strong cleavage to develop a technique, unique among point defect studies, that apparently allows rather direct observation of vacancies in graphite. A crystal is cleaved and then etched in an oxygen–chlorine mixture. The etchant removes carbon atoms in the basal plane surrounding each vacancy exposed in the cleaved surface, leaving a roughly circular depression a single atom deep in the otherwise flat cleavage surface. The edges of the depression are decorated with gold, and the surface examined by replication electron microscopy. The vacancies can be counted as a function of distance from the edge of the cleavage plane and also of the depth in the crystal along the c axis, by repeated cleaving. The sensitivity is high, and vacancy concentrations as low as $\chi_v = 10^{-10}$ can be detected.

Using this technique, Hennig[151] studied good crystals of graphite quenched from 3100°C. With care to avoid impurities, strain, and oxidation during the prequench anneal, he found that χ_v in the bulk of the crystal was less than 10^{-10}. Assuming that this represented an upper limit to the vacancy concentration at 3100°C, he concluded that the vacancy formation enthalpy is greater than 6.6 eV. While vacancies in graphite are probably sufficiently immobile to have been retained by the quench, it appears that interstitials are very mobile in the basal plane. The very low upper limit to χ_v observed in these experiments may therefore reflect, in part at least, annihilation of thermal vacancies by thermal interstitials during the quench.

The stored energy in low-temperature (liquid nitrogen), neutron-irradiated graphite has been measured[152,153] and used to obtain an estimate of about 14 eV for the Frenkel energy ($h_v^F + h_i^F$). Henson and Reynolds[154] have compared the changes in the lattice parameter and the rate of accumulation of stored energy in graphite irradiated with neutrons at temperatures from 200 to 350°C, and have interpreted their results in terms of an enthalpy of formation of the vacancies of 7.3 ± 1 eV. The state of aggregation of the vacancies could not be determined for either experiment. Thus both h_v^F and h_i^F appear to be about 7 eV under irradiation conditions.

High-temperature measurements of the specific heat and thermal conductivity have been interpreted by Rasor and McClelland[155] in terms of vacancy formation with $h_v^F = 7.7$ eV. However, the integrated excess specific heat together with this value for h_v^F [see Eq. (3.6)] required a value of 5×10^{-3} for χ_v at the sublimation temperature, 3650°C. By Eq. (3.1), this in turn implies that the entropy of formation of the vacancies must be of the order of $17k$, an unreasonably large value. It seems probable that there are unresolved experimental problems with sublimation (the heat of sublimation to monatomic carbon is 7.4 eV per carbon atom[156]) or with the difficulty of assessing the ideal-lattice contribution to the specific heat (the law of Dulong and Petit was used), and the correspondence between their value of 7.7 eV for h_v^F and other values may be fortuitous.

Most of the current information about point defects in graphite comes from the study of recovery after irradiation. Goggin and Reynolds[157] observed three recovery stages in the resistivity and thermal conductivity of graphite irradiated with 4-meV electrons at liquid nitrogen temperature, and were able to identify one with the motion of single interstitials with $h_i^M \approx 0.5$ eV, presumably in the basal plane only.

Dislocation loops lying in the basal plane, both interstitial and vacancy, can be formed by various combinations of irradiation and annealing, and information about point defects derived from the kinetics of their growth or dissolution.

At high temperatures and irradiation doses, interstitial loops above a certain size are stable and grow. Turnbull and Stagg[158] have analyzed the kinetics of high-temperature annealing of both vacancy and interstitial loops using the theory of dislocation climb of Friedel[159] as applied to graphite by Baker and Kelly.[160] They find that both kinds anneal by the same process with an activation energy of 8.3 eV and that the kinetics require diffusion both in and perpendicular to the basal plane. This activation energy is given by $h^F + \frac{1}{2}(h_a^M + h_c^M)$, where h_a^M and h_c^M are, respectively, the enthalpies of motion in and perpendicular to the basal plane. Thrower[161] showed that vacancies do not appear to cross the basal plane, so that this result must represent interstitial motion. Thrower[162] had found that interstitial loop formation during 1350°C anneals in heavily irradiated graphite implied *c*-axis interstitial diffusion, with $h_{ic}^M \approx 2.7$ eV.

Thrower[149] put together the results of the last two paragraphs, together with estimates of the entropies of formation and motion of interstitials, to produce the interstitial parameters given in Table XXI.

From the kinetics of growth of vacancy loops and other features involving vacancy motion, a number of researches[154,160,162,163] have resulted in estimates for the enthalpy of motion of vacancies, presumably in the basal plane, of about 3 eV. The mobility perpendicular to the basal plane appears to be negligible.[161]

Self-diffusion experiments also indicate large values for defect formation enthalpies. Kanter[164,165] measured D^* using ^{14}C as a tracer in natural graphite single crystals in the temperature range 2150–2350°C, and found the activation energy $Q = 7.1$ eV. Evans *et al.*[166] found a very similar value (6.8 eV) for basal plane ^{14}C self-diffusion in pyrolytic graphite in the temperature range 2400–2900°C. Feates[167] also studied ^{14}C self-diffusion. Radioactive decay of the ^{14}C produced ^{14}N, which behaved like a vacancy

Table XXI. Defect Parameters in Graphite (from Thrower[149])

Parameter	Interstitial	Vacancy
h^F, eV	6.7	7.0
s^F/k	8.6	—
h_c^M, eV	2.9	≥ 5.5
s_c^M/k	3.3	—
h_a^M, eV	0.4	≥ 3.0
s_a^M/k	0.5	—

when Hennig's technique (above) was used. Feates could section his specimens, and obtain diffusion profiles both in and perpendicular to the c direction. While he did not succeed in obtaining the temperature dependence of the diffusion coefficients, he could show that at 2200°C, the diffusion coefficient along the c axis was about one-tenth that in the basal plane, remarkably large considering the anisotropy of the structure.

Assuming a defect mechanism for diffusion, the activation energy Q should be given by the sum $h^F + h^M$ of the defect formation and motion enthalpies. The observed value, referring primarily to basal plane diffusion, is consistent with the enthalpies of formation and basal plane motion in Table XXI for interstitials, but not for vacancies, and Thrower[149] has concluded that the self-diffusion observed is probably interstitial.

Table XXI contains the estimates of defect formation and motion parameters deduced by Thrower.[149] In a note added in proof, he pointed out that these parameters were incompatible with Feates's[167] observation that self-diffusion at 2200°C occurred perpendicular to the basal plane at a rate as large as one-tenth that in the basal plane. Thrower suggested that diffusion is by a lattice interchange rather than by a defect mechanism, but the comparative isotropy of diffusion in such a highly anisotropic structure remains a problem.

3.4. Ionic Solids

3.4.1. *General References*

Recent reviews of the experimental determination of the free energies of formation of point defects in ionic compounds include the articles by Barr and Lidiard,[168] Lidiard,[169] and Suptitz and Teltow[170] and the book by Adda and Philibert.[171] The older article in the *Handbuch der Physik* by

Lidiard[169a] is still immensely valuable and stands as the basic text in this field. The Barr and Lidiard article, in particular, contains extensive tables of these energies for ionic compounds, from papers published up to 1968. In the present review, these tables will not be repeated, and the reader is referred to them. Rather, the experimental situation will be summarized here and illustrated with the most recent data available at the time of writing. The data presented here will supplement, not reproduce, the data given by Barr and Lidiard.

3.4.2. *Experimental Methods for Ionic Crystals*

The method most often used to obtain values for the free energies of formation, etc., for ionic crystals is the interpretation of the ionic conductivity and its temperature dependence in terms of the ionic model. The conductivity of a cubic crystal on this model is given by

$$\sigma = N \sum_i q_i \chi_i \mu_i \tag{3.43}$$

where N is the number of formula units per unit volume in the crystal, q_i is the charge on the mobile species (usually simple vacancies or interstitials) of the ith kind, and χ_i and μ_i are the corresponding mole fraction and mobility. Where only one type of jump is possible for each carrier, the mobility is given by

$$\mu_i = (y_i r_i^2 q_i \nu_i / kT) R_i \exp(-g_i^M / kT) \tag{3.44}$$

with r_i the distance per jump in the field direction, ν_i the frequency with which the ith mobile species approaches the free-energy barrier of height g_i^M separating the site presently occupied from an adjacent site, and R_i the number of equivalent downfield jumps. The attempt frequency ν_i is usually approximated with the Debye frequency ν_D of the crystal. Coulomb (Debye--Hückel) interactions introduce the mobility drag factor y_i, as discussed by Lidiard.[169a]

The mole fraction of the ith species is governed by the appropriate equilibrium relations [as in Eq. (2.23)], taking into account the requirements of charge balance [Eq. (2.18)] and the possibility of complex formation [Eqs. (2.11) and (2.21)] with mole fractions of the unassociated defects multiplied by the corresponding activity coefficients γ_i given by Eq. (2.22). Since the conductivity, through the mobility given by Eq. (3.44), depends in part upon the free energy of motion g_i^M of the various mobile species, some way must be found to evaluate these motional energies in order that the free energies of formation and of association of the point defects may be found.

Lidiard[169a] has given a useful method of analysis that has been widely employed to eliminate the motional quantities. This analysis makes use of the conductivity isotherm, the variation of the ionic conductivity with concentration of impurity at a given temperature. Assume, for the sake of illustration, a monovalent MX compound with either Schottky or Frenkel defects of M or X predominating, and with divalent cation impurities of mole fraction c. The basic equations governing the ionic conductivity are

$$\chi_+\chi_- = \chi_0^2 = \gamma^{-2}\exp(-g^F/kT) \tag{3.45}$$

$$\sigma = \mathrm{Ne}(\chi_+\mu_+ + \chi_-\mu_-) \tag{3.46}$$

$$\chi_- = \chi_+ + (c - \chi_A) \tag{3.47}$$

$$\chi_A/\chi_-(c - \chi_A) = K_A = Z_A\gamma^2\exp(g^A/kT) \tag{3.48}$$

where χ_- (χ_+) is the mole fraction of the defect bearing the negative (positive) effective charge and χ_A is the mole fraction of the complexes formed with the negatively charged defect and the divalent cation impurity ion. If ϕ denotes ratio μ_+/μ_- and σ_0 the value of σ when $\chi_- \approx \chi_+ \approx \chi_0 \gg c$, these equations can be rewritten in parametric form:

$$\sigma/\sigma_0 = (\varepsilon + \phi\varepsilon^{-1})/(1 + \phi) \tag{3.49}$$

$$c/\chi_0 = (\varepsilon - \varepsilon^{-1})(1 + H\varepsilon) \tag{3.50}$$

where $\varepsilon = \chi_-/\chi_0$ and $H = \chi_0 K_A$. If, as often occurs, $\phi > 1$, the plot of σ/σ_0 against c exhibits a minimum. The mobility ratio ϕ is obtained from

$$(\sigma/\sigma_0)_{\min} = 2\phi^{1/2}/(1 + \phi) \tag{3.51}$$

This result is independent of the magnitude of K_A, and holds whether or not association occurs. Equation (3.49) can now be solved for ε, and the data expressed as a plot of $c/(\varepsilon - \varepsilon^{-1})$ against ε. The intercept is χ_0 and the slope is $\chi_0 H = \chi_0^2 K_A$. Conductivity data at a number of temperatures and for several concentrations can therefore be used to obtain ϕ, χ_0, and K_A as a function of temperature, and hence values of $g_-^M - g_+^M$, g^F, and g^A. By invoking Eq. (3.46) as well, separate values of g_-^M and g_+^M can be obtained.

The motional parameters can also be obtained from tracer diffusion data. In cubic crystals, the tracer diffusion coefficient for an ion with only one kind of jump is

$$D^* = (1/6)\Gamma f r^2 \tag{3.52}$$

where Γ is the number of jumps per second a given ion makes on the average, f is the correlation factor, and r is the jump distance. Γ is given by, e.g.,

$$\Gamma = Z\omega\chi \tag{3.53}$$

where Z is the number of available sites into which equivalent jumps can be made, χ is the mole fraction of the mobile defect species involved in the motion of the ion in question, and ω is the probability per second that a mobile defect will make a given one of the Z jumps available to it. The mole fractions, as before, are governed by the appropriate equilibrium relations. The jump probabilities are given by

$$\omega = \nu \exp(-g^M/kT) \tag{3.54}$$

Hence the tracer diffusion coefficient of a given ion contains the same combination of formation, association, and motional free energies that the conductivity does, for a single ionic species. The two kinds of data, taken together, can be used to sort out the simultaneous contributions to the conductivity made by the several mobile species that may be present. It should be noted, however, that uncharged species (e.g., vacancy pairs; cf. Friauf[171a]) may contribute to the tracer diffusion, but not to the conductivity. This contribution must therefore be subtracted from the diffusion data to obtain that provided by the charged species alone.

In recent years, diffusion data have also been obtained from the relaxation times (T_1, spin–lattice, and T_2, spin–spin) of the nuclear magnetic spin system of some crystals. At sufficiently high temperatures, diffusion of the ions is the principal mechanism by which relaxation of the spin system is induced. Where quadrupole moments are small, so that dipole–dipole interactions provide the main nuclear coupling, a fairly detailed theory[172] can be worked out relating T_1 and T_2 for the ions in the crystal to the probabilities per second, Γ_i, that an ion of the ith kind will make a jump.

In principle, a direct determination of the excess vacancy over the interstitial mole fractions is possible by comparing the change in density Δd to the change in unit-cell size Δa^3 when the crystal is heated from a low temperature, where vacancy concentrations are negligible, to higher temperatures. The difference between the numbers of vacancies and interstitials represents new unit cells, and the change Δn in the number n of unit cells is, for cubic crystals,

$$\frac{\Delta n}{n} = \frac{\Delta d}{d} - \frac{3\,\Delta a}{a} = \chi^T_{V_M} - \chi^T_{I_M} \tag{3.55}$$

where $\chi^T_{V_M}$ is the mole fraction of all thermally produced cation vacancies in a compound of formula MX and $\chi^T_{I_M}$ is the mole fraction of thermally produced cation interstitials. It should be noted that $\chi^T_{V_M}$ includes not only free vacancies, but also all cation vacancies bound up in vacancy pairs and in all valence states.

3.4.3. *Alkali Halides, Mainly Rocksalt Structure*

Four main defect species are included in the complete model for alkali halides in thermal equilibrium: cation and anion vacancies and multivalent substitutional cation and anion impurities. In addition, the model also includes vacancy pairs and impurity–vacancy complexes of both kinds. It is observed that alkali halides with the rocksalt structure usually contain an excess of divalent cation impurities, probably because the solubility of divalent cations such as Ca^{2+}, Cd^{2+}, Sr^{2+}, and Mn^{2+} is considerably higher than that of the common divalent anions such as CO_2^{3-}, O^{2-}, and S^{2-}. However, by careful manipulation, an excess of the divalent anions can be introduced, at least into some of the alkali halides.

The equations governing the ionic conductivity [Eqs. (3.43)–(3.48)] can be written, denoting cation vacancies by the subscript 1 and anion vacancies by 2,

$$\sigma = Ne(\chi_1\mu_1 + \chi_2\mu_2) \tag{3.56}$$

$$\chi_1\chi_2 = \chi_0{}^2 = \gamma^{-2}\exp(-g_s{}^F/kT) \tag{3.57}$$

$$\chi_A/[\chi_i(c - \chi_A)] = 6\gamma^2\exp(g_i{}^A/kT) = K_A \tag{3.58}$$

$$\mu_i T = (ya^2 e v_i/k)\exp(-g_i{}^M/kT) \tag{3.59}$$

where $g_s{}^F$ denotes the sum of the free energies of formation of an isolated cation vacancy and an isolated anion vacancy (Schottky pair), a is the lattice parameter (twice the shortest cation-anion distance), and c is the excess mole fraction of divalent impurity ions. Equation (3.59) holds for both kinds of vacancies, and Eq. (3.58) for the vacancies occupying the same sublattice as the excess impurity ions ("compensating vacancies"). For these compensating vaciancies, Eq. (3.47) becomes

$$\chi_i = (\chi_0{}^2/\chi_i) + [c/(1 + K_A\chi_i)] \tag{3.60}$$

At high temperatures ("intrinsic range"), $\chi_0 \gg c$, and $\chi_1 \approx \chi_2 \approx \chi_0 = \gamma^{-1} \exp(-g_s^F/2kT)$, so that

$$\sigma T = \frac{yNa^2e^2}{k}\left[\nu_1 \exp\left(\frac{\frac{1}{2}s_s^F + s_1^M}{k}\right)\exp\left(-\frac{\frac{1}{2}h_s^F + h_1^M}{kT}\right) + \nu_2 \exp\left(\frac{\frac{1}{2}s_s^F + s_2^M}{k}\right)\exp\left(-\frac{\frac{1}{2}h_s^F + h_2^M}{kT}\right)\right]\gamma^{-1} \quad (3.61)$$

where use has been made of the thermodynamic relation $g = h - Ts$. Since $h_2^M > h_1^M$ in these compounds, a plot of $k \ln(\sigma T)$ against T^{-1} at the lower end of the intrinsic range (see Fig. 8) tends to have a slope close to the quantity $-(\frac{1}{2}h_s^F + h_1^M)$. On the other hand, at higher temperatures, the anion vacancies also make an important contribution to the conductivity, and the (negative) slope increases toward $-h_s^F/2$ plus some mean of h_1^M and h_2^M.

In most experiments, divalent cation impurities are in excess. At temperatures below the intrinsic range, in the "extrinsic range", $\gamma \approx 1$, $\chi_2 \ll \chi_1$, and

$$\begin{aligned} \chi_1 &= (1/2K_A)[(1 + 4K_Ac)^{1/2} - 1] \\ \sigma T &= (yNa^2e^2/k)\nu_1\chi_1[\exp(s_1^M/kT)]\exp(-h_1^M/kT) \end{aligned} \quad (3.62)$$

For small K_A, or at the upper end of this extrinsic range, the degree of association between the cation vacancies and the divalent impurity ions can be very small and χ_1 can approach c. Then,

$$\sigma T = (yNa^2e^2/k)c\nu_1[\exp(s_1^M/k)]\exp(-h_1^M/kT) \quad (3.63)$$

and the slope of the plot of $k \ln(\sigma T)$ against T^{-1} gives h_1^M directly. On the other hand, for large K_A (as the temperature decreases), χ_1 approaches $c/K_A^{1/2}$ and the slope increases again (in an absolute sense) toward $-(\frac{1}{2}h^A + h_1^M)$. At still lower temperatures, precipitation of the divalent impurity can occur, the slope of the plot then reflecting the heat of solution.

These details of the plot of $\ln(\sigma T)$ versus T^{-1} are illustrated in Fig. 8, which shows the data of Kirk and Pratt[173] for "pure" LiF. Stage I is the part of the intrinsic region dominated by cation motion, while at higher temperatures (Stage I′), the anions make a noticeable contribution. Stage II is the upper part of the extrinsic range, and by its slope, indicates that divalent cation impurities are present in excess. At lower temperatures (Stage III), association between the cation vacancies and the divalent

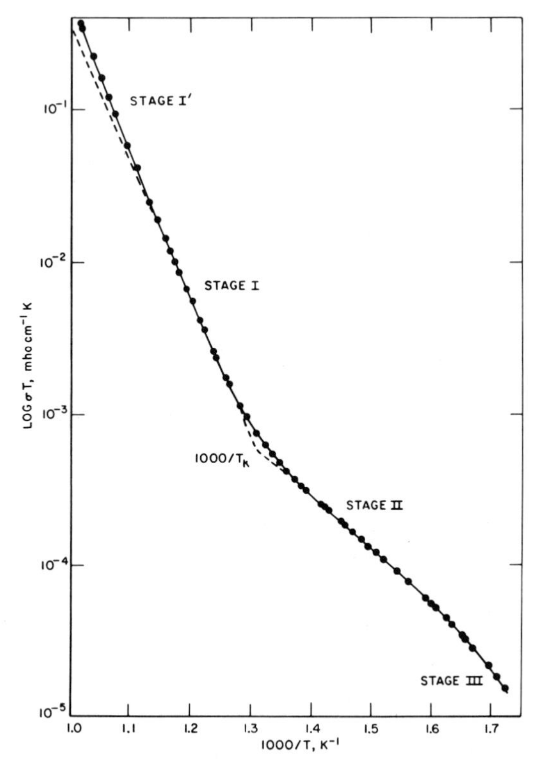

Fig. 8. Temperature dependence of ionic conductivity of pure LiF. The dotted line shows straight line fits to stages I and II. From Kirk and Pratt.[173]

cation impurities becomes important and the slope reflects in part the enthalpy of association.

It has been the practice to plot $k \ln(\sigma T)$ against T^{-1} and to fit various portions of this plot with straight lines whose slopes were then interpreted in the manner described above. In the earlier work, the anion contribution to the ionic conductivity in the intrinsic range was ignored and the slope in this portion of the plot equated with $-(\frac{1}{2}h_s^F + h_1^M)$. Evaluation of h_1^M from the upper part of the extrinsic portion of the plot then allowed h_s^F to be derived. The enthalpy of association h^A could be determined from the lower part of the extrinsic range plot, provided precipitation was avoided.

In recent years, this procedure has been modified. As shown, for instance, by Allnatt and Jacobs[174] for KCl and Rolfe[175] for KBr, the anion contribution to the ionic conductivity is not negligible in every case, particularly at the higher temperatures. Rolfe also showed that this contribution could not be adequately assessed from the intrinsic conductivity data alone. An independent measurement of at least the relative mobilities of the anion and cation vacancies must also be available. Anion diffusion data or conductivity data with excess divalent anions present have provided these independent measurements.

It has also been recognized [176] that the graphical fitting procedure is subject to greater error than was at first realized. The plot of $k \ln(\sigma T)$ against T^{-1} is generally curved over most of its length, and fitting portions with straight line segments involves arbitrary decisions and introduces unknown errors. Beaumont and Jacobs[177] instead fitted the entire curve (for pure and also Sr-doped KCl) to Eqs. (3.56)–(3.60), taking γ and y to be unity, by a computer-operated least-squares technique. In this process, they allowed eight variable parameters to be determined by the best-fit criterion: h_s^F, s_s^F, h_1^M, s_1^M, h_2^M, s_2^M, c, and h^A. It was assumed that $s^A \approx 0$.

This computer curve-fitting technique appears necessary for obtaining accurate values for the parameters, but must be used with caution. Dawson and Barr[178] found that the range of temperature must be wide enough to include both intrinsic and extrinsic regions. Even so, the parameters so determined do not appear to be unique[168] unless, as pointed out by Chandra and Rolfe,[179] some of the data are dominated by anion contributions and some by cation contributions (to yield reliable estimates of g_1^M and g_2^M) plus intrinsic range data as well, if good values for g_s^F are to be obtained.

Table XXII lists some of the most recent results, obtained in studies that were more or less complete in the sense discussed above. Where the comparison can be made, the agreement on h_1^M is better than it is on h_2^M, reflecting the ready doping of these crystals by divalent cations. Those computer-fitted results incorporating independent anion-dominated data and using the Lidiard–Debye–Hückel correction for Coulombic interactions would appear to be the most reliable. Direct comparison of two such sets of results, the work on KCl by Fuller *et al.*[11] and by Chandra and Rolfe,[180] shows relatively good agreement. These results also include Schottky formation enthalpies (2.5–2.6 eV) considerably above the previously accepted range given by Barr and Lidiard,[168] namely 2.26–2.31 from experiment and 1.75–2.26 from theory. This high value for h_s^F seems to be a feature of those studies in which an independent determination of the anion parameters was made. An exception to this statement occurs in the work of Jain and Parashar[181] and may result from the graphical fitting procedure used for data that gave an almost continuously curving plot.

Table XXII. Defect Energies in the Alkali Halides

Substance	Ref.	h_s^F, eV	s_s^F/k	h_1^M, eV	s_1^M/k	h_2^M, eV	s_2^M/k	h^A, eV	s^A/k	LDH[a]	Dopants	Notes
LiF	182	2.34	—	0.70	—	1.0	—	0.4	—	No	Mg, Mn	h_2^M from NMR data
	260	2.4	—	0.73	—	—	—	0.36	—	Partly	Mg	Anion neglected; low-temp. intrinsic only
NaF	269	2.42	7.5	0.95	0.9	1.46	11.7	0.26	—	Yes	Ca	Computer fit, no independent anion data
	271	2.7	—	0.86	—	—	—	—	—	—	—	—
NaCl	261	2.12	6.2	0.80	3.1	—	—	0.3–0.7	—	No	Various	Anion neglected
	262	2.30	—	0.68	—	—	—	0.3	—	No	Mn	Anion neglected; low-temp. intrinsic only
	263	2.17	2.4	0.66	3.2	1.2	9.4	—	—	Yes	Unknown	Computer fit, no independent anion data
	173	—	—	0.74	—	—	—	0.28	—	No	Ca, Cd	Extrinsic range only
NaBr	271	2.16	—	0.63	—	—	—	—	—	—	—	—
KCl	261	2.22	7.1	0.84	3.2	—	—	—	—	No	Various	Anion neglected
	264	2.26	5.4	0.71	1.8	1.04	6.3	0.42	—	No	Sr	Computer fit, no independent anion data
	265	2.49	7.6	0.76	2.6	0.89	4.0	0.57	1.9	Yes	Sr	Computer fit, h_2^M from Cl diffusion data

	177	2.26	5.4	0.71	1.8	1.04	6.6	0.64	4.4	No	Ca	Most parameters adopted from Ref. 264
	11	2.59	9.6	0.73	2.7	0.99	4.1	0.58	1.3	Yes	Sr	Computer fit, h_2^M from CO_3^{2-}-doped crystals
KBr	266	2.53	—	0.67	—	0.87	—	0.46	—	No	Ca	h_2^M from CO_3^{2-}-doped crystals
	180	1.7	—	0.64	—	0.95	—	0.22	—	No	Cd	No independent anion data
KI	175	1.6	—	0.72	—	1.50	—	0.26	—	No	Ba	h_2^M from CO_3^{2-}-doped crystals
	267	2.21	8.9	0.63	1.5	1.29	9.3	0.54	2.2	Yes	Sr	Computer fit, h_2^M from CO_3^{2-}-doped crystals
	271	2.06	—	0.67	—	—	—	—	—	—	—	—
RbCl	181	2.04	—	0.54	—	1.45	—	—	—	Yes	Sr	Computer fit, no independent anion data
CsBr	179	1.80	—	0.36	1.0	0.51	4.3	—	—	Yes	Unknown	Computer fit, no independent cation data
CsCl	268	1.9	—	0.6	—	0.3	—	—	—	No	—	h_2^M from Cl diffusion, h_1^M from Cs diffusion data
KN_3	270	1.25	2.8[b]	0.76	13.4[b]	—	—	0.21	—	No	Ba	Graphical fit, no independent anion data

[a] Indicates whether or not Coulomb interactions between "free" defects have been taken into account, using the version of the Debye–Hückel theory given by Lidiard.[4,5]

[b] There is some numerical confusion in Ref. 270, and it seems possible the s_s^F and s_1^M might have been interchanged.

A number of tracer diffusion studies have been made on the alkali halides. The tracer diffusion coefficients are given by

$$D^* = D_V^* + D_P^* \tag{3.64}$$

where D_P^* is the contribution from vacancy pairs and D_V^* is the contribution from single vacancies:

$$D_V^* = (fa^2\nu_i)\chi_i \exp(-g_i^M/kT) \tag{3.65}$$

By Eq. (2.24), the concentration of vacancy pairs, and therefore D_P^*, is independent of the impurities, but the anion vacancies are suppressed by the presence of divalent cation impurities. Accordingly, D_P^* for anions can be obtained as D^*_{anion} in the limit of large concentration of divalent cation impurities. Since the motion of a vacancy pair involves both cations and anions, the pairs make a similar but not equal contribution to D^*_{anion} and D^*_{cation}. However, there are usually so many cation vacancies present that the pair contribution to D^*_{cation} is ignored.

Once D_P^* is available for the anion, Eq. (3.64) can be used to obtain the free vacancy part, given by Eq. (3.65) together with the solutions to Eqs. (3.57), (3.58), and (3.60) for χ_1 and χ_2. In the intrinsic range, where impurity effects are small, and assuming $\gamma \approx 1$, a plot of $k \ln D_V^*$ against T^{-1} will give a straight line of slope $-(\frac{1}{2}h_s^F + h^M)$. For cation tracer diffusion in the extrinsic range, with divalent cation impurities, the slopes are $-h_1^M$ and $-(\frac{1}{2}h_1^A + h_1^M)$ above and in the association region, respectively. For anions in the extrinsic range (divalent cation impurity) but above the temperature at which association becomes important, the slope should approximate $-(h_s^F + h_2^M)$.

Table XXIII. Single-Vacancy Activation Energies for Diffusion in Alkali Halides

Substance	Ion	Q(intrinsic), eV	Ref.	$\frac{1}{2}h_s^F + h_i^M$, eV	Ref.
NaCl	Anion, $i = 2$	1.92	184	—	—
	Cation, $i = 1$	1.95	272[a]	1.83	173
KCl	Anion, $i = 2$	2.10	273	2.14	11
KBr	Anion, $i = 2$	2.22	178	2.14	175
	Cation, $i = 1$	1.94	178	1.94	175

[a] Reference 272 gives 1.97 eV for Q. This has been corrected roughly for the vacancy pair contribution by assuming the latter is the same for the cation as for the anion, and using the data of Ref. 184.

Results of some of the recent tracer diffusion studies in the alkali halides are collected in Table XXIII. The anion tracer diffusion coefficients were corrected to give the single-vacancy values by subtracting the pair contribution determined as discussed above. In column 3, the experimental activation energies in the intrinsic range are listed, to be compared to expected values, drawn from the data in Table XXII that appear to be most reliable, in column 5. For KCl, the two numbers are not really independent, the diffusion data having been used to obtain the parameters listed for Ref. 11 in Table XXII. For NaCl and KBr, the two sets of numbers are independent. The agreement is good for KBr but not as good for NaCl. It is interesting to note that Q is about the same for anion and cation single-vacancy self-diffusion in NaCl, implying that $h_1^M \approx h_2^M$, contrary to the usual expectation and to what is found in the other alkali halides so far examined, where $h_1^M < h_2^M$.

Nuclear magnetic resonance (NMR) studies have also been used to obtain diffusion coefficients in alkali halides. For LiF, a particularly good case since the nuclear quadrupole moments are very small, Eisenstadt and Redfield[172] give equations of the form

$$T_1^{-1}(i) = \alpha_i \Gamma_i + \beta_i(\Gamma_i + \Gamma_j)$$
$$T_2^{-1}(i) = \gamma_i \Gamma_i^{-1} + \delta_i(\Gamma_i + \Gamma_j)^{-1}$$

where T_1 is the spin–lattice and T_2 the spin–spin relaxation time, Γ_i is the probability per second, as used in Eq. (3.52), that an ion of type i (Li or F) will make a jump, and the α_i, β_i, γ_i, and δ_i are (dimensioned) coefficients that depend upon the orientation of the dc magnetic field with respect to the crystallographic axes. Experimental determination of T_1 and T_2 allows Γ_1 and Γ_2 to be determined, and these, in turn, using Eqs. (3.53) and (3.54), provide estimates of combinations of h_s^F and h_i^M.

The results of several such determinations are contained in Table XXIV. LiF is a particularly favorable case, and Eisenstadt[274] was able to determine both D_{Li}^* and D_{F}^*, the tracer diffusion coefficients, in the intrinsic region. Stoebe and Pratt[182] found that the sum of these two, through the Nernst–Einstein equation,

$$\sigma = (Ne^2/kTf)(D_{\mathrm{Li}}^* + D_{\mathrm{F}}^*) \tag{3.66}$$

accounted very well for the ionic conductivity in the intrinsic range, implying that there is negligible vacancy-pair contribution to the tracer diffusion of either ion. This result allows the Schottky energy to be calculated from the Li diffusion data in the intrinsic and extrinsic ranges. The values listed

Table XXIV. Motional Enthalpies in Alkali Halides by NMR

Substance	Ref.	h_1^M, eV	h_2^M, eV	h_s^F, eV
LiF	274	0.71	1.1	2.2
	275	0.66	—	2.42
NaCl	276	0.75	—	—
	277	0.8	—	—
	278	0.74	—	—

in Table XXIV, found entirely by NMR techniques, agree reasonably well with those obtained from ionic conductivity, Table XXII.

For NaCl, the large nuclear quadrupole moments dominate the relaxation processes, and the theory is not as well developed as for LiF. However, in doped crystals, a fairly extensive extrinsic range exists in which diffusion of the host lattice ions appears to control the relaxation times, and the temperature dependence of T_1 provides the estimates of h_1^M given in Table XXIV. These also agree reasonably well with the conductivity values in Table XXII.

The direct measurement of the thermally produced Schottky defect concentration in alkali halides, using Eq. (3.55), is in principle possible. Since interstitials are present in negligible quantities, $\Delta n/n$ gives directly the mole fraction of Schottky defects, including both unassociated defects and defects in vacancy pairs:

$$\Delta n/n = (\Delta d/d) - (3\Delta a/a) = \chi_1 + \chi_P$$

where n is the number of unit cells of lattice parameter a, d is the density, and χ_1 and χ_P are the mole fractions of thermally produced unassociated vacancies and vacancy pairs, respectively. In the intrinsic range, $\chi_1 \approx \chi_0$, and [Eqs. (2.23) and (2.24)]

$$\Delta n/n = \gamma^{-1}[\exp(-g_s^F/2kT)]\{1 + 12\gamma \exp[\tfrac{1}{2}(\Delta g_v^P - g_s^F)/kT]\} \quad (3.67)$$

with Δg_v^P the binding free energy of the vacancy pairs. If $\gamma \approx 1$ and the vacancy pairs are neglected, a plot of $k \ln(\Delta n/n)$ against T^{-1} would be a straight line of slope $E = -h_s^F/2$. If the vacancy pairs are present in significant concentrations and if $h_v^P \gtrsim h_s^F/2$, then a nearly straight line with negative slope of magnitude $E \gtrsim h_s^F/2$ can result. Some recent data of this kind are shown in Table XXV.

Table XXV. X-Ray Density Schottky Formation Enthalpies in Alkali Halides

Substance	Ref.	$-2E$,[a] eV	h_s^F, eV (Table XXII)
LiF	279	2.45	2.3–2.4
NaCl	280	1.9	2.1–2.3
KCl	279	2.41	2.4–2.6
	281	1.86	—

[a] E is the slope of a plot of $k \ln(\Delta n/n)$ against T^{-1}, where Δn is the thermally produced number of unit cells from combined density and X-ray measurements. Cf. Eq. (3.67).

The data from Ref. 279 and 280 in Table XXV were obtained from samples quenched from near the melting point, while the data from Ref. 281 were obtained at temperature. The agreement between the two sets of data on KCl is not very good. The comparison between $-2E$, twice the temperature slope, and the h_s^F values drawn from Table XXII shown in the fourth column suggests that the vacancy pair binding energy is of the same order as or a bit larger than $h_s^F/2$. This would make it of the order of 1.2 eV, in agreement with the theoretical value found by Tharmalingam and Lidiard[183] and also consistent with the values needed to account for the vacancy-pair contributions to anion diffusion coefficients in KCl[11] and NaCl.[184] The comparison of densities and lattice parameters has been used by Mandel'[184a] with an equation analogous to (3.55) to show that the introduction of small amounts of $CaCl_2$ into NaCl is accompanied by cation vacancies. With 0.034 mol.% $CaCl_2$, no change in the lattice parameter was noted, but a decrease in density corresponding to 1.05 cation vacancy per calcium ion was observed.

Attempts have been made to assess the separate cation and anion vacancy formation energies by way of the dependence of the isoelectric temperature upon concentration of divalent cation impurities and the use of Eq. (2.33). In the rocksalt-structure alkali halides, since $g_1^F < g_2^F$, this procedure gives g_1^F. Some results are listed in Table XXVI. There is fair agreement in the later work on NaCl among the results of Spencer and Plint, [185] Kliewer and Koehler,[186] and Davidge.[187] Moreover, Plint and Brieg[188] obtained estimates of $g_2^F - g_1^F$ by study of the light scattering from dislocations. When these are combined with the estimates for g_1^F for NaCl and KCl, both about 0.7 eV, they produce estimates for g_s^F of about 2.2 eV for NaCl and 2.5 eV for KCl. These numbers are somewhat larger than the values of about 2 eV to be inferred from the data in Table XXII

Table XXVI. Single-Ion Vacancy Formation Energies in Alkali Halides

Substance	Ref.	g_1^F, eV	Temp. range, °K	g_2^F, eV	Notes
LiF	185	0.74	680	—	Dislocation charges
NaCl	23	~0.4	—	—	Yield stress minimum
	282	0.53	—	—	Yield stress minimum
	186	1.07–6.2kT	—	—	Extrema in elastic stiffness and internal friction
	185	0.75	761	—	Dislocation charges
		0.72	627	—	
	188	$g_2^F - g_1^F = 0.8$	—	1.5[a]	Dislocation light scattering
	187	0.95–2.0kT	—	—	Dislocation charges
	189	$g_2^F - g_1^F = 0.26$	—	1.0[a]	Dislocation charges
KCl	188	$g_2^F - g_1^F = 1.10$	—	1.8[a]	Dislocation light scattering
	283	0.7	—	—	Dislocation-enhanced conductivity

[a] Combining $g_2^F - g_1^F$ value from quoted reference with estimate for g_1^F of ~0.7 eV.

for a temperature of about 700°K. The much smaller value of Strumane and deBatist[189] for $g_2^F - g_1^F$ would result in an estimate of g_s^F for NaCl of about 1.8 eV, probably somewhat too small.

3.4.4. *Silver Halides*, AgCl *and* AgBr

The predominant defects in the silver halides with the rocksalt structure are cation Frenkel defects, consisting of cation vacancies and interstitials. This has been most convincingly demonstrated by Fouchaux and Simmons[190] by very careful measurements of the changes with temperature of the lattice parameter a and macroscopic dimensions, l of the crystal. They found that

$$|(\Delta l/l) - (\Delta a/a)| \leq 3 \times 10^{-5}$$

right up to the melting point. The upper limit on the mole fraction of Schottky defects is 9×10^{-5}, whereas at the melting point, the mole fraction of cation Frenkel defects present is about 3×10^{-3}.

Fouchaux and Simmons also found an anomalously high thermal expansion at high temperatures, which they attributed to the formation of the Frenkel defects. The high-temperature thermal expansion may be considered to consist of a lattice contribution, independent of the presence of point defects, plus a contribution from the defects. The point defect contribution, at least for dilute solutions, should vary as the defect concentration. If the lattice contribution can be obtained by extrapolation from lower temperatures, and if the contribution per defect is independent of temperature, then the excess thermal expansion over the extrapolated lattice contribution can be used to estimate the defect contribution and its temperature dependence. Following this reasoning, Fouchaux and Simmons obtained a concentration of Frenkel defects that behaved with temperature according to the mass action law, producing a value of 1.4 eV for h_f^F, the enthalpy of formation of the Frenkel pair in AgCl, as listed in Table XXVII.

In a similar way, as Lidiard[169a] showed, there is a defect contribution to the isobaric specific heat [Eq. (3.6)] given by

$$\Delta C_P = \chi_f (h_f^F)^2 / 2kT^2 \tag{3.68}$$

In the case of AgBr, this contribution is rather large, and a simple linear extrapolation of C_P from lower temperature serves to provide an estimate of the defect-free lattice specific heat. Subtraction from the measured value at high temperatures provides the estimates of ΔC_P. A plot of $\ln(T^2 \Delta C_P)$ was found to be linear. From its slope, given according to Eq. (2.17) by $h_f^F/2k$, the value of 1.27 eV given in Table XXVII was deduced by Lidiard for the cation Frenkel enthalpy h_f^F in AgBr.

The value for h_f^F for AgCl has been substantiated by subsequent work on transport processes. Table XXVII lists the results from some of the more ambitious of the recent efforts. The column headed LDH indicates whether or not Coulombic interactions between "free" defects werc included, using Lidiard's version of the Debye–Hückel theory.[4,5] The predominance of cation Frenkel defects accounts for the observations, first by Tubandt,[191] that electrical conduction is overwhelmingly by cation transport. Since the cation interstitials appear to be more mobile than the cation vacancies, doping with divalent cation impurities like Cd^{2+} is necessary to provide excess vacancies so that the transport properties of the vacancies can be examined.

There appears to be general agreement on the value of h_f^F (1.3–1.5 eV) and h_v^M, the enthalpy of motion of the cation vacancy (about 0.3 eV). For the silver halides, the detailed nature of the interstitial and its motion have not yet been fully worked out. On energetic grounds,[192] it seems un-

Table XXVII. Defect Energies in Silver Halides[a]

Substance	Ref.	h_f^F, eV	s_f^F/k	h_{iC}^M, eV	h_{iN}^M, eV	h_v^M, eV	h^A, eV	LDH	Notes
AgCl	190	1.4	—	—	—	—	—	No	Excess thermal expansion
	284	1.25	—	0.16[b]	—	0.35	0.2–0.3	No	σT, graphical fit, doped with Cd^{2+}, Fe^{2+}, S^{2-}
	9	1.44	9.4	0.055[b]	—	0.27	0.47	Yes	σT, computer fit to conduction isotherms, Cd^{2+} doping
	195	—	12.2	0.15[b]	—	0.32	0.16	No	σT, graphical fit
	193	1.55	—	0.008	0.132	0.26	—	Yes	σT and D^*_{Ag}, graphical fit
AgBr	284	1.06	—	0.15[b]	—	0.34	0.16	No	σT, graphical fit, doping with Cd^{2+}, Fe^{2+}, S^{2-}
	193	1.13	—	0.058	0.27	0.30	—	Yes	σT and D^*_{Ag}, graphical fit
	14	1.27	—	—	—	—	—	No	Excess specific heat

[a] h_{iC}^M and h_{iN}^M refer to the collinear and noncollinear interstitialcy jumps, described in the text, respectively.
[b] Refers to data interpreted without differentiating among the possible interstitial motion processes.

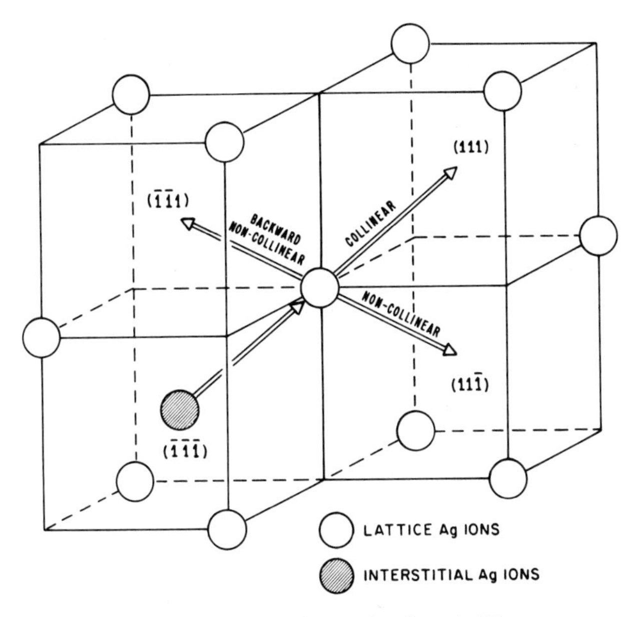

Fig. 9. Interstitialcy jumps in silver halides.

likely that the major transport process is a direct jump of the ion from one interstitial site to a nearby one. Instead, it seems probable that an interstitialcy mechanism is involved, in which the interstitial displaces one of the neighboring lattice silver ions, which then becomes the interstitial. In the rocksalt structure, taking the lattice site of the displaced ion as origin and the interstitial site before the jump takes place as $\bar{1}\bar{1}\bar{1}$, the displaced ion can move into one of three sets of interstitial sites. If it moves to 111, the jump is known as collinear. A jump to $11\bar{1}$, $1\bar{1}1$, or $\bar{1}11$ is called a noncollinear jump, and a jump to $1\bar{1}\bar{1}$, $\bar{1}1\bar{1}$, or $\bar{1}\bar{1}1$ is a backward, noncollinear jump (see Fig. 9).

Weber and Friauf[193] argue that the backward jump is also unlikely on energetic grounds, and express their results in terms of the first two possibilities. Since these have different correlation coefficients, they use the correlation coefficient of the interstitialcy motion, derived from the measured values of D^*_{Ag} and σ in a somewhat involved way, to sort out the mobility ratio of the two kinds of interstitialcy jumps.

Little is known of Schottky energies in the silver halides. Layer *et al.*,[194] in an ingenious experiment, have studied at low temperatures the enhanced ionic conductivity in AgCl crystals quenched from high temperature. They suppose cation-vacancy–anion-vacancy pairs to exist at the high tempera-

Table XXVIII. Activation Volumes in AgCl[a]

Process	Symbol	Value, cm³/mole
Formation of cation Frenkel defects	Δv_f^F	16.7
Motion of cation vacancy	Δv_v^M	4.7
Motion of cation interstitial (ave)	Δv_i^M	3.25
Cation-vacancy–divalent-impurity association	Δv^A	0.98

[a] Results of Abey and Tomizuka.[195]

ture, and to be trapped during the quench process. They further suppose that at the measurement temperature, equilibrium is established between these pairs and dissociated "free" vacancies. The immobility of the anion vacancies holds this metastable situation in a quasiequilibrium state. The temperature dependence of this enhanced conductivity then reflects the combination $(h_v^M + \frac{1}{2}h_v^P)$, where h_v^P is the binding energy of the pair (the enthalpy required to separate the partners in the pair). They find a value of about 0.6 eV for the combination, which yields a value of about 0.6 also for h_v^P.

Abey and Tomizuka[195] have studied the pressure dependence of the ionic conductivity of AgCl containing a low concentration of what appears to be a divalent cation impurity. Using Eq. (2.34), they have derived the volume changes for various processes shown in Table XXVIII.

Estimates have also been obtained for the separate free energies of formation of cation vacancies and cation interstitials in AgCl, shown in Table XXIX. The two references report results of quite different techniques, so that the agreement between them is impressive. Kliewer[196] determined the difference between the entropies of formation of the vacancy and the interstitial from the data of Wakabayasi[197] for the temperature dependence

Table XXIX. Vacancy and Interstitial Formation Parameters, AgCl

Quantity	Symbol	Ref. 196	Ref. 199
Cation vacancy formation, enthalpy	h_v^F, eV	0.47	0.58
Cation vacancy formation, entropy	s_v^F/k	7.7	8
Cation interstitial formation, enthalpy	h_i^F, eV	0.97	0.86
Cation interstitial formation, entropy	s_i^F/k	1.7	1.4

of the surface potential, and the vacancy formation energy from Snavely's[198] measurements of the range of photoelectrons near the surface. Slifkin *et al.*[199] found the vacancy formation enthalpy and entropy from the dependence of the isoelectric temperature of dislocations on the divalent cation impurity content. Experiments involved deformation-induced voltages and internal friction. In each case, the values determined have been combined with the Frenkel formation values of Abbink and Martin[9] (of Table XXVII) to complete the set of values given in Table XXIX.

3.4.5. *Alkaline Earth Halides*

Several of the alkaline earth halides (i.e., CaF_2, SrF_2, BaF_2, and $SrCl_2$) possessing the fluorite structure have been studied fairly extensively. Early work[200] on the X-ray lattice parameter and density of CaF_2 containing Y^{3+} or various trivalent rare earth ions indicated that the aliovalent cations occupied Ca^{2+} sites, and were accompanied by interstitial fluorine ions. Mass transport data[201–203] have shown that Ca^{2+} transport is very much slower than is F^- transport. It has become customary to interpret point-defect-controlled behavior in these materials in terms of anion Frenkel defects as the overwhelmingly predominant defects. On the other hand, cation mobility can be observed, both directly[203,204] in diffusion experiments, and indirectly in such phenomena as dislocation climb.

Mass transport experiments with the alkaline earth halides have not yet achieved the level of sophistication they have with the alkali halides. Graphical fitting of straight lines to portions of the Arrhenius plots for ionic conductivity and diffusion data is universally used, and only in the recent work of Barsis and Taylor[205,206] have corrections been made for Coulomb interactions between isolated defects with the Lidiard version of the Debye–Hückel theory.[4,5]

These crystals also readily incorporate oxygen, most rapidly as a result of hydrolysis (which proceeds at remarkably low temperature, down to 100°C).[207] Direct reaction with oxygen also occurs, but is apparently much slower.[208,209] Oxygen behaves like a substitutional aliovalent impurity and has a strong influence on the ionic conductivity.[210–213] It is doubtful if this strong influence was sufficiently recognized in the earlier work.

EPR experiments on alkaline earth fluorides and related studies have fully confirmed the presence of fluorine interstitials in crystals containing trivalent impurities. When oxygen is excluded, trivalent rare earth ions exhibit either the cubic symmetry of isolated cation sites or a tetragonal symmetry such as might be produced in an F^- ion occupied a nearby interstitial site. Good reviews of much of this evidence are contained in the articles by Weber and Bierig[214] and Fong.[215] Recent ENDOR experiments of Baker *et al.*[216] have demonstrated quite unequivocally the correctness of this description of the tetragonal centers.

In Table XXX, we collect the results of some of the more ambitious attempts to fit the ionic conductivity data with the anion Frenkel defect model. The agreement among the various measurements is reasonably good for the Frenkel defect formation energies and the interstitial motion energies. The studies using oxygen doping appear to be particularly reliable for the energies of anion vacancy motion and association (with oxygen).

The equilibrium equations for MX_2 compounds are somewhat different than for MX compounds. Neglecting cation Frenkel defects, which seem to be unlikely on energetic grounds,[217] the following equations should hold for concentrations of impurities and defects sufficiently low that ideal solution theory can be used:

$$\text{Schottky} \qquad \chi_{V_M''}\chi_{V_X^{\cdot}}^2 = 4\exp(-g_s^F/kT) \tag{3.69}$$

$$\text{Anion Frenkel} \qquad \chi_{I_X'}\chi_{V_X^{\cdot}} = 2\exp(-g_f^F/kT) \tag{3.70}$$

$$\text{Charge balance} \qquad c + \chi_{V_X^{\cdot}} = \chi_{I_X'} + 2\chi_{V_M''} \tag{3.71}$$

In these equations, g_s^F is the free energy expended when a trio of vacancies (one cation and two anion) is formed, isolated from each other and in an otherwise perfect crystal, and g_f^F is the same for the combination of an anion vacancy and an anion interstitial. The quantity c is the excess atom fraction of positive over negative impurity charges introduced by substitutional aliovalent impurities. These equations as written do not contemplate any association among the defects and impurities. Some of the possible associated complexes (e.g., a cation vacancy with a single anion vacancy) would be charged, and their atom fractions should appear on the appropriate side of the charge balance equation (3.71). Equations (3.69)–(3.71) call for the atom fractions of the unassociated defects; an additional equilibrium equation would be added for each type of complex formed, and a conservation equation would be needed for each impurity that entered into association reactions.

These equations produce interesting effects not encountered in the MX compounds like the alkali and silver halides. For the pure crystal, $c = 0$. Solution to Eqs. (3.69)–(3.71) produces a result expressible in terms of the inequality

$$g_s^F \gtrless \tfrac{3}{2}g_f^F + kT\ln 2$$

Table XXX. Defect Energies in Alkaline Earth Fluorides from Ionic Conductivity Studies

Substance	Ref.	$h_F{}^F$, eV	$s_F{}^F/k$	$h^M_{V_X^{\cdot}}$, eV	$s^M_{V_X^{\cdot}}/k$	$h^M_{I_X'}$, eV	$s^M_{I_X'}/k$	$h^A_{V_X^{\cdot}}$, eV	$h^A_{I_X'}$, eV	Dopants
CaF_2	202	2.8	13.5	0.5–0.9	1.3	1.7	10.5	0.07	1.4	Na, Y
	285	2.6–2.2	—	0.8–1.0	—	—	—	—	—	—
	211	2.78	—	0.56	—	1.56	—	0.52	0.48	O, Y
SrF_2	285	2.3	—	—	—	1.0	—	—	—	—
BaF_2	285	1.5–1.8	—	0.85	—	0.78	—	0.32	—	Gd, K
	205	1.9	—	—	—	0.79	—	—	0.44	Gd
	206	—	—	0.56	—	—	—	0.56	—	O

With equality, the number of anion interstitials just equals twice the number of cation vacancies, or charge balance is equally divided between Schottky and Frenkel defects. If the > sign holds, anion Frenkel defects predominate, while with the < sign, the Schottky defects are in the majority. Theoretical estimates for h_s^F and h_f^F have been obtained by Franklin[217] for the alkaline earth fluorides. These h_f^F values are in reasonable accord with the experimental values in Table XXX, while, if anything, the estimates for h_s^F are probably too low.[218] These estimates rather unequivocally lead to the conclusion that the predominant thermal defects are anion Frenkel defects, at all temperatures up to the melting points.

With MX compounds, it can be shown that the thermal defects are also the ones produced as charge balance for aliovalent impurity ions, but for MX_2 compounds, this is not necessarily so. For the usual concentrations of impurity in doping experiments, c in Eq. (3.71) is much larger than the native defects. For $c > 0$ (e.g., with R^{3+} ions on cation sites), the anion vacancy concentration $\chi_{V_X^{\cdot}}$ can be neglected. The criterion determining the relative importance of anion interstitials and cation vacancies in charge balance,

$$\chi_{I_X'} \gtrless 2\chi_{V_M''}$$

becomes, through solution of Eqs. (3.69)–(3.71),

$$kT \ln c^{-1} \gtrless 2g_f^F - g_s^F$$

Since c is an atom fraction and therefore always less than unity, the left-hand side of this inequality is always positive. If $g_s^F > 2g_f^F$, then the > sign always holds and anion interstitials predominate at all temperatures and concentrations of impurity. On the other hand, if $g_s^F < 2g_f^F$, as Franklin's earlier calculations[217] suggest may be possible for CaF_2, then for low temperatures and high impurity concentrations, cation vacancies may become important. When $c < 0$ (e.g., O^{2-} ions on anion sites), the situation is unambiguous, anion vacancies always being formed to balance the impurity charges.

Some diffusion data have been obtained for some of the alkaline earth halides. The most revealing results are gathered together in Table XXXI, which also shows that these data can be fitted reasonably well into the model, particularly the tracer data. In calculating the activation energies for diffusion Q in columns 8 and 9, use was made of the theoretical values for h_s^F obtained by Franklin[217] and the experimentally determined values for the other parameters drawn from the work of Rossing[211] for CaF_2 and

Table XXXI. Diffusion Parameters in Alkaline Earth Fluorides

Substance	Ref.	Diffusing species	Probable range	Dopant	Observed		Calculated Q, eV		Measurement
					Q, eV	D_0, cm^2/sec	Cation vacancy	Anion interstitial	
CaF_2	204	Ca^{2+}	Extrinsic	?	1.7	1×10^{-4}	$h^M_{V_M''} = 1.7$	$h_s^F - 2h_f^F + h^M_{V_M''}$ or $h^M_{V_M''} = 2.2$	Tracer
	203	Ca^{2+}	Intrinsic	—	3.8	130	$h_s^F - h_f^F + h^M_{V_M''} = 4.0$	$h_s^F - h_f^F + h^M_{V_M''} = 4.5$	Tracer
	286	Sr^{2+}	Intrinsic	—	4.4	1.7×10^5	$h_s^F - h_f^F + h^M_{V_M''} \approx 4.0$	$h_s^F - h_f^F + h^M_{V_M''} \approx 4.5$	Tracer
	287	F^-	Intrinsic	—	2.3–2.4	—	$\frac{1}{2}h_f^F + h^M_{V_X^\cdot} = 1.95$	$\frac{1}{2}h_f^F + h^M_{V_X^\cdot} = 1.95$	NMR
	287	F^-	Extrinsic	?	0.36	—	?	—	NMR
	287	F^-	Extrinsic	Sm	0.53	—	$h_f^F - \frac{1}{2}h_s^F + h^M_{I_x'} = 1.8$	$h^M_{I_x'} = 1.56$	NMR
	219	F^-	Intrinsic	—	2.0	50	$\frac{1}{2}h_f^F + h^M_{V_X^\cdot} = 1.95$	$\frac{1}{2}h_f^F + h^M_{V_X^\cdot} = 1.95$	Tracer
	219	F^-	Extrinsic	O?	0.9	1×10^{-4}	$\frac{1}{2}h^A_{V_X^\cdot} + h^M_{V_X^\cdot} = 0.82$	$\frac{1}{2}h^A_{V_X^\cdot} + h^M_{V_X^\cdot} = 0.82$	Tracer
SrF_2	286	Sr	Intrinsic	—	4.3	1.1×10^4	—	—	Tracer
BaF_2	286	Sr	Intrinsic	—	3.8	4.5×10^3	—	—	Tracer
	219	F^-	Intrinsic	—	1.6	3.1	$\frac{1}{2}h_f^F + h^M_{V_X^\cdot} = 1.5$	$\frac{1}{2}h_f^F + h^M_{V_X^\cdot} = 1.5$	Tracer
	288	F^-	Intrinsic	—	1.35	—	$\frac{1}{2}h_f^F + h^M_{V_X^\cdot} = 1.5$	$\frac{1}{2}h_f^F + h^M_{V_X^\cdot} = 1.5$	NMR
	288	F^-	Extrinsic	Eu	0.62	—	$h_f^F - \frac{1}{2}h_s^F + h^M_{I_x'} = 0.2$	$h^M_{I_x'} = 0.79$	NMR

of Barsis and Taylor[205,206] for BaF_2. It was assumed that association was unimportant expect for Matzke's[219] tracer diffusion data for F^- in CaF_2, where association between anion vacancies and the oxygen ions presumably present was assumed. The values for Q listed in column 8 were calculated on the assumption that cation vacancies predominated in the presence of R^{3+} impurity ions, but that otherwise, anion Frenkel defects were the only defects of importance. Column 9 shows the results when only anion Frenkel defects are considered under all circumstances.

A comparison of experimental and calculated values tends to favor slightly the model with cation vacancies as the predominant charge balance for R^{3+} ions in CaF_2, but definitely, in the one comparison possible, to favor anion interstitials in BaF_2. This is what might be expected on the basis of the theoretical values for h_s^F and h_f^F. However, the evidence is not very conclusive for the choice of cation vacancies for the CaF_2 diffusion experiments, and more study of this point is needed.[220]

3.4.6. *Oxides*

Little is known of the energetics of point defects in oxides. A few commonly studied oxides, such as Al_2O_3, BeO, MgO, and CaO, appear to be basically stoichiometric compounds. However, crystals have not yet been prepared with sufficient purity to exhibit the intrinsic behavior necessary for determination of defect formation energies. For instance, Barr and Lidiard[168] have observed a linear relationship between the melting points of a number of ionic compounds and the energy of formation of Schottky defects. Applying their relationship to BaO, which is the lowest melting (approximately 1900°C) of the alkaline earth oxides normally studied, we find an estimate for the Schottky formation enthalpy of 4.6 eV. If we assume that the entropy of formation is similar to those for the alkali halides given in Table XXII, approximately $9k$, then we can use Eq. (3.57) with $\gamma = 1$ for the Schottky equilibrium to estimate the temperature at which the thermally produced vacancy concentration would just equal some chosen impurity concentration. To reach the intrinsic range, experiments would have to be performed at higher temperatures than this. The results of this calculation are given in Table XXXII.

It follows that, with experimental techniques for diffusion and ionic conductivity limited at present to temperatures below about 1700°C, it is necessary to have crystals containing less than about 100 ppm of impurity even to reach the intrinsic range. This is about the limit of purity in presently available oxide crystals. On the other hand, Copeland and Swalin[221] have

Table XXXII

Impurity concentration, ppm	Temperature, °C
100	1670
10	1390
1	1180

interpreted electrical conductivity and other data for SrO in terms of anion Frenkel rather than Schottky defects, and have derived a value of 3.0 eV for the formation enthalpy. A temperature of about 1000°C, comfortably in the middle of their range, would be high enough for the concentration of intrinsic defects to exceed 100 ppm. This result should be checked by other experiments.

Many oxides where several valence states of the cation lie within small energy values of each other exhibit a wide range of composition, and are nonstoichiometric. The various transition metal oxides form an often-studied group, as do the fluorite-structure oxides of uranium and other actinide ions and of the rare earth ions. Ternary oxides containing transition metal ions, such as ferrite and the perovskite-structure titanates, also exhibit extensive departures from stoichiometry.

In the case of the nonstoichiometric oxides, the defect concentration depends upon the partial pressure of oxygen over the crystal, as well as on the temperature. It is not difficult, by varying the oxygen pressure, to vary the defect concentrations. One can write equilibrium reactions involving the increase or decrease in oxygen content of the crystal, and from the temperature dependence of the equilibrium constant, derive free energies, and therefore enthalpies and entropies, for the reaction.

However, two difficulties in practice prevent the assignment of these free energies to individual defects or sets of defects. In the first place, the defect concentrations are usually so high that interactions among them are very important. As we have seen, however, the statistical mechanics of interacting defects has not been satisfactorily solved as yet. The second problem is that the unit reaction involves a change in the valence state of the cation, and the theory of the electronic energy levels in these compounds has not been well enough worked out so that the formation free energy can be separated into the defect and electronic parts. Thus, reliable estimates of defect formation energies are not available for the nonstoichiometric oxides.

REFERENCES

1. F. A. Kroger and H. J. Vink, in *Solid State Physics*, Vol. 3, Ed. by F. Seitz and D. Turnbull (Academic Press, New York, 1956), p. 307.
2. J. E. Mayer and M. G. Mayer, *Statistical Mechanics* (Wiley, New York, 1940), p. 219.
3. A. R. Allnatt and M. H. Cohen, *J. Chem. Phys.* **40**, 1860 (1964).
4. A. B. Lidiard, *Phys. Rev.* **94**, 29 (1954).
5. A. B. Lidiard, *Phys. Rev.* **112**, 54 (1958).
6. F. K. Fong, *Phys. Rev.* **187**, 1099 (1969).
7. R. E. Howard and A. B. Lidiard, *Rept. Progr. Phys.* **27**, 161 (1964).
8. F. A. Kroger, *The Chemistry of Imperfect Crystals* (North-Holland Publishing co., Amsterdam, 1964).
9. H. C. Abbink and D. S. Martin, *J. Phys. Solids* **27**, 205 (1966).
10. R. G. Fuller, M. H. Reilly, C. L. Marquardt, and J. C. Wells, Jr., *Phys. Rev. Letters* **20**, 662 (1968).
11. R. G. Fuller, C. L. Marquardt, M. H. Reilly, and J. C. Wells, Jr., *Phys. Rev.* **176,** 1036 (1968).
12. F. K. Fong, *Phys. Rev. B* **2** (Oct. 1) (1970).
13. M. P. Tosi and G. Airoldi, *Nuovo Cimento* **8**, 584 (1958).
14. A. B. Lidiard, private communication (1966); notes on lectures delivered at the University of Milan, March 1966.
15. J. S. Anderson, *Proc. Roy. Soc. (London)* **A185**, 69 (1946).
16. K. Hagemark, *Kjeller Reports* KR-48 and KR-67 (1964).
17. R. H. Fowler and E. A. Guggenheim, *Statistical Thermodynamics* (University Press, Cambridge, 1939), Chapter 9.
18. L. M. Atlas, *J. Phys. Chem. Solids* **29**, 91 (1968).
19. L. M. Atlas, *J. Phys. Chem. Solids* **29**, 1349 (1968).
20. A. R. Allnatt and M. H. Cohen, *J. Chem. Phys.* **40**, 1871 (1964).
21. J. Frenkel, *Kinetic Theory of Liquids* (Clarendon Press, Oxford, 1946), p. 37.
22. K. Lehovec, *J. Chem. Phys.* **21**, 1123 (1953).
23. J. D. Eshelby, C. W. A. Newey, P. L. Pratt, and A. B. Lidiard, *Phil. Mag.* **3**, 75 (1958).
24. K. L. Kliewer and J. S. Koehler, *Phys. Rev.* **140**, A1226 (1965).
25. K. L. Kliewer, *Phys. Rev.* **140**, A1241 (1965).
26. G. Boato, *Cryogenics* **4**, 65 (1964).
27. G. L. Pollack, *Rev. Mod. Phys.* **36**, 748 (1964).
28. G. K. Horton, *Am. J. Phys.* **36**, 93 (1968).
29. R. M. J. Cotterill and M. Doyama, *Phys. Letters* **25A**, 35 (1967).
30. R. O. Simmons and R. W. Balluffi, *Phys. Rev.* **117**, 52 (1960).
31. O. G. Peterson, D. N. Batchelder, and R. O. Simmons, *Phil. Mag.* **12**, 1193 (1965).
32. B. L. Smith and J. A. Chapman, *Phil. Mag.* **15**, 739 (1967).
33. W. van Witzenburg, *Phys. Letters* **25A**, 293 (1967).
34. J. D. Eshelby, *Acta Met.* **3**, 487 (1955).
35. D. L. Losec and R. O. Simmons, *Phys. Rev. Letters* **18**, 451 (1967); *Phys. Rev.* **172**, 934 (1968).
36. A. J. E. Foreman and A. B. Lidiard, *Phil. Mag.* **8**, 97 (1963).
37. L. S. Salter, *Trans. Faraday Soc.* **59**, 657 (1963).

38. R. H. Beaumont, H. Chihara, and J. A. Morrison, *Proc. Phys. Soc.* (*London*) **78**, 1462 (1961).
39. I. H. Hillier and J. Walkley, *J. Chem. Phys.* **43**, 3713 (1965).
40. P. Flubacher, A. J. Leadbetter, and J. A. Morrison, *Proc. Phys. Soc.* (*London*) **78**, 1449 (1961).
41. J. Kuebler and M. P. Tosi, *Phys. Rev.* **137**, A1617 (1965).
42. A. O. Urvas, D. L. Losee, and R. O. Simmons, *J. Phys. Chem. Solids* **28**, 2269 (1967).
43. D. L. Losee and R. O. Simmons, *Phys. Rev.* **172**, 944 (1968).
44. V. G. Manzhelii, V. G. Gavrilko, and V. I. Kuchner, *Phys. Stat. Solidi* **34**, K55 (1969).
45. A. C. Damask and G. J. Dienes, *Point Defects in Metals* (Gordon and Breach, New York, 1963).
46. L. A. Girifalco and D. O. Welch, *Point Defects and Diffusion in Strained Metals* (Gordon and Breach, New York, 1967).
47. M. W. Thompson, *Defects and Radiation Damage in Metals* (Cambridge University Press, Cambridge, 1969).
48. A. Seeger and D. Schumacher, in *Lattice Defects in Quenched Metals*, Ed. by R. M. J. Cotterill, M. Doyama, J. J. Jackson, and M. Meshii (Academic Press, New York, 1965), p. 15.
49. A. Seeger, in *Theory of Crystal Defects*, Ed. by B. Gruber (Academic Press, New York, 1966), p. 38.
49a. A. Seeger, D. Schumacher, W. Schilling, and J. Diehl (eds.), *Proc. Int. Conf. on Vacancies and Interstitials in Metals, Julich, 1968* (North-Holland Publishing Co., Amsterdam, 1970).
50. N. H. Nachtrieb, J. A. Weil, E. Catalano, and A. W. Lawson, *J. Chem. Phys.* **20**, 1189 (1952).
51. G. V. Kidson and R. Ross, in *Proc. of the Conference on Radioisotopes in Scientific Research*, Vol. I, Ed. by R. C. Extermann (Pergamon, New York, 1958), p. 185.
52. A. Seeger, G. Schottky, and D. Schumacher, *Phys. Stat. Solidi* **11**, 363 (1965).
53. A. Seeger and D. Schumacher, in *Materials Science and Engineering*, Vol. 2 (1967), p. 31.
53a. R. E. Howard, *Phys. Rev.* **144**, 650 (1966); **154**, 561 (1967).
54. H. Mehrer and A. Seeger, *Phys. Stat. Solidi* **35**, 313 (1969).
55. H. Mehrer and A. Seeger, *Phys. Stat. Solidi* **39**, 647 (1970).
56. A. Seeger and H. Mehrer, *Phys. Stat. Solidi* **29**, 231 (1968).
57. D. Schumacher, A. Seeger, and O. Härlin, *Phys. Stat. Solidi* **25**, 359 (1968).
58. A. D. LeClaire, "Correlation Effects in Diffusion in Solids," in *Physical Chemistry, An Advanced Treatise*, Vol. X, Ed. by W. Jost (Academic Press, New York, 1970), p. 261.
59. S. J. Rothman and N. L. Peterson, *Phys. Stat. Solidi* **35**, 305 (1969).
60. L. W. Barr and A. D. LeClaire, *Proc. Brit. Ceram. Soc.* **1**, 109 (1964).
61. D. P. Gregory, *Acta Met.* **11**, 623 (1963).
62. A. C. Damask and G. J. Dienes, *Phys. Rev.* **120**, 99 (1960).
63. T. Mori, M. Meshii, and J. W. Kauffman, *J. Appl. Phys.* **33**, 2776 (1962).
64. C. P. Flynn, J. Bass, and D. Lazarus, *Lattice Defects in Quenched Metals*, Ed. by R. M. J. Cotterill, M. Doyama, J. J. Jackson, and M. Meshii (Academic Press, New York, 1965), p. 639.

65. R. O. Simmons, *J. Phys. Soc. Japan* **18** (Suppl. II), 172 (1963).
66. N. F. Mott and H. Jones, *The Theory of the Properties of Metals and Alloys* (Clarendon Press, Oxford, 1936).
67. R. O. Simmons, and R. W. Balluffi, *Phys. Rev.* **119**, 600 (1960).
68. F. J. Bradshaw and S. Pearson, *Proc. Phys. Soc.* (*London*) **69B**, 441 (1956).
69. K. Misek and J. Polák, *J. Phys. Soc. Japan* **18** (Suppl. II), 179 (1963).
70. Y. Fukai, *Phil. Mag.* **20**, 1277 (1969).
71. R. J. Berry and J. L. G. LaMarche, *Phys. Letters* **31A**, 319 (1970).
72. R. R. Conte and J. Dural, *Phys. Letters* **27A**, 368 (1968).
73. K. P. Chik, in *International Conference on Vacancies and Interstitials in Metals*, Julich Conference 2, Vols. I, II (obtainable from Zentralbibliothek der Kernforschungsanlage Julich GmbH, Julich, Germany); 1970, *Vacancies and Interstitials in Metals*, Ed. by A. Seeger, D. Schumacher, W. Schilling, and J. Diehl (North-Holland Publishing Co., Amsterdam, 1968), p. 183.
74. C. J. Meechan and J. A. Brinkman, *Phys. Rev.* **103**, 1193 (1956).
75. H. B. Huntington and F. Seitz, *Phys. Rev.* **61**, 315 (1942).
76. R. O. Simmons and R. W. Balluffi, *Phys. Rev.* **125**, 862 (1962).
77. B. Okkerse, *Phys. Rev.* **103**, 1246 (1956).
78. S. M. Makin, A. H. Rowe, and A. D. LeClaire, *Proc. Phys. Soc.* (*London*) **B70**, 545 (1957).
79. D. Duhl, K.-I. Hirano, and M. Cohen, *Acta Met.* **11**, 1 (1963).
80. H. M. Gilder and D. Lazarus, *J. Phys. Chem. Solids* **26**, 2081 (1965).
81. A. Gianotti and L. Zecchina, *Nuovo Cimento* **40B**, 295 (1965).
82. U. Ermert, W. Rupp, and R. Sizmann, in *International Conference on Vacancies and Interstitials in Metals*, Julich Conference 2, Vols. I, II (obtainable from Zentralbibliothek der Kernforschungsanlage Julich GmbH, Julich, Germany, 1968), p. 30.
83. W. De Sorbo, *Phys. Rev.* **117**, 444 (1960).
84. D. Jeannotte, and E. S. Machlin, *Phil. Mag.* **8**, 1835 (1963).
85. J. A. Ytterhus and R. W. Balluffi, *Phil. Mag.* **11**, 707 (1965).
85. F. J. Kedves and P. deChatel, *Phys. Stat. Solidi* **4**, 55 (1964).
86. T. Kino and J. S. Koehler, *Phys. Rev.* **162**, 632 (1967).
87. J. Burton and D. Lazarus, *Phys. Rev.* **B2**, 787 (1970).
88. C. Lee and J. S. Koehler, *Phys. Rev.* **176**, 813 (1968).
89. A. Camanzi, N. A. Mancini, E. Rimini, and G. Schianchi, in *International Conference on Vacancies and Interstitials in Metals*, Julich Conference 2, Vols. I, II (obtainable from Zentralbibliothek der Kernforschungsanlage Julich GmbH, Julich, Germany, 1968), p. 154.
90. R. P. Heubener and C. G. Homan, *Phys. Rev.* **129**, 1162 (1963).
91. J.-I. Takamura, in *Lattice Defects in Quenched Metals*, Ed. by R. M. J. Cotterill, M. Doyama, J. J. Jackson, and M. Meshii (Academic Press, New York, 1965), p. 521.
92. S. D. Gertsricken and N. N. Novikov, *Phys. Metals Metallogr.* **9** (2), 54 (1960).
93. Y. Quéré, *Compt. Rend.* **252**, 2399 (1961).
94. M. Doyama and J. S. Koehler, *Phys. Rev.* **127**, 21 (1962).
95. F. Ramsteiner, W. Schule, and A. Seeger, *Phys. Stat. Solidi* **2**, 1005 (1962).
96. C. T. Tomizuka and E. Sonder, *Phys. Rev.* **103**, 1182 (1956).
97. S. J. Rothman, N. L. Peterson, and J. T. Robinson, Phys. Stat. Solidi **39**, 635.

98. M. Beyeler and Y. Adda, in *Physics of Solids at High Pressures*, Ed. by C. T. Tomizuka and R. M. Emrick (Academic Press, New York, 1965), p. 349.
99. R. H. Dickerson, R. C. Lowell, and C. T. Tomizuka, *Phys. Rev.* **137**, A613 (1965).
100. C. T. Tomizuka, in *Progress in Very High Pressure Research* (Proc. Intern. Conf., Bolton Landing), Ed. by F. B. Bundy, W. R. Hibbard, Jr., and H. M. Strong (Wiley and Sons, New York, 1961), p. 266.
101. R. O. Simmons and R. W. Balluffi, *Phys. Rev.* **129**, 1533 (1963).
102. J. Bass, *Phil. Mag.* **15**, 717 (1967).
103. R. O. Simmons and R. W. Balluffi, *Phys. Rev.* **117**, 62 (1960).
104. R. R. Bourassa, D. Lazarus, and D. A. Blackburn, *Phys. Rev.* **165**, 853 (1968).
105. T. S. Lundy and J. F. Murdoch, *J. Appl. Phys.* **33**, 1671 (1962).
106. J. J. Spokas and C. P. Slichter, *Phys. Rev.* **113**, 1462 (1959).
107. B. M. Butcher, H. Hutto, and A. L. Ruoff, *Appl. Phys. Letters* **7**, 34 (1965).
108. M. Doyama and J. S. Koehler, *Phys. Rev.* **134**, A522 (1964).
109. Ya. A. Kraftmakher and P. G. Strelkov, *Soviet Phys.—Solid State* **8**, 460 (1966).
110. A. Ascoli, M. Asdente, E. Germagnoli, and A. Manara, *J. Phys. Chem. Solids* **6**, 59 (1958).
111. G. L. Bacchella, E. Germagnoli, and S. Granata, *J. Appl. Phys.* **30**, 748 (1959).
112. J. J. Jackson, *Lattice Defects in Quenched Metals*, Ed. by R. M. J. Cotterill, M. Doyama, J. J. Jackson, and M. Meshii (Academic Press, New York, 1965).
113. F. Cattaneo, E. Germagnoli, and F. Grasso, *Phil. Mag.* **7**, 1373 (1962).
114. V. S. Kopan and M. J. Skorochod, *Isledovanie nesovershenstv kristallicheskovo stroeniya* (Naukovaya dumka, Kiev, 1965), p. 99.
115. J. Polák, *Phys. Stat. Solidi* **21**, 581 (1967).
116. J. Polák, *Phys. Letters* **24A**, 649 (1967).
116a. R. Feder and A. S. Nowick, *Phil. Mag.* **15**, 805 (1967).
117. F. M. d'Heurle, R. Feder, and A. S. Nowick, *J. Phys. Soc. Japan* **18** (Suppl. II), 184 (1963).
118. J. B. Hudson and R. E. Hoffman, *Trans. AIME* **221**, 761 (1961).
119. N. H. Nachtrieb, H. A. Resing, and S. A. Rice, *J. Chem. Phys.* **31**, 135 (1959).
120. H. Mehrer, H. Kronmüller, and A. Seeger, *Phys. Stat. Solidi* **10**, 725 (1965).
120a. R. C. Brown, J. Worster, N. H. March, R. C. Perrin, and R. Bullough, *Diffusion Processes* (Proc. Thomas Graham Mem. Symp., Univ. Strathclyde) (Gordon and Breach, New York, 1970); R. E. Hoffman, W. Pikus, and R. E. Ward, *Trans. AIME* **206**, 483 (1956).
121. D. L. Martin, *J. Phys. Chem. Solids, Supplement 1, Lattice Dynamics, Proc. of an International Conference, Copenhagen*, Ed. by R. F. Wallis (Pergamon Press, Oxford, 1963), p. 255.
122. A. Lodding, J. N. Mundy, and A. Ott, *Phys. Stat. Solidi* **38**, 559 (1970).
123. J. N. Mundy, L. W. Barr, and F. A. Smith, *Phil. Mag.* **14**, 785 (1966).
124. R. Feder and H. P. Charbnau, *Phys. Rev.* **149**, 464 (1966).
125. R. Feder, *Phys. Rev.* **B2**, 828 (1970).
126. D. K. C. MacDonald, *J. Chem. Phys.* **21**, 177 (1953).
127. R. A. Hultsch and R. G. Barnes, *Phys. Rev.* **125**, 1832 (1962).
128. C. R. Kohler and A. L. Ruoff, *J. Appl. Phys.* **36**, 2444 (1965).
129. I. M. Torrens and M. Gerl, *Phys. Rev.* **187**, 912 (1969).
129a. R. C. Brown, J. Worster, N. H. March, R. C. Perrin, and R. Bullough, to be published.

130. A. Ott, A. Lodding, and D. Lazarus, *Phys. Rev.* **188**, 1088 (1969).
131. D. Jeannotte and J. M. Galligan, *Phys. Rev. Letters* **19**, 232 (1967).
132. H. Schultz, *Lattice Defects in Quenched Metals*, Ed. by R. M. J. Cotterill, M. Doyama, J. J. Jackson, and M. Meshii (Academic Press, New York, 1965), p. 761.
133. R. L. Andelin, J. D. Knight, and M. Kahn, *Trans. AIME* **233**, 19 (1965).
134. A. S. Nowick, *Comments on Solid State Physics* **2**, 30 (1969).
135. C. J. Beevers, *Acta Met.* **11**, 1029 (1963).
136. C. Mairy, J. Hillairet, and D. Schumacher, *Acta Met.* **15**, 1258 (1967).
137. C. Janot, G. Bianchi, and B. George, *Compt. Rend.* **267B**, 336 (1968).
138. P. G. Shewmon, *J. Metals* **8**, 918 (1956).
139. A. Seeger and K.-P. Chik, *Phys. Stat. Solidi* **29**, 455 (1968).
140. A. Seeger, *Comments on Solid State Physics* **1**, 157; **2**, 55 (1969).
141. A. Seeger and M. L. Swanson, *Lattice Defects in Semiconductors*, Ed. by R. R. Hasiguti (University of Tokyo Press, 1968).
142. G. D. Watkins, *J. Phys. Soc. Japan* **18** (Suppl. II), 33 (1963).
143. G. D. Watkins, *Radiation Damage in Semiconductors*, Ed. by P. Baruch (Dunod, Paris, 1965).
144. G. D. Watkins, *Radiation Effects in Semiconductors*, Ed. by F. L. Vook (Plenum Press, New York), 1968.
145. R. E. Whan, *Appl. Phys. Letters* **6**, 221 (1965).
146. R. E. Whan, *Phys. Rev.* **140**, A690 (1965).
147. A. Scholz and A. Seeger, *Radiation Damage in Semiconductors*, Ed. by P. Baruch (Dunod, Paris, 1965), p. 315.
148. M. M. Valenta and C. Ramasastry, *Phys. Rev.* **106**, 73 (1957).
149. P. A. Thrower, *Chemistry and Physics of Carbon*, Ed. by P. L. Walker, Jr. (Marcel Dekker, New York, 1967), p. 217.
150. G. R. Hennig, *J. Chem. Phys.* **40**, 2877 (1964).
151. G. R. Hennig, *J. Appl. Phys.* **36**, 1482 (1965).
152. J. H. W. Simmons, *Radiation Damage in Graphite* (Pergamon Press, New York, 1965), p. 133.
153. L. Bochirol and E. Bonjour, *Carbon* **6**, 661 (1968).
154. R. W. Henson and W. N. Reynolds, *Carbon* **3**, 277 (1965).
155. N. S. Rasor and J. D. McClelland, *J. Phys. Chem. Solids* **15**, 17 (1960).
156. R. J. Thorn and G. H. Winslow, *J. Chem. Phys.* **26**, 186 (1957).
157. P. R. Goggin and W. N. Reynolds, *Phil. Mag.* **8**, 265 (1963).
158. J. A. Turnbull and M. S. Stagg, *Phil. Mag.* **14**, 1049 (1966).
159. J. Friedel, *Les Dislocations* (Gauthier Villiers, Paris, 1956).
160. C. Baker and A. Kelly, *Nature* **193**, 235 (1962).
161. P. A. Thrower, *Phil. Mag.* **18**, 697 (1969).
162. P. A. Thrower, *Carbon* **6**, 687 (1968).
163. G. R. Hennig, *Second Conference on Industrial Carbon and Graphite* (Society of Chemical Industry, London, 1966), p. 109.
164. M. Kanter, *Phys. Rev.* **107**, 655 (1957).
165. M. Kanter, *Kinetics of High Temperature Processes*, Ed. by W. D. Kingery, Jr. (Technology Press, MIT, Cambridge, Mass., 1959), p. 61.
166. R. B. Evans, L. D. Love, and E. H. Kobisk, *J. Appl. Phys.* **40**, 3058 (1969).
167. F. S. Feates, *J. Nucl. Materials* **27**, 325 (1968).

168. L. W. Barr and A. B. Lidiard, "Defects in Ionic Solids," in Vol. 10, *Physical Chemistry—An Advanced Treatise* (Academic Press, New York, 1970).
169. A. B. Lidiard, *Proc. Brit. Ceram. Soc.* **1967** (9), 1.
169a. A. B. Lidiard, Handbuch der Physik, Vol. 20 (Springer-Verlag, Berlin, 1957), p. 246.
170. P. Süptitz and J. Teltow, *Phys. Stat. Solidi* **23**, 9 (1967).
171. Y. Adda and J. Philibert, *La Diffusion dans les Solides* (Presses Universitaires de France, 1966).
171a. R. J. Friauf, *J. Appl. Phys.* **33**, 494 (1962).
172. M. Eisenstadt and A. G. Redfield, *Phys. Rev.* **132**, 635 (1963).
173. D. L. Kirk and P. L. Pratt, *Proc. Brit. Ceram. Soc.* **1967** (9), 215.
174. A. R. Allnatt and P. W. M. Jacobs, *Trans. Faraday Soc.* **58**, 116 (1962).
175. J. Rolfe, *Can. J. Phys.* **42**, 2195 (1964).
176. S. C. Jain and S. L. Dahake, *Indian J. Pure Appl. Phys.* **2**, 71 (1964).
177. J. B. Beaumont and P. W. M. Jacobs, *J. Chem. Phys.* **45**, 1496 (1966).
178. D. K. Dawson, and L. W. Barr, *Phys. Rev. Letters* **19**, 844 (1967).
179. S. Chandra and J. Rolfe, *Can. J. Phys.* **48**, 397 (1970).
180. S. Chandra and J. Rolfe, *Can. J. Phys.* **48**, 412 (1970).
181. S. C. Jain and D. C. Parashar, *J. Phys. C* (*Solid State Phys.*) *Ser. 2*, **2**, 167 (1969).
182. T. G. Stoebe and P. L. Pratt, *Proc. Brit. Ceram. Soc.*, **1967** (9), 181.
183. K. Tharmalingam and A. B. Lidiard, *Phil. Mag.* **6**, 1157 (1961).
184. L. W. Barr, J. A. Morrison, and P. A. Schroeder, *J. Appl. Phys.* **36**, 624 (1965).
184a. V. S. Mandel', *Soviet Phys.—Solid State* **10**, 2530 (1969).
185. O. S. Spencer and C. A. Plint, *J. Appl. Phys.* **40**, 168 (1969).
186. K. L. Kliewer and J. S. Koehler, *Phys. Rev.* **157**, 685 (1967).
187. R. W. Davidge, *Phys. Stat. Solidi* **3**, 1851 (1963).
188. C. A. Plint and M. L. Breig, *J. Appl. Phys.* **35**, 2745 (1964).
189. R. Strumane and R. DeBatist, *Phys. Stat. Solidi* **6**, 817 (1964).
190. R. D. Fouchaux and R. O. Simmons, *Phys. Rev.* **136**, A1664 (1964).
191. C. Tubandt, *Z. Anorg. allgem. Chem.* **115**, 113 (1920).
192. J. E. Hove, *Phys. Rev.* **102**, 915 (1956).
193. M. D. Weber and R. J. Friauf, *J. Phys. Chem. Solids* **30**, 407 (1969).
194. H. Layer, M. G. Miller, and L. Slifkin, *J. Appl. Phys.* **33**, 478 (1962).
195. A. E. Abey and C. T. Tomizuka, *J. Phys. Chem. Solids* **27**, 1149 (1966).
196. K. L. Kliewer, *J. Phys. Chem. Solids* **27**, 705 (1966).
197. H. Wakabayasi, *J. Phys. Soc. Japan* **15**, 2000 (1960).
198. B. B. Snavely, Ph. D. Thesis, Cornell University, Ithaca, New York (1962).
199. L. Slifkin, W. McGowan, A. Fukai, and J.-S. Kim, *Phot. Sci. and Eng.* **11**, 79 (1967).
200. E. Zintl and A. Udgard, *Z. Anorg. allgem. Chem.* **240**, 150 (1939).
201. C. Tubandt, H. Reinhold, and G. Liebold, *A. Anorg. allgem. Chem.* **197**, 225 (1931).
202. R. W. Ure, Jr., *J. Chem. Phys.* **26**, 1363 (1957).
203. V. H. J. Matzke and R. Lindner, *Z. Naturforsch.* **19A**, 1178 (1964).
204. J. Short and R. Roy, *J. Phys. Chem.* **68**, 3077 (1964).
205. E. Barsis and A. Taylor, *J. Chem. Phys.* **48**, 4362 (1968).
206. E. Barsis and A. Taylor, *J. Chem. Phys.* **48**, 4357 (1968).
207. D. C. Stockbarger, *J. Opt. Soc. Am.* **39**, 731 (1949).
208. W. Bontinck, *Physica* **24**, 650 (1958).
209. K. Muto and K. Awazu, *J. Phys. Chem. Solids* **29**, 1269 (1968).

210. J. A. Champion, *Brit. J. Appl. Phys.* **16**, 805 (1965).
211. B. R. Rossing, D. Sc. Thesis, MIT (1966).
212. A. D. Franklin, S. Marzullo, and J. B. Wachtman, Jr., *J. Res. Nat. Bur. Std.* **71A**, 355 (1967).
213. H. B. Johnson, G. R. Miller, and I. B. Cutler, *J. Am. Ceram. Soc.* **50**, 526 (1967).
214. M. J. Weber and R. W. Bierig, *Phys. Rev.* **134**, A1492 (1964).
215. F. K. Fong, *Progr. Solid State Chem.* **3**, 135 (1966).
216. J. M. Baker, E. R. Davies, and J. P. Hurrell, *Phys. Letters* **26A**, 352 (1968); *Proc. Roy. Soc.* (*London*) **A308**, 403 (1968).
217. A. D. Franklin, *Proc. Brit. Ceram. Soc.*, **1967** (9), 15.
218. A. D. Franklin, *J. Phys. Chem. Solids* **29**, 823 (1968).
219. Hj. Matzke, *J. Materials Science* **5**, 831 (1920).
220. A. D. Franklin and S. M. Marzullo, in *Proc. Brit. Ceram. Soc.*, *Conference on Mass Transport in Non-Metallic Solids, Dec. 17 and 18, 1969.*
221. W. D. Copeland and R. A. Swalin, *J. Phys. Chem. Solids* **29**, 313 (1968).
222. I. E. Leksina and S. I. Novikova, *Soviet Phys.—Solid State* **5**, 798 (1963).
223. R. M. Emrick, *Phys. Rev.* **122**, 1720 (1961).
224. H. H. Grimes, *J. Phys. Chem. Solids* **26**, 509 (1965).
225. M. Doyama and J. S. Koehler, *Phys. Rev.* **119**, 939 (1960).
226. Y. Quéré, *Compt. Rend.* **251**, 367 (1960).
227. O. N. Ovcharenko, *Phys. Metals Metallogr.* **11**, 3, 78 (1961).
228. L. J. Cuddy and E. S. Machlin, *Phil. Mag.* **7**, 745 (1962).
229. R. R. Hasiguti, Y. Nakao, and H. Kimura, *J. Phys. Soc. Japan* **20**, 553 (1965).
230. Ya. A. Kraftmakher, *Soviet Phys.—Solid State* **9**, 1458 (1967).
231. W. de Sorbo and D. Turnbull, *Acta Met.* **7**, 83 (1959).
232. W. de Sorbo and D. Turnbull, *Phys. Rev.* **115**, 560 (1959).
233. K. Detert and I. Ständer, *Z. Metallk.* **52**, 677 (1961).
234. D. Locati and T. Federighi, reported by T. Federighi, *Lattice Defects in Quenched Metals*, Ed. by R. M. J. Cotterill, M. Doyama, J. J. Jackson, and M. Meshii (Academic Press, New York, 1965), p. 217.
235. G. Bianchi, D. Mallejac, C. Janot, and G. Champier, *Compt. Rend.* **263B**, 1404 (1966).
236. A. D. King, A. J. Cornish, and J. Burke, *J. Appl. Phys.* **37**, 4717 (1966).
237. T. Federighi, in *Lattice Defects in Quenched Metals*, Ed. by R. M. J. Cotterill, M. Doyama, J. J. Jackson, and M. Meshii (Academic Press, New York, 1965), p. 217.
238. Y. N. Lwin, M. Doyama, and J. S. Koehler, *Phys. Rev.* 165, 787 (1968).
239. Ya. A. Kraftmakher and F. B. Lanina, *Soviet Phys.—Solid State* **7**, 92 (1965).
240. G. R. Piercy, *Phil. Mag.* **5**, 201 (1960).
241. W. Bauer and A. Sosin, *Phys. Rev.* **147**, 482 (1966).
242. G. A. Sullivan and J. W. Weymouth, *Phys. Rev.* **136**, A1141 (1964).
243. J. D. Filby and D. L. Martin, *Proc. Roy. Soc.* (*London*) **284**, 83 (1965).
244. D. F. Holcomb and R. E. Norberg, *Phys. Rev.* **98**, 1074 (1955).
245. A. N. Naumov and G. Ya. Ryskin, *Zh. Tekhn. Fiz.* **29**, 189 (1959).
246. D. C. Ailion and C. P. Slichter, *Phys. Rev.* **137**, A235 (1965).
247. A. Ott, J. N. Mundy, L. Löwenberg, and A. Lodding, *Z. Naturforsch.* **23A**, 771 (1968).
248. N. H. Nachtrieb, E. Catalano, and J. A. Weil, *J. Chem. Phys.* **20**, 1185 (1952).

249. J. N. Mundy, L. W. Barr, and F. A. Smith, *Phil. Mag.* **15**, 411 (1967).
250. Ya. A. Kraftmakher, *Soviet Phys.—Solid State* **5**, 696 (1963).
251. Ya. A. Kraftmakher, *Soviet Phys.—Solid State* **6**, 396 (1964).
252. J. D. Meakin, A. Lawley, and R. C. Koo, *Lattice Defects in Quenched Metals*, Ed. by R. M. J. Cotterill, M. Doyama, J. J. Jackson, and M. Meshii (Academic Press, New York, 1965), p. 767.
253. V. Ya. Chekhovskoy, and V. A. Petrov, *International Conference on Vacancies and Interstitials in Metals*, Julich Conference 2, Vols. I, II (obtainable from Zentralbibliothek der Kernforschungsanlage Julich GmbH, Julich, Germany, 1968), p. 6.
254. W. Glaeser and H. Wever, *Phys. Stat. Solidi* **35**, 367 (1969).
255. M. L. Swanson, G. R. Piercy, G. V. Kidson, and A. F. Quenneville, *J. Nucl. Materials* **34**, 340 (1970).
256. D. Lazarus, *Diffusion in Body-Centered Metals*, Ed. by J. A. Wheeler, Jr., and F. R. Winslow (Am. Soc. for Metals, Metals Park, Ohio, 1965), p. 155.
257. C. Janot, D. Mallejac, and B. George, *Compt. Rend.* **270B**, 404 (1970).
258. H. Letaw, L. Slifkin, and W. M. Portnoy, *Phys. Rev.* **102**, 636 (1956).
259. R. F. Peart, *Phys. Stat. Solidi* **15**, K119 (1966).
260. E. Barsis, E. Lilley, and A. Taylor, *Proc. Brit. Ceram. Soc.* **1967**, (9), 203 (1967).
261. C. F. Bauer and D. H. Whitmore, *Phys. Stat. Solidi* **37**, 585 (1970).
262. L. W. Barr, in *Proc. Brit. Ceram. Soc., Conference on Mass Transport in Non-Metallic Solids, Dec. 17 and 18, 1969.*
263. R. W. Dreyfus and A. S. Nowick, *J. Appl. Phys.* **33**, 473 (1962).
264. A. R. Allnatt and P. Pantelis, *Solid State Comm.* **6**, 309 (1968).
265. V. Trnovcova, *Czech. J. Phys.* **B19**, 663 (1969).
266. E. Rzepka and J.-P. Chapelle, *Compt. Rend.* **268B**, 1770 (1969).
267. V. K. Jain, *Indian J. Pure Appl. Phys.* **7**, 330 (1969).
268. R. G. Fuller, *Bull. Am. Phys. Soc.* **15**, 384 (1970).
269. A. V. Chadwick, B. D. McNicol, and A. R. Allnatt, *Phys. Stat. Solidi* **33**, 301 (1969).
270. P. J. Harvey and I. M. Hoodless, *Phil. Mag.* **16**, 543 (1967).
271. J. N. Maycock, V. R. Pai Verneker, and C. S. Gorzynski, Jr., *Phys. Stat. Solidi* **37**, 857 (1970).
272. F. Bénière and M. Chemla, *Compt. Rend.* **266C**, 660 (1968).
273. R. G. Fuller, *Phys. Rev.* **142**, 524 (1966).
274. M. Eisenstadt, *Phys. Rev.* **132**, 630 (1963).
275. T. G. Stoebe and R. A. Huggins, *J. Mat. Sci.* **1**, 117 (1966).
276. J. Itoh, M. Satoh, and A. Hiraki, *J. Phys. Soc. Japan* **16**, 343 (1960).
277. M. Eisenstadt, *Phys. Rev.* **133**, A191 (1964).
278. M. Satoh, *J. Phys. Soc. Japan* **20**, 1008 (1965).
279. G. Pellegrini and J. Pelsmaekers, *J. Chem. Phys.* **51**, 5190 (1969).
280. J. Pelsmaekers, G. Pellegrini, and S. Amelinckx, *Solid State Comm.* **1**, 92 (1963).
281. B. von Guerard, H. Peisl, and W. Waidelich, *Phys. Stat. Solidi* **29**, K59 (1968).
282. J. S. Koehler, D. Langreth, and B. von Turkovich, *Phys. Rev.* **128**, 573 (1962).
283. H. Kanzaki, K. Kido, and T. Ninomiya, *J. Appl. Phys.* **33**, 482 (1962).
284. P. Muller, *Phys. Stat. Solidi* **12**, 775 (1965).
285. E. Barsis and A. Taylor, *J. Chem. Phys.* **45**, 1154 (1966).
286. M. Baker and A. Taylor, *J. Phys. Chem. Solids* **30**, 1003 (1969).
287. R. J. Lysiak and P. P. Mahendroo, *J. Chem. Phys.* **44**, 4025 (1966).
288. J. R. Miller and P. P. Mahendroo, *Phys. Rev.* **174**, 369 (1968).

Chapter 2

IONIC CONDUCTIVITY (INCLUDING SELF-DIFFUSION)

Robert G. Fuller

Department of Physics
University of Nebraska
Lincoln, Nebraska

1. INTRODUCTION

Ionic conductivity has become relevant. The recent revival of interest in electric automobiles and solid-state batteries has led to production of high-conductivity solid electrolytes. These new high-conductivity materials, such as $RbAg_4I_5$,[1,2] have greatly expanded the range over which ionic transport phenomena in solids have been observed. In Fig. 1, we show the conductivity of $RbAg_4I_5$ in comparison with the conductivity of the more common alkali and silver halides. We find grouped together on the left-hand, or high-temperature, side of the figure the alkali halide crystals. These materials have been extensively studied.* They are excellent insulators at room temperature and only have significant electrical conductivity within a few hundred degrees of their melting temperatures. Furthermore, in this high-temperature region, the conductivity of the alkali halides is strongly temperature-dependent, the conductivity changes by about 3% per degree Celsius, and the activation energy for conduction is about 2 eV. To the right in Fig. 1, we move to lower temperatures and to materials less well characterized than the alkali halides. The cesium and ammonium halides have ionic conductivities that are about equal in magnitude to the alkali halides but are less strongly temperature-dependent, with activation energies

* A recent review of this work is given by Barr and Lidiard.[3]

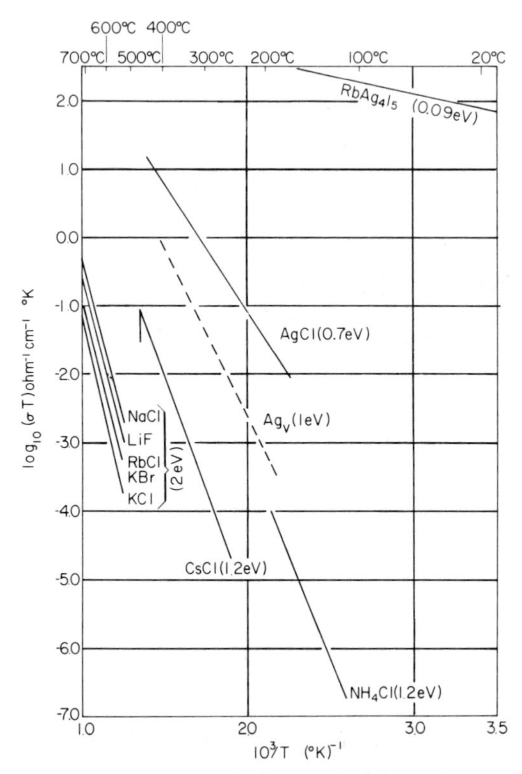

Fig. 1. Electrical conductivity of halide crystals. This figure illustrates the wide range of ionic properties to be found in this group of materials. Activation energies are indicated in parentheses.

of 1.2 eV. With a conductivity higher by three orders of magnitude, we find silver chloride, with an activation energy for conduction of less than 1 eV. In the far upper right-hand corner of the figure, near room temperature, we find the new high-conductivity solid, some four or five orders of magnitude higher than AgCl. The compound $RbAg_4I_5$ shows a small temperature variation of its conductivity, with an activation energy of only 0.09 eV. The ionic transport phenomena shown in Fig. 1 indicate the importance of the following questions to which we will address ourselves: What is currently known about the well-characterized materials such as potassium bromide, potassium chloride, and sodium chloride? How can that knowledge be used to better understand the less well-characterized materials? What does our present knowledge suggest about the future directions for research?

In an attempt to deal with these questions, Section 2 presents an introduction to the conceptual models and phenomenological equations used to understand ionic conductivity and self-diffusion in alkali halide crystals. Section 3 relates these phenomenological equations to the ionic properties of the crystal lattice by means of a simple theory for noninteracting particles. The simple theory is extended to include Coulomb interactions between the particles in Section 4. A review of possible experiments and their techniques begins Section 5, which continues with a discussion of the results of the experiments performed on the fully characterized alkali halides, KBr, KCl, KI, RbCl, and NaCl. A brief summary of the results of work on other halide crystals ends Section 5. Section 6 concludes this chapter with suggestions about the future directions of research in these areas of ionic conductivity and self-diffusion.

2. IONIC TRANSPORT EQUATIONS

The foundations upon which our understanding of ionic conductivity are built were laid down before 1940 by the early work of Schottky,[4] Wagner,[5] and Mott and Littleton.[6] It was found that the transfer of mass and charge occurred in alkali halide crystals by means of an ionic process. The ions in these materials were mobile because of the presence of intrinsic point defects, now called Schottky defects. Figure 2 shows a two-dimensional representation of an alkali halide crystal. A Schottky defect is an isolated pair of cation and anion vacant lattice sites. The concentration of Schottky defects in a nominally pure alkali halide is small, remaining less than 1% even at the melting temperature of the crystal. As a consequence, conduction and diffusion are usually described in terms of the concentration and mobility of the small number of vacancies. Such a description can be made equivalent to motion of the large number of ions.

The ionic processes of the transport phenomena are related to the macroscopic quantities measured in the laboratory by means of the phenomenological law of steady-state diffusion and the theory of random walk. Fick's second law of diffusion is given by*

$$\partial c/\partial t = \mathrm{div}(D \,\mathrm{grad}\, C) \tag{1}$$

where C is the concentration of diffusing particles (cm^{-3}), D is the diffusion coefficient ($\mathrm{cm}^2/\mathrm{sec}$), and $\partial/\partial t$ is the time derivative. An expression for the

* For a general introductory text on matter transport in solids, see Shewmon.[7]

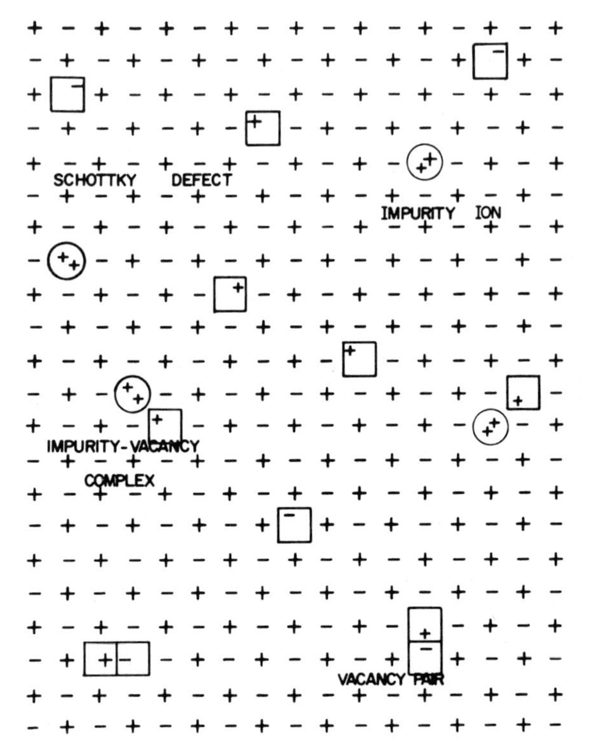

Fig. 2. Model of the perfect alkali halide crystal perturbed by four kinds of point defects: Schottky defects, impurity ions, divalent-cation-impurity–cation-vacancy complexes, and vacancy pairs.

mean-square displacement of the diffusing particles can be obtained from the general solution of Eq. (1). The result is

$$\overline{R^2(t)} = 6Dt \tag{2}$$

where t is the time.

The mean-square displacement of the diffusing particles can also be calculated from an analysis of the motion of the individual particles. If a particle makes Γ random jumps per unit time, and if all the jumps are of the same length r, the random walk theory is applicable and gives the result that the mean-square displacement is equal to the product of the total number of jumps times the square of the length of each jump,

$$\overline{R^2(t)} = \Gamma t r^2 \tag{3}$$

An expression for D, a macroscopic measurable quantity, in terms of the microscopic behavior of the individual particles, or vacancies, in the crystal is obtained by combining Eqs. (2) and (3), i.e.,

$$D = \tfrac{1}{6}\Gamma r^2 \tag{4}$$

In a NaCl-type lattice, if the mole fraction of Schottky defects is x and the probability per unit time that a vacancy will jump from one position to another is w, then the jump frequency Γ is

$$\Gamma = 12xw \tag{5}$$

If the anion–cation separation is represented by a, the jump distance r is equal to $\sqrt{2}a$, so that the expression for the diffusion coefficient becomes

$$D = 4a^2(xw) \tag{6}$$

The parentheses on the right-hand side of Eq. (6) is there to emphasize the fact that the diffusion coefficient is proportional to the product of the vacancy concentration and the vacancy jump probability. A great deal of experimental effort goes into attempting to manipulate the vacancy concentration in a known way so that the properties of the jump probability can be examined. The importance of this manipulation becomes even more apparent when it is realized that if the electrical conductivity and the diffusion occur by the same mechanism, the conductivity times the absolute temperature is directly proportional to the diffusion coefficient.* Hence

* This is known as the Nernst–Einstein relation and a general derivation of it was given by Mott and Gurney[8] as follows. For a system of charged particles in equilibrium in an electric field, the sum of the diffusion current given by Fick's first law, $J_D = -D$ grad C, plus the electric current, must be equal to zero,

$$-qD \operatorname{grad} C + \sigma E = 0 \tag{A}$$

where q is the electric charge of each particle, σ is the conductivity of the system, and E is the electric field. By Maxwell–Boltzmann statistics, the distribution of particles in an electric field will satisfy the following equation:

$$\operatorname{grad} C = CqE/kT \tag{B}$$

where k is the Boltzmann constant and T is the absolute temperature. Combining Eqs. (A) and (B) yields the Nernst–Einstein relation,

$$\sigma T = (Cq^2/k)D \tag{C}$$

For cation diffusion in NaCl, this becomes

$$\begin{aligned} \sigma_c T &= (Ne^2/k)D_c = 4.1\times 10^7 D_c \\ (\text{ohm}^{-1}\,\text{cm}^{-1}\,{}^\circ\text{K}) &= (\text{C}^2\,\text{cm}\,\text{J}^{-1}\,{}^\circ\text{K})(\text{cm}^2\,\text{sec}^{-1}) \end{aligned} \tag{D}$$

since N is the number of cations per cubic centimeter and e is the electronic charge.

the interpretation of both conductivity and diffusion experiments is dependent upon extracting information from the (xw) product.

At the present time, the qualitative features of the experimental results from conductivity and diffusion measurements on alkali halide crystals require a model of the perfect crystal perturbed by the four kinds of point defects shown in Fig. 2: Schottky defects, divalent impurity ions, divalent-impurity–vacancy complexes (vacancies bound to the impurity ions and thereby removed from the conduction process), and vacancy pairs (an anion vacancy and a cation vacancy bound together to form a neutral entity that does not participate in the conduction of electricity).[3] Hence, the electrical conductivity of alkali halide crystals is the sum of the anion and cation contributions and can be written as follows:

$$\sigma = \sigma_a + \sigma_c = (4Ne^2a^2/kT)(x_aw_a + x_cw_c) \tag{7}$$

where x_a and x_c are the mole fractions of isolated anion and cation vacancies, respectively, and w_a and w_c are the anion and cation vacancy jump probabilities. The diffusion coefficient is experimentally determined from the self-diffusion of radioactive tracer ions. Hence, the diffusion equations must be modified to include a correlation factor f which accounts for the nonrandom motion of the tracer ions.* Furthermore, the vacancy pairs make a contribution to the diffusion of radioactive cations. So, the diffusion coefficient of the radioactive ion is equal to the sum of the isolated vacancy contribution and the vacancy pair contribution,

$$D_c^* = D_c + D_p{}^c \tag{8}$$

D_c^* is the diffusion coefficient of the cation tracer ion. The contribution of isolated cation vacancies to the cation tracer diffusion D_c is given by

$$D_c = 4fa^2x_cw_c \tag{9}$$

where f is the correlation factor for isolated vacancy diffusion[9] and is 0.782 for the NaCl-type lattice. The vacancy pair contribution to the cation tracer diffusion $D_p{}^c$ is given by

$$D_p{}^c = \tfrac{4}{3}a^2f_p{}^cw_p{}^cx_p \tag{10}$$

* Consider a tracer which has just exchanged places with a vacancy. The direction of the next jump of the tracer ion is not random, since the vacancy is available for the reverse of the first jump. Hence, consecutive pairs of jumps in opposite directions are more probable than consecutive pairs of jumps in the same direction. This effect requires the modification of the diffusion equations by the inclusion of a correlation factor. See Ref. 9.

where f_p^c is the correlation factor for the vacancy pair contribution to cation diffusion[9] and w_p^c is the probability for a cation to jump into a vacancy pair. Similar equations can be written for the diffusion of radioactive anions:

$$D_a^* = D_a + D_p^a \tag{11}$$

$$D_a = 4fa^2x_a w_a \tag{12}$$

$$D_p^a = \tfrac{4}{3}a^2 f_p^a w_p^a x_p \tag{13}$$

Now, let us turn our attention to the explicit equations for the defect concentrations and the jump probabilities.

3. THE SIMPLE THEORY

3.1. Concentrations of Point Defects

Let us choose as a typical system a potassium chloride crystal containing strontium chloride, where N_+ is the number of KCl molecules in the crystal and N_i ($\ll N_+$) is the number of $SrCl_2$ molecules. The Sr ions are located at K^+ sites in the cation lattice as shown in Fig. 2. Let us define the standard state of the system as a crystal with $N_+ + 2N_i$ sites in each of the anion and cation sublattices. Then, although every site in the anion sublattice is occupied by a chlorine ion, the cation sublattice consists of N_+ potassium ions, N_i strontium ions, and N_i vacant lattice sites. To a first approximation, we assume that the presence of the impurity species does not alter the vibrational frequencies of the lattice. The crystal is permitted to go from the standard state to equilibrium, and in so doing it obtains n_s Schottky defects. Also, there are n_k divalent-cation-impurity–cation-vacancy complexes and n_p vacancy pairs in the sense that an applied electric field will not separate them into their components, which is the case if the components are near neighbors. The simplest assumption to make about the isolated vacancies and impurity ions is that they are distributed at random; their interaction is therefore zero on the average, as is the interaction of the isolated vacancies and impurity ions with the vacancy pairs and impurity–vacancy complexes. There is, however, an average Gibbs free energy of binding associated with each pair or complex, denoted by g_p and g_k, respectively. Since the concentrations of these species is very small, the total binding energy is $n_p g_p + n_k g_k$.

The most probable, or equilibrium, configuration is found by minimizing the Gibbs free energy of the system.[10] From this configuration, the defect concentrations are determined. The Gibbs energy of the crystal relative to the crystal in the standard state is

$$G = n_s g_s - n_k g_k - n_p g_p + T s_c \tag{14}$$

where g_s is the average Gibbs free energy required to create a Schottky defect, T is the absolute temperature, and s_c is the configurational entropy of the system. The Gibbs free energies g_s, g_k, and g_p are not the complete thermodynamic variables since the contribution of the configurational entropy to each Gibbs energy term has been displayed explicitly in the term Ts_c.

The configurational entropy of the system is given[10] by $k \log_e P$, where P is the number of distinguishable arrangements for a configuration specified by the independent variables n_s, n_p, n_k, N_i, and N_+. Let us now calculate P. First, we count the number of arrangements for the complexes, then the number of possibilities left for the pairs, and finally the arrangements left for everything else in the system.

There are $Z_k(N_+ + 2N_i + n_s)$ ways of placing the first complex on the cation lattice, where Z_k is the number of equivalent orientations of the complex and is equal to 12 for KCl. For the second complex, there are $(N_+ + 2N_i + n_s - 2)$ positions for one component and for each of these, Z_k possibilities for the other component, each of these having a probability $(N_+ + 2N_i + n_s - 2)/(N_+ + 2N_i + n_s)$ of being available. Hence, the number of ways of putting the second complex into the cation sublattice is $Z_k(N_+ + 2N_i + n_s - 2)^2/(N_+ + 2N_i + n_s)$. Similarly, the number of ways for the $(j + 1)$th complex is $Z_k(N_+ + 2N_i + n_s - 2j)^2/(N_+ + 2N_i + n_s)$. So, the total number of distinguishable ways of putting n_k complexes into the cation lattice is

$$\left(\frac{Z_k}{N_+ + 2N_i + n_s}\right)^{n_k} \prod_{j=0}^{n_k-1} \frac{(N_+ + 2N_i + n_s - 2j)^2}{n_k!}$$

which can be reduced to

$$\left(\frac{4Z_k}{N_+ + 2N_i + n_s}\right)^{n_k} \left[\left(\frac{\frac{1}{2}(N_+ + 2N_i + n_s)}{n_k}\right)\right]^2 n_k!$$

Next, we put the n_p pairs into the lattice and by exactly analogous reasoning, we find the number of arrangements for pairs. The number of ways of putting in the first pair is $Z_p(N_+ + 2N_i + n_s - 2n_k)$, where Z_p is the number of equivalent orientations for a pair and is equal to six for

KCl. The number of available positions for the cation vacancy of the second pair is $(N_+ + 2N_i + n_s - 2n_k - 1)$ and for each of these, the number of positions available for the anion vacancy component is

$$Z_p(N_+ + 2N_i + n_s - 1)/(N_+ + 2N_i + n_s).$$

The number of ways, then, of placing the $(j + 1)$th pair into the lattice is

$$Z_p(N_+ + 2N_i + n_s - 2n_k - j)(N_+ + 2N_i + n_s - j)/(N_+ + 2N_i + n_s)$$

Hence, the number of distinguishable ways of placing the n_p pairs is

$$\left(\frac{Z_p}{N_+ + 2N_i + n_s}\right)^{n_p} \prod_{j=0}^{n_p-1} \frac{(N_+ + 2N_i + n_s - 2n_k - j)(N_+ + 2N_i + n_s - j)}{n_p!}$$

which can be reduced to

$$\left(\frac{Z_p}{N_+ + 2N_i + n_s}\right)^{n_p} \binom{N_+ + 2N_i + n_s - 2n_k}{n_p} \binom{N_+ + 2N_i + n_s}{n_p} n_p!$$

This leaves $N_+ + 2N_i + n_s - 2n_k - n_p$ cation lattice sites and $N_+ + 2N_i + n_s - n_p$ anion lattice sites for arranging the remaining isolated impurity ions, isolated vacancies, and normal K^+ and Cl^- ions. The number of distinguishable ways this can be done is

$$\frac{(N_+ + 2N_i + n_s - 2n_k - n_p)!(N_+ + 2N_i + n_s - n_p)!}{N_+!(N_i - n_k)!(N_i + n_s - n_k - n_p)!(N_+ + 2N_i)!(n_s - n_p)!}$$

In summary, P is given by the expression

$$P = \left(\frac{4Z_k}{N_+ + 2N_i + n_s}\right)^{n_k} \left(\frac{Z_p}{N_+ + 2N_i + n_s}\right)^{n_p} n_k!\, n_p! \left[\binom{\frac{1}{2}(N_+ + 2N_i + n_s)}{n_k}\right]^2$$
$$\times \binom{N_+ + 2N_i + n_s - 2n_k}{n_p} \binom{N_+ + 2N_i + n_s}{n_p}$$
$$\times \frac{(N_+ + 2N_i + n_s - 2n_k - n_p)!(N_+ + 2N_i + n_s - n_p)!}{N_+!(N_i - n_k)!(N_i + n_s - n_k - n_p)!(N_+ + 2N_i)!(n_s - n_p)!} \quad (15)$$

Next, we find $\log_e P$ using Stirling's approximation for $\log_e(a!)$,

$$\log_e(a!) = a \log_e a - a \quad (16)$$

The logarithm of the binomial coefficients can be similarly approximated,

$$\log_e \binom{a}{b} = a \log_e a - b \log_e b - (a - b) \log_e (a - b) \tag{17}$$

It is straightforward to find the partial derivatives of $\log_e P$ where N equals $N_+ + 2N_i + n_s$ and is much larger than N_i, n_s, n_k, or n_p:

$$\frac{\partial(\log_e P)}{\partial n_s} = -\log_e \left[\frac{(n_s - n_p)(N_i + n_s - n_k - n_p)}{N^2} \right] \tag{18}$$

$$\frac{\partial(\log_e P)}{\partial n_k} = -\log_e \left[\frac{n_k N}{Z_k (N_i - n_k)(N_i + n_s - n_k - n_p)} \right] \tag{19}$$

$$\frac{\partial(\log_e P)}{\partial n_p} = -\log_e \left[\frac{n_p N}{Z_p (N_i + n_s - n_k - n_p)(n_s - n_p)} \right] \tag{20}$$

For determining the most probable equilibrium configuration, the incremental change in the Gibbs free energy is set equal to zero,

$$\begin{aligned} \delta G = 0 = [-g_k - kT\, \partial(\log_e P)/\partial n_k]\, \delta n_k + [-g_s - kT\, \partial(\log_e P)/\partial n_s]\, \delta n_s \\ + [-g_p - kT\, \partial(\log_e P)/\partial n_p]\, \delta n_p \end{aligned} \tag{21}$$

and the following expressions are obtained:

$$n_k N/(N_i - n_k)(N_i + n_s - n_k - n_p) = Z_k \exp(g_k/kT) \tag{22}$$

$$(n_s - n_p)(N_i + n_s - n_k - n_p) = N^2 \exp(-g_s/kT) \tag{23}$$

$$n_p N/(n_s - n_p)(N_i + n_s - n_k - n_p) = Z_p \exp(g_p/kT) \tag{24}$$

These equations can be put into their more familiar form in terms of mole fractions (or fractional concentrations) of defects where the mole fractions of isolated anion vacancies x_a, isolated cation vacancies x_c, divalent-impurity-cation-vacancy complexes x_k, vacancy pairs x_p, and impurity ions c are defined as follows:

$$x_a = (n_s - n_p)/N \tag{25}$$

$$x_c = (N_i + n_s - n_k - n_p)/N \tag{26}$$

$$x_k = n_k/N \tag{27}$$

$$x_p = n_p/N \tag{28}$$

$$c = N_i/N \tag{29}$$

These definitions allow us to reduce Eqs. (22)–(24) to the following simplified forms:

$$x_k/x_c(c - x_k) = Z_k \exp(g_k/kT) \quad (30)$$

$$x_a x_c = \exp(-g_s/kT) \equiv x_s^2 \quad (31)$$

$$x_p = Z_p x_a x_c \exp(g_p/kT) = Z_p \exp[-(g_s - g_p)/kT] \quad (32)$$

From Eq. (32), it is apparent that the concentration of vacancy pairs is independent of the impurity concentration. This fact has been used experimentally to separate the vacancy pair contribution to self-diffusion from the isolated vacancy contribution. The electrical neutrality of vacancy pairs plus the complete independence of Eqs. (30) and (31) from the vacancy pair concentration permits the analysis of electrical conductivity data to be carried out without including vacancy pairs.

When the electroneutrality equation, which assures an equal number of lattice sites in the anion and cation sublattices,

$$x_c = x_a + c - x_k \quad (33)$$

is combined with Eqs. (30) and (31), the various vacancy concentrations can be calculated as functions of temperature and impurity concentration. The concentration of Schottky defects, $x_s = (x_a x_c)^{1/2}$, and fractional concentrations of anion and cation vacancies for various impurity concentrations are plotted as functions of $10^3/T$ in Fig. 3. These curves were computed using typical values for g_s and g_c of 2.26 eV − 5.4kT and 0.42 eV, respectively.[11] The following features of the simple theory can be noted in Fig. 3:

1. The concentration of Schottky defects, given by $x_s = \exp(-g_s/2kT)$, is independent of the impurity concentration, and is shown by the line (- - -).

2. The additional of a small amount, say 10 ppm, of a divalent cation impurity greatly enhances the cation vacancy concentration (——) and reduces the anion vacancy concentration (— —). The product $x_a x_c$ remains equal to x_s^2, but the ratio x_c/x_a becomes much larger than 1 over a wide range of temperatures.

3. The transition from the intrinsic region of vacancy concentration, where the number of vacancies is thermally controlled, to the extrinsic region, where the number of vacancies is controlled by the impurity concentration, occurs in the temperature range where $x_s \approx c$, and the material is essentially extrinsic at temperatures where $5x_s < c$.

4. In the extrinsic region, the cation vacancy concentration versus $10^3/T$ curves have negative curvatures which increase with increasing impurity concentration.

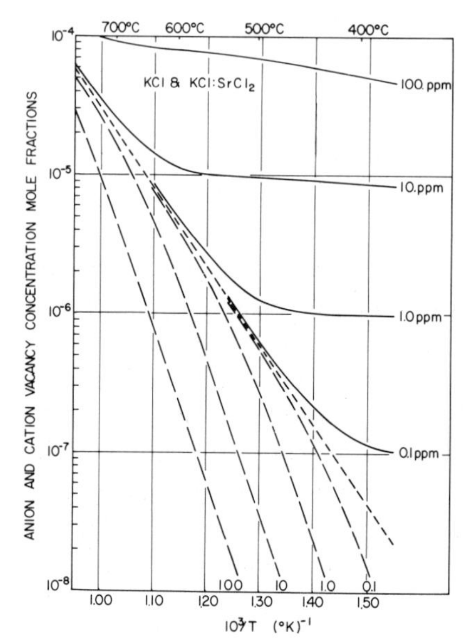

Fig. 3. The simple theory: fractional concentrations of Schottky defects, anion vacancies, and cation vacancies plotted as functions of $10^3/T$ for various divalent cation impurity concentrations. Schottky defect concentration (- - -), cation vacancy concentration (——), anion vacancy concentration (— —). The calculations were made for $g_s = 2.26$ eV $- 5.4kT$ and $g_k = 0.42$ eV.[11]

5. In the extrinsic region, a straight line which would approximate a $\log x_c$ versus $10^3/T$ curve would not have a zero slope. In fact, such lines have negative slopes that increase with increasing impurity concentration, e.g., the slope is approximately -0.01 eV for c equal to 1.0 ppm and -0.1 eV for c equal to 100 ppm.

The influence of divalent-impurity-cation-vacancy complex formation is indicated in the negative curvature discussed in item 4 above. However, the number of complexes is more explicitly displayed by a graph of the per cent association, which is given by $100(x_k/c)$, versus $10^3/T$, and is shown in Fig. 4. The number of complexes is not a monotonically increasing function of reciprocal temperature, but passes through a minimum at temperatures for which $x_s \approx c$. This feature of the simple theory has not been previously discussed in the literature* because those discussions have concen-

* For a review of these discussions, see Lidiard.[12]

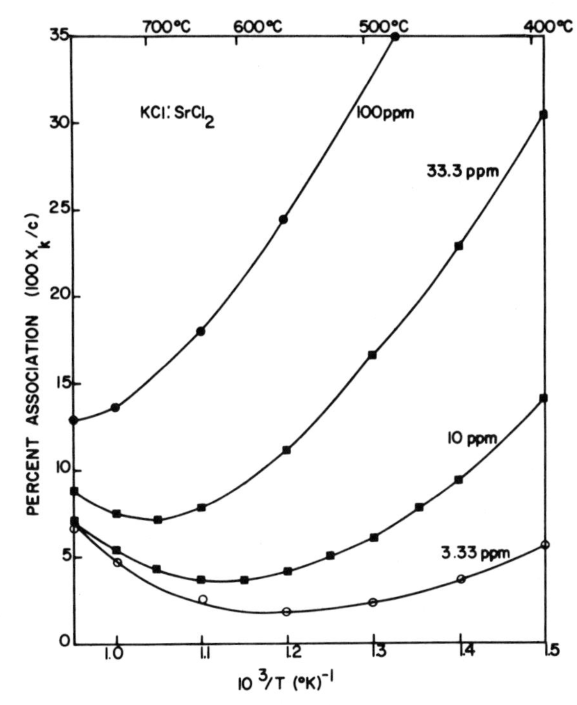

Fig. 4. The simple theory: percent of divalent cation ion impurities associated with cation vacancies to form complexes plotted as a function of $10^3/T$ for various impurity concentrations. Input parameters the same as for Fig. 3.

trated on the extrinsic region, where c is much greater than x_s and where x_c is well approximated by $c - x_k$. If such an approximation is used in Eq. (30), the minima in Fig. 4 disappear. However, for the values of g_s and g_c used above, this approximation, $x_c \approx c - x_k$, is not a good one for KCl crystals containing less than 50 ppm of $SrCl_2$ when the crystals are at high temperatures. The surprising fact is that the number of complexes increases with increasing temperature even though the probability for impurity–vacancy association is decreasing in proportion to the function $\exp(g_k/kT)$.* However, the number of vacancies available for binding to a

* This fact was discovered in an analysis of the experimental results for the $KCl{:}SrCl_2$ system in 1968.[13] The discovery was discussed with Drs. H. B. Rosenstock and M. H. Reilly at the Naval Research Laboratory. Subsequently, the theoretical details were worked out by Dr. Reilly and circulated in an interbranch memo from which Section 1.1 of this chapter is taken. A detailed analysis of the equilibrium concentration of complexes was performed by Dr. Rosenstock.[14]

Sr^{2+} ion increases more rapidly with temperature, so that the number of complexes will approach $12c \exp[-(\frac{1}{2}g_s - g_k)/kT]$ or, using the values from above, $190c \exp(-0.71 \text{ eV}/kT)$ on the high-temperature side of the minima.

3.2. Jump Probability for Point Defects

An expression for the jump probability for a point defect in ionic solids has been derived using several different approaches,[15] e.g., absolute rate theory, many-body theory of equilibrium statistics, and the dynamical theory of diffusion. All of these derivations have resulted in an expression of the following form:

$$w = \nu \exp(-\Delta g/kT) \tag{34}$$

where Δg is the height of the free-energy barrier which an ion must surmount in order to pass to the adjacent point defect and ν is an effective vibration frequency of the ion in the initial site. The derivation of Eq. (34) given by Glyde[15] follows.

Let us define the positions and velocities of N ions in the lattice by the vectors $r_1, r_2, \ldots, r_N$ and $\dot{r}_1, \dot{r}_2, \ldots, \dot{r}_N$, respectively. We wish to observe ion 1 jumping in the x direction to position 2, which we assume to be vacant. If the ion reaches the saddle point, say at (x', y', z'), with an x component of velocity greater than zero, $\dot{x}_1 > 0$, then it will move into the vacant lattice site. Let us denote the probability of such an occurrence by $P(r_1', \dot{x}_1 > 0, r_3, r_4, \ldots, r_N, \dot{r}_3, \ldots, \dot{r}_N)$. Then, the jump probability is given by the rate of ionic flow into an adjacent vacancy,

$$w = \int_0^\infty P(\cdots)\dot{x}_1 \, d\dot{x}_1 \tag{35}$$

We divide the lattice of N ions into identical sublattices containing m ions, so that we can treat the lattice as a canonical ensemble of identical sublattices. We then use the canonical distribution function for the probability P,

$$P(r_1', \dot{x}_1 > 0, r_3, \ldots, r_m) = \frac{\int \cdots \int \exp(-H_m/kT) \, d\dot{y}_1 \, d\dot{z}_1 \, d\dot{r}_3 \cdots d\dot{r}_m}{\int \cdots \int \exp(-H_m/kT) \, d\dot{r}_1 \cdots d\dot{r}_m \, dr_1 \cdots dr_m} \tag{36}$$

where H_m is the Hamiltonian for each sublattice and is given by

$$H_m = \tfrac{1}{2} \sum_{i=1}^{m} m_i \dot{r}_i^2 + V(r_1, \ldots, r_m) \tag{37}$$

The integration of the velocity terms can be done exactly and gives a factor $(2\pi kT/m_i)^{1/2}$ for each one. We are left with

$$P(\cdots) = \frac{[\exp(-m_1\dot{x}_1^2/2kT)]\exp[-V(r_1', r_3, \ldots, r_m)/kT]}{(2\pi kT/m_1)^{1/2}\int\cdots\int\exp[-V(r_1, r_3, \ldots, r_m)/kT]\,dr_1\,dr_3\cdots dr_m} \tag{38}$$

Putting Eq. (38) into Eq. (35) and performing the integrations gives

$$w = \frac{1}{2m^{1/2}}\,\frac{\exp[-V(r_1', r_3, \ldots, r_m)/kT]}{\int\cdots\int\exp[-V(r_1, r_3, \ldots, r_m)/kT]\,dr_1\cdots dr_m} \tag{39}$$

If we postulate a potential minimum at the saddle point, then both of the potentials are expandable in a Taylor series with no first-order terms and can be written as

$$V(r_1', r_3, \ldots, r_m) = V_{sp} + \tfrac{1}{2}\sum_{i,j\neq 1}^{m-1}(\partial^2 V/\partial r_i\,\partial r_j)_{r_i' r_j'}\,\delta r_i\,\delta r_j \tag{40}$$

$$V(r_1, r_3, \ldots, r_m) = V_0 + \tfrac{1}{2}\sum_{i,j=1}^{m}(\partial^2 V/\partial r_i\,\partial r_j)_{r_i^0 r_j^0}\,\delta r_i\,\delta r_j \tag{41}$$

It follows that w is expressible in the usual form,

$$w = \bar{\nu}\exp[-(V_{sp} - V_0)/kT] = \bar{\nu}\exp(-\Delta V/kT) \tag{42}$$

where ΔV is given by the difference in energy for an ion at the saddle point and at the equilibrium position of the static lattice and $\bar{\nu}$ is given in terms of the force constants of the lattice,

$$\bar{\nu} = [1/(2\pi m_1)^{1/2}](K_m^{\,0})^{1/2}/(K_{m-1}^{sp})^{1/2} \tag{43}$$

where

$$\frac{(K_m^{\,0})^{1/2}}{(K_{m-1}^{sp})^{1/2}} = (2\pi kT)^{1/2} \times \frac{\int\cdots\int\exp[-\{\sum_{i,j\neq 1}^{m-1}(\partial^2 V/\partial r_i\,\partial r_j)_{r_i' r_j'}\,\delta r_i\,\delta r_j\}/kT]\,d\delta r_3\cdots d\delta r_m}{\int\cdots\int\exp[-\{\sum_{i,j=1}^{m}(\partial^2 V/\partial r_i\,\partial r_j)_{r_i^0 r_j^0}\,\delta r_i\,\delta r_j\}/kT]\,d\delta r_i\,d\delta r_3\cdots d\delta r_m} \tag{44}$$

If ΔV depends upon the temperature, then with ΔV equal to the change in Gibbs free energy Δg, the usual thermodynamic relations are valid.[15]

In summary, the above formalism results in the usual equation of the form shown in Eq. (34), gives the dependence of jump probability on the mass of the jumping ion by an inverse square-root relationship, and provides an expression for the calculation of w from a knowledge of the crystal force constants.[15]

4. COULOMB INTERACTIONS BETWEEN POINT DEFECTS

The simple theory discussed in Section 3 treats the near-neighbor defects as strongly interacting to form complexes or pairs, and treats the isolated defects as noninteracting. Such a description in an ionic crystal where all the defects have an electric charge or an electric dipole moment is a gross oversimplification. In 1954, Lidiard[16] improved the theory by including the Coulomb interactions between the isolated defects using the Debye–Hückel approximation. More recently, the statistical mechanics of interactions between point defects in solids has been treated by the cluster formalism.* The cluster formalism permits a very concise development of the theory and allows the recent developments in the theory of liquids to be applied to solids. Unfortunately, from the point of view of the experimentalist trying to analyze his data, the cluster solutions are in the form of infinite series not easily used in data analysis. So, in spite of the limitations of the Lidiard theory, such as the arbitrary concept of a complex and the low-concentration-limit validity of the Debye–Hückel approximation, it is the best we have available for the analysis of experimental data and is to be preferred over the simple theory. The derivation of the appropriate equations including the Coulomb interactions between the isolated defects, following Lidiard,[16] is given below.

4.1. Concentrations of Point Defects

The simple theory (Section 3.1) can be modified to include Coulomb interactions by adding the energy due to mutual electrical interactions G_{el} to the equation for the Gibbs free energy, Eq. (14). This additional energy can be calculated by reversibly charging a neutral defect in a field of potential ψ_i.[18] The energy expended during the charging process can be represented by

$$G_{el} = \sum_i \int_0^1 M_i \psi_i z_i e \, d\lambda \tag{45}$$

* For a recent review of this topic, see Allnatt.[17]

where λ denotes the fraction of final charge attained by the defect, z_i is the valence of the ith type of defect, M_i is the number of ith type defects, and ψ_i is the potential at the lattice site of the ith type defect. In the simple case, there are only three types of defects considered, i.e., the strontium impurity ion, the cation vacancy, and the anion vacancy. Furthermore, in the simple association model, if two defects are closer together than one lattice distance d, they are bound together to form a neutral entity; otherwise, they are assumed to interact with a purely Coulombic attraction in a medium of dielectric constant ε. It follows that the potential in Eq. (45) is the potential inside a sphere of radius d centered on the lattice site of a particular ith type defect due to all the charged defects outside of that sphere. The value of ψ_i can be found by solving the Poisson–Boltzmann equation, which is given by

$$\begin{aligned} \nabla^2\psi(r) &= -(4\pi/\varepsilon)\varrho(r) \\ &= -(4\pi/\varepsilon)\left[\left(\sum_i M_i z_i e\right)\Big/ V\right] \exp[-z_i e\psi(r)/kT] \end{aligned} \tag{46}$$

where $\varrho(r)$ is the electrostatic charge density at a distance r from the center of the sphere chosen above, and V is the volume of the crystal. The Debye–Hückel approximation assumes $z_i e\psi(r)/kT$ is much less than one, so that Eq. (46) can be written as

$$\nabla^2\psi(r) = -(4\pi/\varepsilon)\left[\sum_i (M_i z_i e/V)\right]\{1 - [z_i e\psi(r)/kT]\} \tag{47}$$

or, since $M_i z_i = 0$ by the condition of electrical neutrality,

$$\nabla^2\psi(r) = \left[\left(4\pi \sum M_i z_i^2 e^2\right)/\varepsilon VkT\right]\psi(r) = \varkappa^2\psi(r) \tag{48}$$

Solving this equation for the potential inside a sphere of radius* $r = d$ due to the remaining charged defects yields the following result:

$$\psi_i = -z_i e\varkappa/\varepsilon(1 + \varkappa d) \tag{49}$$

In the charging process described by Eq. (45), the defects outside the sphere are given a charge of λz_i. Therefore the combination of Eqs. (45)

* It is in the choosing of the size of d that the arbitrariness of the definition of a complex enters. In this chapter, we have assumed that defects of opposite charge within a distance d ($= 2a$, where a is the anion–cation separation) of one another were bound together to form either a vacancy pair or an impurity–vacancy complex, depending upon the nature of the defects involved.

and (49) gives

$$G_{\text{el}} = -\int_0^1 \sum_i (M_i z_i^2 e^2/\varepsilon)[\varkappa\lambda^2/(1 + \lambda\varkappa d)]\, d\lambda \tag{50}$$

$$G_{\text{el}} = -\left[\left(\sum_i M_i z_i^2 e^2 \varkappa\right)\Big/\varepsilon\right][1/(\varkappa d)^3][\log(1 + \varkappa d) - \varkappa d + \tfrac{1}{2}\varkappa^2 d^2] \tag{51}$$

Since z_i is $+1$ for anion vacancies and strontium ions and z_i is -1 for cation vacancies, $\sum_i M_i z_i^2$ becomes $\sum_i M_i$, which, in the notation of Section 2, is as follows:

$$\begin{aligned}\sum M_i &= (N_i - n_k) + (n_s - n_p) + (N_i + n_s + n_k - n_p) \\ &= 2(N_i + n_s - n_k - n_p) = 2Nx_c\end{aligned} \tag{52}$$

The total Gibbs free energy then becomes

$$\begin{aligned}G_{\text{tot}} = G + G_{\text{el}} &= n_s g_s - n_k g_k - n_p g_p + Ts_c - [2e^2\varkappa(N_i + n_s - n_k - n_p)/\varepsilon] \\ &\times \{[1/(\varkappa d)^3][\log(1 + \varkappa d) - \varkappa d + \tfrac{1}{2}\varkappa^2 d^2]\}\end{aligned} \tag{53}$$

where

$$\varkappa^2 = \left(4\pi \sum M_i z_i^2 e^2\right)/\varepsilon VkT = 8\pi Ne^2 x_c/\varepsilon VkT = 32\pi e^2 x_c/\varepsilon d^3 kT \tag{54}$$

According to the assumptions of this theory, the configurational entropy s_c is the same as given in Section 3.1. The total Gibbs free energy is minimized with respect to n_s, n_k, and n_p as before and instead of Eqs. (22)–(24) of Section 3.1, one obtains the following equations for the numbers of defects:

$$n_k N/(N_i - n_k)(N_i + n_s - n_k - n_p) = z_k \exp[(g_k - \gamma)/kT] \tag{55}$$

$$(n_s - n_p)(N_i + n_s - n_k - n_p) = N^2 \exp[-(g_s - \gamma)/kT] \tag{56}$$

$$n_p N/(n_s - n_p)(N_i + n_s - n_k - n_p) = z_p \exp[(g_p - \gamma)/kT] \tag{57}$$

where

$$\gamma = e^2\varkappa/\varepsilon(1 + \varkappa d) \tag{58}$$

These equations can be expressed in terms of the fractional concentrations of defects, analogous to Eqs. (30)–(32), as follows:

$$x_k/x_c(c - x_k) = z_k \exp[(g_k - \gamma)/kT] \tag{59}$$

$$x_a x_c = \exp[-(g_s - \gamma)/kT] \tag{60}$$

$$x_p = z_p x_a x_c \exp[(g_p - \gamma)/kT] = z_p \exp[-(g_s - g_p)/kT] \tag{61}$$

The concentration of vacancy pairs is independent of the impurity concentration as in the previous theory, and is independent of the Coulomb interaction between the isolated defects; Eq. (61) is identical to Eq. (32).

The Schottky defect concentration, cation vacancy concentration, and per cent association were calculated from the Lidiard–Debye–Hückel equations (33), (59), and (60) using the previous values for g_s and g_k, ε as 5.2, and d as one lattice spacing (6.28 Å for KCl). The results of these calculations are shown and compared with the results from Section 3.1 in Fig. 5–7.

The Lidiard–Debye–Hückel (LDH) equations result in a Schottky defect concentration x_s that is dependent upon the defect concentration. In a pure crystal, the calculated vacancy concentration is enhanced by the inclusion of the Coulomb interactions. This enhancement amounts to about 30% in KCl near the melting point. The addition of divalent impurities

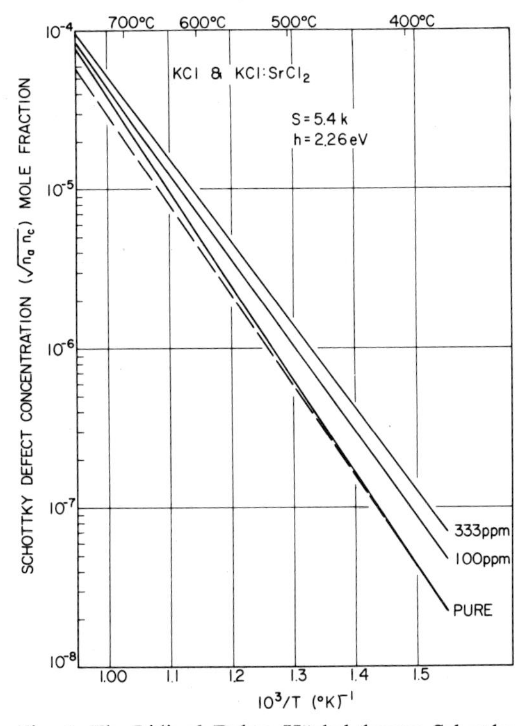

Fig. 5. The Lidiard–Debye–Hückel theory: Schottky defect concentration plotted as a function of $10^3/T$ for various impurity concentrations. The calculations were made with $g_s = 2.26\text{ eV} - 5.4kT$,[11] $g_k = 0.42$ eV,[11] $\varepsilon = 5.2$, and $d = 6.28$ Å. The results of the simple theory are shown by a dashed line.

into a crystal increases the cation vacancy concentration, and as a result [see Eq. (54)], the Schottky defect concentration may be increased by more than a factor of two over that predicted by the simple theory. (See Fig. 5).

The cation vacancy concentrations predicted by the LDH equations are plotted as a function of $10^3/T$ for various impurity concentrations in Fig. 6. The LDH predictions (solid lines) show increased cation vacancy concentrations for each impurity concentration when compared with the predictions of the simple theory (dashed lines). Furthermore, the LDH curves have less curvature than the results from the simple theory.

The most striking effect of the Coulomb interactions is shown in the graph of the per cent association versus $10^3/T$, Fig. 7. The number of complexes to be expected in the crystal is greatly reduced when the Coulomb

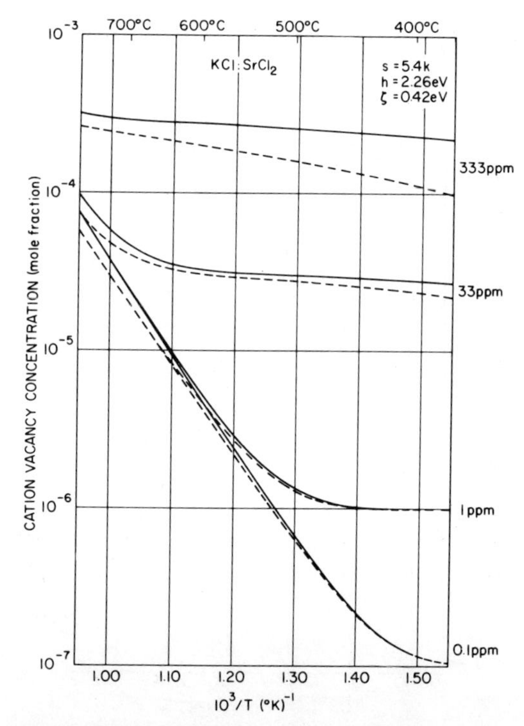

Fig. 6. The Lidiard–Debye–Hückel theory: cation vacancy concentration plotted as a function of $10^3/T$ for various impurity concentrations. Input parameters the same as for Fig. 5. The results of the simple theory, from Fig. 3, are shown by the dashed lines for comparison.

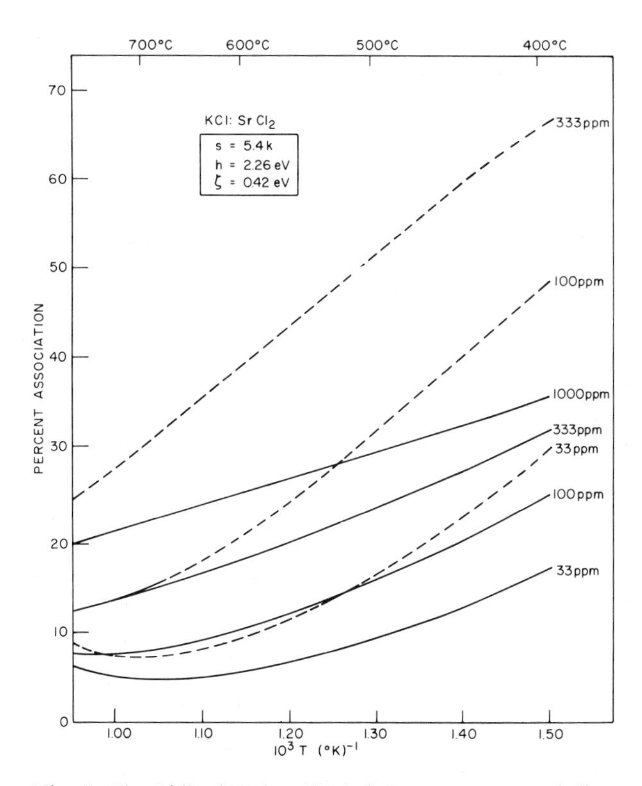

Fig. 7. The Lidiard–Debye–Hückel theory: per cent of divalent cation impurities associated with cation vacancies to form complexes plotted as a function of $10^3/T$ for various impurity concentrations. Input parameters the same as for Fig. 5. The results of the simple theory are shown for comparison by the dashed lines.

interactions between the isolated defects are taken into consideration. It should be noted that the previously discussed minima in these curves are still present in the LDH theory.

4.2. Defect Mobility

The inclusion of the Coulomb interactions not only changes the equations for the defect concentrations, but also the mobility of the defects. In the Debye–Hückel picture of an electrolyte, a charged particle is surrounded by a Debye–Hückel charge cloud of opposite sign. This charge cloud produces a mobility drag effect as a charged particle tends to move through the crystal. Hence, the jump probability as calculated in the simple theory would be reduced by the tendency of the charge cloud to oppose the motion of the jumping ion. Several calculations of the mobility drag effect have been made.[12] An extension of the theory of the conductivity of electrolyte solutions to include Coulomb interactions in the Debye–Hückel approximation was made by Pitts[19] and his result has become the most widely used in the analysis of experimental results of transport measurements in alkali halide crystals. We simply quote Pitts's result here and refer the reader to his paper for the details of the derivation. The mobility drag factor g is given by

$$g = 1 - [e^2\varkappa/7.243\varepsilon kT(1 + 4.828\varkappa a + 4\varkappa^2 a^2)] \tag{62}$$

where $\varkappa$ is given in Eq. (54) and a is the anion–cation separation.

The mobility drag factor g is not very strongly temperature-dependent nor very sensitive to the impurity concentration. For example, for KCl with a divalent impurity concentration of 375 ppm, g varies from 0.91 (at $10^3/T = 1.00$) to 0.84 (at $10^3/T = 0.54$). For pure KCl, g varies from 0.96 to 0.99 as $10^3/T$ goes from 0.98 to 1.22. As far as data analysis is concerned, the inclusion of the mobility drag factor in the transport equations is more esthetic than substantive.

In concluding this section, let us write out completely the equation for ionic conductivity and display its dependence upon the various thermodynamic properties of the crystal. It is common to express the various Gibbs free energies by the usual thermodynamic relationship $g = h - Ts$, where h and s are the enthalpy and entropy, respectively. The ionic conductivity of a pure alkali halide crystal or one containing divalent impurity cations can then be written, using the simple theory, by putting together Eqs. (7), (30), (31), (33), and (34):

$$\begin{aligned}\sigma T = (4Ne^2a^2\nu/k)\{&(x_a/x_s)\exp[(\tfrac{1}{2}s_s + \Delta s_a)]\exp[-(\tfrac{1}{2}h_s + \Delta h_a)/kT] \\ &+ (x_c/x_s)\exp[(\tfrac{1}{2}s_s + \Delta s_c)/k]\exp[-(\tfrac{1}{2}h_s + \Delta h_c)/kT]\}\end{aligned} \tag{63}$$

where

$$(x_a/x_s)^3 + (x_a/x_s)^2[(c/x_s) + K_1(T)x_s] - (x_a/x_s) - K_1(T)x_s = 0 \tag{64}$$

$$x_a x_c = x_s^2 \quad \text{and} \quad K_1(T) = 12\exp[(-\eta/k) + (\chi/kT)] \tag{65}$$

The eight thermodynamic variables are defined as follows: s_s is the entropy of Schottky defect formation, h_s is the enthalpy of Schottky defect formation, Δs_a is the entropy of anion migration, Δh_a is the enthalpy of anion migration, Δs_c is the entropy of cation migration, Δh_c is the enthalpy of cation migration, η is the entropy of divalent-cation-impurity–cation-vacancy association, and χ is the enthalpy of divalent-cation-impurity–cation-vacancy association.

The LDH theory can be incorporated in conductivity equations by replacing $\frac{1}{2}h$ and χ above by $\frac{1}{2}h - \gamma$ and $\chi - \gamma$, and by multiplying the right-hand side of Eq. (63) by the mobility drag factor g, Eq. (62).

Self-diffusion equations can be written in a similar way. In the formulation of the simple theory, we can put together Eqs. (11)–(13) and (30)–(34) to write the expression for anion tracer diffusion,

$$D_a^* = 4fa^2\nu(x_a/x_s)\exp[(\tfrac{1}{2}s_s + \Delta s_a)/k]\exp[-(\tfrac{1}{2}h_s + \Delta h_a)/kT] + D_0^a \exp(-Q_a/kT) \qquad (66)$$

where D_0^a is the preexponential factor for the vacancy pair contribution to anion tracer diffusion and includes the entropy terms as well as the correlation factor, and Q_a is the activation enthalpy for the vacancy pair contribution to anion tracer diffusion and includes both the enthalpy of vacancy pair formation and the enthalpy of vacancy pair migration. The incorporation of the LDH theory into Eq. (66) then proceeds quite straightforwardly.

5. EXPERIMENTAL RESULTS

Experiments to measure the electrical conductivity of the alkali halides began more than 40 years ago. Yet, the characterization of these materials in terms of the thermodynamic variables discussed in the previous two sections is still not precisely accomplished. Our determination of values for the thermodynamic variables depends upon our ability to extract information from conductivity and diffusion experiments. But nature seems to have hidden that information in the (xw) product we discussed earlier. In order to describe with some precision the thermodynamic variables of defect formation and migration for an alkali halide crystal, it is necessary to overdetermine its transport properties. For example, in Table I we have listed various kinds of measurements that have been, or can be, made on an alkali halide crystal to provide information about the intrinsic defects in the crystal. The measurements in Table I are all made at standard pressure, a crystal medium is chosen, either pure or containing a known amount of impurity, and the measurements are performed as a function of temperature. The conductivity of a pure crystal is the sum of the anion and cation contributions (Fig. 8). The cation contribution to the conductivity can

Table I. Intrinsic Defects[a]

Measurement and medium	Property obtained
1. Conductivity: pure crystal	$\sigma_a + \sigma_c$
2. Conductivity: cation impurity crystal	σ_c
3. Conductivity: anion impurity crystal	σ_a
4. Diffusion—cation tracer: pure crystal	$D_c + D_p{}^c$
5. Diffusion—anion tracer: pure crystal	$D_a + D_p{}^a$
6. Diffusion—anion tracer: cation impurity crystal	$D_p{}^a + D$(extrinsic)
7. Diffusion—cation tracer: electric field	D_c
8. Diffusion—anion tracer: electric field	D_a
9. Diffusion—isotope effect: cation impurity crystal	$D_p{}^a$
10. Hall effect	?
11. Diffusion—cation tracer: anion impurity crystal	$D_p{}^c$

[a] Experimental measurements needed to obtain information about the transport properties of alkali halide crystals. Measurements 1–7 have been performed on at least one material. Measurements 8–11 have yet to be performed in such a manner as to further our knowledge of these materials.

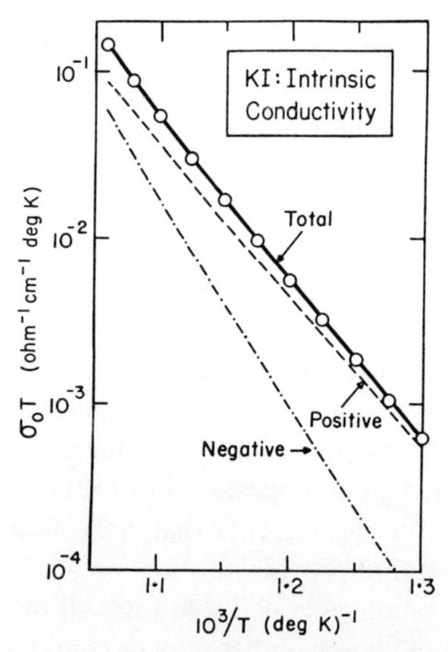

Fig. 8. The conductivity of pure potassium iodide. Plot of log(σT) versus $10^3/T$, showing the total conductivity as the sum of the anion and cation contributions.[20]

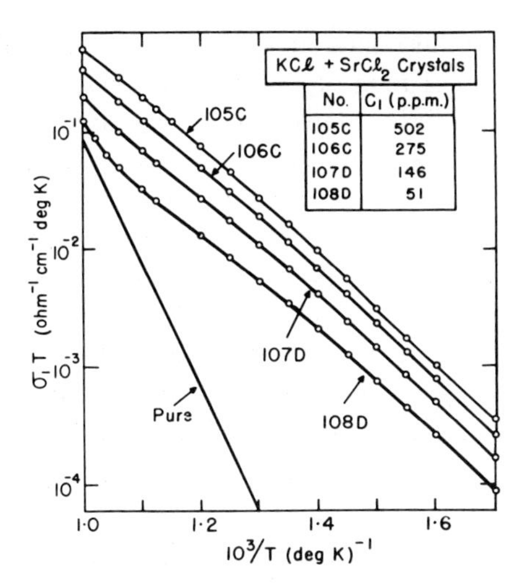

Fig. 9. The conductivity of KCl:$SrCl_2$ crystals. The conductivity is enhanced because the number of cation vacancies is increased to compensate for the presence of the Sr^{2+} impurities.[21]

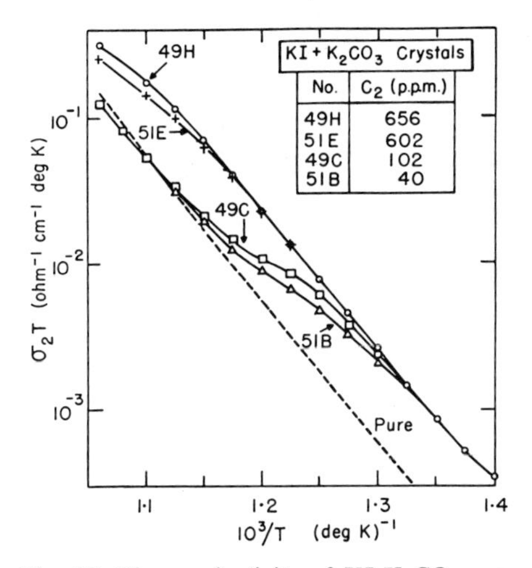

Fig. 10. The conductivity of KI:K_2CO_3 crystals. The conductivity is enhanced because the number of anion vacancies is increased to compensate for the presence of the CO_3^{2-} impurities.[20] A comparison with Fig. 9 shows how the low solubility of K_2CO_3 in KI complicates the analysis of these data.

be measured in a crystal containing a known amount of divalent cation impurity if the degree of impurity–vacancy association can be determined (Fig. 9). Similarly, the anion contribution to the conductivity can be found from the conductivity of a crystal containing a known amount of a divalent anion impurity if the degree of impurity–vacancy association can be determined. (Fig. 10). The diffusion of radioactive cations in a pure crystal gives the sum of the isolated cation vacancy contribution and the vacancy pair contribution to cation diffusion (Fig. 11). The diffusion of radioactive anions in a pure crystal gives the sum of the isolated anion vacancy contribution and the vacancy pair contribution to anion diffusion (Fig. 12). Unfortunately for the analysis of experimental results, the pair contribution to anion diffusion $D_p{}^a$ is not, in general, equal to the pair contribution to

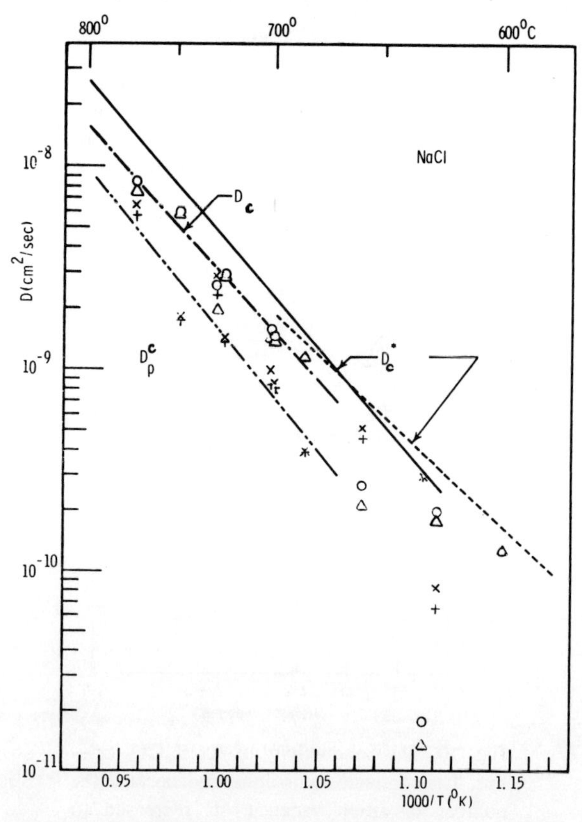

Fig. 11. The diffusion of radioactive sodium ions in sodium chloride crystals. The cation tracer diffusion D_c* is shown as the sum of the isolated vacancy contribution D_c and the vacancy pair contribution $D_p{}^c$.[22]

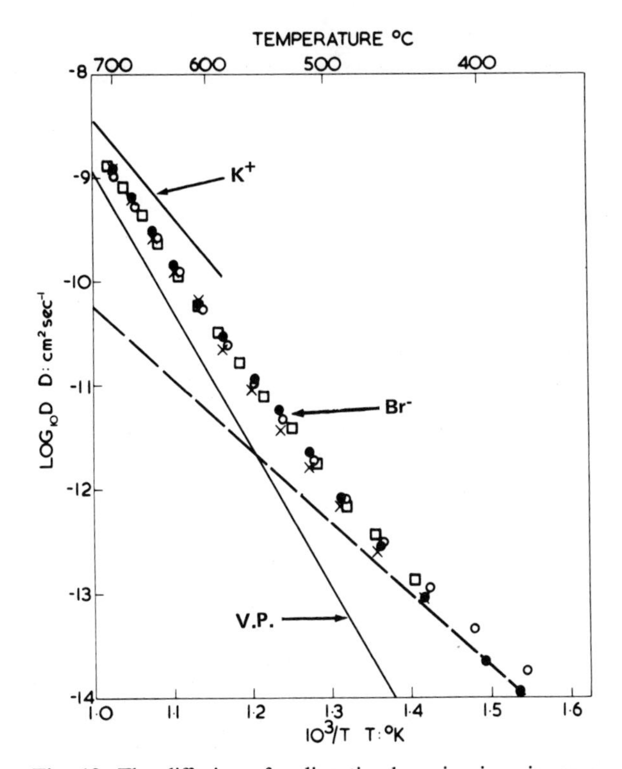

Fig. 12. The diffusion of radioactive bromine ions in pure potassium bromide. The vacancy pair contribution to the anion tracer diffusion is shown by the solid line V.P. The K^+ line represents the diffusion of cation tracer in pure KBr. The dashed line is the extrinsic, or dislocation, contribution to the anion tracer diffusion.[23]

cation diffusion D_p^c (Fig. 13). The diffusion of radioactive anions in a crystal containing a sufficient number of divalent cation impurities is dominated by the vacancy pair contribution (Fig. 14). Under such conditions, $c \gg x_s$, the anion diffusion is independent of the cation impurity concentration, as predicted in Eq. (61). However, in such crystals at low temperatures, the anion diffusion is influenced by an extrinsic, or dislocation, diffusion which makes the interpretation of the low temperature data difficult (Fig. 15). Because the vacancy pair is electrically neutral, the motion of a cation tracer in an electric field will show an enhancement which is entirely derived from the isolated cation vacancy contribution (Fig. 16).

None of the other measurements listed in Table I have actually been made except for the Hall effect. The Hall effect measurements in NaCl

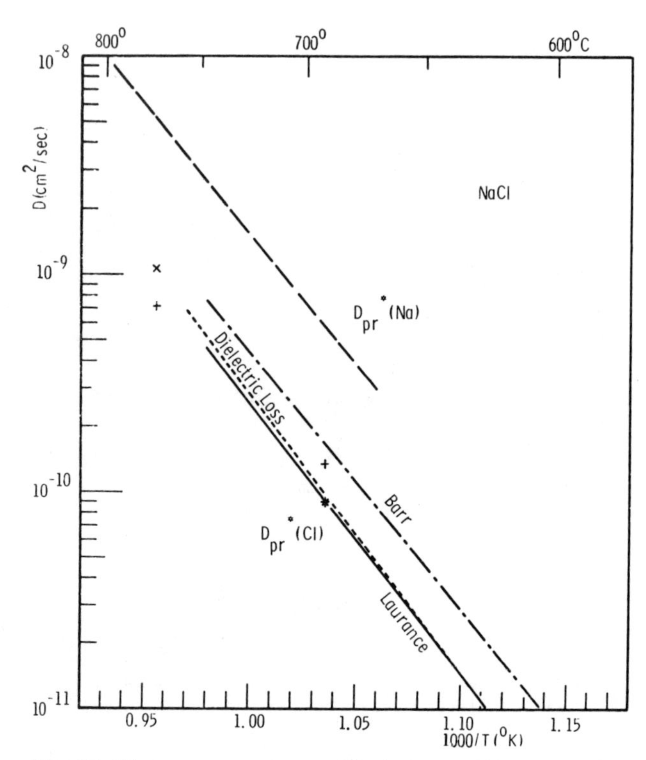

Fig. 13. The vacancy pair contributions to anion and cation diffusion in sodium chloride.[22]

have been reported,[24] but the theoretical interpretation of these results has been the source of discussion. Is a Hall effect possible in an ionic crystal where the conducting particles have limited directions in which they can move, i.e., they must jump into an adjacent vacant lattice site?*

There are two other measurements that one might think to add to Table I. They are cation tracer diffusion in a crystal containing divalent cation impurities and anion tracer diffusion in a crystal containing divalent anion impurities. However, these experiments would give results that include a large contribution to diffusion from the motion of impurity–vacancy complexes. Hence, these measurements are rightly left out of a table that focuses our attention on the properties of defects intrinsic to pure crystals.

* A preliminary theoretical investigation of the Hall effect in ionic crystals was made by Dr. M. Gomez at the Naval Research Laboratory. He treated the ions as harmonic oscillators in perpendicular electric and magnetic fields. His results indicated a transverse component of electric current.

The experimental procedures used in measuring the transport properties of alkali halide crystals (Table I) are easy to describe. The alkali halides are good insulators at most temperatures and a small experimental sample (1 mm × 7 mm × 7 mm) placed between two platinum electrodes makes a capacitor with a small amount of resistive loss. Hence, the resistance of these materials is measured at low frequencies using a standard ac capacitance bridge. The temperature dependence of the conductivity is determined by placing the sample in a furnace and measuring its resistance at various temperatures. The diffusion measurements are made using radioactive isotope methods. There are two different techniques now in use, the sectioning technique[25] and the isotopic exchange technique.[26] The sectioning technique proceeds by evaporating a thin layer of the radioactive material on one surface of the sample, putting the sample in a furnace, holding it at constant temperature for a measured length of time, removing the sample from the furnace, slicing it into thin sections, and measuring the radioactivity of each section. A graph of the radioactivity versus penetration can be made. The diffusion coefficient is then determined from this graph. In this technique, the total process must be repeated at many different temperatures and with many different samples to obtain a graph of the diffusion coefficient as a function of temperature and impurity concentra-

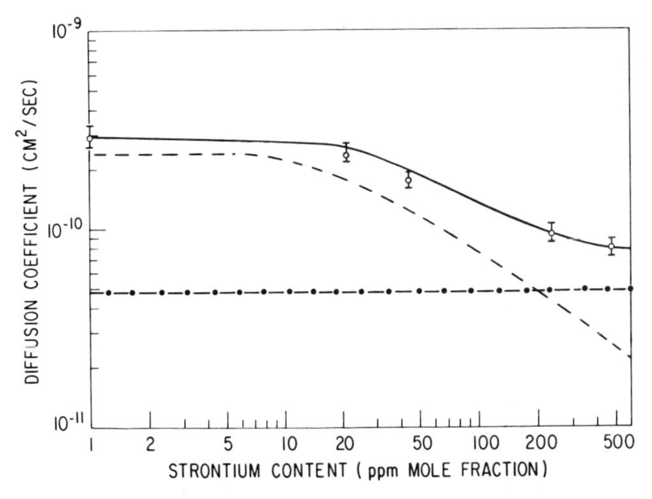

Fig. 14. The diffusion coefficient for chlorine ion diffusion in KCl and KCl:$SrCl_2$ single crystals as a function of the strontium concentration. (- - -) The isolated anion vacancy contribution to the chlorine diffusion; (-··-) the vacancy pair contribution to the chlorine ion diffusion.[13]

tion. The isotopic exchange technique uses a sample containing a uniform distribution of the radioactive ions and measures the rate of ion exchange at the surface of the sample. For example, a $K^{36}Cl$ sample is placed in a furnace and chlorine gas is passed over the surface of the sample. The radioactivity of the gas is then simply related to the rate at which the radioactive chlorine ions diffuse to the surface of the $K^{36}Cl$ sample and exchange with the chlorine gas ions. In this technique, a single sample can be used to obtain the diffusion rate as a function of temperature. The recent work of Dawson and Barr[23] has found that the sectioning and exchange techniques give equivalent values for the diffusion coefficient.

Since both the conductivity and the diffusion rate are strongly temperature-dependent, it is essential that precise measurements of the sample temperature be made. The use of a large, low-thermal-gradient furnace with calibrated thermocouples in proximity to the sample is necessary.

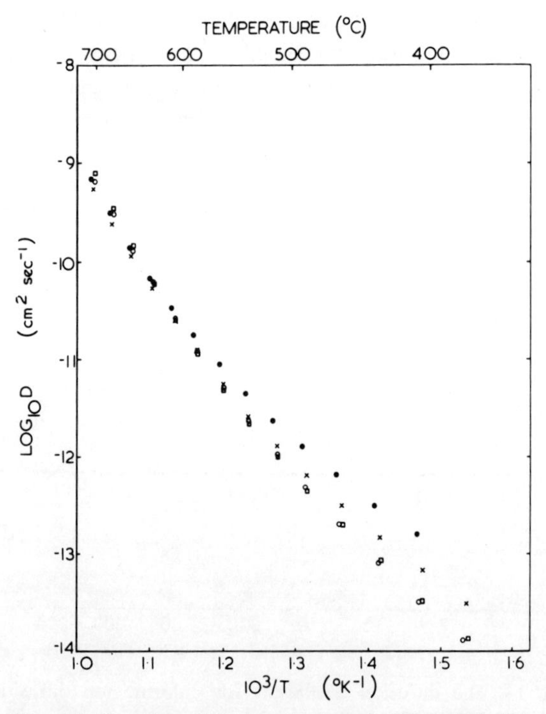

Fig. 15. Bromine diffusion in doped KBr. (●) 135 ppm $SrBr_2$; (×) 240 ppm; (○) 84 ppm; and (□) 25 ppm $CaBr_2$. Note the variation in the diffusion coefficient below 550°C.

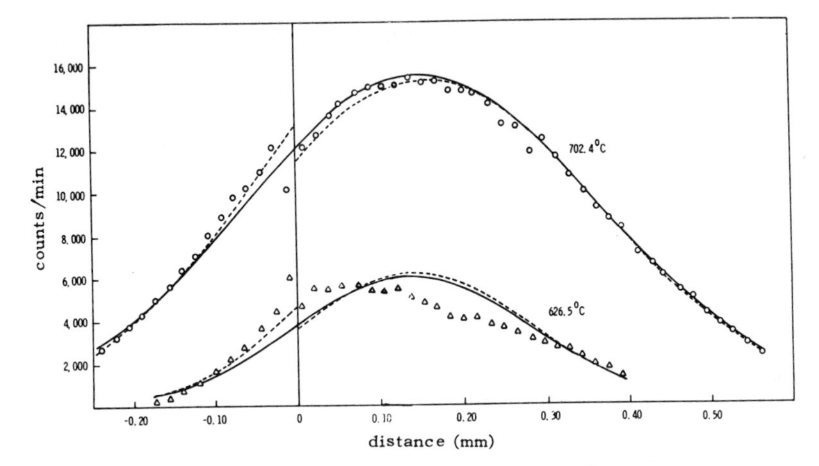

Fig. 16. Diffusion profiles for Na in NaCl with a dc electric field. The open circles and triangles give actual data points obtained by sectioning two different runs. The solid curves are least-squares fit without an interface barrier and the dotted curves are the fit assuming a barrier to diffusion at the interface. The initial interface is at $x = 0$, and the field is applied along the positive axis.[22]

Furthermore, the transport properties of these materials are very sensitive to the presence of small amounts of aliovalent impurities, and the precise control and determination of impurity concentrations are important in obtaining reliable results.

5.1. Fully Characterized Alkali Halide Crystals

Each of measurements 1–7 listed in Table I has been performed on at least one alkali halide. Because it is sometimes difficult to compare experiments done in different laboratories by different experimental groups, a well-characterized alkali halide would be one on which all seven measurements had been performed by the same group and the results properly analyzed. By this definition, there are no well-characterized alkali halides. So, we have decided to group together as fully characterized* those materials for which the thermodynamic variables needed to specify its ionic conductivity, Eqs. (63)–(65), have been determined from more than one of the

* It seemed appropriate for this chapter to give special attention to the work which had been carefully done and the analysis of which had taken into account the importance of the Coulomb interactions between the defects. Hence, the term "fully characterized" is used to denote experimental results of sufficient accuracy and consistency that the scientists were motivated to perform the more complex LDH analysis of their data.

measurements shown in Table I. Furthermore, the analyses of the experimental results have been done using the LDH theory.

5.1.1. *Potassium Chloride*

Ionic transport in KCl as determined from the measurements of the conductivity and anion tracer diffusion in single crystals of KCl and KCl:$SrCl_2$ was reported in 1968.[13] Subsequently, Chandra and Rolfe[21] reported on the ionic conductivity of single crystals of KCl, KCl:$SrCl_2$, (Fig. 9), and KCl:K_2CO_3. These two papers gave results in good agreement with one another, and the values obtained for the thermodynamic variables for ionic transport in KCl are given in the first column of Table II. The values shown there are "best" values estimated from the results and uncertainties reported in the above two papers. Of measurements 1–7 in Table I, only those involving cation diffusion have not been carefully done on KCl.

5.1.2. *Potassium Bromide*

Recently, Dawson and Barr[23] have completed a monumental work on KBr. They have measured the bromine diffusion in pure and divalent cation doped KBr crystals (Figs. 12 and 15). They have measured the ionic conductivity in the same materials and have also measured the cation diffusion in pure KBr. They have used the LDH theory in their analyses and have obtained the values for the various enthalpies and entropies as shown in the second column of Table II. These results are in general agreement with the earlier conductivity results of Rolfe,[27] which were analyzed by the use of the simple theory.

5.1.3. *Potassium Iodide*

The ionic conductivity of single crystals of KI (Fig. 8), KI:SrI_2, and KI:K_2CO_3 (Fig. 10) was reported by Chandra and Rolfe[20] in 1970. Unfortunately, the low solubility of the carbonate ions in KI makes it difficult to extract the anion contribution to the conductivity from these data. Nevertheless, a complete analysis was done and the values are given in the third column of Table II.

5.1.4. *Rubidium Chloride*

The amount of experimental data available on the transport properties of RbCl is very limited.[28,29] However, a recent analysis of the conductivity

Thermodynamic variable	Alkali halide crystal				
	KCl[a]	KBr[b]	KI[c]	RbCl[d]	NaCl[e]
Schottky defect formation					
Enthalpy h_s, eV	$2.52^{+0.07}_{-0.03}$	2.37 ± 0.06	2.21	2.04	2.45 ± 0.05
Entropy s_s/k	$8.4^{+1.2}_{-0.8}$	7.18 ± 0.30	8.87	2.47	9.3 ± 1.2
Cation vacancy migration					
Enthalpy Δh_c, eV	0.74 ± 0.02	0.667 ± 0.02	0.63 ± 0.03	0.54	0.65 ± 0.01
Entropy $\Delta s_c/k$	2.88 ± 0.18	2.54 ± 0.30	1.58 ± 0.34	1.73	1.75 ± 0.1
Anion vacancy migration					
Enthalpy Δh_a, eV	$0.90^{+0.09}_{-0.01}$	0.92 ± 0.04	1.29	1.45	0.86 ± 0.1
Entropy $\Delta s_a/k$	$4.34^{+0.10}_{-0.20}$	3.95 ± 0.30	9.35	13.4	2.2 ± 0.5
Divalent-cation-impurity vacancy association					
Enthalpy χ, eV	0.58 ± 0.01 (Sr)	0.57 ± 0.003 (Ca)	0.54 ± 0.03 (Sr)	0.64 (Sr)	0.55 ± 0.05 (Sr), 0.34^f (Cd), 0.31^g (Ca)
Entropy η/k	$1.50^{+0.38}_{-0.20}$ (Sr)	1.82 ± 0.01 (Ca)	2.20 ± 0.35 (Sr)	0.94 (Sr)	1.1 ± 0.5 (Sr)
Jump attempt frequency ν, sec^{-1}	4.25×10^{12}	—	2.8×10^{12}	1×10^{12}	4.91×10^{12}
Vacancy pair contribution to anion diffusion					
Activation energy Q_a, eV	$2.62^{+0.01}_{-0.10}$	2.60 ± 0.03	—	—	$2.50\pm0.02^{h,j}$
Preexponential constant, $D_0{}^a$, cm^2/sec	5150^{+150}_{-3460}	$(1.5\pm0.5)\times10^4$	—	—	990 ± 90^j
Vacancy pair contribution to cation diffusion					
Activation energy Q_c, eV	—	—	—	—	$2.35\pm0.22^{i,j}$
Preexponential constant $D_0{}^c$, cm^2/sec	—	—	—	—	1130 ± 310^j

[a] The weighted average of the values given in Refs. 13 and 21. [b] Ref. 23. [c] Ref. 20. [d] Ref. 30. [e] "Best" values taken from Refs. 31–33. [f] Analysis of NaCl:$CdCl_2$ conductivity data using the LDH theory.[16] [g] Analysis of NaCl:$CaCl_2$ conductivity data using the LDH theory.[34] [h] Footnote 9, Ref. 28. [i] Ref. 22. [j] The analyses were done using the simple theory, Section 2.

of pure RbCl and RbCl:$SrCl_2$ single crystals has been reported.[30] These results are shown in the fourth column of Table II. Even though no uncertainties are given for the values of the thermodynamic variables, these values are much less certain than any of the others reported for the fully characterized materials.

5.1.5. *Sodium Chloride*

The amount of experimental data available on the transport properties of NaCl is horrendous. In a recent review paper,[3] the results of 14 different experiments (dated from 1950 to 1968) reported values for the enthalpy of cation migration in NaCl ranging from 0.65 to 0.85 eV. Since 1968, several additional articles[22,31–33] have been added to the literature on NaCl. Unfortunately, much of the NaCl data has been analyzed by a linearized version of the simple theory, i.e., drawing straight lines through the experimental conductivity data points when plotted in the usual $\log(\sigma T)$ versus $10^3/T$ form. Such a procedure for analysis is repudiated by the transport theory discussed in Section 3.

The results from the measurement of the conductivity of pure NaCl and NaCl:$SrCl_2$ crystals have been carefully analyzed by Allnatt *et al.*[33] using the LDH theory. Their results are in general agreement with the other recent results of conductivity[32] and diffusion measurements.[22,33] Therefore, the "best" values for the thermodynamic variables for ionic transport in NaCl, given in the fifth column of Table II, are rather heavily weighed in favor of the Allnatt *et al.*[33] values. Their data are not very sensitive to the presence of anion vacancies since the only significant anion contribution to the conductivity of NaCl is in the high-temperature intrinsic region. So the values listed for the anion enthalpy and entropy of migration are taken from the chlorine ion diffusion measurements.[28,31]

Additional experimental results are quoted in the fifth column of Table II. The analyses of NaCl:$CaCl_2$ [34] and NaCl:$CdCl_2$ [16] conductivity data using the LDH theory reported values for Δh_c and g_k of about 0.84 and 0.33 eV, respectively. In the light of more recent experimental work,[32,33] a value of 0.84 for Δh_c for NaCl appears to be about 20% high.

The results for diffusion of radioactive tracer ions in a constant electric field have been reported for Na diffusion in NaCl by Nelson and Friauf.[22] Two of the typical diffusion profiles are shown in Fig. 16. The drift of the center of the Gaussian distribution of the tracer ions is directly related to the mobility of the cation vacancies since the electrically neutral vacancy pairs are unaffected by the electric field. From diffusion profiles of this type, Nelson and Friauf were able to measure the mobility of isolated cation vacancies and to calculate the contribution of cation vacancies to the diffusion of Na ions in NaCl, D_c. Having measured the diffusion coefficient of the cation tracer ions D_c^* and knowing D_c, they quite easily computed the vacancy pair contribution to cation diffusion, $D_p{}^c$ [see Eq. (8)]. The results of their experiments are shown in Figs. 11, 13, and 15 and the values they obtained for the vacancy pair contribution to cation diffusion are given in the fifth column of Table II.

The numerical values shown in Table II are the results of fitting experimental data to the transport equations given in Section 3. The fitting procedures involve using an electronic computer to perform a least-squares analysis. Because the transport equations involve complicated exponential functions of the thermodynamic variables [see Eqs. (63) and (65)], there is no exact way to solve the matrix obtained from the minimization equations of the least-squares method. Hence, the results in Table II are the final products of iterative procedures. These procedures treat the thermodynamic variables as parameters whose values are changed in some systematic way in order to minimize the value of the function W, where W is the sum of the squares of the differences between the experimental and computed values of the transport properties.[13]

There are at least two dangers in these procedures. First, there is the possibility of local minima that will be mistaken for the true minimum value of W. Second, there is the possibility that W will have a minimum that is relatively flat in parameter space; i.e., large variations in the values of the parameters will make very little change in the value of W. Furthermore, computer techniques require a large number of data points free from any systematic errors. These techniques of computer analysis must be carefully undertaken. While a poor fit of the transport equations to the experimental data might indicate some inadequacy in the present transport equations, the existence of additional transport mechanisms must be verified by direct experimental evidence.

With these warnings given in advance, let us examine the values in Table II in detail. In particular, the enthalpies of formation and migration have been subject to numerous theoretical calculations[3] and it is of interest to compare experimental and theoretical values for these quantities. The five fully characterized materials have been divided into two families on the basis of a common cation, the potassium family, and a common anion, the chlorine family. It is noticed that for the potassium family, the Schottky defect formation enthalpies are proportional to the melting temperature of the compound, as suggested by Barr and Lidiard[3] (Fig. 17). However, the

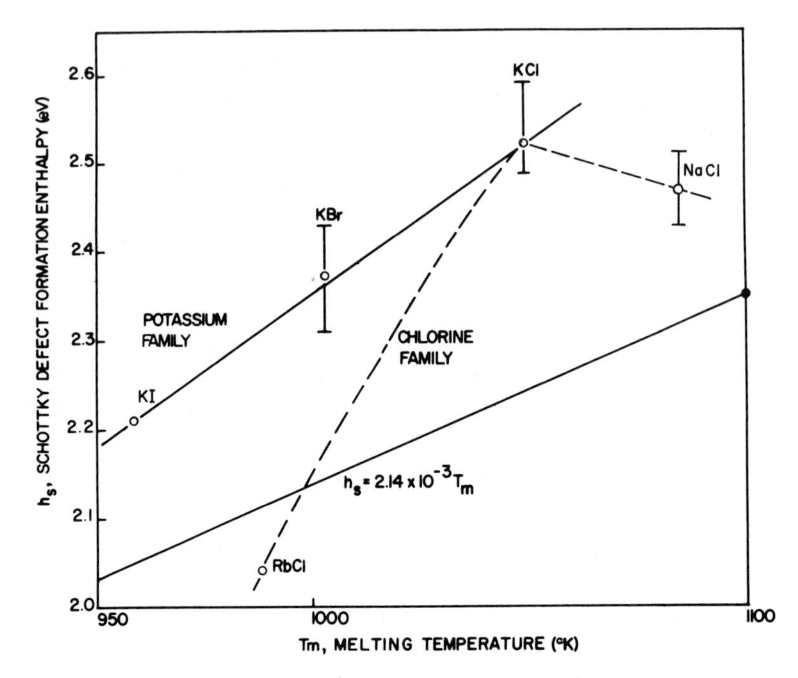

Fig. 17. Experimental values for the enthalpies of Schottky defect formation plotted as a function of the melting temperature of the compound. The $h_s = 2.14\times10^{-3}T_m$ line was suggested by Barr and Lidiard from a similar graph for 17 ionic crystals.[3]

formation enthalpies in the chlorine family seem to follow a more complex pattern as shown by the dashed curve in Fig. 17.

In an attempt to examine in more detail the transport properties of these two families, the experimental values for the enthalpies of formation and migration have been collected into Table III along with some typical theoretical values[35–37] for these same enthalpies. The potassium family seems to be particularly well behaved. Just as the theoretical values for the Schottky defect formation enthalpies decrease as the anion size increases, so do the experimental values. It seems quite reasonable on the basis of a simple rubber-sphere model for the ions that the migration enthalpies of the anions should increase with increasing anion size. This follows because, for a constant cation size,[38] the volume through which the anion must be squeezed as it jumps from one site to another would seem to decrease with increasing anion size. Hence, the anion migration enthalpies in the potassium family should increase from Cl to Br to I. This is observed experimentally. Conversely, as the anion size is increased, the small cation ion

Table III. Enthalpies of Schottky Defect Formation and Cation and Anion Migration for the Fully Characterized Alkali Halides

	Schottky defect formation h_s, eV		Anion migration Δh_a, eV		Cation migration Δh_c, eV	
	Experiment[a]	Theory	Experiment[a]	Theory[d]	Experiment[a]	Theory[d]
KCl	$2.52^{+0.07}_{-0.03}$	2.232[b] 1.935[c]	$0.90^{+0.09}_{-0.01}$	0.702	0.74 ± 0.02	0.592
KBr	2.36 ± 0.06	2.080[b] 1.811[c]	0.92 ± 0.04	0.712	0.67 ± 0.02	0.491
KI	2.21	1.866[b] 1.621[c]	1.29	—	0.63 ± 0.03	—
NaCl	2.45 ± 0.05	2.138[b] 1.904[c]	0.85 ± 0.1	0.706	0.65 ± 0.01	0.426
KCl	$2.52^{+0.07}_{-0.03}$	2.232[b] 1.935[c]	$0.90^{+0.09}_{-0.01}$	0.702	0.74 ± 0.02	0.592
KbCl	2.04	2.198[b] 1.900[c]	1.45	0.720	0.54	0.565

[a] From Table II.
[b] Ref. 35.
[c] Ref. 36.
[d] Ref. 37.

should find it easier to squeeze between its large neighbors as it jumps from one site to another, so the cation migration enthalpy should decrease as the anion size increases. Again, this is observed for KCl, KBr, and KI. No doubt it gives theoretical physicists comfort to note that the results of their calculations exhibit trends that are explicable on the basis of such a simple model.*

The chlorine family does not fit any simple model. The enthalpies of Schottky defect formation seem to follow no predictable trend. Why is the defect formation enthalpy largest in KCl? The migration enthalpies seem to be contrary to common sense. How can it be that the anion finds it less energetically favorable to move to a neighboring vacant site in RbCl than in NaCl when its companion cation is 25% smaller in the latter case? These peculiar trends within the chlorine family at least seem to receive some support from the theoretical values, which predict the largest formation enthalpy for Schottky defects in KCl. But even the most recent theoretical calculations[37] have difficulty in matching the behavior of the ions moving through the lattices of the crystals in the alkali chloride family.

* Some care should be taken in the criticism of this type of model. It was used quite effectively by Rabin and Klick[39] to predict the energy of formation of *F* centers in alkali halides.

Further evidence for the peculiarity of the alkali chlorides may be indicated by the measurements of the pressure effects on the ionic conductivity of some alkali halide crystals.[40] It was found that the Schottky defect activation volumes were 50–80% larger than the simple molar volumes of the defects in NaCl and KCl, whereas the experimentally determined activation volumes for KBr and NaBr were within 20% of the calculated molar volumes of the defects. The significance of these results is confused by the fact that the experimental data were analyzed as if all of the intrinsic conductivity arises from the motion of cation vacancies. Since it is known that the anion vacancies account for a considerable fraction of the high-temperature intrinsic conductivity in NaCl, KCl, and KBr, the implications that may be drawn from these analyses of the pressure data are uncertain.

Perhaps in the light of this discussion it is not surprising that the computer analyses of the conductivity of KCl and NaCl have shown some nonrandom deviations between the transport equations and the experimental data[13,33] (Fig. 18). As a result, additional mechanisms for charge transport in the alkali chlorides have been suggested, such as trivacancies.[41]

Trivacancies could exist in the crystal in two types, the cation type $(+ - +)$ or the anion type $(- + -)$. Each type of trivacancy could exist in either of two shapes, a collinear shape, with all the vacancies in a row, or an elbow shape, with the two end vacancies bent around the central vacancy. Simple point-ion calculations predict that the collinear cation trivacancies are more energetically favorable in the alkali chlorides.[41] It follows that the ratio of the amount of electrical current carried via a trivacancy mechanism to that carried by an isolated cation vacancy mechanism is independent of the divalent cation impurity concentration but has an exponential temperature dependence. Therefore, in the high-temperature extrinsic regions of conductivity, the presence of trivacancies might be detected. For example, in a typical case for KCl:$SrCl_2$ crystals at 750°C, it has been suggested that 23% of the electrical current could be carried by trivacancies. However, recent measurements of the conductivity of NaCl:$SrCl_2$ crystals have not shown any evidence of the presence of trivacancies.[32,33]

The transport properties of NaCl seem to have special puzzles of their own. Two different groups have made extensive measurements of the electrical conductivity of NaCl:$SrCl_2$ crystals. Both groups have performed analyses of their experimental data using the LDH theory. They have found precise values for the enthalpy of cation migration of 0.75 ± 0.02 eV[32] and 0.65 ± 0.01 eV.[33] A third group has measured both diffusion and

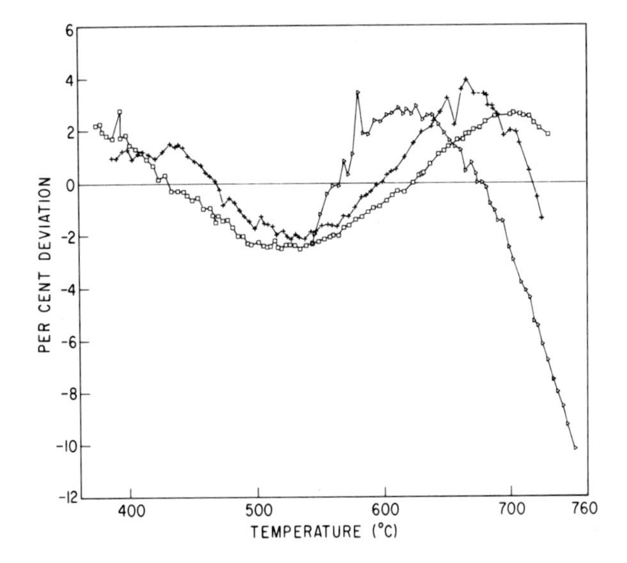

Fig. 18. Per cent deviation as a function of temperature using the Lidiard–Debye–Hückel theory. (△) The per cent deviation between the calculated and measured values of conductivity for the Harshaw sample (0.5 ppm computed divalent impurity concentration); (+) the per cent deviation between the calculated and measured values of conductivity for sample *D* (15 ppm computed divalent cation impurity concentration); (□) the percent deviation between the calculated and measured values of conductivity for the sample containing 375 ppm Sr by atomic absorption analysis.[13]

conductivity in NaCl and NaCl:$SrCl_2$ crystals.[31] They performed their analyses with a linearized form of the simple theory. They found that the Nernst–Einstein relationship is satisfied for NaCl without including the correlation factor (see footnotes to pp. 107 and 108). This implies that the obviously correlated jumps of a tracer ion are not correlated in NaCl! A possible way out of such a dilemma is to suggest that the concept of vacancies that are fixed in time and space is meaningless and to propose alternative concepts for the explanation of the transport properties of NaCl. Such a procedure has not yet been received with much enthusiasm.

In an attempt to make what we have learned about the fully characterized materials useful in understanding other materials, let us summarize the qualitative features of the transport properties of alkali halide crystals. We shall thereby overlook the numerous quantitative discrepancies for which we have no satisfactory explanation.

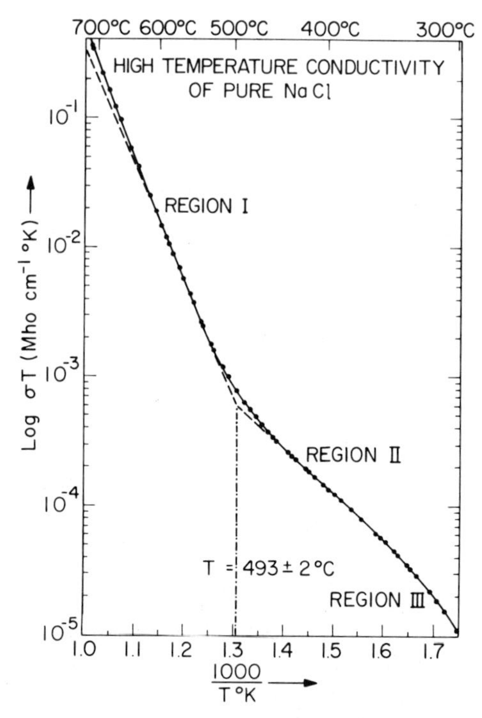

Fig. 19. High-temperature conductivity of "pure" NaCl from Kirk and Pratt.[54]

The following features are suggested by the theory of Section 3 and/or supported by the experimental data represented in Table II:

1. The ionic conductivity of an alkali halide crystal is such a complicated function of temperature and impurity concentration that the $\log(\sigma T)$ versus $10^3/T$ plot of conductivity data is a curve of continuously changing slope (Fig. 19).

2. The anion vacancies make a significant contribution to the intrinsic conductivity of these materials.

3. The cation vacancy is the most mobile intrinsic defect in these materials, with the cation enthalpy of migration less than the anion migration enthalpy by fractional amounts varying from 18% in KCl to 53% in RbCl.

4. The vacancy pair contribution to tracer diffusion is approximately half of the isolated vacancy contribution near the melting temperatures of these materials, but the pair contribution decreases rapidly with decreasing

temperature; e.g., the vacancy pair contribution to chlorine ion diffusion in KCl decreases from 50% of the isolated vacancy contribution at 775°C to about 8% at 525°C.

5. The enthalpy of Schottky defect formation is of the order of $2\frac{1}{4}$ eV in these materials and the enthalpy of association for divalent-cation-impurities and cation vacancies is of the order of $\frac{1}{2}$ eV.

5.2. Other Halide Crystals

Recent review articles have supplied a catalog of experimental values for the transport properties of the ionic crystals.[3,42] Rather than attempt to update these reviews, we will only discuss three of the less well-characterized halide crystals, each of which exhibits a characteristic not in agreement with the qualitative features listed above.

5.2.1. *Lithium Fluoride*

The results from the conductivity and diffusion measurements of Stoebe and Pratt,[43] Stoebe and Huggins,[44] and Eisenstadt[45] show that the

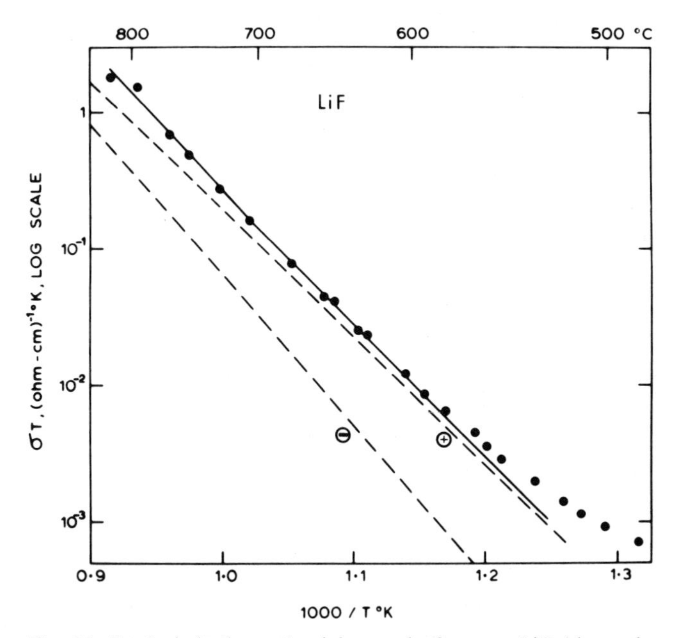

Fig. 20. Intrinsic ionic conductivity results for pure LiF (shown by the points) compared with the summation of anion and cation contributions determined from diffusivities.[43]

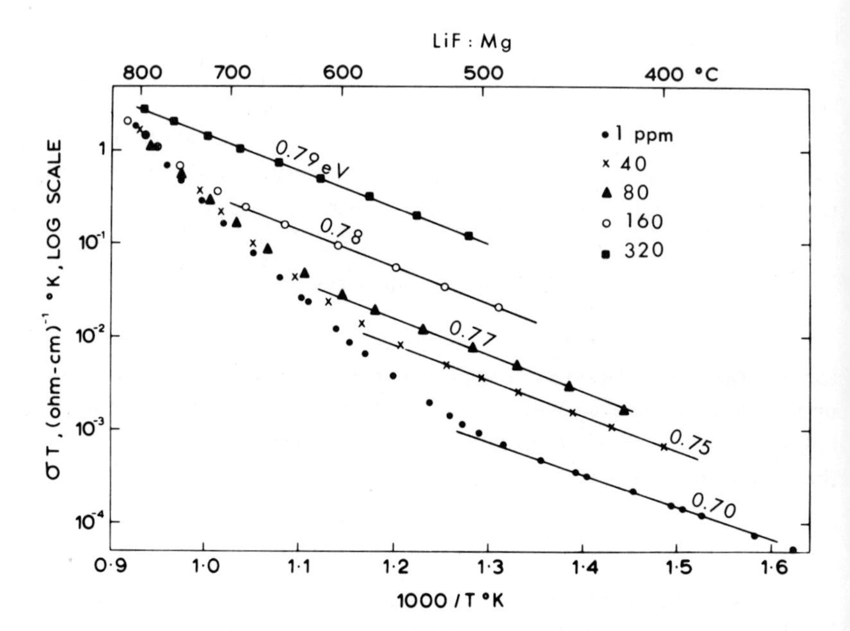

Fig. 21. Ionic conductivity of LiF:MgF_2 crystals.[43] Slope of the extrinsic conductivity curve increases with increasing impurity concentration.

intrinsic conductivity of LiF is exactly proportional to the sum of the cation and anion tracer diffusivities in LiF. This means that there is no uncharged mobile defect making a sizeable contribution to the diffusion of tracer ions in LiF. So, the vacancy pair which appears to make important contributions to the tracer diffusion in the other alkali halides may be omitted from LiF analyses, as shown in Fig. 20.

Stoebe and Pratt also showed the dangers in using the linearized simple theory to find the cation migration enthalpy from extrinsic conductivity data. Their results, shown in Fig. 21, show that the linearized slope of the extrinsic conductivity curve is a function of the impurity concentration, as we noted in Section 2, and can best be used to give a least upper bound to the value for the cation migration enthalpy.

5.2.2. *Cesium Chloride*

The cesium halides appear to be the only members of the alkali halide group in which the anions are consistently more mobile than the cations. The results of Harvey and Hoodless[46] from Cs and Cl diffusion in CsCl (Fig. 22) show that the anion diffusion rate is more than four times the cation diffusion rate over the whole intrinsic range. Their results also seem to indicate that vacancy pairs make contributions of greater than 35% to the tracer diffusivities over the whole temperature range from 280 to 460°C.

5.2.3. *Ammonium Chloride*

The ammonium halides have the cesium chloride crystal structure. So, from the beginning, their ionic transport properties have been compared to those of the cesium halides. However, the first measurements of the conductivity of the ammonium halides showed that ammonium chloride, in particular, is a significantly better intrinsic conductor than any of the cesium halides.[47] Furthermore, when single crystals of NH_4Cl were doped with either divalent cation or divalent anion substitutional impurities, the conductivity was enhanced by about the same factor, implying that anions and cations make comparable contributions to the conductivity. These facts, plus others, lead to the postulation of a conduction mechanism controlled by the motion of the hydrogen ion, or proton.[47] The results of the measurements of the electrical conductivity in NH_4Cl, ND_4Cl, and

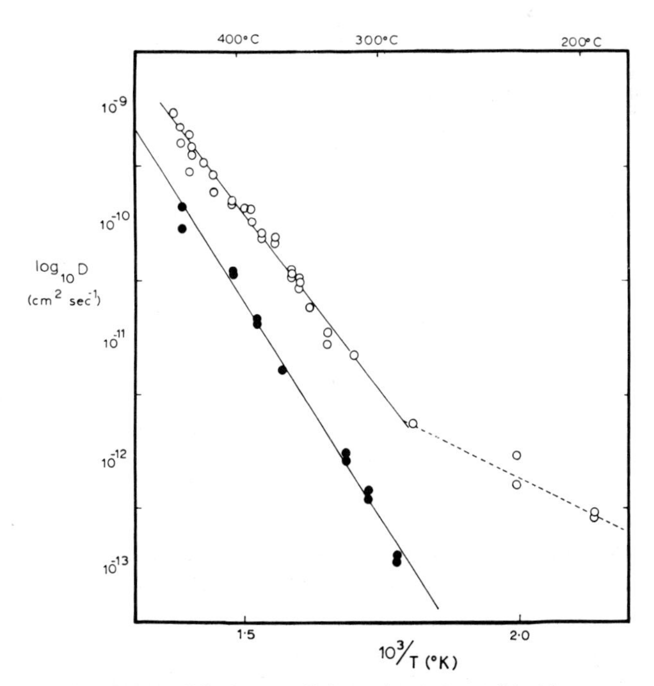

Fig. 22. Tracer diffusion coefficients in cesium chloride versus reciprocal temperature; (●) ^{137}Cs, (○) ^{36}Cl.[44]

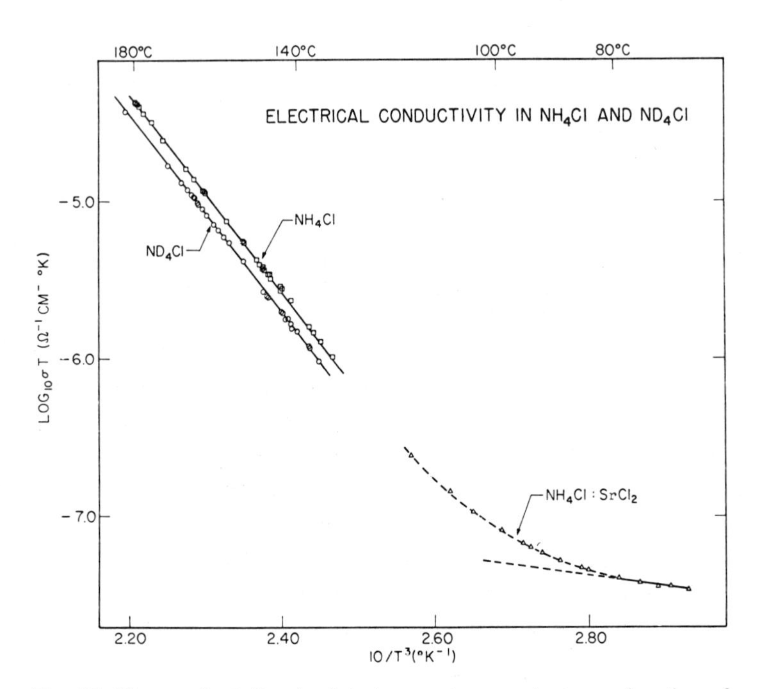

Fig. 23. The conductivity–absolute temperature product as a function of reciprocal temperature for NH_4Cl, ND_4Cl, and NH_4Cl:$SrCl_2$ single crystals. The solid lines are a result of a least-squares computer fit of a first-degree polynomial to the data (only five data points in the low-temperature extreme of NH_4Cl:$SrCl_2$ were fitted).[48]

NH_4Cl:$SrCl_2$ single crystals are shown in Fig. 23.[48] These data were analyzed using the dependence of the jump probability on the inverse square root of the mass of the moving ion shown in Eq. (43). The conduction mechanism in NH_4Cl was found to be dominated by the motion of the proton. The extrinsic conductivity region showed an activation energy of only 0.15 eV, indicating the high mobility of the proton in the NH_4Cl lattice. The comparison of the enthalpies of formation and motion for NH_4Cl and CsCl can be made from the values shown in Table IV.

6. COMMENTS ON FUTURE EXPERIMENTAL AND THEORETICAL WORK

At this time, our civilization is confronted by enormous technological and social problems. In such a setting, it is only natural that the fruits of basic research be made as plain and as useful as possible. Clearly, then, the

Table IV. Experimental Values of Enthalpies of Formation and Migration

Material	Schottky defect formation h_s, eV	Cation migration Δh_c, eV	Anion migration Δh_a, eV
LiF[a]	2.34	0.70±0.02	1.1
CsCl[b]	1.86	0.60	0.34
NH_4Cl[c]	2.2	0.15[d]	—

[a] Ref. 43.
[b] Ref. 46.
[c] Ref. 48.
[d] This is not a true cation migration enthalpy since the present model of the conduction mechanism proposes that the NH_4^+ ion dissociates into NH_3^0 and H^+ ions and the conduction process is dominated by the migration of the H^+ ions with an activation enthalpy of 0.15 eV.[48]

future of research in ionic transport properties lies not in just further honing of our considerable theoretical and experimental tools on the alkali halides, but also in applying these tools to a wide range of materials about which very little is known, but which show promise of future applications. Of course, the alkali halides have not been the only ionic materials to receive attention. A considerable body of experimental data has been accumulated on the silver halides[42] and the alkaline earth halides[42] as well. The transport properties of the oxides[49] and azides[50] are becoming more widely studied.

Before considering some new materials for ionic transport research, let us summarize some problem areas that remain in the alkali halide group. For future experimental work, there are at least the following:

1. The present state of the data for NaCl needs to be improved by a complete set of careful chlorine ion diffusion data analyzed with the LDH theory.

2. The potassium diffusion in KCl needs to be measured over a wide temperature range.

3. Further conductivity and diffusion work for RbCl is required.

4. Additional crystals, such as NaBr, RbBr, NaI, and RbI, need to be brought into the fully characterized group. Then, the sodium, rubidium, bromine, and iodine families could be subjected to the kind of scrutiny presently possible for the potassium and chlorine families.

5. Perhaps a search for additional transport mechanisms, such as suggested for trivacancies,[41] should be carried out.

6. The vacancy pair contributions to the diffusivities in LiF and the cesium halides could stand some investigation.

7. Additional experimental techniques, such as the last four listed in Table I, could be applied to the transport properties in alkali halide crystals. For example, the use of the isotope effect* in sodium diffusion in NaCl might be an experimental way of determining the temperature dependence of the vacancy pair correlation factor.[9]

For future theoretical work, there are at least the following:

1. Perform calculations of enthalpies of formation and migration that not only explain the trends shown in Figs. 20 and 22, but also predict what might be found among the other cation and anion families in the alkali halide group.

2. Develop an explanation of the apparent temperature dependence of the anion migration enthalpy in KCl reported[52] as 0.6 below room temperature and as 0.9 at 650°C.[13]

3. Improve the transport equations of Section 3 by refining the treatment of defect interactions, or by including impurity vibrational effects,[14] or anharmonic lattice vibration theory.[53]

It now seems clear that the work that led to the discovery of the high-conductivity electrolyte $RbAg_4I_5$ had its beginning in the knowledge of the transport processes in the silver halides. It is known that only a small part of the electric current that flows in AgCl is carried by the motion of vacancies (shown by the dashed line labeled Ag_v in Fig. 1) and that the Ag ion is a very mobile interstitial charge carrier in AgCl. In a similar fashion, the recent work on NH_4Cl seems to point us toward research on a different group of materials. Since the conductivity of NH_4Cl seems to arise from the motion of the proton, then perhaps proton-rich halides offer us the possibilities of high ionic conductivity. If so, the organic halides such as tetraethyl-ammonium bromide, $(C_2H_5)_4NBr$, and tetramethyl-ammonium chloride, $(C_2H_3)_4NCl$, may be good materials for future research on ionic transport properties.

ACKNOWLEDGMENTS

I wish to thank the people who have shared their data, drawings, and insights with me via correspondence and conversation, especially my colleagues at the Naval Research Laboratory, C. L. Marquardt, M. H. Reilly, H. B. Rosenstock, M. Gomez, M. N. Kabler, and C. C. Klick.

* For a discussion of this technique as applied to sodium diffusion in RbCl, see Peterson and Rothman.[51]

REFERENCES

1. B. B. Owens and G. R. Argue, *Science* **157**, 308 (1967).
2. G. G. Bentle, *J. Appl. Phys.* **39**, 4037 (1968).
3. L. W. Barr and A. B. Lidiard, "Defects in Ionic Crystals," in *Physical Chemistry—An Advanced Treatise* (Academic Press, New York, 1970), Vol. X.
4. W. Schottky, *Z. Phys. Chem. Abt. B* **29**, 335 (1935).
5. C. Wagner, *Z. Phys. Chem. Abt. B* **38**, 325 (1938).
6. N. F. Mott and M. J. Littleton, *Trans. Faraday Soc.* **34**, 485 (1938).
7. P. G. Shewmon, *Diffusion in Solids* (McGraw-Hill, New York, 1963).
8. N. F. Mott and R. W. Gurney, *Electronic Processes in Ionic Crystals*, 2nd Ed. (Oxford Press, London, 1948).
9. A. D. LeClaire, "Correlation Effects in Diffusion in Solids," Chapter 6 in *Physical Chemistry—An Advanced Treatise* (Academic Press, New York, 1970), Vol. X.
10. C. Kittel, *Elementary Statistical Physics* (Wiley, New York, 1958).
11. J. H. Beaumont and P. W. M. Jacobs, *J. Chem. Phys.* **45**, 1496 (1966).
12. A. B. Lidiard, *Handbuch der Physik*, Vol. 20, p. 246 (1957).
13. R. G. Fuller, C. L. Marquardt, M. H. Reilly, and J. C. Wells, Jr., *Phys. Rev.* **176**, 1036 (1968).
14. R. G. Fuller and H. B. Rosenstock, *J. Phys. Chem. Solids* **30**, 2105 (1969).
15. H. R. Glyde, *Rev. Mod. Phys.* **39**, 373 (1967).
16. A. B. Lidiard, *Phys. Rev.* **94**, 29 (1954).
17. A. R. Allnatt, "Statistical Mechanics of Point-Defect Interactions in Solids," in *Advances in Chemical Physics* (Interscience, 1967), Vol. XI.
18. R. H. Fowler and E. A. Guggenheim, *Statistical Thermodynamics* (Cambridge, 1949), Chapter 9.
19. E. Pitts, *Proc. Roy. Soc. London* **A217**, 43 (1953).
20. S. Chandra and J. Rolfe, *Can. J. Phys.* **48**, 397 (1970).
21. S. Chandra and J. Rolfe, *Can. J. Phys.* **48**, 412 (1970).
22. V. C. Nelson and R. J. Friauf, *J. Phys. Chem. Solids* **31**, 825 (1970).
23. D. K. Dawson and L. W. Barr, *Phys. Rev. Letters* **19**, 844 (1967); *Proc. Brit. Ceram. Soc.*, No. 9, 171 (1967); and to be published.
24. P. L. Read and E. Katz, *Phys. Rev. Letters* **5**, 466 (1960).
25. D. Mapother, H. N. Crooks, and R. J. Maurer, *J. Chem. Phys.* **18**, 1231 (1950).
26. D. Patterson, J. A. Morrison, and G. S. Rose, *Phil. Mag.* **1**, 393 (1956).
27. J. Rolfe, *Can. J. Phys.* **42**, 2195 (1964).
28. R. G. Fuller and M. H. Reilly, *Phys. Rev. Letters* **19**, 113 (1967).
29. G. Arai and J. G. Mullen, *Phys. Rev.* **143**, 663 (1966).
30. R. G. Fuller, *Bull. Am. Phys. Soc.* **15**, 384 (1970).
31. F. Bénière, M. Bénière, and M. Chemla, *J. Phys. Chem. Solids* **31**, 1205 (1970).
32. E. Laredo and E. Dartyge, *J. Chem. Phys.* **53**, 2214 (1970).
33. A. R. Allnatt, P. Pantelis, and S. J. Sime, *J. Phys. C: Solid St. Phys.* **4**, 1778 (1971).
34. H. Kanzaki, K. Kido, S. Tamura, and S. Oki, *J. Phys. Soc. Japan* **20**, 2305 (1965).
35. I. Boswarva and A. B. Lidiard, *Phil. Mag.* **16**, 805 (1967).
36. A. M. Karo and J. R. Hardy, *Phys. Rev.* **B3**, 3418 (1971).
37. P. D. Schulze and J. R. Hardy (to be published).
38. F. G. Fumi and M. P. Tosi, *J. Phys. Chem. Solids* **25**, 31 (1964).
39. H. Rabin and C. C. Klick, *Phys. Rev.* **117**, 1005 (1960).

40. D. Lazarus, D. N. Yoom, and R. N. Jeffery, *Z. Naturforsch.* **26a**, 56 (1971).
41. R. G. Fuller and M. H. Reilly, *J. Phys. Chem. Solids* **30**, 457 (1969).
42. P. Süptitz and J. Teltow, *Phys. Stat. Sol.* **23**, 9 (1967).
43. T. G. Stoebe and P. L. Pratt, *Proc. Brit. Ceram. Soc.* **9**, 171 (1967).
44. T. G. Stoebe and R. A. Huggins, *J. Metals Sci.* **1**, 117 (1966).
45. M. Eisenstadt, *Phys. Rev.* **132**, 630 (1963).
46. P. J. Harvey and I. M. Hoodless, *Phil. Mag.* **16**, 545 (1967).
47. T. M. Herrington and L. A. K. Staveley, *J. Phys. Chem. Solids* **25**, 921 (1964).
48. R. G. Fuller and F. W. Patten, *J. Phys. Chem. Solids* **31**, 1539 (1970).
49. Y. Adda and J. Philibert, *La Diffusion dans des Solids* (Presses Universitaires de France, Paris, 1966), Vol. II.
50. A. L. Laskar and J. Sharma, *Bull. Am. Phys. Soc.* **15**, 390 (1970).
51. N. L. Peterson and S. J. Rothman, *Phys. Rev.* **177**, 1329 (1969).
52. F. Lüty, Chapter 3 in *Physics of Color Centers*, edited by W. B. Fowler (Academic Press, New York, 1968).
53. W. Franklin, *Phys. Rev.* **180**, 682 (1969).
54. D. L. Kirk and P. L. Pratt, *Proc. Brit. Ceram. Soc.* **9**, 215 (1967).

Chapter 3

DEFECT MOBILITIES IN IONIC CRYSTALS CONTAINING ALIOVALENT IONS

A. S. Nowick

Henry Krumb School of Mines
Columbia University
New York, New York

1. INTRODUCTION

In ionic crystals, there is an opportunity to create a rather unique type of defect complex by the introduction of aliovalent impurities, i.e., impurity ions which differ in charge from the corresponding solvent ion. When such impurities are introduced, additional defects (either vacancies or interstitial ions) must accompany the aliovalent ions in order to achieve charge compensation, the defect possessing an effective charge equal and opposite to that of the impurity ion. Accordingly, the two entities will experience a Coulomb attraction tending to produce, if the temperature is not too high, a dipole or defect complex which possesses a relatively strong binding energy. The existence of such complexes profoundly affects many of the properties of ionic crystals, including mechanical as well as electrical properties. This chapter is concerned mainly with information about ionic motions which can be obtained by studying such doped crystals. On the one hand, diffusion of the aliovalent ion may be very different from self-diffusion, due to the presence of the compensating defect. On the other hand, the methods of dielectric and anelastic relaxation may be used to observe the reorientation of the complex in an appropriate externally applied field (electric or stress field, respectively). These measurements, therefore, give direct information on the kinetics of the rate-controlling steps in the reorientation process. Combining this information with corresponding information on kinetics of migration (from diffusion measurements or the study of the dipole aggregation), can give a rather complete picture of the various ionic motions which take place in the presence of aliovalent impurity ions.

A major portion (Section 4) of the present chapter will be devoted to the theory and methods involved in the study of relaxation processes, since these questions are not covered in detail elsewhere in this treatise. The effect of aliovalent impurities on ionic conductivity will not be discussed here, since it is covered in the chapter by Fuller.

2. INTERACTION BETWEEN ALIOVALENT IMPURITIES AND OTHER POINT DEFECTS

As already mentioned, the introduction of aliovalent impurity ions into an ionic crystal results in the presence of additional defects (particularly, vacancies or interstitial ions) required for charge compensation. Probably the best-known example is that of an alkali halide AX doped with a divalent halide MX_2. In this case, the M^{2+} ion is accompanied by a cation vacancy, V_A. The M^{2+} ion provides an excess of one unit of positive charge (relative to the perfect crystal), while the A^+ vacancy introduces a unit negative charge. In other examples, to be discussed later, an aliovalent ion may be compensated by an interstitial ion, or by a vacancy on the opposite sublattice. If the crystal is treated as a continuum, the Coulomb interaction between such a pair of point defects at a separation R_j is

$$E_j = -q^2/\varkappa R_j \tag{1}$$

where $\varkappa$ is the dielectric constant of the crystal and q is the charge. At relatively low temperatures, the defects may be expected to form a nearest-neighbor (nn) complex, which may be thought of as the "ground-state" of the defect pair. At higher temperatures, the complex may take on the next-nearest-neighbor (nnn) configuration, or may involve even greater separations out to complete dissociation. Fig. 1 shows the nn and nnn configurations for the example of the M^{2+}–V_A pair in the sodium chloride lattice. (Also shown, for later use, are the possible types of vacancy jumps.) At close distances, Eq. (1), which treats the crystal as a continuum, tends to overestimate the association energy. Detailed atomistic calculations[1,2] carried out using the well-known Born–Mayer model show that, for the case of AX:MX_2, the nnn complex has about the same binding energy as

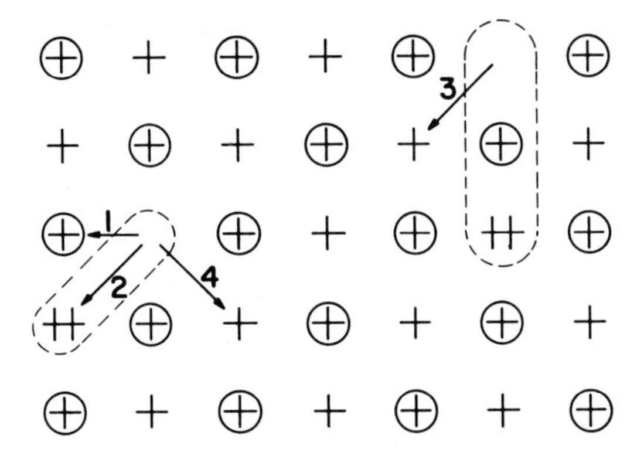

Fig. 1. The cation lattice of NaCl showing nearest-neighbor and next-nearest-neighbor vacancy–impurity pairs, and the four vacancy jumps discussed in the text. Circled positive charges represent Na^+ ions lying above and below the plane of the page by one-half of the lattice parameter. The impurity ions (indicated by the double charges) lie in the plane of the uncircled Na^+ ions. The rates corresponding to the vacancy jumps 1, 2, 3, and 4 are w_1, w_2, w_3, and w_4, respectively.

the nn complex (both about 0.3–0.4 eV) and in some cases (e.g. Sr^{2+} in KCl), the nnn value may even be slightly greater than the nn value. For larger separations, however, Eq. (1) may be used. When the temperature and binding energies are such that only nn and nnn complexes need to be considered, we may write the quasichemical reaction

$$\alpha \rightleftarrows \beta \rightleftarrows \text{dissociated pair} \tag{2}$$

where α denotes the nn and β the nnn complex. From the dissociation "reaction" of the α complex and the condition for conservation of aliovalent impurities, the following mass action equation may be written:

$$x_\alpha / x_d(c_I - x_\alpha - x_\beta) = z_\alpha \exp(\Delta g_\alpha / kT) \equiv K_\alpha \tag{3}$$

A similar equation applies to the β complex. Here, x_α, x_β, and x_d are, respectively, the mole fractions of nn, nnn, and the free (dissociated) compensating defect, c_I is the *total* aliovalent impurity concentration, and z_α is the number of equivalent orientations of the nn complex. The quantity K_α is defined by Eq. (3), while the quantity Δg_α is the Gibbs free energy of association of the nn complex, which may be expressed in terms of the corresponding enthalpy Δh_α and entropy Δs_α by

$$\Delta g_\alpha = \Delta h_\alpha - T\,\Delta s_\alpha \tag{4}$$

A simplification occurs if the only charged defects present in significant amounts are the dissociation products (i.e., the dissociated aliovalent ions and the compensating defect d). In this case, the charge neutrality condition is

$$x_d = c_I - x_\alpha - x_\beta \tag{5}$$

With this condition, a solution of Eq. (3) and the corresponding equation for β is readily obtained. For the even simpler case where the β (nnn) complex may be ignored (often referred to as the "Stasiw–Teltow model"), Eqs. (3) and (5) yield

$$p/(1-p)^2 = K_\alpha c_I \tag{6}$$

in which $p \equiv x_\alpha/c_I$ is the degree of association. This equation shows that at low impurity concentrations c_I, $p \propto c_I$, while for high concentrations, $p \rightarrow 1$. It is convenient to note that when $\Delta g_\alpha/kT = 10$, $p \approx 1$ for $c_I \gtrsim 10^{-3}$.

As already indicated, however, at elevated temperatures, it is probably necessary to take into account the existence of a whole series of bound states, in accordance with Eq. (1). Various theoretical refinements have been introduced to consider this situation, beginning with the adaptation to this problem of the Debye–Hückel theory of strong electrolytes by Lidiard,[3] as well as the more detailed statistical mechanical treatments of Fong[4] and of Allnatt and Cohen.[5] Since these approaches have already been discussed in the chapter by Franklin, we will not review them again. In any case, much of the interest of the present chapter centers on situations in which consideration of only nn and nnn pairs is sufficient.

3. DIFFUSION OF ALIOVALENT IMPURITIES

An aliovalent impurity ion which is compensated by a vacancy on the same sublattice will spend a relatively large fraction p of its time as part of an nn complex with that vacancy, p being the degree of association [see Eq. (6)]. When diffusion occurs by means of a vacancy mechanism, such an ion is expected to have a much greater diffusion coefficient than one which only makes chance encounters with a vacancy, as in the case of ions with the same valence as the solvent ions, or of aliovalent ions which are compensated by interstitials or by vacancies on the opposite sublattice. The most interesting case that meets this criterion, and by far the most explored one, is that of M^{2+}-doped alkali halides, i.e., of $AX:MX_2$, for which the compensating defect is the alkali-ion vacancy V_A. We will first review the theory for diffusion of such aliovalent ions, due to Lidiard,[3,6] and will then present some of the results of such diffusion experiments.

3.1. Theory of Diffusion of M^{2+} in AX

Since, in terms of the vacancy mechanism, an M^{2+} ion can only diffuse when it is associated with a vacancy, it is convenient to regard the α (nn) complex as the diffusing entity.* We consider a one-dimensional diffusion problem, for which the diffusion coefficient D_M is given by

$$J_M = -D_M\, dn_M/dx = -D_\alpha\, d(pn_M)/dx \tag{7}$$

where J_M is the flux of M ions and dn_M/dx the gradient of concentration of M (expressed as number/volume) in direction x. The last equality in Eq. (7) treats the problem as a diffusion of complexes α. Since

$$d(pn_M)/dx = [d(pn_M)/dn_M]\, dn_M/dx \tag{8}$$

it follows that

$$D_M = D_\alpha\, d(pn_M)/dn_M = D_\alpha\, d(pc_M)/dc_M \tag{9}$$

where we have changed over to expressing the concentration as a mole fraction c_M. Now, D_α is evaluated in the usual way as $\frac{1}{6}\Gamma f r^2$ [see Chapter 1, Eq. (3.52)], to obtain

$$D_\alpha = \tfrac{1}{12}a^2 f w_2 \tag{10}$$

where a is the lattice parameter and f the correlation factor given by

$$f = (2w_1 + 7w_4)/(2w_1 + 2w_2 + 7w_4) \tag{11}$$

Here, w_1, w_2, and w_4 are, respectively, the jump frequencies for the interchange of a vacancy adjacent to an M^{2+} ion with an nn A^+ ion, an M^{2+} impurity ion, and an nnn A^+ ion, as shown in Fig. 1.† Each of these jump

* This assumption is not always valid when all of the Onsager coefficients of the thermodynamics of irreversible processes are taken into account, as shown by Howard and Lidiard.[6a]

† The reader should be cautioned that in much of the literature on diffusion, the definitions of w_3 and w_4 are the reverse of those given here.

frequencies depends on the temperature through an expression of the form

$$w_i = w_{i0} \exp(-\Delta h_i/kT)$$

where Δh_i is the corresponding enthalpy of activation. The importance of correlations is best seen by considering the case in which w_2 is very large. In this case, the impurity ion merely jumps back and forth into the vacancy many times without migrating, and here, $f \ll 1$. On the other hand, from Eq. (11), it is clear that if $w_2 \ll w_1, w_4$ correlations are unimportant, i.e., $f \approx 1$. This case is the one in which jumps of alkali ions into a vacancy take place much more readily than those of the impurity ion. Owing to the double charge on the M^{2+} ion, one may actually expect a higher potential barrier opposing its motion, except perhaps when the radius of the impurity ion is very small. As we shall see later, the condition of a relatively small w_2 and, therefore, of minor importance of the correlation factor, is in fact generally valid. Nevertheless, for generality, we keep the correlation factor f in the equations.

In order to obtain an explicit expression for D_M from Eq. (9), we use the mass action equation (6) for p, which gives

$$d(pc_M)/dc_M = p + c_M(dp/dc_M) = 2p/(1+p) \tag{12}$$

and then, substituting for p from Eq. (6), obtain

$$D_M(c_M) = (a^2 f w_2/12)[1 - (1 + 4K_\alpha c_M)^{-1/2}] \tag{13}$$

which is referred to as Lidiard's equation. Note that Eq. (13) expresses D_M as a function of the total concentration of divalent impurity c_M, as well as of the temperature through the terms w_2, f, and K_α. Two limiting cases of this equation are of interest. In the low-concentration, high-temperature limit ($K_\alpha c_M \ll 1$, but the crystal not hot enough to be intrinsic), where $p \to 0$, we obtain

$$D_M(c_M) \approx 2a^2 f w_2[\exp(\Delta g_\alpha/kT)]c_M \qquad (p \to 0) \tag{14}$$

where the explicit expression for K_α from Eq. (3) has been inserted, with $z_\alpha = 12$ for this fcc lattice. Thus, in this limit, $D_M \propto c_M$. In the opposite case, the high-concentration, low-temperature limit ($K_\alpha c_M \gg 1$), where $p \to 1$, $D_M(c_M)$ becomes independent of c_M, approaching the saturation value $D_M{}^s$ given by

$$D_M{}^s = a^2 f w_2/12 \qquad (p \to 1) \tag{15}$$

It is noteworthy that the effective activation energy E for the diffusion coefficient in the low-concentration limit is

$$E \approx \Delta h_2 - \Delta h_\alpha \tag{16}$$

where Δh_2 is the enthalpy of activation for the impurity–vacancy interchange, Δh_α is given by Eq. (4), and the temperature dependence of f is taken to be negligible. On the other hand, in the saturation limit, the activation energy is given by

$$E_s = \Delta h_2 \tag{17}$$

In neither case does a vacancy formation energy appear in the expression for E, as it does in the case of self-diffusion, since, in this model, the vacancies are present as charge compensators rather than as thermally produced defects.

The variation of $D_M(c_M)$ with c_M between these extreme limits, as calculated from Eq. (13), is shown in Fig. 2 for different values of the parameter $\Delta g_\alpha/kT$. Lidiard[3] also used the Debye–Hückel approach to obtain

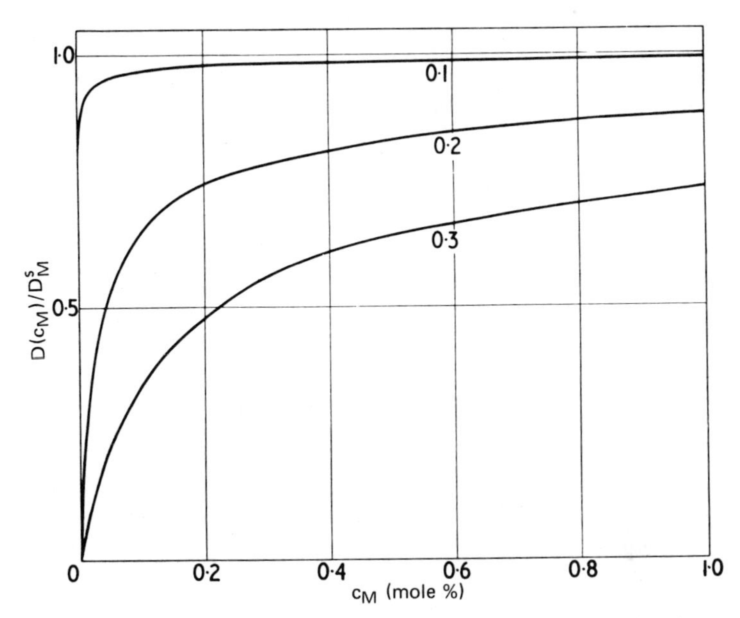

Fig. 2. Dependence of the diffusion coefficient D_M of an M^{2+} ion in an NaCl-type crystal on the concentration c_M as calculated from Eq. (13) for the three different values of the parameter $kT/\Delta g_\alpha$ marked on the curves. (From Lidiard.[6])

a modified form of Eq. (13). Finally, Lidiard has considered the diffusion of a radioactive isotope of the M^{2+} impurity in a solid solution which is already doped uniformly with M^{2+} ions. In this case, p depends on this concentration of M^{2+} ions and hardly at all on the relatively small tracer concentration in the crystal. Accordingly, while there is a gradient in tracer concentration $c_M{}^*$, there is none in the degree of association p. The tracer diffusion coefficient $D_M{}^*$ is then, from Eqs. (9) and (10),

$$D_M{}^*(c_M) = D_\alpha p = a^2 f w_2 p/12 \tag{18}$$

Measurement of $D_M{}^*$ as a function of doping concentration from the linear range to the saturation range then allows for evaluation of both Δh_2 and Δh_α.

3.2. Methods

The methods of studying diffusion will be mentioned here only very briefly. Details on techniques may be found, for example, in a review article by Tomizuka.[7]

The initial conditions for diffusion measurements are chosen so as to obtain simple solutions for Fick's second law. The most common condition is to deposit a thin layer of impurity (normal or radioactive) on one end of a relatively long sample. After diffusing at temperature T for a time t, the concentration as a function of depth x is given by[8]

$$c_I(x, t) = [c_0/(Dt)^{1/2}] \exp(-x^2/4Dt) \tag{19}$$

for the case in which the diffusion coefficient D is independent of concentration, where c_0 is a constant. By plotting $\log c_I$ versus x^2, one obtains a straight line whose slope gives the value of D at that temperature. The dependence of c_I on depth is best determined by sectioning the sample into thin slices, and measuring the chemical concentration or radioactivity of each slice.

Another frequently used boundary condition is that of constant surface composition during the time of diffusion, as realized, for example, by maintaining equilibrium with a vapor containing the impurity. Under the assumption of constant D, the solution is

$$c_I(x, t) \approx \text{erfc}(x/2D^{1/2}t) \tag{20}$$

where erfc refers to the complementary error function.

If D is actually a function of c_I, as, for example, in the case of Eq. (13), the Matano method[8] of analysis of the data may be used to obtain $D(c_I)$ from a single diffusion profile.

3.3. Experimental Results

The first major experimental study of the diffusion of divalent impurity ions in alkali halides was carried out by Chemla,[9] who studied conventional diffusion as well as diffusion in an electric field to obtain both the ionic mobility μ and the diffusion coefficient D. For M^{2+} ions, he consistently found that μ/D was much less than the value e/kT predicted by the Einstein relation. The significance of this result is that a substantial part of the diffusion occurs via neutral pairs which do not contribute to the transport of charge. This result provides strong qualitative support for the Lidiard theory which we have reviewed in Section 3.1.

Subsequent experiments by various other workers have borne out the expectation that the activation energy for diffusion of M^{2+} ions relates primarily to the activation energy for defect migration and not to defect formation energies, and is therefore lower than that for self-diffusion (or monovalent cation impurity diffusion). Thus while $E = 1.7$ eV for tracer diffusion of Na^+ in NaCl, values of $E \lesssim 1.0$ eV are obtained for M^{2+} ions. A tabulation of such values will be given later in this chapter (Table V), for comparison with activation energies obtained from relaxation-type experiments.

Aside from the activation energies, the general form of Lidiard's equation (13) for $D_M(c_M)$ has been verified by a number of different workers. For example, Fig. 3 shows data for diffusion of Pb^{2+} in KCl in comparison with theoretical curves based on Eq. (13). In this work, the crystal is not predoped with Pb^{2+} and, therefore, in each experiment, D is a function of concentration. For this reason, the curves of $D(c_M)$ were obtained by the Matano analysis. Fitting the best theoretical curves to the data of Fig. 3 gives values for both $D_M{}^s$ and for Δg_α at each temperature of measurement. The value of Δg_α obtained in this way for Pb^{2+} in KCl is $\sim$0.45 eV at 450°C and decreases slightly with increasing temperature. The activation energy E_s is 1.18 eV.*

Rothman *et al.*[11] have studied diffusion of Zn^{2+} into NaCl using predoped samples, so that Eq. (18) applies, and D is a constant in each diffusion

* In later work (W. J. Fredericks, private communication), however, this value has been revised to 0.91 eV.

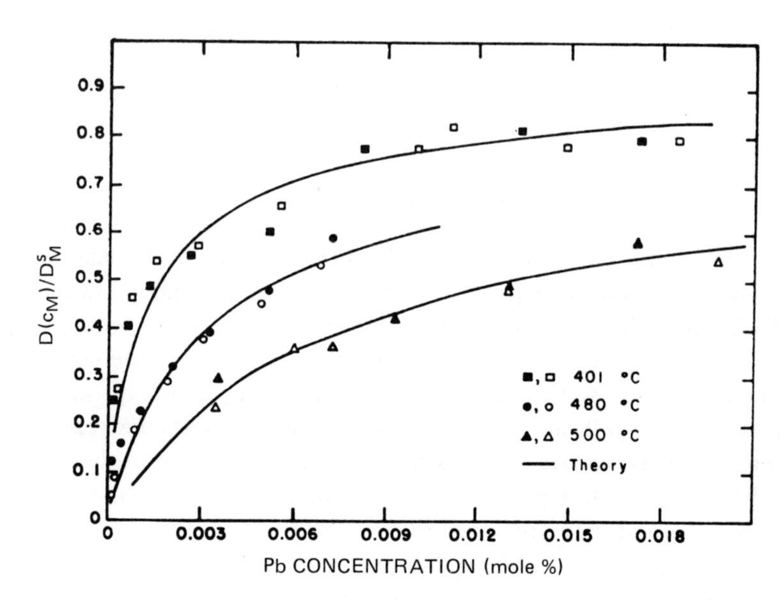

Fig. 3. Data for diffusion coefficient of Pb^{2+} into KCl as a function of Pb^{2+} concentration, as compared with the theoretical curves. (From Keneshea and Fredericks.[10])

experiment. They obtain a very low value of 0.52 eV for the activation energy E_s in the saturation range, as against a value of ~1.0 eV for tracer diffusion into essentially pure crystals. There is good indication that the situation for Zn^{2+} in NaCl may be complicated by the occurrence of precipitation. In such a case, the equilibrium among complexes, dissociated complexes, and precipitate must be taken into account. Also, because of the relatively small ionic radius of Zn^{2+}, it is possible that for this ion, the assumption that w_2 is small, which results in ability to neglect the correlation factor, may not be valid.

In summary, the Lidiard theory does provide a simple approach to understanding the diffusion of divalent cations in alkali halides which, although surely oversimplified, gives quite reasonable results for a wide variety of divalent impurities, especially in NaCl and KCl.

In the case of diffusion of M^{2+} ions in the silver halides, the same theory would be expected to apply, since the crystal structure is the same and the band gap is large, as in the alkali halides, but recent work[12,13] shows the behavior to be quite different. Apparently, there are two modifications which enter into the problem and make the Lidiard theory inapplicable, as follows:

1. At the higher temperatures, and for small amounts of M^{2+} tracers, the concentration of cation vacancies present is not given by that which accompanies the divalent impurity, but rather by the intrinsic value. The latter concentration varies as $\exp(-h_s/2kT)$, where h_s is the enthalpy of formation of a Schottky pair. Since this formation enthalpy is included, the diffusion activation energy of the divalent cation at high temperatures is, therefore, not much different from the self diffusion of Ag^+ ions.

2. There is evidence[13] for a difference in behavior of M^{2+} ions having an outer electron *d* shell, e.g., Mn^{2+}, Cd^{2+}, and Zn^{2+}, which indicates that a fraction of these impurities occupy interstitial sites where they are extremely mobile. At low temperatures, a break then occurs in the plots of $\ln D$ versus T^{-1} below which interstitial diffusion dominates. Ions without a *d* shell, e.g., Ca^{2+} or Sr^{2+}, appear to show no such break.

4. DIELECTRIC AND ANELASTIC RELAXATION: THEORY

Since dielectric and anelastic relaxation methods have been widely used to study defects involving aliovalent impurities in ionic crystals, we will use this and the next section to present in some detail the theoretical background and the methods used to carry out such measurements. For more complete presentations, see Ref. 14–17.

4.1. Formal Theory

In a formal way, dielectric and anelastic relaxation represent a generalization of the behavior of a perfect dielectric or a perfectly elastic material represented by the equations

$$P = \chi F, \qquad \varepsilon = s\sigma \tag{21}$$

Here, P is the polarization per unit volume (henceforth just called the "polarization"), F the electric field, and χ the electric susceptibility, while ε, σ, and s are, respectively, the strain, stress, and elastic compliance constant. In general, P and F are vectors and the others are all tensor quantities, but for the present, it is more convenient to deal with suitable components such that all quantities may be considered as scalars. For relaxation to occur, Eqs. (21) must be generalized so as to include time as a variable. In this way, response of the material to an applied field is no longer instantaneous. Two features of Eqs. (21) are still maintained, however. These are: (1) *linearity*, in the sense that doubling the field (F or σ) doubles the response

(P or ε), and (2) a *unique equilibrium relationship*, meaning that to every value of field, there corresponds a unique value of the response which the material will attain if given sufficient time. Maintaining these two restrictions readily permits the generalization of Eqs. (21) to include time derivatives. (We will use the dielectric behavior to illustrate, but it should be realized that the exactly analogous equations apply in the anelastic case.) The simplest generalization which meets these two restrictions is of the form

$$\chi_R F + \tau \chi_U \dot{F} = P + \tau \dot{P} \tag{22}$$

in terms of three new constants χ_R, χ_U, and τ, the significance of which will soon become apparent. Equation (22) characterizes what we shall term a "standard solid." It may be solved under suitable conditions to give the response of such a material to different experimental situations. Thus, for example, under constant electric field F_0 applied at $t = 0$ (i.e., $F = F_0$ and $\dot{F} = 0$ for $t > 0$), we obtain for the time-dependent susceptibility, defined by

$$\chi(t) \equiv P(t)/F_0 \tag{23}$$

the solution

$$\chi(t) = \chi_U + (\chi_R - \chi_U)[1 - \exp(-t/\tau)] \tag{24}$$

This process, of a time-dependent adjustment of the polarization following the application of a field, is an example of a relaxation process, and is illustrated in Fig. 4. The meaning of the constants χ_U, χ_R, and τ now

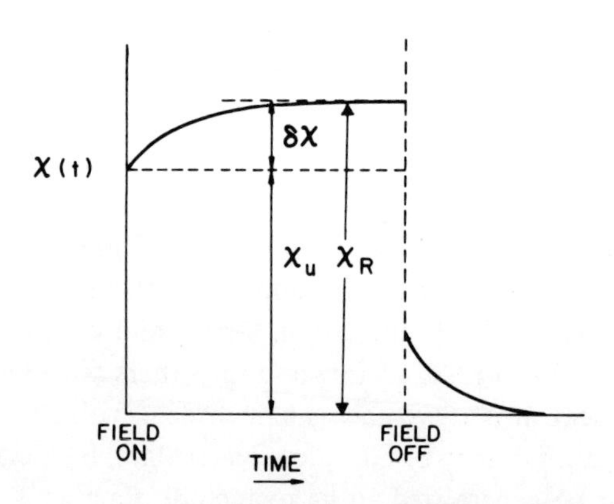

Fig. 4. The time-dependent susceptibility of a standard solid as a function of time after application of a field, and after removal of the field ("aftereffect").

become clear: χ_U, called the "unrelaxed susceptibility," gives the instantaneous response to the applied field;[†] χ_R, called the "relaxed susceptibility," gives the equilibrium value of $\chi(t)$ as $t \to \infty$; τ, which is the time to reach $1/e$ of completion of the process, is called the "relaxation time." The difference

$$\delta\chi \equiv \chi_R - \chi_U \tag{25}$$

called the "relaxation of the susceptibility," measures the total magnitude of the relaxation process. The relaxation process can also be observed as an "aftereffect," which in the present case, is an exponential decay of polarization following release of the field, as also shown in Fig. 4.

The behavior of a solid which obeys Eq. (22) under dynamical conditions is also of great interest. Here, a sinusoidal field is applied; in complex notation,

$$F = F_0 \exp(i\omega t) \tag{26}$$

where F_0 is the amplitude and ω the circular frequency. In view of Eq. (22), P is not in phase with F, but lags behind by a phase angle δ so that

$$P = P_0 \exp i(\omega t - \delta) \equiv (P_1 - iP_2) \exp(i\omega t) \tag{27}$$

where P_0 is the amplitude of P, while P_1 and P_2 are, respectively, the components which are in phase and out of phase with the field. The relationship between P and F can be expressed in the form $P = \chi^* F$, where χ^* is called the "complex susceptibility" and is given by

$$\chi^*(\omega) = \chi_1(\omega) - i\chi_2(\omega) \tag{28}$$

Note that the real part of χ^*, namely $\chi_1 = P_1/F_0$, and the imaginary part, $\chi_2 = P_2/F_0$, are in general both functions of ω. The relationship of the phase angle δ to these quantities is readily seen from Eq. (27) to be

$$\tan \delta = P_2/P_1 = \chi_2/\chi_1 \tag{29}$$

It should be noted that Eqs. (26)–(29) are completely general, in that they apply to any material showing linear behavior, not just to the standard solid of Eq. (22). If we now substitute these equations into the differential

[†] That this assignment is correct can be seen from Eq. (22) when the rate of application of field, $\dot{F}$, and therefore the rate of response $\dot{P}$ are both very large so that the other two terms are negligible by comparison.

equation (22), we obtain

$$\chi_1(\omega) = \chi_U + [\delta\chi/(1 + \omega^2\tau^2)] \tag{30}$$

$$\chi_2(\omega) = \delta\chi\,\omega\tau/(1 + \omega^2\tau^2) \tag{31}$$

These are the well-known "Debye equations." The function $\chi_2(\omega)$ when plotted versus $\log(\omega\tau)$ gives a symmetric peak centered about $\log(\omega\tau) = 0$ (i.e., $\omega\tau = 1$) called the "Debye peak." Figure 5 shows plots of both $\chi_1(\omega)$ and $\chi_2(\omega)$. An important characteristic of the Debye peak is that its width at half-maximum, $\Delta[\log_{10}(\omega\tau)]$, is just over one decade; specifically,

$$\Delta[\log_{10}(\omega\tau)] = 1.144 \tag{32}$$

The phase angle δ is called the "loss angle" since it is a measure of the internal mechanisms that give rise to energy dissipation. In the anelastic case, it is also known as the "internal friction."

Turning now to more complex relaxation behavior than the standard solid, we find that many cases, particularly those related to point defects, can be described by a "relaxation spectrum" made up of a superposition of standard solids. Instead of having a single relaxation time involving a magnitude $\delta\chi$ and a time constant τ, we introduce a set of relaxation magnitudes $\delta\chi_i$ and their corresponding relaxation times τ_i, which may be written $\{\delta\chi_i, \tau_i\}$. The corresponding functions $\chi(t)$, $\chi_1(\omega)$, and $\chi_2(\omega)$ are

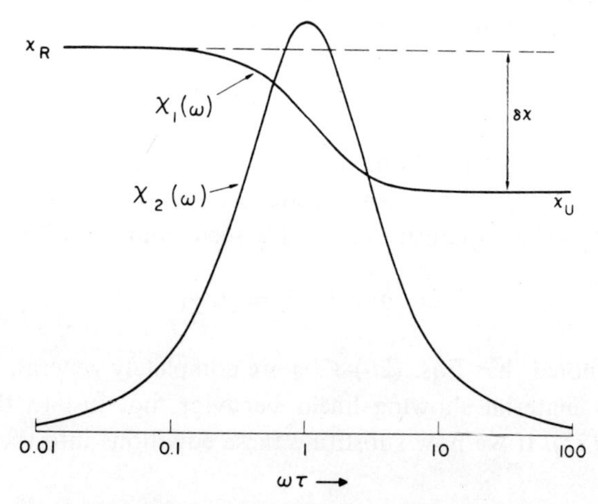

Fig. 5. Plots of the Debye equations for the real and imaginary parts of the complex susceptibility, $\chi_1(\omega)$ and $\chi_2(\omega)$, respectively, as functions of $\log(\omega\tau)$.

then summations on i of expressions of the form given in Eqs. (24), (30), and (31) for a standard solid or single relaxation process. Thus, for example, Eq. (31) for $\chi_2(\omega)$ becomes

$$\chi_2(\omega) = \sum_i \delta\chi_i \, \omega\tau_i/(1 + \omega^2\tau_i^2) \tag{33}$$

We refer to such a material as having a "discrete relaxation spectrum." In the limit of very closely spaced τ values, the spectrum may go into a continuous relaxation spectrum and Eq. (33) becomes an integral. However, for the purpose of point defect studies, the case of a discrete spectrum is of greater interest.

It is also important to note that for most relaxations produced by point defects, the relaxation times τ_i are strongly temperature-dependent, in fact, given by an Arrhenius equation of the form

$$\tau^{-1} = \tau_0^{-1} \exp(-E/kT) \tag{34}$$

This exponential temperature dependence provides the basis for the common ac methods in the study of dielectric and anelastic relaxation (see Sections 5.1 and 5.2).

4.2. Concept of Relaxational Normal Modes

In this section, we consider a species of point defect which may possess a number, say n, of crystallographically equivalent orientations. Most defect pairs of the type involving aliovalent impurities are in this category. For example, the nn M^{2+}–V_A pair in the sodium chloride structure shown in Fig. 1 can occur in 12 equivalent orientations, corresponding to the 12 directions of the type $\langle 110 \rangle$. In the absence of a field (electric or stress), all of the n equivalent orientations have equal probability when the crystal is in equilibrium, so that, if C_0 is the total concentration (expressed as a mole fraction) of defects, the concentration in each orientation will be C_0/n. In the presence of a field, however, the different orientations may no longer be equivalent, and so their equilibrium concentrations will no longer be equal.

To illustrate the principles, we first consider a defect that possesses two equivalent orientations. Each orientation may be assigned a free energy g, which includes the vibrational entropy but not the configurational entropy; i.e., g is the free energy of the defect in a *specific* location and orientation in the crystal. In the absence of a field, the free energies of the

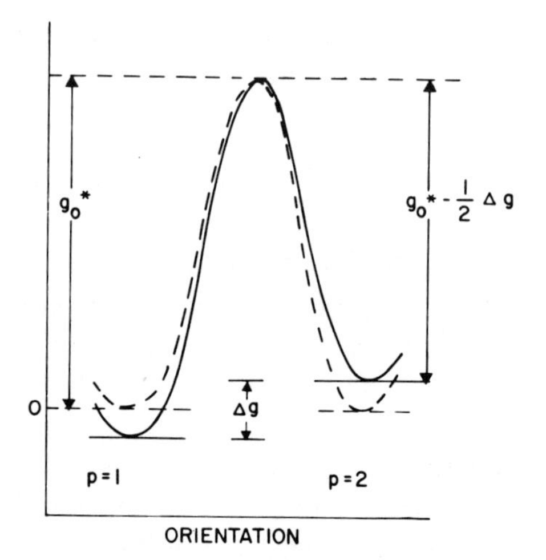

Fig. 6. Illustrating the activation barriers before (dashed curve) and after (solid curve) the splitting by stress of the free energies of a defect in two orientations.

two orientations are equal, but the presence of a field may split the free energy levels by an amount Δg (as shown in Fig. 6) and thereby lead to a time-dependent redistribution of the defect concentrations. The kinetic equation describing the rate of change of C_1, the concentration of defects in orientation number 1, is

$$\dot{C}_1 = -C_1\nu_{12} + C_2\nu_{21} \tag{35}$$

where ν_{pq} is the probability per second for reorientation from p to q. A similar expression may be written for $\dot{C}_2$. The reorientation frequency ν_{12} is given by

$$\begin{aligned}\nu_{12} &= \nu_0 \exp[-(g_0^* + \tfrac{1}{2}\,\Delta g)/kT] \\ &\approx \nu[1 - (\Delta g/2kT)]\end{aligned} \tag{36}$$

where g_0^* is the activation free energy in the absence of a field, as shown in Fig. 6, ν_0 is an appropriate lattice vibration frequency, and

$$\nu \equiv \nu_0 \exp(-g_0^*/kT) \tag{37}$$

is the reorientation frequency in the absence of a field. [The last step in Eq. (36) is obtained by expanding the exponential, assuming that $\Delta g/kT \ll 1$, which is valid for small fields.] Similarly,

$$\nu_{21} \approx \nu[1 + (\Delta g/2kT)] \tag{38}$$

Putting Eqs. (36) and (38) into (35), together with the conservation condition

$$C_0 = C_1 + C_2 \tag{39}$$

gives, to first order,

$$\dot{C}_1 = -2\nu(C_1 - \bar{C}_1) \tag{40}$$

where

$$\bar{C}_1 = \tfrac{1}{2}C_0[1 + (\Delta g/2kT)] \tag{41}$$

is the equilibrium value of C_1 in the presence of the field. Equation (40) shows that C_1 increases with time (and, correspondingly, C_2 decreases) according to an exponential equation with time constant τ given by

$$\tau = 1/2\nu \tag{42}$$

Further, an increase in C_1 gives rise to a net change in the polarization P proportional to the change in C_1; thus $P(t)$ shows the same time dependence as C_1, and with the same relaxation time τ. Comparison with Eq. (24) shows that this case corresponds to standard-solid behavior.

The calculation of the relaxation time could also have been carried out more simply as an aftereffect, i.e., under zero field ($\Delta g = 0$), but starting from an initial condition in which $C_1 \neq C_2$. The reader may easily verify that the resulting equation is the same as Eq. (40), with $\bar{C}_1 = C_0/2$. In this case, $\nu_{12} = \nu_{21} = \nu$, since $\Delta g = 0$. We will shortly see that this approach via the aftereffect is most convenient for generalizing the theory.

Equation (42) expresses the relaxation time in terms of the reorientation frequency at zero field. In an actual defect problem, however, it is desired to express τ^{-1} in terms of specific atom or ion jump frequencies w_i, appropriate to the defect in question. The conversion of reorientation frequencies into jump frequencies is not a difficult calculation. In fact, it can usually be carried out by inspection of a model of the crystal containing the defect.

We now turn to the question of generalizing the above calculation when several different defect species ζ are present, each possessing n_ζ equivalent orientations, and each capable of interconversion into the others via suitable defect reactions. The most important example for our purposes is where there are two species which are nearest-neighbor (nn) and next-nearest-neighbor (nnn) pairs involving an aliovalent ion, each species having several equivalent orientations. The concentration of species ζ in orientation r may be labeled $C_{r(\zeta)}$. It is more convenient, however, to use a single index, say $u = 1, 2, \ldots, n$, obtained by taking the double subscripts in the following order: $1(\alpha), 2(\alpha), \ldots, n_\alpha(\alpha), 1(\beta), 2(\beta), \ldots$ for species $\alpha, \beta, \ldots$. Note that $n = n_\alpha + n_\beta + \cdots$. The kinetic equations in the case of zero field then take the form

$$\dot{C}_u = \sum_v C_v \nu_{vu} \qquad (u = 1, 2, \ldots, n) \tag{43}$$

where ν_{vu} is the frequency of reorientation from v to u in the absence of a field,* and the quantities ν_{uu} are separately defined as

$$\nu_{uu} \equiv -\sum_v{}' \nu_{uv} \tag{44}$$

in which $\sum_v'$ implies the summation over all v except $v = u$. The conservation condition now takes the form

$$\sum_u C_u = C_0 = \text{const} \tag{45}$$

Equations (43) constitute a set of n linear first-order differential equations in n variables. For the simplest relaxation behavior, we seek a solution of the form

$$C_u = A_u \exp(-t/\tau) \tag{46}$$

in which τ is independent of u, i.e., the same for all C_u. Such a solution may be called a "relaxational normal mode" (by analogy with the vibrational normal mode solutions of a vibrating system in n degrees of freedom). Substitution of Eq. (46) into (43) gives a set of n linear homogeneous algebraic equations in n unknowns A_u:

$$\sum_u A_u(\nu_{uv} + \tau^{-1}\delta_{uv}) = 0 \qquad (v = 1, 2, \ldots, n) \tag{47}$$

where δ_{uv} is the Kronecker delta. The eigenvalues τ^{-1} are given by the following "secular equation" obtained by setting the determinant of the coefficients A_u equal to zero:

$$| \nu_{uv} + \tau^{-1}\delta_{uv} | = 0 \tag{48}$$

* Note that, when all defects are not of the same (equivalent) species, ν_{uv} does not form a symmetric matrix, i.e., for indices u and v belonging to different species ζ, $\nu_{uv} \neq \nu_{vu}$.

Equation (48) is a polynomial of degree n in the variable τ^{-1}. From the properties of the matrix ν_{uv}, it can be shown that all roots of this equation are real and positive, so that there are n solutions for the relaxation rate τ^{-1}. For each such solution, a set of amplitudes A_u, constituting a relaxational normal mode, can be found.

While the above constitutes a solution in principle, whenever n is not a small number, solving Eq. (48) may be a formidable problem indeed. In seeking the eigenvalues, it is therefore advantageous to make considerable use of symmetry, i.e., of group-theoretical methods. In referring to symmetry, we mean not only that of the host crystal, but also the local symmetry about the defect itself. Although a familarity with group theory will not be assumed here, the results of the group-theoretical approach will be discussed in terms which, hopefully, can be readily appreciated. It is found that each eigenvalue τ^{-1} is associated with a symmetry designation (called an "irreducible representation" in group theory) denoted by γ. To each eigenvalue, there corresponds at least one normal coordinate $C_s'^{(\gamma)}$ of the form

$$C_s'^{(\gamma)} = \sum_u B_{su}^{(\gamma)} C_u \tag{49}$$

i.e., each being a linear combination of the original concentrations C_u with coefficients determined by solving Eqs. (47).* The reason for the statement, "at least one normal coordinate," is that some designations γ are doubly or triply degenerate, so that τ^{-1} then corresponds, respectively, to two or three normal coordinates. The normal coordinates transform among themselves under the operations of the point group of the crystal in particularly simple ways. For defects of relatively low symmetry, there may be more than one τ value (and, therefore, more than one set of normal coordinates) belonging to a given designation γ. In such a case, τ^{-1} and the normal coordinates must be identified with an additional index which runs over the number of times that the designation γ is repeated.

The complete set of normal coordinates forms an orthogonal set in n-dimensional space, which means that all but one can be set equal to zero at a time. Each normal coordinate, when present only by itself (the others being zero), constitutes a normal mode. The normal modes can be shown diagramatically by a method due to Haven and van Santen.[18] We illustrate for the case of the three most common defect symmetries in a cubic crystal. These are the "tetragonal defect," which may be oriented along one of the

* The coefficients $B_{su}^{(\gamma)}$ of Eq. (49) are simply proportional to the respective amplitudes A_u of Eq. (47), obtained for a particular eigenvalue τ^{-1} of Eq. (48).

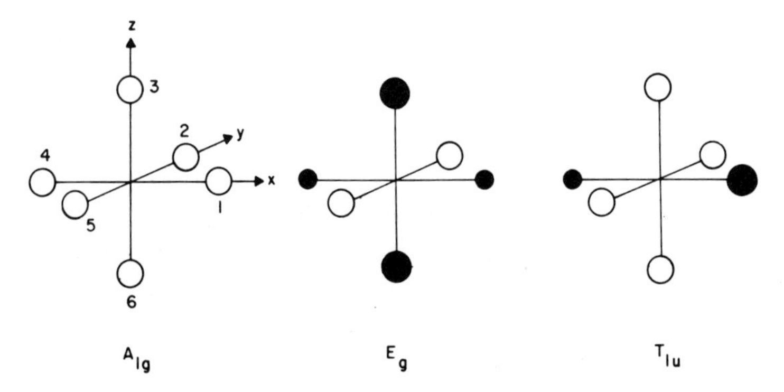

Fig. 7. Haven–van Santen diagrams for the normal modes of a tetragonal defect in a cubic crystal. Open circle means a random probability of occupation (here, a concentration $C_0/6$); large, filled circle means greater than random probability of occupation by amount δ; small, filled circle means less than random probability of occupation by amount δ. Each diagram is labeled with the appropriate symmetry designation (or "irreducible representation").

six $\langle 100 \rangle$ directions,* the "trigonal" defect, which lies along one of the eight $\langle 111 \rangle$ orientations, and the $\langle 110 \rangle$ "orthorhombic" defect, which possesses twelve $\langle 110 \rangle$ directions. The corresponding Haven–van Santen diagrams for each defect species taken alone are shown in Figs. 7–9. These diagrams show the relative probability of occupation of each of the n equivalent sites, with the following convention: An open circle means that the concentration of defects in that site is the average value C_0/n; an enlarged solid circle represents an increased concentration $(C_0/n) + \delta$; while a small solid circle represents a correspondingly decreased concentration $(C_0/n) - \delta$. The lettered designation below each diagram is the conventional name of the symmetry designation γ. Those having the letter T (e.g., T_{1u} and T_{2g}) are triply degenerate. It is easy to see in each such case that three equivalent diagrams can be drawn simply by interchanging the cubic axes. The corresponding relaxation times must, of course, be the same. The diagrams with designation E_g are doubly degenerate. (At first glance, it would seem that in these cases, there are also three ways of drawing equivalent diagrams, but it is easily shown that only two of the corresponding normal coordinates are linearly independent.) The modes with letter A are nondegenerate. In particular, A_{1g} is that mode in which all concentra-

* We regard negative orientations as distinct from the corresponding positive ones because of our interest here in unlike pairs, for which two orientations 180° apart are distinct from each other.

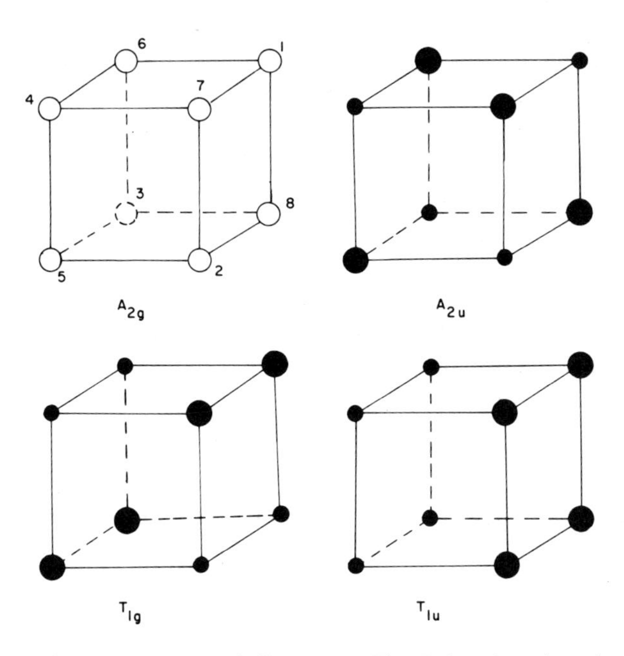

Fig. 8. Same type of diagram as Fig. 7, but for trigonal defect in cubic crystal.

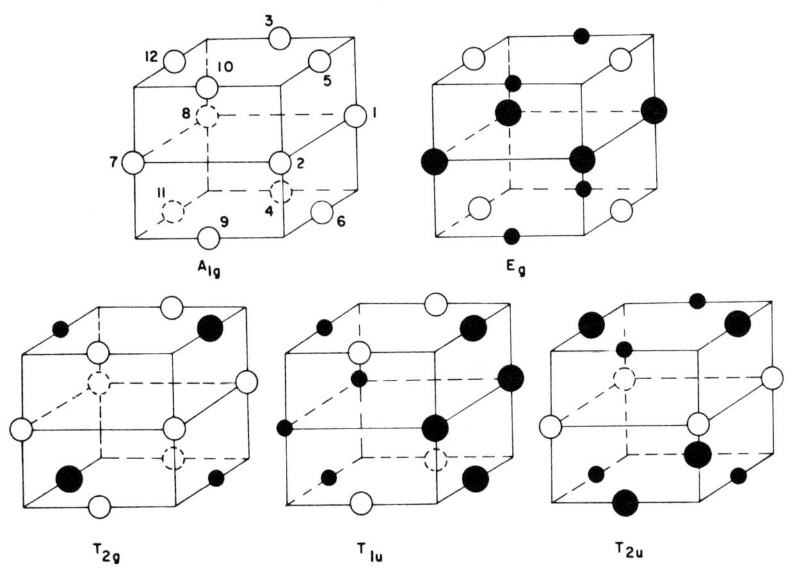

Fig. 9. Same type of diagram as Fig. 7, but for ⟨100⟩ orthorhombic defect in cubic crystal.

tions change by the same amount; however, in view of the conservation condition (45), such a change is not permissible. Accordingly, the A_{1g} mode, though formally counted as one mode, involves no change in concentrations at all, and is therefore called the "equilibrium mode." It should be noted that for these three relatively high symmetry defects, no symmetry designation appears more than once, so that the complexity of introducing an additional index to $C_s'^{(\gamma)}$ (as mentioned above) is unnecessary.

The relaxation rates as eigenvalues of Eq. (48) for those three cases are calculated in a straightforward way with the aid of group theory.[16] (They can also be obtained from the diagrams of Figs. 7–9 by simply setting up a kinetic equation for the decay of the distribution as shown.) The results are as follows (in terms of the numbering convention shown in each figure on the diagram for A_{1g}).

(a) *Tetragonal Defect—Cubic Crystal*

$$\tau^{-1}(E_g) = 6\nu_{12}, \qquad \tau^{-1}(T_{1u}) = 4\nu_{12} + 2\nu_{14} \tag{50}$$

(b) $\langle 110 \rangle$ *Orthorhombic Defect—Cubic Crystal*

$$\begin{aligned} \tau^{-1}(E_g) &= 6(\nu_{13} + \nu_{19}) \\ \tau^{-1}(T_{2g}) &= 4(\nu_{12} + \nu_{13} + \nu_{19}) \\ \tau^{-1}(T_{1u}) &= 2(\nu_{12} + \nu_{13} + \nu_{17} + 3\nu_{19}) \\ \tau^{-1}(T_{2u}) &= 2(\nu_{12} + 3\nu_{13} + \nu_{17} + \nu_{19}) \end{aligned} \tag{51}$$

(c) *Trigonal Defect—Cubic Crystal*

$$\begin{aligned} \tau^{-1}(T_{2g}) &= 4(\nu_{12} + \nu_{16}) \\ \tau^{-1}(A_{2u}) &= 2(\nu_{15} + 3\nu_{16}) \\ \tau^{-1}(T_{1u}) &= 4\nu_{12} + 2\nu_{15} + 2\nu_{16} \end{aligned} \tag{52}$$

In each of the above equations, the result is expressed in terms only of the *independent* values of ν_{uv}. For example, in all cases, $\nu_{uv} = \nu_{vu}$. In addition, for the tetragonal case, it can be seen from Fig. 7 that $\nu_{12} = \nu_{13} = \nu_{15} = \nu_{16}$. Similarly, for the orthorhombic defect (Fig. 9), $\nu_{12} = \nu_{18}$, $\nu_{13} = \nu_{14} = \nu_{15} = \nu_{16}$, and $\nu_{19} = \nu_{1\,10} = \nu_{1\,11} = \nu_{1\,12}$. For the trigonal defect (Fig. 8), $\nu_{12} = \nu_{13} = \nu_{14}$ and $\nu_{16} = \nu_{17} = \nu_{18}$.

4.3. The Selection Rules

Equation (49) expresses the normal coordinates as linear combinations of the concentrations which transform in particularly simple ways under the

Table I. Numbers of Relaxational Normal Modes for Various Defects in High-Symmetry Cubic (O_h) Crystals and Symmetry Coordinates of Field Components $F'^{(\gamma)}$ and Stress Components $\sigma'^{(\gamma)}$

γ	$(m)^a$	$F'^{(\gamma)}$	$\sigma'^{(\gamma)}$	Defect symmetry		
				Tetrag.	⟨110⟩ Ortho.	Trig.
A_{1g}	(1)	—	$\sigma_1 + \sigma_2 + \sigma_3$	1	1	1
E_g	(2)	—	$(2\sigma_1 - \sigma_2 - \sigma_3,\ \sigma_2 - \sigma_3)$	2	2	0
T_{2g}	(3)	—	$(\sigma_4, \sigma_5, \sigma_6)$	0	3	3
T_{1u}	(3)	(F_x, F_y, F_z)	—	3	3	3
A_{2u}	(1)	—	—	0	0	1
T_{2u}	(3)	—	—	0	3	0
	Mechanically active modes			3	6	4
	Electrically active modes			3	3	3
	Inactive modes			0	3	1
	Total modes			6	12	8

[a] Degeneracy.

symmetry operations of the crystal. The three components of electric field F_i and the six components of stress σ_k can also be expressed as linear combinations each of which belongs to one of the designations γ. These are called "symmetry coordinates" of electric field and stress and are denoted, respectively, by the notation $F'^{(\gamma)}$ and $\sigma'^{(\gamma)}$, with suitable subscripts for degenerate symmetry designations and for those that occur more than once. These symmetry coordinates for high-symmetry cubic crystals (i.e., those of O_h symmetry) are given in columns 3 and 4 of Table I. It should be noted that the three components of electric field all belong to the three-fold degenerate designation T_{1u}, while the six symmetry coordinates of stress belong to A_{1g}, E_g, and T_{2g}. In the case of axial crystals (trigonal, hexagonal, or tetragonal), however, the component of electric field F_z along the major crystal axis belongs to a different γ than the components in the basal plane. Table I also shows how many normal modes of each designation occur for each of the three defect types covered by Figs. 7–9. Just as the relaxational normal modes could be shown diagramatically, so too can the symmetry coordinates of electric field and stress. Thus, Fig. 10 gives a diagramatic representation of electric fields of type T_{1u} (a vector lying in one of the

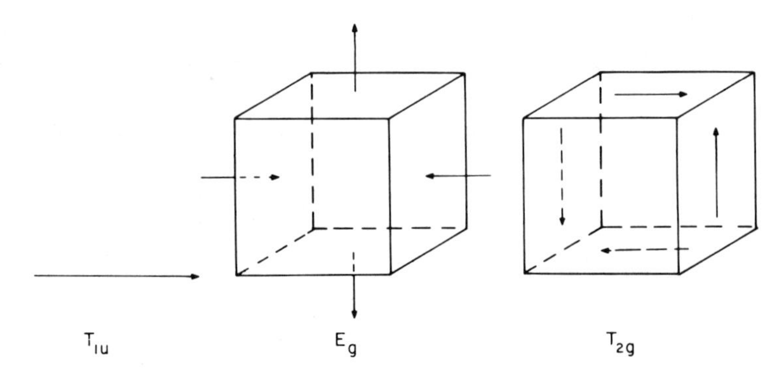

Fig. 10. Diagrams showing symmetry coordinates of electric field (T_{1u}) and of two different types of shear stresses (E_g and T_{2g}) for a cubic (O_h symmetry) crystal.

three cube directions), and the stress patterns, which are of the type designated by E_g and T_{2g}, respectively. By comparing these diagrams with Figs. 7–9, the reader can convince himself that the *diagram* of a symmetry coordinate of a field which belongs to a given designation γ has the same symmetry as that of a corresponding normal mode belonging to γ.

A basic theorem from group theory is that a given normal mode can only be excited by field components which belong to the same γ. This theorem has profound consequences, as follows.

1. Some normal modes can only be excited electrically, others only mechanically, and still others by neither type of field. These are called, respectively, "electrically active," "mechanically active," and "inactive" modes. (In the case of a piezoelectric crystal, there is also the possibility of having normal modes which are *both* mechanically and electrically active.)

2. Some fields do not excite any normal modes. This statement is the basis for the "selection rules." These rules tell which γ's do not show normal modes, and therefore, which field components do not give rise to relaxation, for the particular crystal and defect species. Since the absence of normal modes belonging to certain γ's is characteristic of the defect symmetry, the application of the selection rules permits one to obtain information about the symmetry of an unknown defect.

As examples of these statements, we note from Table I that, in a cubic crystal of class O_h, the E_g and T_{2g} modes are mechanically active, the T_{1u} modes are electrically active, while A_{2u} and T_{2u} modes are inactive. Also, the absence of T_{2g} modes characterizes a tetragonal defect, while the ab-

Table II. Symmetry Coordinates of Fields and the Numbers of Relaxational Normal Modes for Various Defects in Trigonal Crystals (D_3 symmetry)

γ	(m)	$F'^{(\gamma)}$	$\sigma'^{(\gamma)}$	Defect symmetry		
				Trig.	Monocl.	Tricl.
A_1	(1)	—	$\left\{ \begin{matrix} \sigma_3 \\ \sigma_1 + \sigma_2 \end{matrix} \right\}$	1	1	1
A_2	(1)	F_z	—	1	0	1
E	(2)	(F_x, F_y)	$\left\{ \begin{matrix} (\sigma_4, \sigma_5) \\ (\sigma_1 - \sigma_2, \sigma_6) \end{matrix} \right\}$	0	2	4
Total modes				2	3	6

sence of E_g modes characterizes the trigonal defect in such a cubic crystal. Since all defects considered in Table I are dipolar, the electrically active mode T_{1u} always appears; if the defect were not dipolar (for example, a pair consisting of two identical defects), there would be no T_{1u} modes. For crystals of lower than cubic symmetry, the electrically active modes belong to more than one γ; therefore, there is a greater variety of possibilities. Table II illustrates this by classifying the relaxational normal modes and the field components for the case of a trigonal crystal of D_3 symmetry, to which the crystal α-quartz belongs. Note here that the E modes are both electrically and mechanically active. The field components belonging to this designation are F_x and F_y, as well as the four pure shear components of stress. It is these electric and stress components which interact with each other to produce piezoelectric behavior. A full listing of the selection rules for all crystal and defect symmetries has been given elsewhere.[15]

It is important to note that is not usually necessary to apply only a pure symmetrized stress $\sigma'^{(\gamma)}$ in order to verify the selection rules. Such combinations, which are usually pure shear stresses, are often difficult to apply in an actual experiment. Instead, uniaxial tension or compression is easier to work with. In a cubic crystal, for example, a uniaxial stress $\sigma\langle 100\rangle$ along a cube axis may be resolved into a combination of hydrostatic (A_{1g} type) and shear of E_g type. Similarly, uniaxial stress along $\langle 111\rangle$ resolves itself into A_{1g} plus shear of T_{2g} type. Symbolically, we may write

$$\sigma\langle 100\rangle \rightarrow \sigma'^{(A_{1g})} + \sigma'^{(E_g)} \tag{53}$$

$$\sigma\langle 111\rangle \rightarrow \sigma'^{(A_{1g})} + \sigma'^{(T_{2g})} \tag{54}$$

We may, therefore, distinguish tetragonal defects by the absence of relaxation under $\sigma\langle 111\rangle$ and trigonal defects by the absence of relaxation under $\sigma\langle 100\rangle$. These results may be seen simply enough by the fact that a stress along $\langle 111\rangle$ is equally inclined toward all six $\langle 100\rangle$ orientations of a tetragonal defect, and, therefore, cannot split their free energy levels, and similarly for $\sigma\langle 100\rangle$ in relation to trigonal defects.

When more than one defect species capable of "reaction" [i.e., of interconversion—see Eq. (2), for example] are present simultaneously, the number of normal modes for each γ is the sum of those for the separate species. The normal modes themselves, however, are now, in general, linear combinations of the modes of the individual defects which belong to the same γ. Similarly, while the number of relaxation times is additive, the specific expressions for τ^{-1} are more complex than those given in Section 4.2, since they now also involve jump rates of the type $\nu_{p(\alpha)r(\beta)}$ between one orientation of species α and another of species β. There is even an important change, however, with respect to the number of active relaxational modes, specifically those belonging to A_{1g}. The reader should recall that for a single defect species, the A_{1g} relaxation was excluded on the grounds of the auxiliary condition that the total concentration C_0 of defects must be conserved, Eq. (45). In the case of a reaction between species α and β, however, only the total concentration of α and β defects together is conserved. Thus, a relaxational mode of type A_{1g} is now possible, in which there is a net change in concentration of the α defects at the expense of the β defects, but the concentrations in the various orientations of each species remain equal, as required for the A_{1g} mode. Such a normal mode is called a "reaction mode."

Although general expressions for the relaxation rates have been obtained for the case of reacting defects,[17] the equations become rather complicated. In this chapter, we will therefore quote only results when we come to specific defect reactions of interest.

4.4. Relaxation Magnitudes

In addition to the calculation of the kinetics of relaxation in the form of relaxation rates τ^{-1}, application of thermodynamic concepts permits the working out of expressions for the relaxation magnitudes $\delta\chi$ and δs [see Eq. (25)] in terms of defect properties.[15] The form of the expression for the relaxation of the electric susceptibility $\delta\chi$ is

$$\delta\chi = \alpha C_0 \mu^2 / v_0 kT \tag{55}$$

where α is a geometric factor, C_0 is the total defect concentration, v_0 is the molecular volume, and μ is an appropriate electric dipole moment associated with the defect in question. For a cubic crystal, $\alpha = 1/3$.

The corresponding expression for the relaxation of the elastic compliance δs is of the form

$$\delta s = \beta v_0 C_0 (\Delta\lambda)^2/kT \tag{56}$$

where β is a numerical constant and $\Delta\lambda$ is an appropriate difference among components of the "elastic dipole tensor" which characterizes the strain about a defect.[19]

5. DIELECTRIC AND ANELASTIC RELAXATION: METHODS AND RESULTS

5.1. Electrical Methods

5.1.1. *AC Dielectric Loss*

In this method, one obtains the real and imaginary parts of the electric susceptibility (χ_1 and χ_2) or of the dielectric constant $(1 + 4\pi\chi)$ under an alternating electric field. Actually, the loss tangent, $\tan\delta$, given by Eq. (29) is usually measured directly with a conventional ac capacitance bridge, in which the equivalent circuit of the sample is a lossy condenser (i.e., one with a parallel resistance). The real part of the susceptibility, $\chi_1(\omega)$, is usually also obtained. Since $\delta\chi$ is usually small, the quantity $\tan\delta$ has the same dependence on $\omega\tau$ as χ_2. For example, in the case of a single relaxation process, $\tan\delta \propto \omega\tau/(1 + \omega^2\tau^2)$, i.e., it shows a Debye peak of the type given by Eq. (31) and plotted in Fig. 5. In the earlier discussion, it was implied that such a peak is traced out by varying ω while keeping τ constant. Often, it is not convenient to vary the frequency continuously over approximately two decades, as required to trace out a Debye peak. In practice, such a peak is most often studied not by changing ω, but by varying τ while keeping ω constant. The basis for this approach is the fact that τ^{-1} usually obeys an Arrhenius equation of the form (34). This makes it possible to vary τ over a wide range simply by changing the temperature. From Eq. (34), we may write

$$\ln(\omega\tau) = \ln(\omega\tau_0) + (E/k)(1/T) \tag{57}$$

which shows that there is a linear relation between $\ln(\omega\tau)$ and T^{-1}. Thus, the curves of χ_1, χ_2, or $\tan\delta$ plotted versus T^{-1} take on the same form as those plotted versus $\ln(\omega\tau)$ (as, for example, in Fig. 5), the only difference being a scale factor of E/k. The condition that $\omega\tau = 1$ at the maximum of a Debye peak now becomes

$$\ln(\omega\tau_0) + (E/k)(1/T_m) = 0 \tag{58}$$

where T_m is the temperature at the maximum. One may therefore obtain a series of peaks by choosing two or more different frequencies. Plotting $\ln \omega$ versus $1/T_m$ then gives a straight line whose slope is E/k and whose intercept gives τ_0.

Combining Eqs. (32) and (57), we obtain for the width at half-maximum of a Debye peak when plotted versus T^{-1}

$$\Delta(T^{-1}) = (1.144)(2.303)k/E = 2.635k/E \tag{59}$$

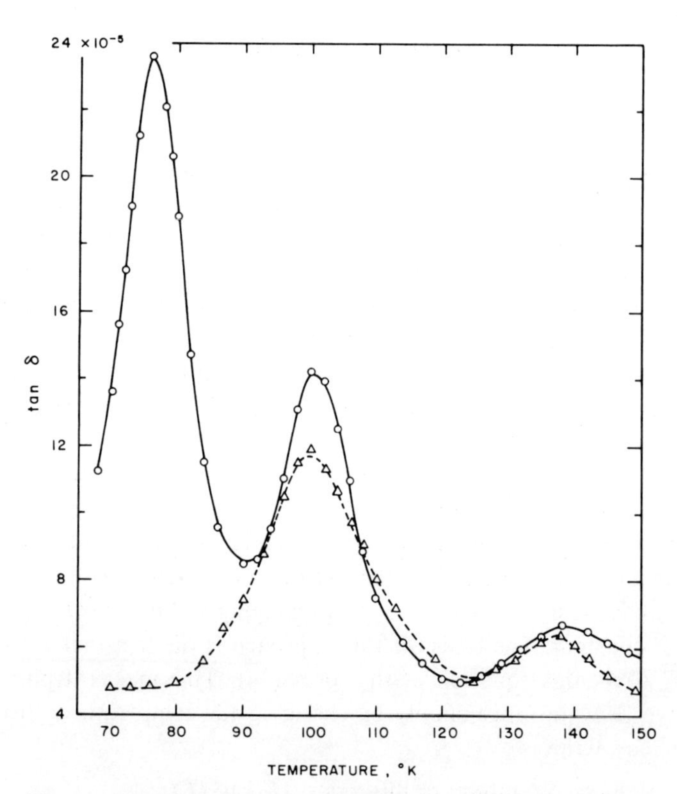

Fig. 11. Dielectric loss of a synthetic crystal of α-quartz as a function of temperature at 2 kHz. Solid curve is from measurements with electric field parallel to the z axis; dashed curve for field perpendicular to the z axis. (From Park[20]).

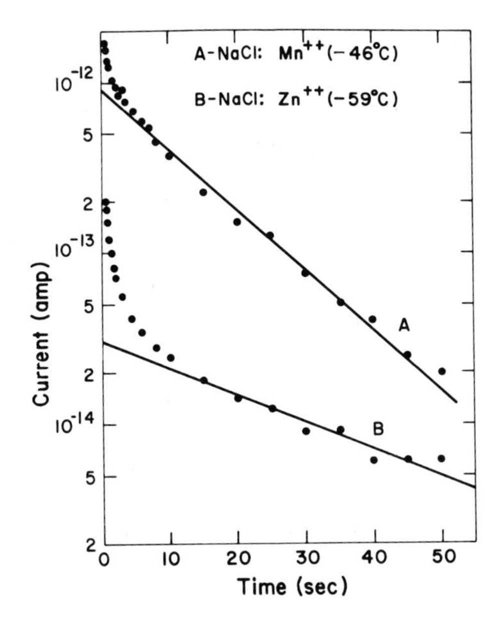

Fig. 12. The polarization discharge current as a function of time for NaCl:$MnCl_2$ and NaCl:$ZnCl_2$ showing exponential behavior at relatively long times (from Dreyfus[21]).

Figure 11 provides an illustration of dielectric loss data as a function of temperature for the case of α-quartz containing suitable impurities. The data (which are plotted here versus T rather than T^{-1}) show the presence of three distinct Debye peaks. The origin of such peaks in α-quartz will be discussed later (see Section 5.4.5).

The detection limit of the ac dielectric loss method is typically $\sim 10^{-5}$ dipoles, expressed as a mole fraction.

5.1.2. *Time-Dependent Polarization*

This is essentially a dc method, based on Eq. (24), in which an abrupt change in the electric field is made at $t = 0$, and a transient current density ($i = dP/dt$) is observed immediately thereafter. By integrating the transient current, one obtains the polarization $P(t)$. Measurements may be made upon application of a constant field, giving a charging current, or upon removal of the field to produce a discharging current or "aftereffect." The transient current, which may be measured with a vacuum tube electrometer, can usually be analyzed into one or more exponentials of the form $\exp(-t/\tau)$. Figure 12 provides an example of a discharging curve, showing exponential behavior at relatively long times. The residual early part of the curve can also be fitted to an exponential.

The sensitivity of the equipment is such that a mole fraction close to 10^{-7} dipoles can be detected by this method.

5.1.3. *Ionic Thermocurrent* (*ITC*)

This method, first used by Bucci and co-workers,[22,23] is also a dc method, but it makes use of the depolarizing current released by slowly warming up a crystal from a relatively low temperature T_0 (at which τ is very long). The steps are as follows:

1. Dipoles are preferentially oriented at a "polarizing temperature" T_p by applying an electric field F for a time much greater than $\tau(T_p)$.
2. Keeping the field on, the sample is cooled to a temperature T_0, at which $\tau(T_0)$ is several hours or longer. The dipoles now remain oriented as they were at T_p (actually, at a temperature somewhat below T_p due to the finite cooling rate).
3. The field F is removed and the electrodes across the sample are now switched to a current detector, which consists of a resistor, the potential across which is measured with a vibrating reed electrometer.
4. The sample is warmed at a *constant rate* $b = dT/dt$, and the current density $i(T)$ is observed as a function of temperature in the range where the dipoles return to a random orientation. For a single relaxation process, $i(T)$ first increases exponentially, then goes through a maximum, and finally drops to zero.

Figure 13 illustrates the type of data that can be obtained in an ITC experiment. At the start of an ITC peak, the current takes the form

$$i(T) = \text{const} \times \exp(-E/kT) \tag{60}$$

from which the activation energy E is most readily obtained. The maximum occurs at a temperature T_m given by

$$T_m = [bE\tau(T_m)/k]^{1/2} \tag{61}$$

from which τ_0 is obtained once E has been determined. (The heating rate b is, of course, known experimentally.) Finally, the area under a peak directly gives the magnitude of the relaxation, i.e.,

$$\delta\chi = (1/F) \int_{\text{peak}} i(T)\, dT \tag{62}$$

In accordance with Eq. (55), this area is then proportional to the number

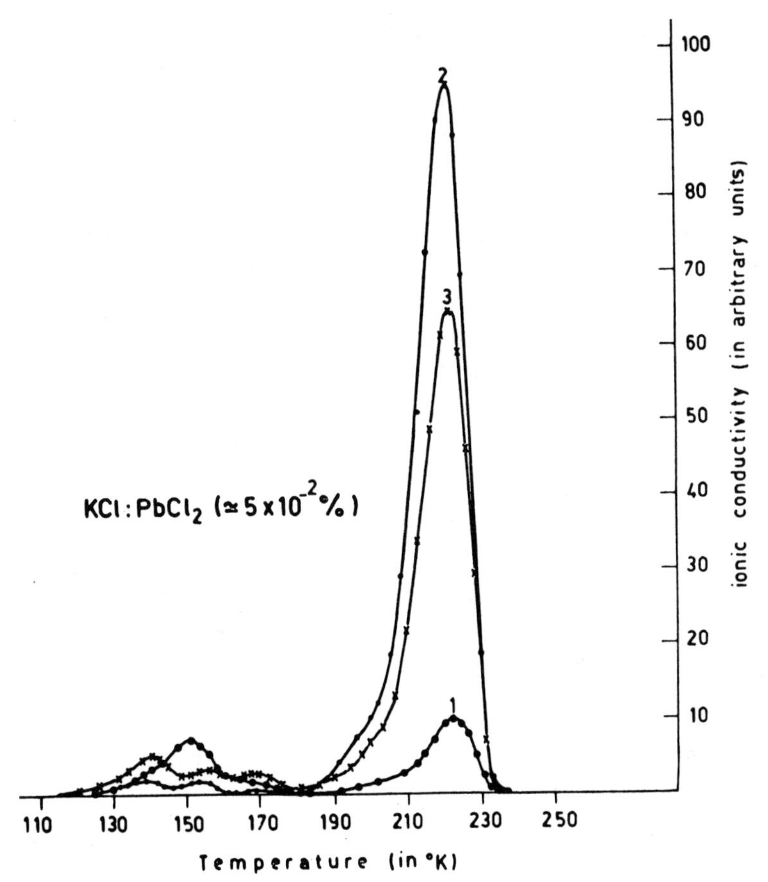

Fig. 13. Some examples of ITC curves for KCl:$PbCl_2$ which had been given three different aging treatments prior to the polarization treatment (from Capelletti and Fieschi[24]).

of dipoles C_0. The ITC method is remarkably sensitive, approaching a detection limit of $\sim 10^{-7}$ mole fraction of dipoles.

When two relaxation times τ_1 and τ_2 are so close that the corresponding ITC peaks overlap, the method can be modified so as to resolve them. To do this, one polarizes at a temperature T_p which falls between the two maxima T_{m1} and T_{m2} for a time $t_p \sim \tau_1(T_p) \ll \tau_2(T_p)$. The dipoles of the first kind are then preferentially oriented while those of the second kind are almost random. The subsequent ITC curve then shows only the lower-temperature peak. The other peak can be obtained by polarizing at a higher temperature, then discharging only the lower peak and dropping the temperature again. On the second reheating, the upper peak will be obtained almost alone.

5.2. Internal Friction

This method is the mechanical analog of the ac dielectric loss treated in Section 5.1.1. Here, one sets a sample into mechanical vibration (in a longitudinal, a flexural, or a torsional mode) at one of its resonant frequencies. The sample may be driven by applying an alternating electrostatic or magnetic field, or by attaching to it a piezoelectric transducer or some other mechanical drive. The sample vibrations can be detected by similar methods, either while the driving force is being applied (i.e., in forced vibration), or in free vibration after the driving force is removed. Such measurements give the resonant frequency ω_r of the sample and the loss tangent $\tan \delta$.* The latter may be obtained from the width at half-maximum, $\Delta\omega$, of the resonance peak in forced vibration,

$$Q^{-1} \equiv \Delta\omega/\sqrt{3}\omega_r \tag{63}$$

(called Q^{-1} by analogy to electrical resonant circuits), or from the logarithmic decrement, log dec, in free decay. For small $\tan \delta$, the relations connecting these different measures of internal friction are

$$\tan \delta = Q^{-1} = (\log \text{dec})/\pi \tag{64}$$

The quantity ω_r itself may be related to the real part s_1 of the compliance by a relation of the form

$$\omega_r^2 = \gamma/s_1 \tag{65}$$

where γ is a factor containing both the geometric and inertial features of the mechanical system.[14] Since, by analogy with Eq. (29), $\tan \delta = s_2/s_1$, a knowledge of both ω_r and $\tan \delta$ gives both the real and imaginary parts of the complex compliance. As in the dielectric case, $\tan \delta$ obeys a Debye peak, or a sum of Debye peaks, which may be traced out by varying the temperature rather than the frequency. Equation (59) thus applies as well to internal friction as to dielectric loss.

Typical internal friction measurements are shown in Fig. 14 for NaCl doped with $CaCl_2$. The significance of these results will be discussed in Section 5.4.3.

* In addition to resonance methods, high-frequency (MHz) ultrasonic wave propagation methods may also be used. In this case, the velocity v and attenuation α of the wave are measured. These are related to s_1 and $\tan \delta$ by the relations $v^2 \propto s_1^{-1}$ and $\alpha = (\omega/2v) \tan \delta$.

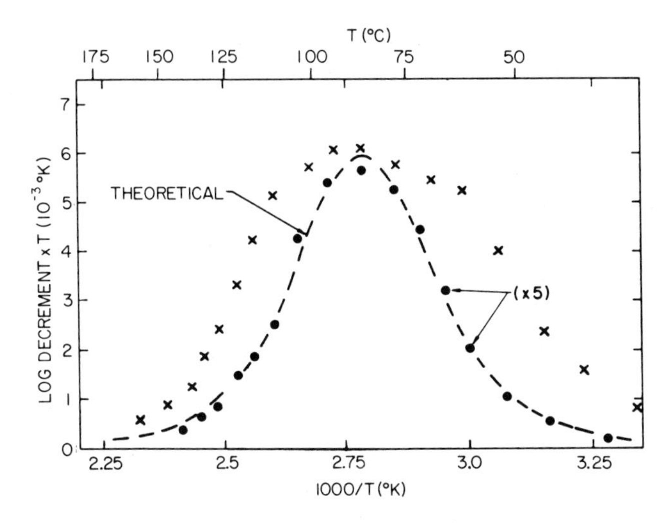

Fig. 14. Internal friction versus reciprocal temperature for NaCl: 200 ppm $CaCl_2$ crystals oriented in the ⟨100⟩ and ⟨111⟩ directions. The dashed curve represents a Debye peak for an activation energy of 0.70 eV. ×, ⟨100⟩ at 6.6 kHz; •, ⟨111⟩ at 10 kHz. (From Dreyfus and Laibowitz.[25])

5.3. Defect Aggregation Studies

Until now, we have shown how kinetic data could be obtained from the study of relaxation processes via the relaxation time τ. In a number of investigations, however, use has been made of the fact that the magnitude of a dielectric loss peak is proportional to the defect concentration, i.e., $\tan \delta_{max} \propto \delta\chi \propto C_0$ [Eq. (55)]. If the defects undergo aggregation to form higher clusters which do not contribute to the peak in question, then the relative decrease of concentration with time $C_0(t)$ can be obtained merely by observing the peak height as a function of time. The same method may, of course, be applied in the case of internal friction; thus far, however, such internal friction studies have only been carried out on metallic systems. Actually, this method is not really a new one, in that the same quantities are measured as in the usual ac relaxation methods, but it has turned out to be important enough to treat separately. In addition, when the defect involves a suitable ion (notably Mn^{2+}) which gives an EPR (electron paramagnetic resonance) signal, the aggregation process can also be observed by EPR as well as by relaxation measurements.

Defect aggregation is actually a long-range migrational process, and may therefore be controlled by a different ion jump than the reorientation process. For example, reorientation of a vacancy–impurity pair in an AX crystal often involves vacancy motion to another site adjacent to the impurity, i.e., a vacancy–A-ion interchange. This constitutes what may be considered a *rotation* of the vacancy about the impurity. On the other hand, for migration of the defect to take place, *both* rotation and impurity–vacancy interchange must occur. Since the impurity–vacancy interchange is usually the slower process, the activation energy for aggregation may, in general, be anticipated to be higher than that for relaxation.

5.4. Examples of Defect Relaxations

As was mentioned earlier, the best-studied examples of defect relaxations involve cubic crystals. We will illustrate with the examples of defects in crystals which possess the CaF_2 and NaCl structures. Finally, to demonstrate a noncubic case, we will consider the example of a defect in α-quartz.

5.4.1. CaF_2:NaF *and Analogs*

When NaF is introduced into CaF_2, the lower-valent Na^+ ion substitutes for a Ca^{2+} and is compensated by a fluorine vacancy V_F to form a Na^+–V_F pair. Since the next-nearest-neighbor (nnn) separation is very much greater than the nearest-neighbor (nn) distance, we are led to expect primarily an nn species of defect. This nn pair has trigonal symmetry, in which the V_F occupies one of eight positions about the Na^+ as shown in Fig. 15. (This figure shows only one-fourth of the unit cell of the fluorite structure, since the body-centered position is unfilled in alternate cubes of F^- ions.) These eight positions are numbered in accordance with the same numbering convention shown in Fig. 8.

A completely analogous situation exists for ThO_2:CaO and also for CeO_2:CaO. Both the thorium and cerium dioxides have the fluorite structure, but here, all valences are double those of the CaF_2:NaF case (Th^{4+}, Ca^{2+}, O^{2-}). Thus, the same type of defect pair is formed, namely Ca^{2+}–V_O. These oxides were, in fact, studied before the CaF_2 system, but only in polycrystalline form, due to the difficulty in growing large crystals of these refractory materials.

Because of the trigonal symmetry of the defect, we see from the selection rules (Table I) that anelastic relaxation of the T_{2g} type may be expected as well as dielectric relaxation (T_{1u} modes), but that the anelastic E_g relaxation is absent. From Eqs. (53) and (54), this implies that anelastic relaxation will be observed for a $\langle 111\rangle$ stress axis but not for a $\langle 100\rangle$ stress axis.

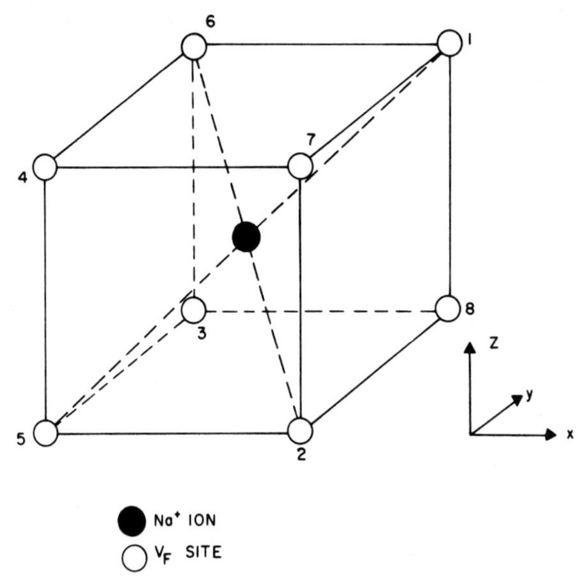

Fig. 15. Part of the unit cell of the CaF_2 lattice showing the site of a Na^+ impurity and the eight surrounding sites on which a charge-compensating fluorine vacancy may reside.

This prediction was verified for single crystals of CaF_2:NaF.[26] In addition, we have expressions for the relaxation rates of the various relaxational modes for a trigonal defect in a cubic crystal, given by Eqs. (52). It is more useful, however, to convert the independent reorientation frequencies ν_{12}, ν_{15}, and ν_{16} into specific ion jump frequencies. Examination of Fig. 15 shows that ν_{16} is equal to the nearest-neighbor jump frequency of the vacancy, while ν_{12} and ν_{15} are, respectively, the second and third nearest-neighbor jump frequencies. Usually, the latter two frequencies will be negligible relative to the first one. Accordingly, if we denote by w the nearest-neighbor jump frequency between two specific sites, we have $\nu_{16} = w \gg \nu_{15}, \nu_{12}$ so that Eqs. (52) become

$$\begin{aligned} \tau^{-1}(T_{2g}) &= 4w \qquad \text{(anelastic, } \sigma\langle 111\rangle) \\ \tau^{-1}(T_{1u}) &= 2w \qquad \text{(dielectric)} \end{aligned} \tag{66}$$

for the mechanically and electrically active modes, while the relaxation frequency of the inactive mode is $\tau^{-1}(A_{2u}) = 6w$. This result, which may be expressed as $\tau_{\text{diel}}/\tau_{\text{anel}} = 2$, was first verified for the case of ThO_2:CaO by Wachtman,[27] later for CeO_2:CaO by Lay and Whitmore,[28] and more recently, for CaF_2:NaF by Johnson *et al.*[26] (see Table III). These results,

Table III. Relaxation Data on Crystals of Fluorite Structure Doped with Lower-Valent Cation Impurities

Material	$\tau_{\text{diel}}/\tau_{\text{anel}}$	E, eV	Ref.
ThO_2:CaO	2 ± 0.2	0.95	27
CeO_2:CaO	2 ± 0.2	0.86	28
CaF_2:NaF	2.2	0.53	26

taken together with the verification of the selection rules (absence of a $\sigma\langle 100\rangle$ relaxation) by the latter authors, indicate that the model which considers only nn pairs and $w = \nu_{16}$ as the dominant jump frequency completely accounts for the relaxation behavior. The activation energies obtained from the relaxation experiments are also listed in Table III, representing the jump which takes an anion vacancy from one nn position of the aliovalent impurity to another. Since this jump bears no relation to that involved in the migration of the impurity ion, it is clear that such relaxation measurements do not give any direct information on the activation energy for impurity diffusion. In general, we can expect that the latter process will involve a considerably higher activation energy. On the other hand, the activation energies in Table III should not be too different from the corresponding values for the migration of free anion vacancies in the same crystals obtained from conductivity studies (see the chapter by Fuller). In the case of CaF_2, for example, the activation energy for migration of a free fluorine vacancy is 0.55 eV,[29] which is very close to the value of 0.53 eV from the relaxation measurements.

In the CaF_2:NaF system, studies were also made of the effect of increasing the NaF concentration. It is found that the relaxation peaks become higher, but also that they become broader than a Debye peak, undoubtedly due to dipole–dipole interaction effects. A detailed study of these interaction effects by ITC methods was carried out by Shelley and Miller.[30]

5.4.2. CaF_2:MF_3

When a trivalent ion like Y^{3+} replaces Ca^{2+} in CaF_2, charge compensation is accomplished by the introduction of an interstitial F^-, as shown by EPR and ENDOR as well as by density measurements.[31,31a,32] The model for the associated defect pair is shown in Fig. 16, in which both nn and nnn positions are included. The nn defect is tetragonal, while the nnn defect,

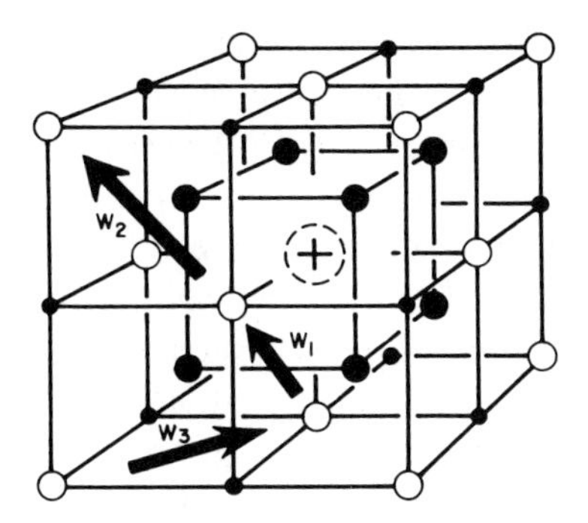

Fig. 16. The unit cell of CaF_2 with a Y^{3+} ion replacing a Ca^{2+} at the center (net + charge). The small, filled circles are Ca^{2+} ions, and the large ones are F^- ions. Sites for charge compensating F^- interstitials are shown as open circles. The three jumps of the interstitial, having frequencies w_1, w_2, and w_3, are shown. (From Southgate.[33])

having a separation which is $\sqrt{3}$ times greater, has trigonal symmetry. It we consider only the nn defects, then from Table I, only E_g and T_{1u} relaxational modes are obtained. The relaxation rates would be controlled by the nn → nn jump frequency w_1 of the F^- interstitial, as shown in Fig. 16. Actually, in anelastic experiments on CaF_2:YF_3 shown in Fig. 17,

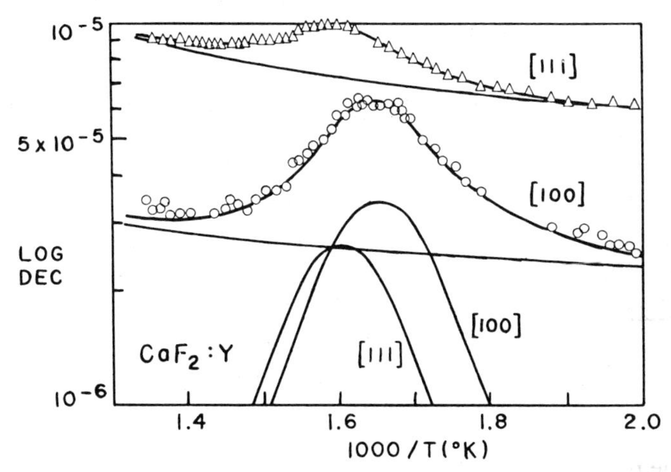

Fig. 17. Internal friction data (at 38 kHz) from single crystals of CaF_2:YF_3 oriented in the ⟨100⟩ and ⟨111⟩ directions. The two lower curves show the peaks with the background removed. (From Southgate.[33])

Southgate[33] has found both $\sigma\langle 111\rangle$ and $\sigma\langle 100\rangle$ relaxations, which led him to conclude that defects are present in both nn and nnn configurations. (This conclusion is also supported by EPR measurements, which show that an appreciable concentration of nnn pairs is present.) With both the tetragonal nn defects and trigonal nnn defects present, we expect (from Table I) to obtain one set of each of the mechanically active E_g and T_{2g} modes as well as two sets of dielectric T_{1u} modes. In addition, there will be a reaction mode of type A_{1g}, which is mechanically active.* In terms of the jump frequencies w_1, w_2, and w_3 defined in Fig. 16 (w_2 being the frequency for nn $\to$ nnn and w_3 for nnn $\to$ nn), the relaxation rates turn out to be

$$\tau^{-1}(E_g) = 6w_1 + 4w_2 \tag{67}$$

$$\tau^{-1}(T_{2g}) = 3w_3 \tag{68}$$

$$\tau^{-1}(A_{1g}) = 4w_2 + 3w_3 \tag{69}$$

$$\tau^{-1}(T_{1u}) = 2w_1 + 2w_2 + \tfrac{3}{2}w_3 \pm [(2w_1 + 2w_2 - \tfrac{3}{2}w_3)^2 + 4w_2w_3]^{1/2} \tag{70}$$

In order to simplify the problem, Southgate makes the assumption that $w_1 = w_2 = w_3$. Then, $\tau^{-1}(E_g)/\tau^{-1}(T_{2g}) = 10/3$, which accounts for the fact that the E_g peak falls at a slightly lower temperature than the T_{2g} peak, as shown in Fig. 17. However, this shift can also be due to small differences in the activation energies of the controlling jump process. Since, from the Arrhenius equation, a 10% difference in activation energy corresponds to an order-of-magnitude change in the corresponding jump frequency w, there seems to be enough reason to question the validity of Southgate's assumption.

Dielectric measurements on this system have been reported, but their interpretation is still unclear. Chen and McDonough[34] found a simple Debye loss peak whose activation energy (1.17 eV) is very close to that obtained by Southgate, but the preexponential of $\tau_0^{-1} = 3.3\times 10^{13}$ sec^{-1} is more than a decade smaller than the corresponding anelastic values. They attribute the peak to Y^{3+}–interstitial-F^- dipoles. This interpretation, however, is seriously challenged by Barsis and Taylor,[35] who point out that the magnitude of this dielectric peak is much too large to permit it to be attributable to the aliovalent Y^{3+} ion. In the past two years, evidence has rapidly accumulated,[35a] primarily through ITC measurements, that CaF_2 doped with

* It is probable that the magnitude of this relaxation will be too small to be observable in an experiment which employs uniaxial stress, since then the major components are shear stresses.

Er^{3+}, Gd^{3+}, Ce^{3+}, and Y^{3+} shows two low-temperature dielectric relaxation peaks. The main relaxation falls in the range of activation energy 0.38–0.46 eV, and a smaller relaxation at still lower temperatures gives 0.15–0.18 eV. These are attributed to the presence of both nn and nnn impurity–F^- interstitial pairs, in accordance with EPR and ENDOR observations. The relaxation rates should then be interpreted in terms of Eq. (70), and the wide split between the two observed peaks suggests that $w_3 \gg w_1, w_2$. The major problem, however, is that these results cannot be reconciled with Southgate's much higher activation energies. Clearly, it would be very desirable to conduct anelastic measurements on CaF_2:MF_3 crystals at low temperatures, to see if relaxations consistent with these dielectric measurements and corresponding to Eqs. (67)–(69) can be observed.*

5.4.3. NaCl:$M$$Cl_2$

Here, M^{2+} is a divalent cation, e.g., Ca^{2+}, Mn^{2+}, or Zn^{2+}. As already mentioned, charge compensation takes place by the introduction of cation vacancies to form pairs of the type M^{2+}–V_{Na}. These defects and the corresponding vacancy jumps are shown in Fig. 1. The nn pair constitutes a $\langle 110 \rangle$ orthorhombic defect. As in the case of CaF_2:YF_3, we will see that here, too, the anelastic results cannot be explained by nn pairs alone. Further support comes from theoretical calculations[1,2] which show that the nnn binding energy is not much different from the nn binding energy, both being ~0.4 eV. The symmetry of the nnn defect is tetragonal, so that from Table I, if both nn and nnn are present together, the relaxational modes include the following sets:

$$1\,A_{1g}, \quad 2\,E_g, \quad 1\,T_{2g}, \quad \text{and} \quad 2\,T_{1u}$$

as well as inactive modes. The A_{1g} mode is the reaction mode involving interconversion between nn and nnn species. Dreyfus and Laibowitz[25] and also Franklin[36] first worked out expressions for the relaxation rates in terms of the four jump frequencies w_1 through w_4 defined in Fig. 1. Specifically, w_1 is the jump frequency for a vacancy interchange with a Na^+ ion that takes nn $\leftrightarrow$ nn, w_3 is the same for nnn $\rightarrow$ nn, and w_4 is that for nn $\rightarrow$ nnn. Finally, w_2 is the jump frequency for the interchange of the impurity and the vacancy, causing 180° rotation of the dipole. The results obtained for the various types of relaxational modes which are mechanically

* A preliminary investigation of this type has been reported very recently by Franklin and Crissman.[35b]

and electrically active are

$$\tau^{-1}(A_{1g}) = 4w_3 + 2w_4 \tag{71}$$

$$\tau^{-1}(T_{2g}) = 4w_1 + 2w_4 \tag{72}$$

$$\tau_{\pm}^{-1}(E_g) = (3w_1 + 2w_3 + w_4) \pm [(3w_1 - 2w_3 + w_4)^2 + 2w_3w_4]^{1/2} \tag{73}$$

$$\tau_{\pm}^{-1}(T_{1u}) = (w_0 + 2w_3 + w_4) \pm [(w_0 - 2w_3 + w_4)^2 + 4w_3w_4]^{1/2} \tag{74}$$

where

$$w_0 \equiv w_1 + w_2 \tag{75}$$

Note that the frequencies of the two sets of E_g modes are given by Eq. (73) with the plus and minus signs, respectively, and similarly for $\tau^{-1}(T_{1u})$ in (74). Also note that only the dielectric relaxational modes, of the T_{1u} type, involve the jump frequency w_2. This is because stress is a centrosymmetric quantity and therefore cannot differentiate between defect orientations which differ by 180°.

While dielectric relaxation on NaCl:MCl$_2$-type materials has been studied by numerous workers over the years (see Dryden and Meakins[37] for an early review), an experimental study of anelastic relaxation in this type of material was conducted only more recently by Dreyfus and Laibowitz,[25] particularly for the cases of $CaCl_2$- and $MnCl_2$-doped NaCl. The data given in Fig. 14 show that under $\sigma\langle 100\rangle$, a multiple peak is obtained which can, in fact, be analyzed into three Debye peaks, as in Fig. 18. This

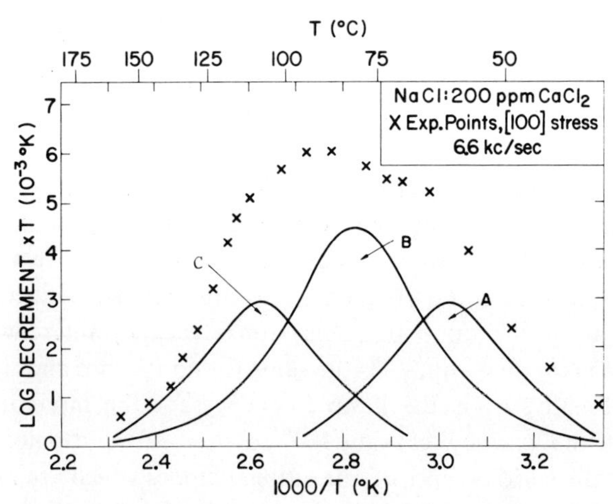

Fig. 18. Decomposition of the data for the $\langle 100\rangle$ sample in Fig. 14 into three Debye peaks (from Dreyfus and Laibowitz[25]).

Table IV. Numerical Values for Jump Rates and Activation Energies Obtained for NaCl:Ca^{2+} and NaCl:Mn^{2+} by Dreyfus and Laibowitz[25]

w_i	Value of w_i at 87°C (units of 10^3 sec^{-1})		Activation energy, eV	
	Ca^{2+}	Mn^{2+}	Ca^{2+}	Mn^{2+}
w_1	2	6	0.76	>0.71
w_3	114	79	0.63	0.65
w_4	27	41	0.68	0.67

result shows that nn defects alone cannot give rise to the anelastic relaxation. Rather, it is necessary to introduce both nn and nnn defects, leading to the relaxation rates given by Eqs. (71)–(74). On the other hand, the data for $\sigma\langle 111\rangle$ follow a single Debye peak. This is consistent with the above equations, provided that the reaction mode (A_{1g}) relaxation is too weak to be detected, a result which is often found in uniaxial stress experiments. Thus, the $\sigma\langle 111\rangle$ peak in Fig. 14 is the T_{2g} relaxation, while two of the three $\sigma\langle 100\rangle$ peaks are the two E_g relaxations, and the third (peak C of Fig. 18) is unaccounted for. Presumably, it is due to a higher-order complex.

From the three identified anelastic peaks (A and B for $\sigma\langle 100\rangle$ and the single peak for $\sigma\langle 111\rangle$), Dreyfus and Laibowitz have solved Eqs. (72) and (73) for the three unknown jump frequencies w_1, w_3, and w_4 and obtained numerical values for these frequencies. Table IV shows the values obtained for both Ca^{2+} and Mn^{2+} doping, at a single temperature, as well as the calculated activation energies. These results for both dopants led to the unexpected conclusion that $w_3 > w_4 > w_1$. This means that reorientation among nearest neighbors is more rapidly accomplished by passing through the nnn site and back again than via a direct jump of the w_1 type.

Turning now to dielectric measurements on the same materials, Dreyfus[21] has observed the two T_{1u} relaxations predicted by Eq. (74), especially for M^{2+} ions with smaller atomic radii, e.g., Mn^{2+}, Zn^{2+}, and Mg^{2+}. (See, for example, Fig. 12.)* The dielectric relaxation rates involve not only

* More recently, Burton and Dryden,[52] using ac dielectric loss measurements, resolved the data for NaCl:Mn^{2+} into two peaks. Although apparently unaware of the anelastic studies, they arrived at essentially the same conclusion as did Dreyfus and Laibowitz (see Table IV), namely that $w_3 > w_1$ with an activation energy difference of about 0.05 eV.

Table V. Comparison of Activation Energies for Dielectric Relaxation E_R, for Dipole Aggregation E_{AG}, and for Impurity Diffusion in the Saturation Limit E_s for Divalent Impurities of Different Radii in NaCl[a]

Impurity, M^{2+}	$r_{M^{2+}}/r_{Na^+}$	E_R, eV	E_{AG}, eV	E_s, eV
Zn^{2+}	0.78	0.66	—	0.52[c]
Co^{2+}	0.82	0.72[b]	0.61	(1.1)[d]
Mn^{2+}	0.84	0.68	0.78	0.95
Cd^{2+}	1.02	0.67	0.50	0.86[f]
Ca^{2+}	1.04	0.70	0.92	0.96
Pb^{2+}	1.26	0.69[e]	0.80	0.98[f]

[a] Data shown are obtained from summaries in Refs. 21 and 38 unless otherwise noted.
[b] Ref. 39.
[c] Ref. 11.
[d] At low concentration rather than at saturation limit; Ref. 40.
[e] From data communicated by J. S. Dryden.
[f] Ref. 51, from a new method involving simultaneous diffusion of two divalent ions, which utilizes common ion interaction.

w_1, w_3, and w_4, but also the rate w_2 of impurity–vacancy interchange. It would appear, therefore, that the dielectric measurements, taken together with the anelastic results, offer the opportunity to determine w_2. On the other hand, if w_2 is much smaller than the other jump frequencies, Eq. (74) can only be used to provide a check on the values of w_1, w_3, and w_4 already obtained from the anelastic measurements. Dreyfus and Laibowitz have verified that the latter situation is indeed the case for Ca^{2+}- and Mn^{2+}-doped NaCl, indicating that w_2 involves a relatively high activation energy. Experiments on impurity diffusion generally verify this conclusion, as can be seen from Table V, which compares activation energies obtained from dielectric relaxation E_R, from impurity diffusion, and from dipole aggregation (which will be discussed below). Clearly, without impurity–vacancy interchange, diffusion of the impurity cannot take place. If w_2 is much less than the other jump frequencies, this means that the jump frequency w_2 must be rate-determining in the diffusion process.

The ratio w_3/w_4 provides an experimental value for the relative populations of nn and nnn configurations, and therefore, for the difference in association free energies. Specifically,

$$\Delta g_\alpha - \Delta g_\beta = kT \ln(w_3/w_4)$$

where Δg_α and Δg_β are the nn and nnn association free energies, respectively. From the results of Table IV, Dreyfus and Laibowitz obtain $\Delta g_\alpha - \Delta g_\beta = 0.044$ and 0.02 for NaCl:$CaCl_2$ and NaCl:$MnCl_2$, respectively.

In the specific case of Mn^{2+}-doped NaCl, electron spin resonance provides a powerful tool for the study of the local configuration about the Mn^{2+} ion, and therefore of the relative populations in nn and nnn configurations. The first such study was by Watkins,[40a] who observed both the nn and nnn configurations of the Mn^{2+}–V_{Na} dipoles and obtained values of w_3/w_4 in reasonable agreement with those quoted above. Symmons[40b] has extended the use of EPR methods to investigate the kinetics of the establishment of the equilibrium population distribution between nn and nnn dipoles. His results for w_3 and w_4 are in good agreement with those of Dreyfus and Laibowitz.

From dielectric studies on NaCl:MCl_2 for a wide variety of M^{2+} ions, Dreyfus[21] concludes that, as the ionic radius of M^{2+} decreases, the fraction of vacancies occupying nnn sites increases. On the other hand, the activation energy for the major dielectric relaxation* E_R is strikingly independent of ionic radius, as shown by Table V. This result lends further support to the conclusion, drawn earlier, that w_2 is unimportant compared to the other jump rates, since surely w_2 is expected to be strongly dependent on the radius of the impurity. In addition, it is interesting to compare these values of E_R with the activation energy for migration of a free vacancy, the best value for which appears to be 0.71 ± 0.04 eV.[41] Clearly, then, the activation energy for interchange of a Na^+ and a vacancy is not much different whether or not the vacancy is adjacent to an M^{2+} impurity.

For NaCl and KCl doped with the very small Be^{2+} ion, Bucci,[42] using ITC methods, found very different results. A relaxation which gives a very low activation energy ($\sim$0.25 eV) and an unusually low frequency factor τ_0^{-1} is obtained for which the dominant jump is said to be w_2. However, because of the very small size of the Be^{2+} ion (only about one-third the size of Na^+), the application of the previous theory to this material is questionable. In particular, the Be^{2+} ion may occupy an off-center position, as Bucci suggests, or may even enter the lattice interstitially.

Similar dielectric relaxations have been studied in other alkali halides, especially in KCl:MCl_2, for which the results are not much different than the case of NaCl. Also, dielectric relaxation in AgCl:MCl_2 has been studied for a variety of M^{2+} ions by Kunze and Müller[43] using ITC methods. In the silver halide, a double loss peak is usually observed at relatively low temperatures, giving activation energies in the range 0.30–0.35 eV. Although

* This is the slower T_{1u} relaxation of Eq. (74), which is dominated by w_4.

anelastic measurements have not been carried out for these crystals, the interpretation given for these relaxations is very much the same as that for doped NaCl crystals.

5.4.4. *Dipole Aggregation in Alkali Halides*

The phenomenon of dipole aggregation (see Section 5.3) has been studied most extensively using dielectric loss measurements by Dryden and co-workers[44–46] for NaCl or KCl doped with divalent cation impurities. The most striking result is the observation of third-order kinetics, suggesting that dipoles come together, three at a time, to form "trimers." These unusual kinetics have been confirmed by several other workers using EPR, ITC as well as dielectric loss. Crawford[38] interpreted this kinetic behavior as due to a two-step process involving (1) the formation of a dimer, and then (2) the addition of a third dipole to form a trimer. From the present viewpoint, however, the most interesting quantity is the activation energy for the aggregation process, E_{AG}. Although this quantity is not measured to as high a precision as the activation energy for relaxation, E_R, reasonably good values of E_{AG} have been obtained for various systems, some examples of which are included in Table V.

In terms of Crawford's two-step model, the activation energy E_{AG} which governs the kinetics of aggregation is not the same as the activation energy of motion E_s of the defect pair, but is smaller than E_s by an amount E_B which is the binding energy of the precursor pair of dipoles. The quantity E_s can be determined from measurement of diffusion of the divalent impurity in the high-concentration (or saturation) limit, i.e., under conditions favoring impurity-vacancy association [see Eq. (17)]. Values of E_s are included in the final column of Table V for the various divalent cations in NaCl. In general, the results are consistent with the idea that E_{AG} is less than E_s by a small binding energy. Taken together with the relaxation results, they also suggest, for ions of radii larger than Zn^{2+}, that the diffusional motion is governed by the jump frequency w_2, while the relaxational motion is governed by the frequencies (w_1, w_3, and w_4) of the jumps which leave the impurity ion fixed and only produce reorientation of the dipole.

5.4.5. *The* Al^{3+}–Na^+ *Defect in* α-*Quartz*

The α-quartz (SiO_2) structure belongs to the crystal class D_3 and is best pictured as a set of slightly distorted oxygen tetrahedra, each surrounding a Si^{4+} ion. There are three different orientations of the tetrahedra, each possessing one of the three twofold (C_2) axes of symmetry of the crystal. Between the tetrahedra are relatively wide open channels running parallel to the z axis of the crystal. Hydrothermally grown crystals usually contain Al^{3+} as an impurity which apparently enters the lattice substitutionally for the Si^{4+} ion, i.e., at the center of an oxygen tetrahedron. Charge compensation then occurs by incorporation of an interstitial alkali or H^+ ion, which forms an associated pair with the Al^{3+} due to electrostatic attraction. Although relaxations have been observed that were attributed to Li^+ or K^+ paired with Al^{3+}, the most thoroughly studied defect has been the Al^{3+}–Na^+ pair, for which both dielectric and anelastic relaxation were observed. A detailed interpretation of these observations which has been given recently[47,48] will be briefly reviewed here.

The selection rules for a D_3 crystal are given in Table II. Since the threefold axis in α-quartz is a screw axis, there are no sites in the crystal which possess trigonal symmetry. Only monoclinic and triclinic defects are therefore possible in this crystal, referring, respectively, to the cases where a defect pair lies along one of the C_2 axes or in an arbitrary direction. Table II shows that relaxational modes may be of the A_2 type, which are dielectrically active in response to an electric field in the z direction, or of the E type, which are *both* electrically and mechanically active in response to either an electric field in the basal plane or to any kind of shear stress.

Experiments on dielectric relaxation were carried out by Stevels and Volger.[49] Using a frequency of 32 kHz, they found Debye-type relaxation peaks at 38 and 95°K and were able to establish that both are due to a dipolar defect formed from Al^{3+}–Na^+ pairs. Figure 19 shows schematically the dielectric behavior observed for both the A_2 and E modes. The 76°K peak in Fig. 11 is the upper of these two Al^{3+}–Na^+ peaks, occurring at a lower temperature than 95°K because of the lower (2 kHz) frequency. (The remaining two peaks in Fig. 11 were not observed by Stevels and Volger and, therefore, must be related to a different impurity.) Anelastic experiments in shear vibrations, carried out at 5 MHz by Fraser,[50] show two peaks at higher temperatures, but when corrected for the frequency shift, they are found to match the two dielectric peaks[48]. Thus we conclude that two A_2- and two E-type peaks occur. This is why a dashed peak is shown in Fig. 19 for the E mode at 95°, indicating that although the dielectric peak was not observed, it is undoubtedly not of zero magnitude, since the corresponding anelastic peak appears. Examination of Table II shows that the occurrence of two A_2 relaxations can only mean that two different triclinic defect species are present, presumably Al^{3+}–Na^+ pairs in nn and nnn configurations, denoted by α and β species, respectively. Further, at

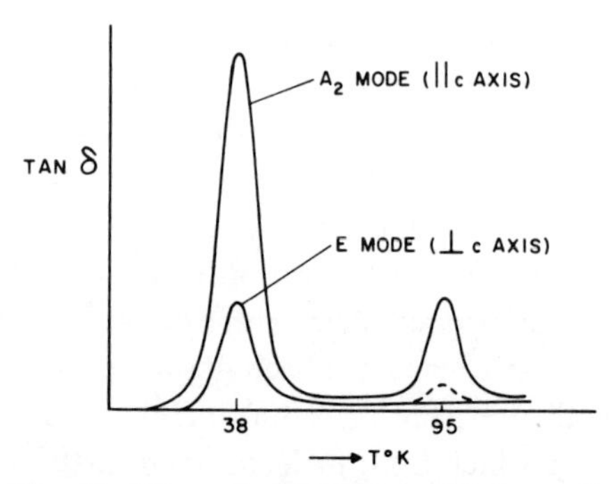

Fig. 19. Schematic illustration of data on dielectric loss versus temperature (at 32 kHz) in which electric field is parallel and perpendicular to the z axis, for α-quartz doped with Al^{3+} and Na^{+}.

the relatively low temperatures of these experiments, it is reasonable to suppose that only jumps of the interstitial Na^{+} ion can take place and not those of the substitutional Al^{3+} ion, since the latter would require too high an activation energy. Accordingly, of the six orientations which a given triclinic defect can possess, a given nn or α defect (say, that in orientation 1) can jump only to the other nn orientation located on the same oxygen tetrahedron (conventionally chosen as number 5). In addition, an interstitial jump may also take it into two orientations of the β (nnn) defect located on the same tetrahedron. The orientations of a set of four such mutually accessible defect orientations, $1(\alpha)$, $5(\alpha)$, $1(\beta)$, and $5(\beta)$, is shown in Fig. 20(a), while the jumps of the Na^{+} ion which most directly take the defect among these four orientations are shown in Fig. 20(b). The respective jump frequencies are denoted by ν_a (for $\alpha \leftrightarrow \alpha$), ν_b ($\alpha \rightarrow \beta$), and ν_c ($\beta \rightarrow \alpha$). The relaxation rates may be worked out in terms of these frequencies to obtain

$$\tau_{\pm}^{-1} = \nu_a + \tfrac{1}{2}(\nu_b + \nu_c) \pm \{[\nu_a + \tfrac{1}{2}(\nu_b + \nu_c)]^2 - 2\nu_a\nu_c\}^{1/2} \tag{76}$$

which applies to either the A_2 or the E modes of relaxation. If we assume that $\nu_a \gg \nu_b$, ν_c, which was shown to be reasonable[48] (since the contrary assumption gives unreasonable relative values for the dipole moment components), then Eq. (76) becomes

$$\tau_{+}^{-1} \approx 2\nu_a; \qquad \tau_{-}^{-1} \approx \nu_c \tag{77}$$

The quantity τ_{+}^{-1} gives the relaxation rate for the lower-temperature peak of Fig. 19, while τ_{-}^{-1} applies to the higher-temperature peak. The corresponding activation energies are 0.060 eV and 0.16 eV, respectively. From the relative magnitudes of the A_2- and E-type dielectric relaxations, one can also conclude that the major components of the motions involved in the jumps whose frequencies are ν_a and ν_c take place along the open channels of the crystal parallel to the z axis. However, the motions are not exactly parallel to the z channels, as Stevels and Volger had originally proposed.

Finally, it should be noted that while the jumps shown in Fig. 20, which allow limited access among the various defect orientations, account for all the τ values of the A_2 relaxations, they do not do so for the E relaxations. Specifically, Table II shows that for *each* triclinic defect, there must be *two* E relaxations, so that a total of four τ values are required, only two of which are given by Eq. (76). The two remaining relaxation frequencies involve jumps of the Al^{3+} ions among the three equivalent substitutional sites. As already mentioned, this jump should require a much higher ac-

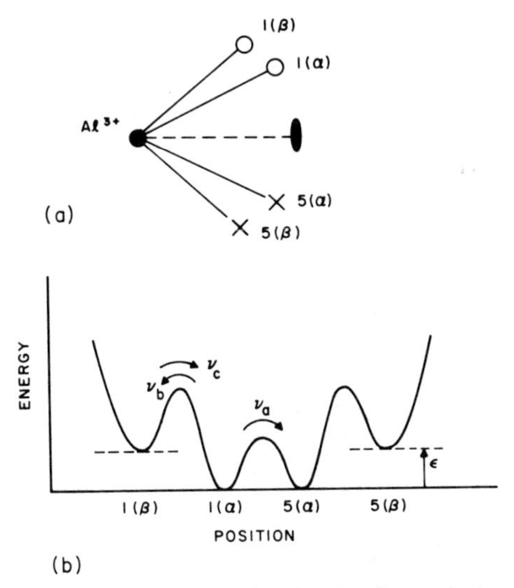

Fig. 20. (a) Diagram showing the four orientations (two of nn species α and two of nnn species β) of a Na^+ ion adjacent to a fixed Al^{3+} ion in α-quartz. The Al^{3+} ion substitutes for a Si^{4+} at the center of an oxygen tetrahedron whose twofold axis of symmetry is shown. Circles and crosses lie, respectively, above and below the plane of the page, which is perpendicular to the trigonal axis. (b) Schematic diagram of energy versus position for the four orientations shown in (a), defining the jump frequencies ν_a, ν_b, and ν_c.

tivation energy, and therefore the corresponding relaxations can only appear at considerably higher temperatures than those at which the relaxation experiments were carried out. In this example, therefore, the relaxation frequencies split into two types: (1) those involving the interstitial jump only and therefore giving rise to rotation of the Al–Na pair, and (2) those involving a migrational or diffusion-type jump. The latter relaxations have, unfortunately, not been observed, nor have experiments on diffusion of Al^{3+} (which would give the same type of information) been carried out.

6. CONCLUDING REMARKS

It is hoped that this chapter will have given the reader some idea of the variety of kinetic processes associated with defects involving aliovalent impurities in ionic crystals, and of the various methods used to separate these processes. In many cases, the combination of relaxation measurements and diffusion or defect aggregation studies has given a great deal of information about the various possible defect jump rates. Surely, the best-studied crystals are those of the type NaCl:MCl$_2$, with data summarized, for example, in Table V. Even for these materials, however, the story is not quite complete. Better diffusion measurements, particularly with radioactive tracers into predoped crystals, would be particularly useful. For other materials, the potentialities of this combination of techniques has barely been tapped.

ACKNOWLEDGMENTS

The author is grateful to the U. S. Atomic Energy Commission for supporting the writing of this review, as well as its support of much of the earlier research described in this chapter. This paper has benefited greatly from comments by J. H. Crawford, Jr., J. S. Dryden, W. J. Fredericks, A. B. Lidiard, S. K. Nenmeli, D. S. Park, and H. F. Symmons.

REFERENCES

1. F. Bassani and F. Fumi, *Nuovo Cim.* **11**, 274 (1954).
2. M. P. Tosi and G. Airoldi, *Nuovo Cim.* **8**, 584 (1958).
3. A. B. Lidiard, in *Encyclopedia of Physics*, Vol. 20, Ed. by S. Flügge (Springer, Berlin, 1957), p. 246.
4. F. K. Fong, *Phys. Rev.* **187**, 1099 (1969).
5. A. R. Allnatt and M. H. Cohen, *J. Chem. Phys.* **40**, 1860, 1871 (1964).

6. A. B. Lidiard, *Phil. Mag.* **46**, 815 (1955).
6a. R. E. Howard and A. B. Lidiard, *Rep. Prog. Phys.* **27**, 161 (1964), § 5.
7. C. Tomizuka, in *Methods of Experimental Physics*, Vol. 6A, Ed. by Lark-Horovitz and Johnson (Academic Press, New York, 1959), p. 364.
8. J. Crank, *The Mathematics of Diffusion* (Clarendon Press, Oxford 1946).
9. M. Chemla, Thesis, University of Paris, 1954.
10. F. J. Keneshea and W. J. Fredericks, *J. Chem. Phys.* **41**, 3271 (1964).
11. S. J. Rothman, L. W. Barr, A. H. Rowe, and P. G. Selwood, *Phil. Mag.* **14**, 501 (1966).
12. P. Süptitz and R. Weidmann, *Phys. Stat. Sol.* **27**, 631 (1968).
13. A. P. Batra, A. L. Laskar, G. Brebec and L. Slifkin, *Radiation Effects* **4**, 257 (1970).
14. A. S. Nowick and B. S. Berry, *Anelastic Relaxation in Crystalline Solids* (Academic Press, New York, 1972).
15. A. S. Nowick and W. R. Heller, *Adv. Phys.* **14**, 101 (1965).
16. A. S. Nowick, *Adv. Phys.* **16**, 1 (1967).
17. A. S. Nowick, *J. Phys. Chem. Solids* **31**, 1819 (1970).
18. Y. Haven and J. H. van Santen, *Nuovo Cim. Suppl.* **7** (2), 605 (1958).
19. A. S. Nowick and W. R. Heller, *Adv. Phys.* **12**, 251 (1963).
20. D. S. Park, Master's Thesis, Columbia University, 1970.
21. R. W. Dreyfus, *Phys. Rev.* **121**, 1675 (1961).
22. C. Bucci and R. Fieschi, *Phys. Rev. Lett.* **12**, 16 (1964).
23. C. Bucci, R. Fieschi, and G. Guidi, *Phys. Rev.* **148**, 816 (1966).
24. R. Cappelletti and R. Fieschi, *Crystal Lattice Defects* **1**, 69 (1969).
25. R. W. Dreyfus and R. B. Laibowitz, *Phys. Rev.* **135**, A1413 (1964).
26. H. B. Johnson, N. J. Tolar, G. R. Miller, and I. B. Cutler, *J. Phys. Chem. Solids* **30**, 31 (1969).
27. J. B. Wachtman, Jr., *Phys. Rev.* **131**, 517 (1963).
28. K. W. Lay and D. H. Whitmore, *Phys. Stat. Sol.* (*b*) **43**, 175 (1971).
29. R. W. Ure, Jr., *J. Phys. Chem.* **26**, 1363 (1957).
30. R. D. Shelley and G. R. Miller, *J. Solid State Chem.* **1**, 218 (1970).
31. B. Bleaney, *J. Appl. Phys.* (*Suppl.*) **33**, 358 (1962).
31a. J. M. Baker, E. R. Davies, and J. P. Hurrell, *Proc. Roy. Soc.* **A308**, 403 (1968).
32. J. Short and R. Roy, *J. Phys. Chem.* **67**, 1860 (1963).
33. P. D. Southgate, *J. Phys. Chem. Solids* **27**, 1623 (1966).
34. J. H. Chen and M. S. McDonough, *Phys. Rev.* **185**, 453 (1969).
35. E. Barsis and A. Taylor, *Phys. Rev. B* **3**, 1506 (1971).
35a. J. P. Stott and J. H. Crawford, Jr., *Phys. Rev. Lett.* **26**, 384 (1971) and references quoted therein. Also, J. H. Crawford, Jr., private communication.
35b. A. D. Franklin and J. Crissman, *J. Phys. C* **4**, L 239.
36. A. D. Franklin, *J. Res. Nat. Bur. Std.* **67A**, 291 (1963).
37. J. S. Dryden and R. J. Meakins, *Disc. Faraday Soc.* **1957** (23), 39.
38. J. H. Crawford, Jr., *J. Phys. Chem. Solids* **31**, 399 (1970).
39. S. C. Jain and K. Lal, *Proc. Phys. Soc.* **92**, 990 (1967).
40. Y. Iida and Y. Tomono, *J. Phys. Soc. Japan* **19**, 1264 (1964).
40a. G. D. Watkins, *Phys. Rev.* **113**, 79 (1959).
40b. H. F. Symmons, *J. Phys. C* **3**, 1846 (1970).
41. L. W. Barr and A. B. Lidiard, in *Physical Chemistry—An Advanced Treatise*, Vol. 10, Ed. by H. Eyring *et al.* (Academic Press, New York, 1970).

42. C. Bucci, *Phys. Rev.* **164**, 1200 (1967).
43. I. Kunze and P. Müller, *Phys. Stat. Sol.* **33**, 91 (1969).
44. J. S. Cook and J. S. Dryden, *Proc. Phys. Soc.* **80**, 479 (1962).
45. J. S. Dryden, *J. Phys. Soc. Japan, Suppl. III*, **18**, 129 (1965).
46. J. S. Dryden and G. G. Harvey, *J. Phys. C* (*Solid State*) **2**, 603 (1969).
47. A. S. Nowick and M. W. Stanley, *J. Appl. Phys.* **40**, 4995 (1969).
48. A. S. Nowick and M. W. Stanley, in *Physics of the Solid State*, Ed. by S. Balakrishna (Academic Press, New York, 1969), p. 183.
49. J. M. Stevels and J. Volger, *Philips Res. Repts.* **17**, 283 (1962).
50. D. B. Fraser, *J. Appl. Phys.* **35**, 2913 (1964).
51. J. L. Krause and W. J. Fredericks, *J. Phys. Chem. Solids*, **32**, 2673 (1971).
52. C. H. Burton and J. S. Dryden, *J. Phys. C.* (*Solid State*) **3**, 523 (1970).

Chapter 4

DEFECT CREATION BY RADIATION IN POLAR CRYSTALS

E. Sonder and W. A. Sibley*

Solid State Division
Oak Ridge National Laboratory †
Oak Ridge, Tennessee

1. INTRODUCTION

Radiation damage studies have contributed significantly in the past to our understanding of defect creation and interaction. In addition, much of the recent progress in the area of defect–lattice interactions has been due indirectly to the experimental background furnished by detailed investigations of the identity and properties of radiation-produced defects in alkali halide crystals carried out over a number of years.

From a practical point of view, the recent search for new materials with exotic magnetic and electrical properties and development of new techniques such as ion implantation for producing materials with specific properties emphasize the importance of understanding how material properties are altered by inadvertent or deliberate irradiation. In fact, the production of materials with "tailor-made" properties by use of radiation may turn out to be quite practical, especially in the area of computer technology.[1] Moreover, materials that will be used in space or near nuclear reactors must be tested to determine how seriously the hostile environments will affect important properties.

* Present address: Department of Physics, Oklahoma State University, Stillwater, Oklahoma.

† Operated by Union Carbide Corporation for the U. S. Atomic Energy Commission. The work going into this chapter was sponsored by the U. S. Atomic Energy Commission under contract with Union Carbide Corporation.

In studies of radiation damage, polar crystals are especially useful. These materials, being insulators, lend themselves to optical spectroscopy, magnetic resonance, and electrooptical-property measurements. Combinations of these measurements permit the identification of many defects and thus simplify studies of the interactions of these defects with impurities and among themselves. Partly because of this, a large base of experimental information has been accumulated, particularly for alkali halide crystals. The configurations for a number of the radiation-produced defects are known in great detail; of course, there are also many speculative models of defects which are extant in the literature. Our procedure will be to emphasize those defects for which there is reasonably good evidence that the model is correct. We will start by describing the primary defects that can be produced by radiation. Next, the nomenclature used to describe these defects will be discussed and a consistent notation applying to defects in both monovalent and divalent crystals will be suggested. Then, a summary of how radiation affects polar crystals will be given. In discussing radiation damage, the total process of displacing an ion from its lattice site in a crystal and the stabilization of this ion so that it cannot return to an empty lattice site will be considered. Finally, a detailed review of our present knowledge of defect production and reactions for alkali halide crystals and polar oxide crystals will be given.

2. RADIATION DEFECTS

When a polar crystal is subjected to radiation, numerous changes can occur in both the indigenous lattice ions and in the impurities present. We will classify the end products of these changes in terms of three categories of defects: (1) electronic defects, which involve changes in valence states; (2) ionic defects, which consist of displaced lattice ions; and (3) gross imperfections, such as dislocation loops and voids. Each of these types of defect has been observed in polar crystals and a more detailed description of each class of defects is given in the following subsections.

2.1. Electronic Defects

2.1.1. *Valence Changes of Impurities in Crystals*

The simplest radiation products arise from the impurities which are present in all samples and whose valence states are changed by trapping electrons or holes created elsewhere in the lattice by radiation. The symbol we will use for a substitutional impurity will be [Y^-] or [Me^{2+}], where Y denotes a negative-ion impurity, Me denotes a metallic impurity, and the superscripts give the valence state of the impurity, e.g., [Ag°] would be a silver atom, whereas [Ca^{2+}] would denote a calcium ion. A change of valence state due to radiation may result in different optical absorption bands or spin resonance spectra characteristic of the defect. Moreover, when the temperature is raised above the radiation temperature, many defects become unstable and release charge which may be detected by measurable electric currents or thermoluminescent glow peaks.

Radiation-induced valence changes have been studied in some detail for specific crystal systems with laser potential.[2–5] In addition, radiation-produced valence changes in transition metal ions and their thermal recovery have been reported in connection with studies of heat treatment,[6] ionization effects,[7,8] and deformation effects[9] in MgO. The valence of transition metal ions has also been changed by irradiation in alkali halides,[10] and the optical absorption and spin resonance spectra of ([Ag°] and [Ag^{2+}])[11] and ([Tl°] and [Tl^{2+}])[12] in KCl have been studied in detail. Charges can also be trapped in the vicinity of impurity anions, as spin resonance experiments[13,14] on alkali halides containing impurity halide ions have shown. Examples of impurities occurring with different charge states are shown in Fig. 1. In summary, a change of charge state of an impurity in an insulator is a common product of ionizing radiation. The capture cross sections of impurities for electrons or holes vary with the type of impurity and the nature of the host lattice, as does the temperature dependence of the stability of the centers.

2.1.2. *Valence Changes of Lattice Defects and Self-Trapped Charges*

Imperfections in crystals can also change their nature by trapping electrons or holes. There are reports in the literature of radiation-produced changes in the charge states of existing defects in alkali halides[15] and in alkaline earth oxides.[7,16–19] More interesting perhaps is the fact that free charges can be trapped even in perfect crystals. In the alkali halides, characteristic spin resonance spectra[20,21] and optical absorption bands[22–24] have been observed due to so-called self-trapped holes. The configuration of these centers shown in Fig. 1(b) is now well known;[25] they consist of two nearest-neighbor ⟨110⟩ halide ions that have given up an electron (trapped a hole) and have moved together to form a halide [X_2^-] molecular ion (X stands for an indigenous negative ion). It should be emphasized that

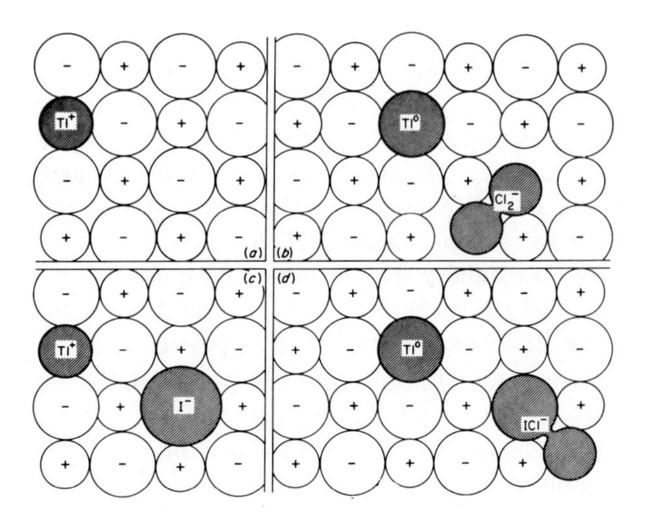

Fig. 1. Typical charge redistribution centers in alkali halides: (a) portion of KCl crystal containing thallium impurity in its normal valence; (b) thallium impurity that has trapped an electron, causing a corresponding self-trapped hole center to be formed; (c) portion of crystal containing both a thallium and an iodine impurity; (d) thallium and iodine impurities with an electron transferred from the iodine to the thallium.

these self-trapped charges are not lattice defects in the usual sense of involving a vacancy or interstitial, but that the two halide ions (atoms) are displaced from their normal lattice sites so that they have a smaller spacing than the normal negatively charged halide ions. Also, it is noted that in order for a self-trapped hole to be produced, the accompanying electron must be trapped at some other defect, such as an impurity (see Fig. 1b).

2.2. Ionic Defects

2.2.1. *Vacancies*

The lattice site of a missing lattice ion or atom is called a vacancy. Vacancies are probably the best-known radiation damage products. In simple elemental crystals (e.g., metals), all vacancies are equivalent. In compounds, there are a number of possible vacancy configurations, depending upon which elemental species is missing, and in more complicated crystal structures, upon which of a number of nonequivalent lattice sites is vacant.

Most of the observations of radiation defects in polar compounds have been made on negative-ion vacancies. As a result, these vacancy defects have been given special names. In singly charged compounds, the anion vacancy that contains an electron is referred to as the F center.[25] The center has the same charge as the anion originally present there and is consequently uncharged *with respect to the perfect lattice.* In doubly charged compounds, there is some confusion in terminology; both the vacancy containing one electron and the vacancy containing two electrons have been referred to as the F center. Henderson and Wertz[7] have recently attempted a rationalization of the system of naming defects in oxides. In agreement with their notation and in order to preserve consistency between singly and doubly charged polar compounds, we shall use the F-center nomenclature *whenever the anion vacancy is uncharged with respect to the lattice*, i.e., when it contains one electron in singly charged compounds and two electrons in doubly charged compounds. We shall use the symbols F^+ and F^- for negative-ion vacancy centers that are respectively positively or negatively charged with respect to the original lattice.

In alkali halides containing impurities, some of the radiation-produced absorption bands with peaks close to that of the F band have recently been found to be due to F centers adjacent to alkali impurities. These centers are now generally referred to as F_A centers.[26]

A number of defects involving positive-ion vacancies have been discovered in MgO.[7] The simplest of these consists of just one hole trapped at the Mg^{2+} ion vacancy, with the hole being shared by the next-neighbor oxygens. This defect has been called V_1, but to be consistent with our notation, we will refer to it as V^-. Two other centers consist of a cation vacancy with O^- (instead of the normal O^{2-} of the lattice) on one side and either an OH^- or an F^- on the other side. These latter two centers have the same *net* charge as does the perfect lattice and have been labeled V_{OH} and V_F, respectively. They will be discussed in detail later (see Fig. 19). The use of the letter V for cation vacancy centers is based on the original V-center nomenclature proposed by Seitz[27] for what were then thought to be positive-ion vacancy centers in the alkali halides.

At this point, it is perhaps necessary to mention that the labeling used in this chapter is for *centers*. This should not be confused with the labeling, often using the same letters, of experimental manifestations of defects such as the various ultraviolet absorption *bands* in alkali halides. The labeling of bands prior to a determination of what defect is responsible for that absorption can lead to confusion when the configuration of the center is finally discovered. Many times, the labeling scheme of the bands and that for the centers does not coincide. This has occurred, for example, in the alkali halides. Some of the ultraviolet absorption bands were called *V* bands since they occurred in the violet region. The centers responsible for these bands were thought to be positive-ion vacancy centers and were called *V* centers. Recent evidence, however, suggests that some, if not all, of the *V bands* in the alkali halides are due to interstitial defects.[28,29] In the meantime, the positive-ion vacancy V^- *center* was found in MgO and its absorption band was labeled V_1. The use of the same *V*-band notation for vacancy centers in oxides and interstitial centers in halides leads to great communication difficulties; we feel, therefore, that it is best to discard the *band* nomenclature and retain only the labels for centers whose configurations are known.

2.2.2. *Vacancy Aggregates*

In many cases, it is possible to produce groups of vacancies by prolonged radiation, heat treatment, or optical bleaching. Some of the definitions that are often used in connection with aggregates (an undefined group) of vacancies are the following: A "vacancy pair" is composed of nearest-neighbor anion and cation vacancies. A "divacancy" consists of two nearest-neighbor anion vacancies or two nearest-neighbor cation vacancies; if there is any question as to which, the defect is designated as a "cation" or "anion" "divacancy." The term "vacancy aggregate" can in general refer to anions and/or cations and can include groups of from two to many.

In recent years, much work has been done on aggregates of *F* centers in alkali halides.[30] When the term "*F*-aggregate center" (rather than vacancy-aggregate center) is used, or when the shorter term "aggregate" refers specifically to an *F*-aggregate center, the meaning is much more restrictive, referring specifically to a center consisting of from two to four *F* centers adjacent to each other. These *F* aggregates have been given specific designations, *M* centers (two *F* centers), *R* centers (three *F* centers), and *N* centers (four *F* centers). The (*M*, *R*, *N*)-center notation was first given to the characteristic absorption bands and the models originally proposed[31] for the centers responsible for these bands have since been modified.[25,32,33] This has led to some confusion; but the notation *M*, *R*, and *N* center respectively for two, three, and four *F*-center aggregates is now generally accepted. Recently, the notation F_2^+ center has been used[30] for a positively charged *M* center. This notation, in which the subscript refers to the number of *F* centers involved in the aggregate and the sign refers to the deviation from the normal crystal charge, is descriptive of the centers and is consistent with the remainder of the notation described here. It is, however, cumbersome in some cases. For that reason, we will use the M and F_2 notation interchangeably for di-F centers.

Few, if any, observations have been made on positive-ion vacancy aggregates and there is no specific notation for them in use. On the other hand, vacancy-pair-type centers have been reported in MgO, CaO and SrO.[7] These centers have been designated F_c^+ centers. Our notation is P^-. There is also some evidence[8,34] that the uncharged vacancy pair has been observed in deformed MgO, CaO, and SrO.

2.2.3. *Interstitials*

When a lattice atom or ion is displaced from its normal site, it either returns to a normal site (annihilates), travels to a surface (a grain boundary or dislocation), or remains in a position that is not a normal lattice site. Such a position is referred to as an "interstitial" site. The atom or ion is then an interstitial and the double defect consisting of the vacancy and interstitial is called a "Frenkel defect." There are many possible types of interstitial centers. In compounds, there may be cation and anion interstitial ions or atoms, and in any but the simplest and most symmetric structures, there exist more than one type of interstitial site which can be populated. Moreover, interstitial ions can sit either at the center of an interstitial site or may be drawn toward one of the lattice ions to form a center of differing configuration such as a molecular ion. In fact, the only interstitial species produced in polar compounds by radiation that has been carefully studied is not located at a normal interstitial site, but is an interstitial halide atom that has bonded itself to a lattice halide ion and shares that ion's lattice site. This molecular-ion defect occurs in irradiated alkali halides and is called the "H center."[25] Let us emphasize that the H center is a real lattice defect, in the sense that there are more ions than corresponding lattice sites. This is to be contrasted to the [X_2^-] hole trap in the perfect lattice, where the two anions have two lattice sites.

The only other interstitial center in polar crystals, the configuration of which appears to be on a firm foundation, is a variation of the H center. This center is an H center adjacent to a foreign alkali cation in an impure crystal.[35,36] For many years, optical absorption bands due to such centers have been observed and have been labeled V_1 *bands*. However, since there is much disagreement concerning the centers giving rise to the various V bands and especially since the V notation should be reserved for positive-ion vacancies, we propose that this defect be called the H_A center in analogy to F_A centers.

2.2.4. *Interstitial Aggregates*

Interstitial-aggregate centers have not been investigated as thoroughly as have the vacancy-aggregate centers, but Itoh and his collaborators[28] have recently made a strong effort to put the concept of interstitial aggregates on a better experimental foundation, and find that some of the ultraviolet absorption bands observed in irradiated alkali halides are probably due to interstitial aggregates. They suggest that the so-called V_4 band in KBr is due to diinterstitial halogen centers. If this is correct, then our notation for this center would be H_2.

2.3. Dislocations and Voids

It has been shown rather conclusively in the case of oxide crystals that irradiation produces large macroscopic defects such as dislocation loops.[37] The evidence for the production of dislocation loops and voids by irradiation in halide crystals is perhaps not as strong, but the observation of the growth of large macroscopic defects in alkali halides during electron irradiation at high temperature in an electron microscope[38] suggests that radiation-produced defects may aggregate in these crystals as well to form voids or dislocation loops. Damm and Suszynska[39] have also reported the effects of radiation-induced mechanical damage such as production of dislocations by radiation-induced strain in alkali halides.

2.4. Nomenclature

We believe that a reasonably consistent nomenclature applicable to all polar crystals is possible. The scheme we propose is a compromise between retaining the notation used in the literature for known defect centers and being somewhat more descriptive and having more universal applicability. Table I lists many of the defect centers in polar crystals that have been observed. The third column gives the notation we use in our discussion, while the fourth column gives other names that have appeared for the same centers in the literature. There is no attempt to include all possible defects. Only those defects whose configuration is reasonably well confirmed are listed. Below, we propose general rules which can be used for naming new defect centers in a consistent fashion whenever their configuration becomes known.

1a. The name F center is reserved for negative-ion vacancies containing the *same number of electrons as the charge of the normal lattice ion*. Thus, in singly charged polar crystals (e.g., alkali halides), the *F* center is a vacancy with one electron, and in doubly charged polar crystals, as, for example, MgO, the *F* center is a vacancy with two electrons.

1b. The name *V* center is reserved for positive-ion vacancies whose neighbors contain the same number of holes as the charge of the normal lattice ion that is missing.

1c. The name *H* center is reserved for a negative-ion interstitial atom that has combined with a lattice ion so that a molecular ion shares a normal lattice site. The *H* center has no net charge as compared with the perfect lattice.

2. If an *F*, *V*, or *H* center is adjacent to an impurity, the composite defect can be specified by subscripts following the *F*, *V*, *H* notation. Thus, an F_A (or F_{Na}) is an *F* center adjacent to an impurity cation (or particularly Na); V_{OH} is a positive-ion vacancy that is next to a substitutional OH^- impurity.

3. An aggregate of single point defects is specified by a number subscript. Thus, for example, three adjacent *F* centers are labeled F_3.

4. The defects specified in 1–3 all have the same charge as does the perfect lattice. If the charge is different from that of the perfect lattice, a superscript follows the defect symbol. For example, a negative-ion vacancy in an alkali halide that does *not* contain an electron would be labeled F^+, since the uncharged vacancy is one electronic charge more positive than a negative-ion that would be there in the perfect halide lattice.

5a. Centers for which the number of ions (atoms) is equal to the number of lattice sites are not ionic defects in the normally accepted sense and are not given letter designation (as, for example, *V* center). They are designated, as has been done in the recent literature,[40] by specifying the ions or molecules involved. Thus, a self-trapped hole in KCl or NaCl is a [Cl_2^-] center and a hole trapped at a lattice Cl^- and a neighboring substitutional Br^- is a [$ClBr^-$] center. This notation can also be used for substitutional impurities; for example, for iron in MgO, the notation would be [Fe^{2+}] or [Fe^{3+}].

5b. Brackets are used around all centers that are written in terms of their chemical species. This is necessary to prevent confusion with defect centers like *F* or *H* centers, which could be confused with fluorine or hydrogen impurities.

5c. The superscript giving the valence of the center should be placed inside the bracket since it is descriptive of the center and does not reflect the charge, with respect to the crystal, of the defect. For example, a [$ClBr^-$]

Table I. Defect Nomenclature

Description of defect	Ionic charge of normal ion at lattice site	Notation	Previously used notation
A. Single defects			
Negative-ion vacancy	-1[a]	F^+	α
	-2[a]	F^{2+}	—
Negative-ion vacancy with one electron	-1	F	F
	-2	F^+	F_1, F^+
Negative-ion vacancy with two electrons	-1	F^-	F'
	-2	F	F, F'
Positive-ion vacancy	$+1$	V^-	—
	$+2$	V^{2-}	—
Positive-ion vacancy with one hole at a near neighbor	$+1$	V	—
	$+2$	V^-	V_1, V'
Interstitial atom centered at interstitial space	0[b]	I	—
Split interstitial (atom) molecular ion at halide site	—	H	H
Interstitial anion (singly charged) centered at interstitial space	—	I^-	I
B. Composite defects			
Di-*F* center	-1	F_2, M[c]	M, F_2
	-2	F_2	—
Tri-*F* center	-1	F_3, R[c]	R, F_3
	-2	F_3	—
Ionized Di-*F* center	-2	F_2^+	M^+
	-4	F_2^+	—
Vacancy pair	0	—	—
Charged vacancy pair	-1	P^-	F_C^+
Diinterstitial atom	0	H_2 or I_2[d]	H'
C. Impurity-defect centers			
F center adjacent to cation impurity of same valence as host	-1	F_A[e]	F_A
	-2	F_A[e]	—
F center adjacent to divalent cation–positive-ion vacancy pair in alkali halide	-1	F_Z[e]	Z_1

Table I (*continued*)

Description of defect	Ionic charge of normal ion at lattice site	Notation	Previously used notation
V center adjacent to impurity	+1	V_A [e]	—
	+2	V_F, V_{OH}	V_F, V_{OH} [f]
H center trapped at impurity of same valence host	—	H_A [e]	—
D. Simple charge defects without missing or extra ions			
Self-trapped hole	−1	$[X_2^-]$[g]	V_K
Hole trapped at halide impurity	−1	$[XY^-]$[g]	$[XY^-]$
Hole trapped at cation impurity	+1	$[Me^\circ]$[g]	—

[a] Alkali halides are singly charged; materials like MgO are doubly charged; for materials as, for example, MgF_2, the cation (Mg) is doubly charged and the anion is singly charged.
[b] An interstitial site carries no charge.
[c] The *M*- and *R*-center notation for F_2 and F_3 centers is generally accepted by workers using alkali halides and is used universally in the literature. It is therefore retained.
[d] It is not known whether the interstitial atoms of a diinterstitial are at interstitial sites or have *H*-centerlike configurations.
[e] The subscript *A* stands for any alkali atom, impurity. The F_A center in sodium-doped KCl, for example, could be written as F_{Na}.
[f] The V_{OH} and V_F defects also include a hole located at a nearby divalent anion in order for charge of the composite defect to be the same as that of the perfect lattice.
[g] The letter *X* stands for a host halide ion; the letter *Y* stands for any other halide species; Me stands for any metal atom.

has a net positive charge in an alkali halide, as does an $[Fe^{3+}]$ in a divalent oxide. If desirable, the charge of the defect with respect to the lattice can be placed outside the bracket (for example, $[ClBr^-]^+$).

3. RADIATION DAMAGE PROCESSES

In this section, the mechanisms by which defects can be created in solids by radiation will be considered. A much more detailed discussion of these processes will follow in Sections 4 and 5, where reference will be made to specific materials, theories, and experimental results. Before any extended discussion, however, it is necessary to define and carefully distinguish three generic classes of radiation damage processes. We define these classes as electronic processes, elastic collisions, and radiolysis. The electronic class includes all processes in which an electronic state is changed or charge is moved about by the absorption of radiant energy, but in which no ionic or atomic defects are formed. Elastic collisions are those in which atoms or ions are displaced due to momentum and energy transfer from irradiating particles. Radiolysis processes are those in which atomic or ionic defects are produced by a series of reactions beginning with an electronic excitation.

3.1. Electronic Processes

The initial step in the production of any electronic defect is the absorption of energy from the radiation field. This absorption occurs somewhat differently for various types of radiation. Therefore, let us first consider how energy is absorbed and then discuss the result of this energy transfer.

A heavy, energetic particle passing through matter is usually stripped of some or all of its electrons and thus represents a rapidly moving point charge which interacts with the crystal's electrons. The scattering of a charged particle by the electrons of the crystal through which it is passing has been calculated[41] and is discussed in detail in nuclear physics textbooks.[42] From such scattering calculations, one obtains the energy loss per centimeter of path of the irradiating particle to be

$$dE/dx = (4\pi e^4 z^2 N_0 Z/mv^2)\log(2mv^2/I) \tag{1}$$

where e and m are the electron charge and mass, respectively, z and v are the irradiating particle charge and velocity, respectively, N_0 is the atomic density of the crystal, Z is the number of electrons per atom, and I is the average excitation potential of the crystal electrons. Equation (1) is valid for irradiating particles moving with nonrelativistic velocities. Obviously, the energy loss of the irradiating particle is equal to the energy gain of the electrons of the crystal. For heavy particles moving with velocities approaching the speed of light, the log term in Eq. (1) is replaced by $\log(2mv^2/I) - \log(1 - \beta^2) - \beta^2$, where $\beta = v/c$, c being the velocity of light.

Similar calculations have been made for fast electrons. The formulas differ slightly from those for the heavy, charged particle because bombarding electrons are not distinguishable from the crystal electrons. Moreover, since bombarding electrons, in contrast to heavy particles, are as light as the crystal electrons, they can lose an appreciable fraction of their energy and can be scattered through large angles in a single collision. Nevertheless, in a qualitative sense, electrons give up energy in a fashion similar to that of heavy particles so that the initial factor term of Eq. (1) also applies for electrons and only the log term differs.[42]

The penetration depth of a particle in a crystal is very important in radiation damage. Since the depth depends on the energy loss, it is of interest to compare the energy loss of heavy particles and electrons. Considering only the initial factor of Eq. (1), we note that for electrons, $z = 1$, whereas for alpha particles and heavier nuclei, it may become larger, and that for a given energy, v is inversely proportional to the square root of the mass of the particle. This suggests that the energy absorbed from a heavy particle is very much greater than that for electrons. Consequently, electrons will penetrate deep into crystals for which heavy particles are stopped near the surface. Table II gives some numbers[43] that show how far particles that are often used in radiation damage experiments can be expected to penetrate materials of the type that are of interest in this discussion. In particular, it might be noted that 1-MeV electrons penetrate at least 1 mm of most of the materials noted.

Another difference between electrons and heavy particles having kinetic energies of the order of 1 MeV is the fact that electrons approach relativistic speeds whereas heavy particles do not, and the relativistic form of Eq. (1) must be used. For electron energies above approximately 0.5 MeV, as can be seen from Fig. 2, the ionization produced is practically independent of energy.[44] Thus, fast-electron irradiation can produce uniform energy deposition throughout relatively thick crystals.

Fast neutrons, since they are not charged, do not excite the crystal electronically as do charged particles. However, when a fast neutron displaces

Table II. Penetration of Irradiating Particles[a]

Particle	Energy, MeV	Approximate penetration, cm			β^2
		Air	KCl	MgO	
Electron	0.5	1.5×10^{2}	9×10^{-2}	5×10^{-2}	0.75
	1	3.8×10^{2}	0.22	0.12	0.88
	5	2×10^{3}	1.2	0.7	0.99
Proton	1	2.2	2×10^{-3}	1×10^{-3}	2.1×10^{-3}
	5	3.4	2×10^{-2}	1×10^{-2}	1.1×10^{-2}
Alpha particle	1	0.5	2×10^{-4}	1×10^{-4}	5.4×10^{-4}
	5	3.6	2×10^{-3}	1×10^{-3}	2.7×10^{-3}

[a] Calculated from data assembled in Ref. 43 and references therein.

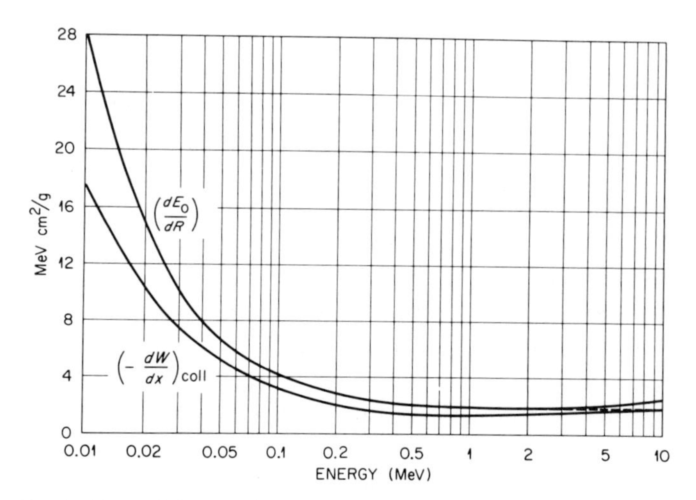

Fig. 2. Rate of change of energy with range (dE_0/dR) computed from an empirical equation, compared with the rate of energy loss for electrons ($-dW/dx$) according to theory.[44]

a crystal ion, the ion gives up some of its kinetic energy to the electronic structure of the crystal. In some materials, thermal neutrons can be quite effective in producing electronic excitation indirectly. This comes about when a thermal neutron is captured by a nucleus and the excited new isotope decays. For example, in LiF, the ^{6}Li has a relatively high (n, α) cross section (950 barns). The alpha particle and the recoiling triton which result from neutron capture move through the crystal with initial energies of 2.05 and 2.73 MeV, respectively, and these heavy charged particles excite the electronic structure of the crystal as described above.

Photons with energies in the range obtainable with X-ray or isotope sources can transfer their energy to the electronic system of a crystal by a number of processes.[42] In the *photoelectric effect*, the full photon energy is transferred into ionization and kinetic energy of one of the crystal electrons. The cross section for absorption of a photon by a crystal atom varies approximately as the fifth power of the atomic number of the absorbing atom and inversely as the 7/2 power of the photon energy, which means the process is most efficient for high-Z materials and low-energy photons. In the case of X-rays, the photoelectric effect is the chief mechanism of energy transfer in low-Z as well as high-Z materials. Energy transfer increases in efficiency as the photon energy decreases until the energy becomes too small to excite K-shell electrons. At that point, there is a sudden drop in efficiency. This drop is quite evident at 33 keV for the case of NaI depicted[45] as an example in Fig. 3. At even lower energies, the L- and M-shell electrons can no longer be ionized, so that a graph of efficiency versus energy has a sawtooth character in the keV energy region.

For photon energies above approximately 100 keV, a second mechanism of energy transfer becomes important. This is the "Compton effect," in which a photon transfers only a portion of its energy to an electron of the crystal. Figure 4 shows that, particularly in the range between 1 and 10 MeV, the energy is fairly evenly distributed between the scattered photon and the electron. The total cross section for Compton scattering, as can be seen by comparing the slopes of the "Compton" and "photoeffect" curves of Fig. 3, varies much more slowly with energy than does that for the photoelectric effect. Also, since Compton scattering is proportional to the number of electrons, it varies as Z, compared to the Z^5 variation of the photoeffect. Thus, particularly for the lighter elements, the Compton effect is already dominant for energies below 1 MeV.

At energies above 1.02 MeV ($2mc^2$), "pair production" becomes important and electron–positron pairs are produced with increasing efficiency as the energy rises. For energies above 10 MeV, nuclear processes can be initiated.

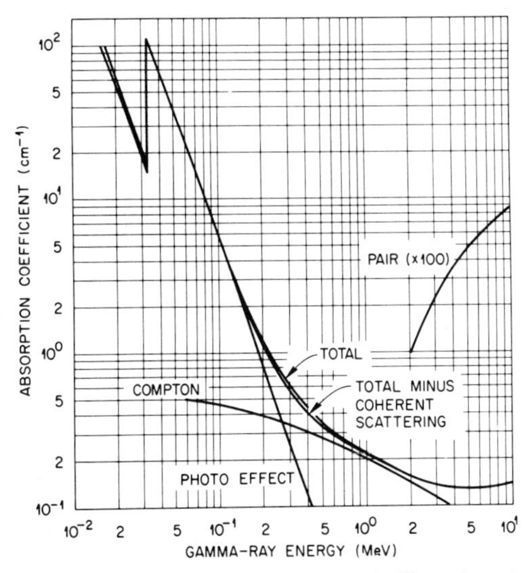

Fig. 3. Absorption coefficient for thallium-doped NaI as a function of gamma-ray energy. Separate curves for the photoelectric effect, Compton scattering, and pair production are shown and the small effect of coherent scattering is indicated on the sum curve.[45]

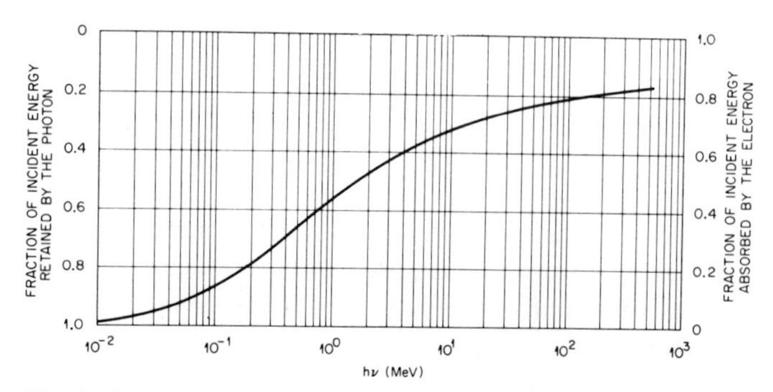

Fig. 4. Average distribution of energy of incident photon between scattered photon and scattering electron in a Compton effect scattering event. [National Bureau of Standards Circular 542, *Graph of the Compton Energy–Angle Relationship and the Klein–Nishina Formula from 10 keV to 500 MeV* (1953).]

As examples of special interest to us, we consider energy transfer from X-rays and gamma rays. The energy transferred to a thin section of crystal by X-rays of a given wavelength (since in the photoelectric process, the X-rays lose all their energy) is proportional to the decrease in the irradiating beam, so that the energy transferred to crystal is

$$dI/dx = \alpha I = I_0 \alpha e^{-\alpha x} \tag{2}$$

where I is the energy flux of the beam. This energy flux depends very strongly on the product of the distance from the crystal surface x and the absorption coefficient α. Since absorption coefficients are of the order of 10–100 cm^{-1}, the variation in energy transferred to the crystal is as much as a factor of 10 in a few millimeters. Moreover, let us recall (see Fig. 3) that α is a strong function of photon energy. Therefore, for a nonmonochromatic X-ray beam, the energy absorbed is the sum of many contributions of the form of Eq. (2) with different values of α, so that the absorbed energy depends on distance from the crystal surface in a complicated manner.

In the case of the higher-energy gamma-ray photons, uniformity of energy deposition is much improved for two reasons. The obvious one is that the absorption coefficient α for photons in the 1-MeV range is about two orders of magnitude below that for X-rays. Also, as was mentioned above, the Compton-scattering cross section is less strongly dependent on initial photon energy. Less obvious and sometimes forgotten is the fact that although the initial photon beam does decrease exponentially with the appropriate absorption coefficient, the scattered photons in Compton scattering (or the positrons and electrons in the case of pair production) cause additional energy to be transferred to the crystal as the initial beam decreases in intensity. Thus, high-energy photons initiate a type of electron–photon cascade that leads to relatively uniform energy deposition throughout fairly thick crystals.

When energy is absorbed in a crystal as described above, it appears in the form of electrons in a normally empty conduction band and holes in the normally occupied valence bands, or in the form of excitons (electron–hole pairs bound to each other) at lattice ions, impurity ions, or defects in the crystal. These excitations are of great interest in their own right and have been studied extensively. In electronic processes, as we define them, the excitation is but the first step and must be followed by processes that lead to observable electronic states. This usually involves separation of the electrons and holes, and trapping of the separated charges at impurities, defects, or in the perfect lattice. Motion of either the electron or the hole must occur following electronic excitation in order for an electronic defect to be produced. A fact that should be remembered is that normally the crystal as a whole must remain neutral, and free electrons and holes are always created in pairs. For every electron trapping center formed, there must also be a corresponding hole center formed. To use one illustrative example, when KCl containing [Tl^+] ions is irradiated at low temperature, [Tl°] atoms are produced due to capture of mobile radiation-produced electrons. However, for every electron trapped at a thallium impurity, a self-trapped hole [Cl_2^-] appears somewhere in the crystal, thus preserving the charge balance (see Fig. 1).

When the temperature of the crystal is raised above that at which a particular set of electronic defects is formed, the defects may become unstable or mobile. The charge carriers will then either recombine, restoring the perfect crystal, or they can be retrapped, forming different sets of electronic defects. Again, in the case of the Tl-containing KCl we considered above, heating the crystal to the vicinity of 200°K after irradiation at a lower temperature permits the self-trapped holes to move. Some of these holes will recombine with electrons that have been trapped, to restore the crystal to its preirradiation condition, but other holes will become trapped at [Tl^+] ions and produce new hole centers, [Tl^{2+}], leaving a corresponding number of [Tl°] in the crystal.[12]

3.2. Elastic Collisions

In order to create defects through elastic collisions, it is necessary for the incident particle to impart sufficient energy to a lattice atom or ion to displace it through the press of its neighbors into an interstitial site. Thus the effectiveness of an incident particle in creating damage depends on the maximum amount of kinetic energy T_m it can transfer to a lattice ion; this depends, in turn, on the energy E and the mass M_1 of the incident particle and the mass M_2 of the lattice ion, as shown by the equation for a non-relativistic head-on collision:

$$T_m = 4M_1M_2E/(M_1 + M_2)^2 \tag{3}$$

In the case of relativistic particles, the equation is altered to

$$T_m = 2(E + 2M_1c^2)E/M_2c^2 \tag{4a}$$

For the specific case of electron bombardment, insertion of numerical values for $M_1 = m$ and c yields

$$T_m = 2147.7E(E + 1.022)/A \quad \text{(eV)} \tag{4b}$$

where E is in MeV and A is the atomic mass of the lattice ion.

There have been several detailed reviews of the creation of defects by elastic or "knock-on" collisions in metals and semiconductors,[46–50] but little on polar crystals. About 30 eV is normally required to displace a lattice atom through the saddle point between its nearest neighbors. The value for a given material is called its displacement energy T_d. If the radiation transfers less energy than the displacement energy ($T_m < T_d$), then no elastic collision radiation damage will result.

In general, five types of radiation may produce displaced atoms or ions by elastic collisions. These are: (1) gamma rays, (2) energetic electrons, (3) thermal neutrons, (4) fast neutrons, and (5) energetic atoms or ions. Table III illustrates the amount of energy T_m transferred to lattice ions of K, Cl, Mg, and O for various types and energies of radiation. It is clear from the table that the heavier particles will be much more effective in displacing lattice ions than the lighter ones. In fact, they are so effective that they produce so-called "cascades." Cascades result when a large amount of kinetic energy is transferred from the incident radiation to one lattice ion and that lattice ion, in moving through the lattice, transfers by elastic collisions sufficient kinetic energy to other lattice ions to dislodge them from their normal positions. If enough energy is available, this process continues until a large number (cascade) of displaced ions is formed. In such processes, large defect clusters are produced. The details of the collisions leading to such clusters have been discussed a number of times in the past,

Table III. Energy Transfer between Various Irradiating Particles and Lattice Ions

Particle	Energy, MeV	T_m, eV			
		K	Cl	Mg	O
^{60}Co γ-ray[a]	1.17	110	123	178	268
Cs γ-ray[a]	0.667	41	46	67	101
Electron	1	110	121	179	269
	2	330	368	537	805
	14	11,500	12,000	19,000	28,000
	30	51,000	56,500	83,000	125,000
Proton	2	195,000	216,000	307,200	443,000
Neutron	2	195,000	216,000	307,200	443,000
Alpha particle	6	2.0×10^6	2.2×10^6	2.9×10^6	3.8×10^6

[a] Gamma rays produce defects in two steps. First, they generate fast electrons through the photoelectric or Compton effects; then, these electrons in turn transfer energy and momentum to lattice ions.

in particular with reference to radiation damage in metals.[46–50] For example, Kinchin and Pease[50] have calculated the average number of displacements Ω produced by an ion to which an energy T has been transferred by the incident radiation. They find that

$$\begin{aligned} \Omega(T) &= 1, \qquad && T_d \le T \le 2T_d \\ \Omega(T) &= T/2T_d, \qquad && T \ge 2T_d \end{aligned} \tag{5}$$

As mentioned in the introduction, one of the great advantages of polar crystals as compared to metals for studying radiation damage is that optical and ESR techniques are so convenient to use. Since these techniques are most useful for studying single defects, most workers have preferred to irradiate their samples in such a way as to produce primarily single defects. This requires that the energy transferred to a lattice ion be between T_d and $2T_d$ (less than approximately 100 eV).

Elastic collision processes are conceptually straightforward and defect production occurring by elastic collisions is relatively well understood. However, in most discussions, the assumption is made that there is a sharp displacement threshold. Recently, considerable evidence has accumulated in metals,[47] III–V compounds,[47,51] and LiF[52] that the displacement threshold energy is strongly dependent on the crystallographic direction of the impinging electron beam, and hence the direction or displacement of the lattice ions. Moreover, recent discoveries that irradiating particles can move through "channels" in a crystal lattice and can penetrate to anomalous depths without transferring much energy to any lattice atom or ion[53] indicate that the study of elastic collision processes is still a very active field of research.

Let us consider how we can use optical absorption measurements of insulators to determine displacement energies T_d and cross sections σ_d. Many insulating crystals develop absorption bands due to the introduction of defects which absorb light at wavelengths at which the perfect crystal is transparent. The intensity of these bands is proportional to the concentration of specific defects such as F centers, F^+ centers, or F-center aggregates. By measuring the optical absorption, it is possible in many cases to determine the identities of the defects produced and in conjunction with types of measurements such as magnetic susceptibility to obtain the absolute concentration of these defects. In fact, once the oscillator strength of a particular center has been found by means of combined experiments, the absolute concentration can be found by simply making optical measurements.[25] Since we can also measure the flux of incident particles ϕ, we can determine the displacement cross section σ_d, which may be defined by the equation

$$N = \phi N_0 \sigma_d(E) \tag{6}$$

where N is the number of defects per cubic centimeter of crystal and N_0 is the atomic density. If the sample is thin enough so that the irradiating particles are of essentially constant energy throughout their traverse of the crystal (see previous section on energy loss due to electronic interaction) the cross section obtained is for the specific particle energy E. Determination of the cross section as a function of energy or crystal orientation is possible and energy thresholds and $\sigma_d(E)$ curves for specific defects can be obtained for various orientations and host crystals.

3.3. Radiolysis Processes

In the more highly ionic polar crystals such as alkali halides, ionic defects are created quite easily with ionizing radiation.[29] A consideration of Eq. (6) in the above discussion of elastic collisions indicates that for an average displacement cross section of about 10 barns, a 2-MeV electron can displace at most *one* lattice ion from its site to form a Frenkel pair. When alkali halide crystals are irradiated with electrons, many ionic defects are produced per electron hit.[29] In fact, it is even possible to produce

Frenkel pairs by excitation with ultraviolet light. This means that in certain ionic materials, defect creation is highly efficient and is most likely due to the conversion of electronic excitation energy into a form capable of manufacturing lattice defects rather than to elastic collisions. Such photochemical processes also occur in insulators other than polar crystals and are most likely involved in the photographic process and photosynthesis.

In radiolysis defect production, there must be at least three steps: (1) an electronic excitation resulting at least momentarily in the creation of a polarized or charged electronic defect in the lattice, (2) the conversion of this energy into kinetic energy of a lattice ion in such a way that the ion moves, and (3) the motion and stabilization of the ion. It appears now after many years of research that at least some of these steps are understood. The steps involved in radiolysis defect production will be discussed in detail in the next section.

4. DEFECT CREATION IN HALIDE CRYSTALS

In metal halide compounds, defect production by radiolysis is extremely efficient. Thus the discussion of radiation damage in metal halides will of necessity be chiefly a description of radiolysis, even though elastic particle collisions also produce defects in halides as they do in metals and semiconductors. The small amount of work with alkali halides that is concerned with defect production by elastic collisions will be mentioned first and then the remainder of this section will be devoted to electronic processes and radiolysis.

4.1. Elastic Collisions

Elastic collision processes are only rarely observed. Farge and his co-workers[52] have studied thermal neutron and electron irradiation of LiF and have attributed an absorption band at 550 nm to interstitial Li produced by elastic collisions. They have also studied the dependence of the production of this band on electron energy and crystallographic direction of the incident beam.

4.2. Electronic Processes

The initial product of radiation is excitation of valence electrons of crystal ions to higher energy levels. These excited levels may be localized, as, for instance, the excitation of an impurity would be, or they may require motion of one of the charges, as is the case when the electron is ionized into a high conduction band state.

Effects of electronic excitation without charge motion can certainly be observed. For instance, radiation luminescence due to direct recombination of excitons or bound electron–hole pairs is well known.* Also, it is possible that charge separation is not necessary for radiolysis.[55] However, in this section, we will be concerned with electrons that are excited into the conduction band by the radiation and that consequently move to a site different from where they originated, causing two sites in the crystal to change their charge state, one becoming more negative and the other more positive.

In doped alkali halides, valence changes of impurity ions can be observed both after a short irradiation at liquid nitrogen temperature and after subsequent warming. In [Tl^+]-doped KCl, the [Tl^+] concentration can be determined readily from the height of an absorption band at 245 nm in the ultraviolet.[56] Delbecq *et al.*[57] have identified a strong band with peak at 380 nm due to Tl° and bands with peaks at 220, 262, 294, and 364 nm due to Tl^{2+} and have demonstrated that Tl° and Tl^{2+} are produced by irradiation of the doped samples. Other heavy-metal impurities in alkali halides also trap electrons and holes when the samples are irradiated at low temperature. Silver in KCl has been studied in detail and recent work[58] has shown that Pb in KCl can trap both holes and electrons. Additional absorption bands which appear to be due to electron or hole trapping at impurities have been reported for other alkali halides.[59,60] However, systematic studies of radiation-induced valence changes in impurities are limited. One reason for this is that ionic defects are produced so rapidly by radiolysis that it is often difficult to determine whether a radiation-produced band is due to trapped charge or an ionic defect. There is no certain way of identifying a particular charge state of a given impurity, so that the study of any single system requires careful sample preparation (in order to know what impurity and how much is present) and a complete accounting of all the charges released by the radiation. This usually requires a combination of experimental techniques, e.g., optical absorption and spin resonance measurements, and often involves not only the impurity being studied, but other possible impurities and defects present in the crystal before irradiation as well as defects produced by the radiation.

From studies of charge trapping, one very important result has emerged. Apparently in alkali halides, only electrons can move freely once they are

* See, e.g., Tables 1–4 of Knox and Teagarden.[54]

Table IV. Contributions to the Self-Trapping Energy for Holes and Electrons in KCl

	Localization energy, eV	+	Polarization energy, eV	+	Binding energy, eV	=	Self-trapping energy, eV
Holes	−0.3		+0.5		+1.3		1.5
Electrons	−1.9		+0.3		+0.3		−1.3

ionized into the conduction band. Holes, by contrast, can be trapped at low temperature in the perfect crystal and are observed to move only at higher temperatures (in KCl, $\gtrsim 200°K$) by a diffusionlike hopping from one lattice space to the next.[29] Gilbert and co-workers have estimated[61,62] the relative stability of localized trapped charges and charges distributed over many lattice volumes for both electrons and holes. They show that the energy necessary to localize a hole, approximately half the width of the valence band, is much less than the energy gained from polarization, lattice relaxation, and Cl_2^- binding, whereas it takes more energy to localize an electron than is gained from polarization and binding. The various energies contributing to the self-trapping energy are given for KCl in Table IV.

The motion of self-trapped holes has been studied in some detail in recent years and the activation energy for motion of trapped holes has been measured in a number of alkali halides. In Table V, we list the temperatures at which self-trapped holes become mobile and the activation energies for a number of the halides. The table is not complete, due to unavailability of data; however, it clearly shows that for all cases noted, holes are self-trapped and cannot move at liquid helium temperatures. The basic diffusive jump involves a shifting of the center a distance of $a\sqrt{2}/4$ (where a is the lattice constant) in a $\langle 110 \rangle$ direction. This is accompanied by a 60° change in orientation of the center axis.

4.3. Radiolysis

The absorption of radiant energy by the electronic system of a crystal as discussed in Section 3.1 occurs in all materials, determines the loss of energy of the irradiating particle, and is the first step in the creation of electronic defects. It is also the first step in the creation of ionic defects

Table V. Stability of Self-Trapped Holes

Material	Property measured	Thermal treatment	Temp. of max. change, °K	Activation energy,[a] eV	Ref.
LiF	Dichroism of opt. abs. of [F_2^-] band	Anneal for 2 min; recool to 80°K for measurement	110	—	22
	Dichroism of opt. abs. of [F_2^-] band	Continuous warming (no rate given)	114	—	23
	Decay of opt. abs. of [F_2^-] band	Continuous warming (no rate given)	138	—	23
	Decay of spin res.	Anneal for 30 min; recool to 80°K for measurement	~120	0.32 (2)	63
NaF	Dichroism of opt. abs.	Continuous warming (no rate given)	155	—	23
	Decay of opt. abs.	Continuous warming (no rate given)	175	—	23
	Decay of spin res.	Anneal for 4 min; recool for measurement	160	—	64
	Growth of [F_2^- Li] centers	Not given	140–160	—	65
NaCl	Decay of spin res.	Anneal for 3 min; recool to 120°K for measurement	160	—	66
	Opt. abs. of [Cl_2^-], [Tl°], and [Tl^+]^2	Isochronal anneal	150–160	—	67
		Continuous warming (3°K/min)	151		
KCl	Decay of spin res.	Anneal for 3 min; recool to 120°K for measurement	200	—	66
	Decay of spin res.	Anneal for 2 min; recool to 80°K for measurement	140	—	68
	Opt. abs. of [Tl^+] band	Anneal for 3 min; recool to 80°K for measurement	208	—	12
	Opt. abs. of [Cl_2^-] band	Anneal and measure at temperature	~200	0.53 (2)	69
KCl	Dichroism of opt. abs. of [Cl_2^-] band	Anneal for 2 min; recool to 80°K for measurement	173	—	22
	Dichroism of opt. abs. and ESR of [Cl_2^-] band	Anneal and measure at temperature	~170	0.54 (1)	70
KBr	Dichroism of opt. abs. of [Br_2^-] band	Anneal for 2 min; recool to 80°K for measurement	140	—	22
NaI	Dichroism of opt. abs. of [I_2^-] band	Continuous warming (no rate given)	52	—	24
	Growth of [Tl^{2+}] bands	Continuous warming (no rate given)	58	0.15	24
KI	Dichroism of opt. abs. of [I_2^-] bands	Anneal and measure at temperature	—	0.273 (1)	71
	Decay of [I_2^-] opt. abs. band and afterglow	Anneal and measure at temperature	~100	0.28 (2)	71
RbI	Dichroism of opt. abs. of [I_2^-] band	Warm at 2°K/min	102	—	24
	Decay of opt. abs. of [I_2^-) band	Warm at 2°K/min	125	0.32 (e)	24

by radiolysis. In this portion, we discuss mechanisms by which electronic energy can be converted into kinetic energy capable of producing and separating ionic defects. It will also be necessary to consider the secondary reactions by which the primary defects recombine or are stabilized.

4.3.1. *Primary Defect Production Processes*

(a) *The Identity of the Produced Defects.* There has been some controversy in recent years concerning the identity of the primary defects produced in alkali halides by radiolysis, in particular, whether there are Frenkel defects in the halide sublattice or Schottky pairs consisting of cation and anion vacancies. The latter possibility stemmed from a suggestion[27] that electronic excitation would deexcite at dislocation jogs, causing positive and negative ion vacancies to be alternately "evaporated" from the dislocation. This mechanism for Schottky-defect production has been repeatedly discussed in the literature[25,29] and will not be covered here, especially since there now exists fairly good evidence that defects are not produced as Schottky pairs.

The initial evidence that Frenkel defects, rather than Schottky pairs, are formed is due to Känzig and Woodruff,[72] who studied the ESR of low-temperature irradiated LiF, KCl, and KBr. They identified one spectrum as being due to a dihalide molecular ion located in one lattice site, and interacting with two additional halide ions at nearest-neighbor sites in the $\langle 110 \rangle$ direction. In short, they discovered the H center, diagrams of which are given in a number of recent review articles.[25,29,73,74] At about the same time, Compton and Klick[75] studied the polarization of one of the main absorption bands produced together with the F band when KCl and KBr are irradiated at liquid He temperature and they found that the centers giving rise to the band, whether in KCl or KBr, could be aligned in a $\langle 110 \rangle$ direction by bleaching with polarized light. Moreover, they found that the growth of the band was proportional to that of the F band, indicating that the centers were probably complements of F centers produced by the radiation. Subsequently, it was shown[76] that these bands are due to H-center absorption. Delbecq *et al.*[76] measured both the thermal disorientation of aligned H centers and the disappearance of these centers, by optical absorption and spin resonance. They found in KCl that the anisotropic absorption of both the spin resonance signal and the polarized optical absorption disappear at the same temperature, namely at 11°K, and that the H-center spin resonance signal disappears together with the optical absorption band between 40 and 70°K.[77]

These results strongly indicate that in alkali halides irradiated at low temperatures, F centers are produced together with interstitial halide atoms, i.e., H centers. Considerable indirect evidence exists which shows that when irradiations are performed at higher temperatures, the damage is also of the Frenkel type. For example, measurements of the flow stress of irradiated KCl, KBr, NaCl, and LiF[78–80] show that for all temperatures, the increase in flow stress that accompanies radiation F-center production is of such a magnitude as theoretically[81] would be expected for interstitials or interstitial clusters, but not for vacancies.

Even more powerful verification that Frenkel rather than Schottky defects are produced at all temperatures up to room temperature comes from simultaneous measurement of the expansion of a crystal and its increase in lattice constant. When a Frenkel pair is formed, the lattice atom or ion must be accommodated interstitially so that the percentage increase in distance between lattice planes is equal to the percentage increase in the crystal macroscopic dimensions. This fractional expansion will be proportional to the mole fraction of defect pairs, N_p/N_0, and the per cent increase in linear dimension produced per defect, or

$$\Delta l/l = \Delta a/a = (N_p/N_0)(\delta_v + \delta_i) \tag{7}$$

where l is the macroscopic dimension, a the lattice constant, δ_v the dilatation due to one vacancy (this might be negative), and δ_i the dilatation produced by one interstitial. If, on the other hand, vacancies are produced and the corresponding atoms or ions are not forced into interstices, new crystal planes will be formed to accommodate these atoms (ions). The percentage increase in lattice constant in this case will not be equal to the percentage increase in macroscopic size. Rather,

$$\begin{aligned} \frac{\Delta l}{l} &= \frac{N_+}{N_0}\left(\frac{b_+}{b_+ + b_-} + \delta_+\right) + \frac{N_-}{N_0}\left(\frac{b_-}{b_+ + b_-} + \delta_-\right) \\ \frac{\Delta a}{a} &= \frac{N_+}{N_0}\,\delta_+ + \frac{N_-}{N_0}\,\delta_- \end{aligned} \tag{8}$$

where the plus and minus subscripts refer to cation and anion vacancies and b is the average linear dimension of a normal lattice ion. Thus the production of Schottky defects, even in the presence of Frenkel defects, will manifest itself by unequal changes in the macroscopic size and the lattice constant of a crystal. Rather careful measurements of the lattice parameter and the density have been made on a number of alkali halides irradiated at different temperatures.[82] The results of these measurements

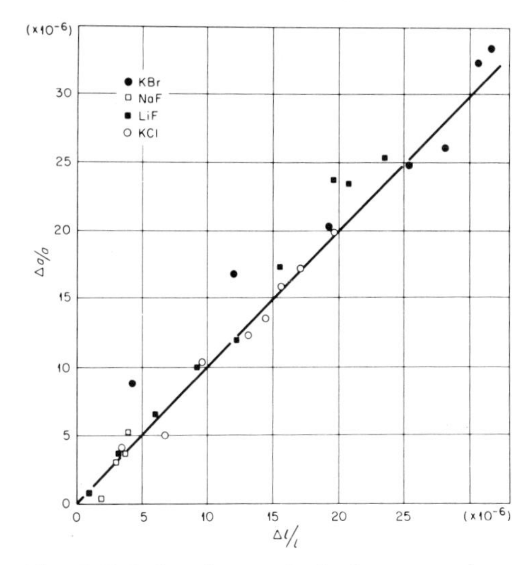

Fig. 5. Relative changes in lattice parameter a and length l of X-irradiated alkali halides.[82]

are exemplified by Fig. 5, which shows that the percentage increase in lattice parameter and length is the same after room temperature irradiation. Other results obtained by the same group demonstrate that the equality is independent of irradiation temperature. Thus the evidence appears to be overwhelming that the damage produced by radiolysis in alkali halides is of the Frenkel type, no matter what the radiation temperature.

(b) *Is Radiolysis a Bulk Process*? There is a second general question that concerns the production process. This is whether the conversion of electronic energy to lattice defects takes place in the perfect lattice or at particular sites, e.g., defects. This question is not yet completely settled and we will periodically return to it in later discussion. However, there is relatively good evidence that the major fraction of defects produced are created in the environment of the perfect crystal. For example, the efficiency of vacancy production at liquid He temperature in KCl does not depend upon impurity content.[83] Moreover, there has been no evidence that it is affected by the dislocation concentration. In addition, it has been shown that more than 10^{19} defects/cm^3 can be produced at room[84] or liquid helium* temperature in most halides. This would make the average separation of defects ten

* Faraday and Compton[85] obtain $\gtrsim 5 \times 10^{18}$ F centers. Ritz[88] shows that for every F center, approximately four F^+ centers are produced, indicating that more than 2×10^{19} vacancies/cm^3 can be produced by low-temperature irradiation.

lattice spaces if they are uniformly distributed. If defects were created at special sites, such as near dislocations, the concentrations in these volumes would be orders of magnitude greater than the average defect concentration and the average spacing would be only a few lattice spaces. Under these circumstances, the effects of interaction of nearby centers would become important and a much larger number of F_2 centers (F centers in adjacent sites) would be seen than is observed experimentally.[86]

(c) *Does Defect Production Involve Multiple Ionization?* We come now to the problem of how electronic excitation can produce vacancy–interstitial pairs in a halide crystal lattice. Varley[87] in 1954 suggested that high-energy radiation as it becomes degraded will produce a number of multiply ionized halide ions and that these ions, if they remain in their multiply charged state long enough, will be electrostatically ejected into interstitial positions some distance away from the original sites. Capture of an electron would then produce the F-center–H-center pairs usually observed.

The production efficiencies originally calculated by Varley[87] were in good agreement with experimentally measured production rates for F centers at liquid helium temperatures. They were, however, much too low when compared with the production rate for all Frenkel defects (that is, including interstitial-ion–F^+-center pairs) at liquid helium temperature[88] or the initial production rate at room temperature.[89] More recent calculations by Varley[90] and Itoh[91] suggest that double ionization may be efficient enough to account for observations in at least some of the alkali halides. There is, however, another difficulty with the double-ionization mechanism. It is questionable whether the double-ionization state is sufficiently long-lived[92] to permit the ions to begin to move before an electron is recaptured. Thus it appears unlikely that double ionization is the predominant process in radiation defect production. Many of the arguments are summarized in other recent reviews.[29,73]

(d) *Proposals for Defect Production by Single Ionization.* There is one characteristic of ionic crystals that may permit the production of directed ionic motion from a single ionization of a halide ion. This characteristic is the rather large amount of ionic relaxation that follows any electronic change. When a hole is self-trapped, the two halide nuclei are quite a bit closer than are two normal halide lattice ions.* A recombining electron will not only make available approximately 7 eV electrostatic recombination

* See, e.g., the calculation of Jette *et al.*[93]

energy, but it will produce an impulse pushing the two ions toward their normal lattice positions.

The foregoing is the basis of the model proposed by Pooley and Runciman[94] to explain the highly efficient defect production in alkali halides. The conclusion that there exists a relation between electron–hole recombination and the production of ionic defects was in actuality arrived at independently by three different groups of investigators, Hersh and co-workers in the U. S., Lushchik and co-workers in the USSR, and Pooley and co-workers in England. The observations that pointed the way were for the most part made in connection with studies of ultraviolet light absorption and luminescence. As early as 1930, Smakula[95] reported that alkali halides could be colored with ultraviolet light. This has been confirmed by more recent measurements.[96,97] However, it was not clear that the ultraviolet-induced coloration was connected with an "intrinsic" defect production mechanism until it was shown that it was possible to create concentrations of *F* centers that were much greater than the impurity concentrations in the samples,[98] and that the shapes of the *F*-center production curves were very similar to those obtained with X-rays.[99]

More definitive evidence connecting *F*-center production with electron–hole recombination at self-trapped holes appeared when the temperature dependence of the fundamental luminescence in KI was compared with the temperature dependence of *F*-center production.[100,101] It turned out that below approximately 100°K, irradiation with ultraviolet light or X-rays produced an emission spectrum peaking at 302 and 380 nm; this so-called fundamental emission could not be observed at higher temperatures, but just in the range where the emission disappeared, the *F*-center production efficiency increased approximately an order of magnitude. Meanwhile, Murray and Keller[102] had demonstrated that the fundamental luminescence could also be observed when electrons were permitted to recombine with self-trapped holes, thus showing that the fundamental luminescence was due to electron–hole recombination. Their results indicated, moreover, that electron–hole recombination *during* irradiation at low temperatures was identical to recombination of electrons with trapped holes.

A few measurements have been made that indicate that what we have described above for KI is more general. The identity between the fundamental luminescence and the electron–self-trapped-hole recombination luminescence has been demonstrate for RbI and NaI;[24] the anticorrelation in the temperature dependence of the fundamental luminescence and *F*-center production has been observed[103] for a number of other alkali halides, although it appears not to hold in every case. The most clear-cut evidence that electron–hole recombination produces Frenkel pairs was obtained in a recent experiment in which trapped electrons were released in KCl at low temperature.[104] The recombination of these electrons with holes trapped as [Cl_2^-] or [$ClBr^-$] centers caused *H* centers to be formed in proportion to the trapped hole centers destroyed.

The Russian workers,[105] in an investigation of the ultraviolet excitation wavelength dependence of *F*-center production in NaCl, found that defects can be produced not only when free electron–hole pairs are created, but even when the ultraviolet photons have only enough energy to excite bound electron–hole pairs, or excitons. It would be an attractive simplification if we could assume that electron–hole recombination always goes by way of a definite bound exciton state, so that some calculations could be made on how Frenkel pairs separate. However, it appears that there may be several recombination paths. When KI is irradiated with light in the first exciton band at low temperature, only one emission band at 371 nm is observed,[106] whereas higher-energy photon irradiation or free electron–hole recombination[102] produces two emission bands, one at 371 and another at 300 nm. Results of this kind indicate at least two electron–hole recombination paths that produce luminescence.

Even though there appears to be general agreement among investigators that radiationless recombination of electrons and holes is the source of defect production, investigators still differ about the details of the process. Pooley's point of view is that in the exciton state, the two halogens can be considered to be in an [X_2^-]-center configuration with the nearby compensating electron producing little perturbation of the trapped hole center. The relaxation of the halides that occurs upon recombination of the exciton can be as much as 40% of a lattice spacing and is considered to be large enough to cause a replacement collision chain to be propagated by halide ions along a close-packed $\langle 110 \rangle$ direction. Pooley's calculations[107] indicate that in most easily colored alkali halides, the replacement threshold is approximately the same as the exciton recombination energy available. However, other calculations[108] indicate that the replacement threshold may be two to three times the energy available from electron–hole recombination. In view of the difficulty of making calculations of this type and the number of assumptions that are inherent in them, the different results are not surprising.

Data exist for more complicated structures than alkali halides which suggest that defect production by radiolysis can go on without a replacement collision sequence. Magnesium fluoride, for example, is tetragonal (rutile structure) and there are no close-packed rows of fluoride ions analogous to the ⟨110⟩ rows in NaCl structures. Thus it is difficult to see how a replacement chain would propagate. Yet MgF_2 can be colored by radiolysis[109] just as can the alkali halides, i.e., *F* and *F*-aggregate centers can be produced in large concentrations. On the other hand, analysis of data taken on mixed alkali halide crystals[110] suggest that focusing collisions do occur and that impurities interfere with the focusing to inhibit the production of defects.

Konitzer and Hersh, in their original paper,[100] suggested that the Frenkel pair separates by diffusion of the atomic (uncharged) halogen atom. Although no quantitative calculation was made, they viewed the exciton state as being similar to, or decaying to, an antibonding configuration of a $[Cl_2^-]$ molecule ion, one of whose members, being smaller in the uncharged state than in the negatively charged state, would move under the influence of the electronic recombination process, leaving the electron behind in the vacancy. Lushchik *et al.*,[55] in a recent discussion of the excitonic production mechanism, also favor the direct production of *F*-center–interstitial-atom pairs rather than F^+-center–interstitial-ion pairs which would result from displacement collisions.

Ueta and co-workers[111] have one piece of experimental evidence that appears to support the point of view that it is the uncharged Frenkel pair that is produced by radiolysis. They found that when a virgin sample of KCl is irradiated with a 30-nsec, 60-MeV electron pulse at liquid nitrogen temperature, *F* centers were formed in times shorter than the pulse width. In samples containing F^+ centers, however, they were able to observe a characteristic slow growth of *F* centers which was due to trapping of electrons by empty vacancies followed by the well-known anomalously slow decay of the electron from the *F*-excited state to its ground state.[25,112] Figure 6 shows the lack of slow growth of *F*-band absorption in a virgin sample (A) compared to one containing F^+ centers (B).[111] These results are not compatible with a production process by which F^+-center–interstitial-ion pairs are produced, since in that case, the slow growth could be observed in a virgin crystal in the same way it is observed in a preirradiated one.

4.3.2. *Defect Mobilities*

From the above discussion, it is evident that there is some disagreement concerning the mode in which the members of the radiation-produced Frenkel pair separate. The two modes that have been discussed, interstitial-ion motion as a replacement collision chain and interstitial-atom diffusion, are clearly not the only possible ways in which defects can move. This is

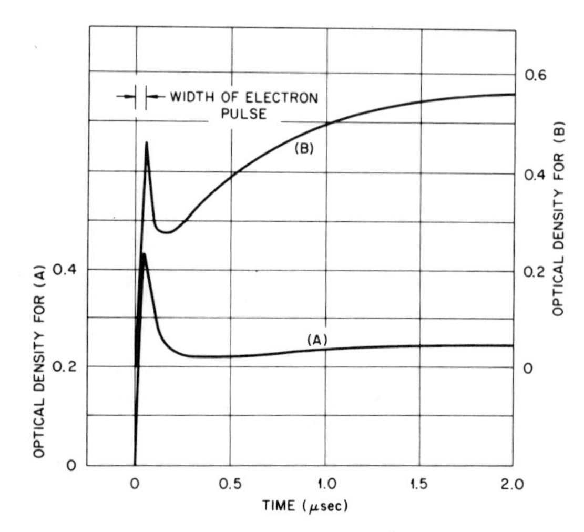

Fig. 6. F-center formation at liquid nitrogen temperature by a single electron pulse in (A) freshly cleaved and (B) previously irradiated KCl. The optical density values were taken from the peak of the F band at 540 nm.[111]

especially true at low temperatures, where motion may well be under the influence of a localized elevated temperature, i.e., so-called thermal spikes, resulting from the electron–hole recombination. Also, there is most likely some recombination of the primary defects during and after the production process. Experimentally, it is extremely difficult to separate the primary production process from these secondary reactions that take place between the time of absorption of the radiant energy and the time of measurement of the defects. This problem is complicated by the fact that the secondary reactions, which involve motion of primary defects, are almost certainly temperature-dependent. In this part, the mobility of defects will be considered; then, the secondary interactions will be discussed.

(a) *Mobility of Interstitial Atoms.* The temperature ranges in which H centers become unstable in LiF, KCl, KBr, and KI are given in Table VI. Not many halides have been investigated and for those that have, only an approximate temperature range of annealing can be extracted from the results. Moreover, a recent investigation of interstitial atom centers in LiF, in which double resonance techniques (ENDOR) were employed, has indicated that the temperature 100–110°K, originally thought to be the annealing temperature for H centers,[63] was probably the annealing temperature for H_A centers[113] (interstitials trapped adjacent to impurity cat-

Table VI. Interstitial-Atom Stability

Material	Property measured	Thermal treatment	Temp. of max. change, °K	Activation energy,[a] eV	Ref.
LiF	Spin res. of ⟨110⟩ *H* center	Anneal for 30 min; recool for measurement	100–110	—	63
	Spin res. of ⟨111⟩ int. atom	Not stated	∼60	—	114
KCl	Spin res. of ⟨110⟩ *H* center	Anneal for "few min"; recool for measurement	42–60	—	72
	Opt. absorption of *H* band	Warm at 1°K/min	33	0.04 (1)	115
			44	0.09 (2)	
KBr	Spin res. of ⟨110⟩ *H* center	Anneal for "few min"; recool for measurement	20–40	—	72
	Opt. absorption of *H* center	Anneal and recool to 6°K for measurement	12	—	116
			25		
			39		
	Opt. absorption at 380 nm	Warm at 0.1°K/min	27	0.1 (1–2)	117
			40		
			52		
			110		
KI	Opt. absorption of *H* band	Warm at ½°K/min	55–70	—	100

[a] The number in parentheses following the activation energy refers to the order of reaction assumed to obtain the given result.

ions). The intrinsic H center in LiF, which is more easily produced in the higher-purity material now available and which appears to be oriented in the $\langle 111 \rangle$ direction, is mobile above approximately 60°K.[114]

When a detailed annealing experiment is performed, the H centers (and, as we shall see below, I^- centers) disappear in steps. For example, the H center in KCl decreases by about 20% near 33°K, with much of the remainder going away near 44°K.[115] These two annealing steps are indicated by the two temperature entries on separate lines in Table VI. (Two temperatures with a dash between them indicate uncertainty; i.e., that annealing takes place somewhere within the range bracketed by the two numbers.)

(b) *Mobility of Interstitial Ions.* Interstitial ions, since they have an even number of electrons, do not give rise to spin resonance spectra. Moreover, there are no well-known interstitial-ion absorption bands. This makes direct experimental observation of the stability of interstitial ions more difficult. However, if one assumes that an interstitial ion which is negatively charged is strongly attracted to a positively charged empty vacancy (F^+ center), then one can obtain information about the motion of interstitial ions by monitoring the concentration of F^+ centers. This can be done by measuring an optical transition of the F^+ center which is known as the α band.[115,118–121] Table VII is a summary of annealing experiments performed. We have also included experiments in which lattice constant, length change, and thermal conductivity were the properties measured. These latter measurements are more indirect in that it is less certain that the effect observed was due to annealing of interstitial ions. However, the authors usually compared the annealing of these indirect properties with changes in the α band and found comparable behavior.

Table VII shows that much more detailed work has been done on interstitial-ion annealing than on interstitial-atom annealing, even though results are limited to KCl and KBr. A number of annealing stages have been observed and for the most part there is rough agreement among different investigators about the temperature of the annealing steps, ~18, ~28, and ~35°K for KCl and ~12 and ~20°K for KBr.

Some of the gross annealing steps can be divided into substages, as has been shown by Itoh *et al.*[119] for the 20°K annealing step in KBr. The explanation for the multiple steps that is generally accepted is that the attraction between close pairs of Frenkel defects lowers the potential barrier so that defect pairs located in specific relative positions recombine at particular temperatures in a first-order process. Only the highest-tem-

Table VII. Interstitial-Ion Mobility

Material	Property measured	Thermal treatment	Temp. of max. change, °K	Activation energy,[a] eV	Ref.
KCl	Opt. absorption of alpha band	Warm at 1°K/min	18	0.025 (1)	115
	Opt. absorption of alpha band	Anneal and recool for measurements	23	—	118
			35		
	Lattice constant	Anneal for 10 min; recool for measurements	29	—	122
	Length change	Anneal for 10 min; recool for measurements	28	—	123
	Thermal conductivity	Anneal and recool for measurements	18	0.06–0.11	124
			36		
KBr	Opt. absorption of alpha band	Warm at 0.6°K/min	11	0.015 (1)	119
			17	0.032 (1)	
			19	0.04 (1)	
			21	0.06 (2)	
	Opt. absorption of alpha band	Warm at $\frac{1}{2}$°K/min	14	0.025–0.04	120
			22	0.07–0.09	
	Opt. absorption of alpha band	Anneal and recool for measurements	12	—	121
			20		
	Thermal conductivity	Anneal and recool for measurements	13	—	124
			22		

[a] The number in parentheses following the activation energy refers to the order of reaction assumed to obtain the given result.

perature stage is considered to involve uncorrelated recombination of the defects. Calculations have been made of the mobility energy of certain close pairs;[125] the results indicate that the temperatures found for KBr (10–20°K) are not unreasonable for near-neighbor interstitial-ion–F^+-center pairs recombining via an interstitialcy mechanism of diffusion.

Comparison of the temperatures shown in Table VII with those in Table VI shows that interstitial-atom centers are stable to higher temperatures than are interstitial ions. Nevertheless, it should be kept in mind that under irradiation conditions, free electrons are released into the lattice, which can change interstitial atoms to ions. This can result in radiation-induced annealing of interstitial atoms at temperatures at which they are normally stable.

(c) *Mobility of Negative-Ion Vacancies.* As is the case in the noble metals,[46,47] vacancy defects are more stable than interstitials. If an estimate of negative-ion vacancy mobility is made from diffusion or ionic conductivity measurements,[126] it is evident that vacancies should not move except at temperatures above about 300°C. However, measurements of *F*-center production and aggregation in the room-temperature range have made it certain that some form of negative-ion vacancy defect must be mobile as much as 50–100° below room temperature. For example, when a KCl sample containing *F* centers is bleached at room temperature, F_2 and F_3 aggregate centers form, as shown in Fig. 7.[127]

One might try to explain the appearance of the F_2 and F_3 bands in terms of electron transfer from *F* centers to empty divacancies that have previously been produced by the radiation; but if that were the case, the F_3 ban would begin to grow with a maximum slope, as does the F_2 band, and would not exhibit the delay (initial positive curvature) which appears to be due to the necessity for building up F_2 centers before F_3 centers can begin to form. Moreover, recent work[128–130] has demonstrated that vacancy aggregates with missing electrons (F_2^+ and F_3^+ centers) give rise to absorption bands in the infrared. Yet these bands cannot be observed after irradiation at room temperature, even though more than 10^{17} F_2 centers/cm^3 form upon bleaching.

The question is, "What are the forms of the negative-ion vacancy that might be mobile near and below room temperature?" The *F* center in the ground state can be eliminated with a fair degree of certainty. In all the alkali halides, it is possible to store an irradiated or additively colored sample at room temperature in the dark for long times without observing any appreciable disappearance of *F* centers. Even the disappearance of *F*

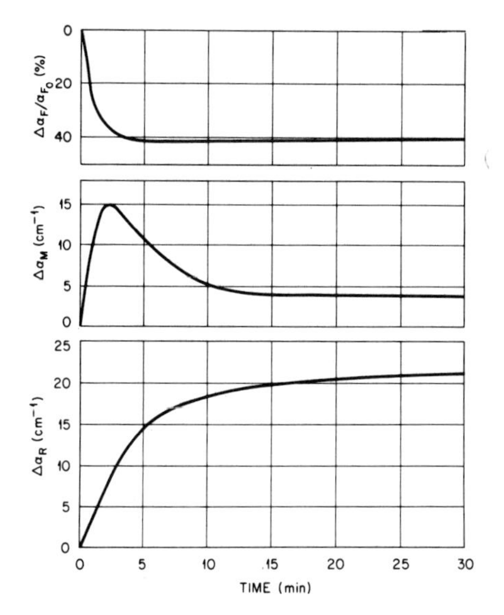

Fig. 7. Growth of F_2 (α_M) and F_3 (α_R) centers and decay of F (α_F) centers resulting from bleaching additively colored KCl with 546-nm light at room temperature.[127]

centers above room temperature appears to be related to thermally released charges. This is supported by experiments which show that in doped or deformed materials, where one would expect large numbers of trapped charges that can be thermally released, the F band anneals at lower temperatures than it does in pure materials.[131–133]

One possible mobile form is the F^- center. Table VIII gives the temperatures at which F^- centers disappear in a number of alkali halides. In the potassium salts, these are well below room temperature, so that if the disappearance of the bands were connected with motion of the defects, we would have good evidence for a mobile vacancy defect. However, the evidence suggests that F^- centers disappear by losing one of the two electrons, thus becoming F centers, rather than by diffusing to sinks. In a number of the references indicated in Table VIII, a thermoluminescent glow peak[135,138,140,142] or a measurable electronic current,[135,140] which was interpreted as being due to released electrons, was observed to accompany the disappearance of the F^- centers. Moreover, as Fig. 8 shows, the F^--center decay is accompanied by F^+-center decay and F-center growth.[142] This is what might be expected if electrons released from F^- centers are attracted by F^+ centers forming uncharged F centers in the process. Also,

Table VIII. F^--Center Stability[a]

Material	Thermal treatment	Temp. of max. change, °K	Ref.
NaF	Warm at $\frac{3}{4}$°K/min	280	134
NaCl	—	360	135
	Anneal for 15 min; measure at temp.	290	136
KCl	Warm at ~1°K/min	<180	137
	Warm at ~5°K/min	198	138
	Warm at $2\frac{1}{2}$°K/min	215	139
	Warm at 5°K/min;	205	140
	Anneal for 15 min; measure at temp.	200	136
KBr	Anneal for 15 min; measure at temp.	130	136
	Warm at 5°K/min; measure at temp.	~145	140
	Warm at $2\frac{1}{2}$°K/min	157	139
KI	Anneal for 15 min; recool to 78°K for measurement	180	141

[a] In all cases, the property measured was the decay of optical absorption of the F^- band. The measurement reported for KBr from Ref. 139 was made at 87°K after F light at 130°K.

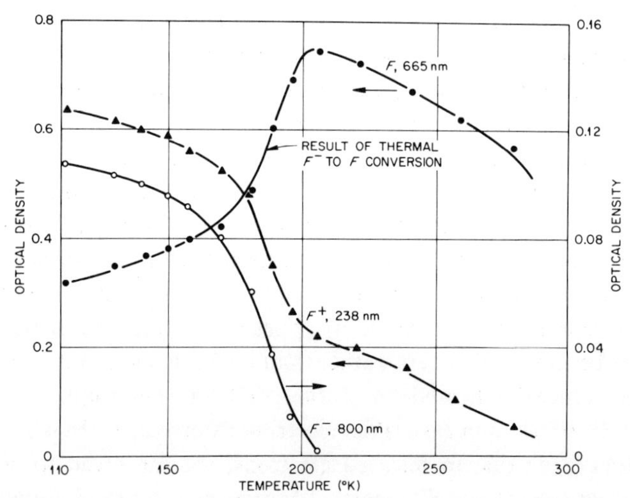

Fig. 8. Decay of F^+ and F^- centers and accompanying growth of the F band resulting from isochronal annealing of irradiated KI. Irradiation and optical measurements have been made at 80°K.[142]

as has been pointed out by Rabin,[137] F-aggregate-center formation does not occur at temperatures at which F^- centers disappear.

The conclusion that F^- centers disappear in all cases because of electron releases is not yet certain. For example, Delbecq *et al.*[12] have found that in Tl-doped KCl at 208°K, a thermoluminescent glow peak arises from mobile holes (see Table V) recombining with [Tl°] centers. This would suggest that the F^- disappearance in pure KCl, which occurs in the same temperature range, is also due to mobile holes. On the other hand, in pure material, there exist many more F^- centers than self-trapped holes after irradiation, so that hole motion should destroy at most a fraction of the F^- centers. Another problem concerns F^- centers that disappear at low temperatures. Sonder *et al.*[138] find in warming KCl irradiated at 77°K that only about 20% of the F^- centers present at liquid nitrogen temperature disappear between 80 and 190°K. But in Rabin's work,[137] where a very large defect concentration was present after irradiation and a slower warming rate was used, 80% disappeared below 120°K. It is not clear yet whether this low-temperature disappearance involves electron tunneling, hole release, or recombination of interstitials with the F^- centers.

Alternatively, F^+ centers could be the mobile vacancy form. Table IX gives temperatures at which F^+ centers disappear and F_2 and $F_2{}^+$ centers form spontaneously in the dark upon warming. Potassium chloride is the only material for which both the decay of the α band and the growth of aggregate centers have been observed. The results listed in the table show that F^+ centers decay at a number of different temperatures in the range 100–300°K and it is only in the vicinity of the highest-temperature F^+-center anneal step (where all remaining centers disappear) that aggregate centers form. For example, Giuliani *et al.*[118,148] found F^+-center annealing in KCl near 113, 170, 223, and 235°K. Divacancy centers form in KCl between 235 and 270°K,[137,138–146] indicating that perhaps the high-temperature step may be connected with vacancy motion, but that the lower-temperature steps almost certainly do not involve long-range motion of vacancies. Most likely, much of the low-temperature F^+-center annealing is due either to electron release by F^- or other traps in the crystal or to interstitial–vacancy recombination. It should be clear from Table IX that very little definitive annealing work has been done from which quantitative information about the mobility of F^+ centers can be obtained.

The results of thermal studies of vacancy centers summarized in Tables VIII and IX do not make it possible to decide unequivocally whether the empty vacancy or the one containing two electrons is the mobile species at room temperature. The fact that F^- centers have in all experiments been

Table IX. F^{+}-Center Stability

Material	Property measured[a]	Thermal treatment	Temp. of max. change, °K	Activation energy,[b] eV	Ref.
LiF	A	Anneal for 13 min; recool to 248°K for measurement	275	0.67 (1)	143
	B	Anneal for 13 min; recool to 248°K for measurement	320	0.75 (1)	143
	A	Anneal and measure at temperature	285	0.74	144
NaF	A	Anneal and measure	280	0.73	144
	A	Warm at $\frac{3}{4}$°K/min	~280	—	134
	B	Warm at $\frac{3}{4}$°K/min measurement	310	—	134
NaCl	C	Anneal for 15 min; recool for measurement	295	0.9(2)	145
	B	Anneal for 15 min; recool for measurement	265	—	146
KF	B	Anneal for 15 min; recool for measurement	245	—	146
KCl	C	Anneal for 15 min; recool for measurement	204	0.37 (2)	145
	C	No details given	128	—	147
	C	No details given	235	—	118
KCl	C	Anneal for 15 min; recool for measurement	113 170 223	—	148
	A	Warm at 1°K/min	235	—	137
	B	Warm at 1°K/min	245	—	137
	B	Warm at 5°K/min	270	—	138
	B	Anneal for 15 min; recool for measurement	250	—	146
	A, B	Max. growth of F_2 band; isothermal annealing	250	0.52 (1)	149
KBr	C	Anneal for 15 min; recool for measurement	135	—	145
	B	Anneal for 15 min; recool for measurement	233	—	146

Table IX (*continued*)

Material	Property measured[a]	Thermal treatment	Temp. of max. change, °K	Activation energy,[b] eV	Ref.
KI	C	Anneal for 10 min; recool for measurement	240	0.5	150
	C	Anneal for 15 min; recool for measurement	180 230	—	142
RbCl	B	Anneal for 15 min; recool for measurement	225	—	146
RbBr	B	Anneal for 15 min; recool for measurement	225	—	146
RbI	B	Anneal for 15 min; recool for measurement	220	—	146

[a] A: Growth of optical absorption of F_2^+ band. B: Growth of optical absorption of F_2 band. C: Decay of optical absorption of F^+ band.
[b] The numbers in parentheses following the activation energy refer to the order of reaction assumed to obtain the given result.

observed to disappear completely at temperatures well below that at which aggregation takes place does suggest, however, that the vacancy containing two electrons is not the mobile vacancy species we are looking for.

A number of recent experiments done with KCl yield stronger evidence that the F^+ center is the mobile vacancy entity. Lüty and co-workers[26,151,152] have used F_A centers to show that it is the empty vacancy that is mobile at temperatures below room temperature. F_A centers, in contrast to spherically symmetric F centers, have an axis of symmetry, $\langle 100 \rangle$, and can therefore be aligned by polarized optical bleaching. The onset of mobility of a vacancy next to an alkali impurity can be observed by a loss of alignment. Link and Lüty[152] used heavily-Na-doped KCl with a large concentration of F_A centers and aligned these centers by bleaching with polarized light at 77°K. Moreover, they removed electrons from some of these centers by bleaching the crystal at 50°K in a strong electric field. In this way, a crystal was prepared which contained about equal numbers of aligned F_A^+, F_A, and F_A^- centers. Although the positively charged centers could not be monitored, it was possible to measure the concentration and dichroism of the F_A and F_A^- species. Isochronal anneals were made to temperatures in the range 130–

170°K; after each heating pulse, the sample was measured at 50°K. In this experiment, no change was observed in either the F_A or F_A^- dichroism. A repeat of the same experiment, but this time converting all three defects to F_A centers after each warming pulse, resulted in a decrease of the F_A-center dichroism. The amount of decrease varied exponentially with the annealing temperature. Link and Lüty concluded from these two experiments that only the positive charge state F_A^+ center is mobile in the range 130–170°K, and that its energy of motion around the neighboring Na impurity is 0.45 eV. This result suggests the F^+ center is mobile in the perfect lattice with a slightly higher activation energy; and, in fact, the 0.6-eV mobility[151] energy that has been attributed to motion of the mobile vacancy species in pure KCl appears to be of the correct order of magnitude to support this argument.

Giuliani[153] demonstrated that the F^+ center is the mobile species in a slightly different way. He used Sr-doped KCl that contained F centers. It is well known that bleaching in the vicinity of room temperature causes F_Z centers (F centers adjacent to Sr^{2+} and empty positive-ion vacancies) to form. Giuliani cooled his sample to 80°K and bleached at that temperature to produce all three charge states of the negative-ion vacancy. During subsequent warming, he found an aggregate center forming near 250°K at the same time the F^+ centers decreased. The new aggregate centers were identified as F_Z^+ species since they could subsequently be converted to F_Z centers at liquid nitrogen temperature (where neither positive holes nor any species of vacancy is mobile) by F-light bleaching.

The evidence appears to be sufficient to conclude that, for KCl at least, the F^+ center can move below room temperature with an activation energy of approximately 0.5 eV. This produces a dilemma since it is clear from diffusion and ionic conductivity measurements that the negative-ion vacancy moves with an activation energy of approximately 1 eV.[126] Additional work is necessary to decide whether the interpretation of the low-temperature experiments is in error, whether there is another way of interpreting the high-temperature transport measurements, or whether some subtle polarization effect permits the vacancy to move with a much lower activation energy below room temperature than in the diffusion-ionic conductivity range.

4.3.3. *Secondary Processes*

(a) *Charge Redistribution among Vacancy Centers at* 4°K. Until recently, it was thought that a different production process operates at liquid-helium than at higher temperatures since the predominant vacancy defect produced at the lowest temperatures was the F^+ center, whereas at higher temperatures, it was the F center. However, Ritz's careful and complete study[88] of F- and F^+-center production at low temperature in KBr has shown that the differently charged vacancies are affected in the same way by changing the energy of the impinging radiation and that, moreover, the ratio of F to F^+ centers depends only upon the F-center absorption coefficient of the sample. These results suggest that only one charge state of the vacancy is produced and that the distribution of charge occurs as a secondary reaction. Further support for this hypothesis comes from the observation[88,154] that the primary radiation-induced luminescence disappears at temperatures near which the F^+ to F production ratio drops from the range 5–10 to approximately 0.2. It is likely that the F-center–interstitial-atom pair is the primary defect, as suggested by the recent results of Ueta *et al.*,[111] and that the fundamental luminescence and, to a lesser extent, radiation excite F centers. The excited state is known to be long-lived.[112] In Frenkel pairs, where the interstitial and vacancy are close together, the excited electron may tunnel to the H center with a high degree of probability to form an interstitial ion. For Frenkel pairs that are widely separated, the excited electron has to return to the ground state and ionizes only due to radiation ionization. Such ionization of widely separated Frenkel pairs is infrequent; therefore for these defects, F^+ centers appear only if the electron is trapped elsewhere, for example, at impurity atoms. Giuliani and Reguzzoni[148] have designated two types of F^+ centers: one type that anneals at low temperature (temperatures where the closely spaced interstitial-ion–F^+-center pairs annihilate) and another that is stable to temperatures above that of liquid nitrogen, *the concentration of which depends upon the impurity content*. This result is in accord with the foregoing proposition.

The negatively charged species of the vacancy (F^- center) has usually been ignored at liquid helium temperatures because it is not generally observed, but Kabler[154] points out that the very fact that it is not observed is in agreement with the idea that the fundamental luminescence causes electron excitation in the vacancy centers when a sample is irradiated at very low temperatures. The F^- center, for which absorption of light ionizes the electron into the conduction band, would clearly be the first species to be destroyed by the luminescence.

(b) *Interstitial Motion at* 4°K. The work summarized in Tables VI and VII shows that both I and H centers disappear above 10°K by close pair annihilation with activation energies between about 0.025 and 0.05 eV and by free motion with energies up to 0.1 eV. One might conclude from this that these defects cannot move under any condition at 4°K, but ESR results[72] indicate that interstitials are produced such that they are removed at least a few lattice spaces from their corresponding vacancies.[25] This requires some type of motion. Moreover, Itoh and Saidoh[28] have shown recently that diinterstitial atoms (H_2 centers) as well as interstitial-atom–impurity pairs (H_A centers) are formed at liquid helium temperature in concentrations much larger than one would expect due to the statistical probability that one interstitial is produced randomly next to an impurity or another interstitial. They interpret their results in terms of interaction volumes for nascent interstitials and impurities of approximately 150 lattice volumes and for two interstitials of approximately 2000 lattice volumes. Such large interaction volumes require that the interstitial be mobile enough within this volume to move a distance of approximately 10 lattice spaces toward another interstitial or an impurity. It might be noted at this point that the *F*-center saturation[84] produced with protons (2×10^{19} F/cm^{-3}) also can be explained by assuming that an interstitial annihilates vacancies already present within approximately 1000 lattice sites of the interstitial production site.

The fact that interstitials form complexes or annihilate vacancies requires that, at least in a radiation field, they move through a large enough distance to pass adjacent to approximately 1000 lattice sites. If diffusive motion is involved, this corresponds to a range of about 10 lattice speces. If the interstitials move only along straight lines, as they would in a replacement chain, then the range would be of the order of 100 lattice spaces.

(c) *Effect of Mobile Interstitials.* As temperatures are raised above the liquid helium range, primary defects become mobile even without the help of the radiation field. As is shown by the results summarized in Tables VI–IX, there is a large range of temperatures in which interstitial defects are thermally mobile, whereas vacancies are not. For most alkali halides, liquid nitrogen temperature is very conveniently near the middle of this temperature range. We discuss now the work, performed chiefly at liquid nitrogen temperature, for which secondary reactions involving mobile interstitials should not be ignored. The pertinent observations that have to be accounted for are as follows:

(1) The rate of production of color centers is much lower at liquid nitrogen temperature than at 4°K. This has been observed for KCl[155] and KBr[156] and is true both for the total vacancy concentration and for the uncharged (*F* center) species. However, a sample irradiated at 4°K and

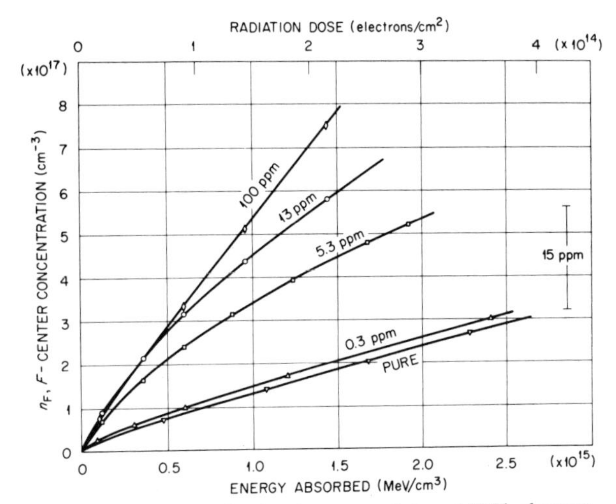

Fig. 9. *F*-center production in pure and doped KCl, electron-irradiated at 80°K. The numbers labeling the curves give concentrations of Pb in ppm.[160]

warmed to 80°K will exhibit very nearly the same concentration of *F* and F^+ centers[155] as an identical sample irradiated at 80°K.

(2) *F*-center production curves are nearly linear[157,158] in dose over a wide range of concentrations. Moreover, there is no dose-rate dependence, i.e., a large increase in dose rate with a corresponding decrease in irradiation times produces no change in the number of *F* centers produced per absorbed energy.

(3) Trace impurities contained in the alkali halides enhance the defect production at liquid nitrogen temperature by a factor of from 2 to 5.[159–161] They also produce a greater curvature in the *F*-center production curves, as shown in Fig. 9.[159]

The observation of H_A centers,[35,36] as well as the impurity enhancement of defect production observed for liquid-hydrogen[162] and liquid-nitrogen temperature irradiation, show that interstitial trapping of some sort takes place. Moreover, the recent discovery of Mn-impurity–interstitial-halide aggregates[163] and the indication that H_2 centers can be produced[28] suggests that interstitial trapping and aggregation may be a general phenomenon which makes it possible for vacancies in the form of *F* centers to remain in the crystal at higher temperatures. Some recent indirect measurements[164] suggest, in fact, that at temperatures approaching room temperature, a distribution of interstitial clusters is formed that becomes larger with increasing irradiation temperatures. For example, the stability of F centers produced at various temperatures between 80 and 300°K was measured by bleaching the samples after warming. It was found that in the higher-temperature irradiated samples, the defect structure was more stable, even though the F centers themselves were identical in all samples. This higher stability was attributed to the presence of larger and more widely separated interstitial clusters. It was also found both for KCl[164] and KBr[165] that the ultraviolet absorption spectra in the samples irradiated at various temperatures and warmed to room temperature varied continuously with irradiation temperature. If one accepts the usual assumption that the ultraviolet bands reflect the complementary defects to the F centers, then this result suggests a continuous variation of interstitial-type centers, which could be a result of an increase of cluster size with irradiation temperature. This concept of an increasing interstitial cluster size as the temperature of irradiation approaches room temperature is also supported by low-temperature thermal conductivity measurements.[166,167] An analysis of these data suggests that after heavy room-temperature irradiation of KCl, clusters of perhaps 100–1000 halide atoms are present.

Now let us consider the impurity enhancement of the coloration as shown in Fig. 9. The simplest explanation is that impurities act as traps for radiation-produced mobile interstitials, thus suppressing interstitial-vacancy recombination and producing, as a result, a greater concentration of vacancies. Such an explanation also can account qualitatively for the increased curvature in doped samples, since impurities may well be able to trap only a finite number of interstitials before losing their effectiveness.

A quantitative version of this model has been applied to F-center production curves in LiF irradiated at room temperature.[168,169] This approach is represented by the rate equations

$$\begin{aligned} di/dt &= \gamma - Biv - CiN \\ dv/dt &= \gamma - Biv \end{aligned} \tag{9}$$

where i and v are respectively the numbers of interstitials and vacancies per cubic centimeter during irradiation, γ is the vacancy production rate, and Biv and CiN are recombination and trapping rates, respectively; B and C are equal to $4\pi rD$, where r is the recombination or trapping radius, respectively, and D is the diffusion coefficient for interstitials; N is the density of trapping sites, which was assumed constant. The solution of these equations at long irradiation times yields a square-root dependence on dose, which is in agreement with the room-temperature experimental data. The theory ought to be especially applicable in the temperature range 100–250°K since, as Tables VI and IX indicate, in this temperature range, interstitial atoms are mobile and vacancies are not. Unfortunately, the *F*-center production curves in LiF deviate from the square-root dependence predicted by Eq. (9) and are more nearly linear at 80°K. This is even more clearly observed in KCl[159] and NaCl[158] and has led to further analysis of the model represented by Eq. (9). A review of the solutions suggested[157] that only for unreasonably small values of the recombination parameter B would the equations yield an approximately linear production curve. Moreover, for such conditions, the solution predicted a dose-rate dependence. This was looked for and not observed. It was therefore concluded[157] that Frenkel defect production and stabilization were so fast that it would be sufficient to consider each single production and stabilization event independent of all previous or subsequent ones. On that basis, Eq. (9) becomes unnecessary because there is no buildup of any significant steady-state concentration of interstitials. The *F*-center production rate is simply the product of the initial Frenkel pair production efficiency, times the probability that the produced center does not recombine.* This conclusion is in accord with the observation (1) listed above, namely that the same end product is obtained for liquid-nitrogen-temperature irradiation as for liquid-helium-temperature irradiation and subsequent warming. It also agrees with the results of the pulsed measurements mentioned earlier,[111] which indicate that the total production process (including stabilization or recombination) is completed in KCl at liquid nitrogen temperature in approximately 40 nsec.

The kinetics of production and stabilization of defects at liquid nitrogen temperature are far from being well understood. In particular, the very rapid completion of the secondary reactions that the pulsed measurements indicate and that is necessary to account for the liquid-nitrogen-temperature curve shapes and absence of dose-rate dependence seems to be in contradiction to the (lower) mobilities for interstitial ions and H centers determined from annealing measurements. As mentioned previously, this discrepancy may be the result of localized high temperatures (thermal spikes) during irradiation or the existence of a species of interstitial that is more mobile than the ion and H center so far observed.

* This stabilization probability may depend upon impurity content and the concentration of already produced F centers, but not upon any intermediate product whose steady-state concentration might depend on dose rate.

(d) *Effects of Mobile Vacancies and Holes.* As the radiation temperature increases, the probability of vacancies and holes being mobile during irradiation, in addition to interstitials, becomes greater. This means that other secondary reactions occur which appear to complicate the whole radiation damage picture. We mentioned in the previous section that interstitial–interstitial and interstitial–impurity trapping can explain the existence of vacancies after irradiation at 77°K. As the radiation temperature increases and vacancies themselves become mobile, some of them will move through the lattice and recombine with interstitial–impurity pairs or interstitial clusters to reduce the amount of *F*-center coloration. Moreover, others will cluster together to form F_2 and F_3 centers, which will further reduce the *F*-center concentration. Thus, we see that "back reactions" can have a marked effect on *F*-center coloration curves and the apparent *F*-center production efficiency.

In part 4.3.1(c) of this section, a discussion of the radiolysis production mechanism in alkali halides was presented. In that discussion, it was pointed out that the most likely explanation for the efficient Frenkel-pair production in these materials had to do with radiationless electron–hole recombination in the perfect lattice. It has been proposed[170] that the production mechanism itself can be short-circuited due to electron–hole recombination at defects. This becomes possible when holes become sufficiently mobile to move to impurities before recombination occurs.

In order to distinguish the effects of secondary reactions and short-circuiting on the colorability, we will first review the experimental observations for temperatures in the vicinity of room temperature and then discuss them in terms of the two possibilities mentioned above. The procedure normally followed in studying the kinetics of defect production is to irradiate a sample under well-determined conditions of temperature, dose rate, etc., and then a few minutes later to measure the defect concentration by means of optical absorption techniques. Dahnke and Thommen[171] have noted that some differences exist between samples whose absorption is measured during irradiation and those irradiated and then measured. More recent work of this nature[172] shows the differences even more clearly.

Experimental curves obtained by plotting the *F*-center concentration versus absorbed energy (dose) exhibit a rapidly saturating early stage, which can consist of a number of substages, followed by a slowly saturating late stage. Even though the growth curves are different from sample to sample, the gross qualitative behavior is reproducible. The same thing can be said for the production of *F*-aggregate centers and the relation between these centers and F centers. Therefore, we can summarize the general behavior of radiation-induced defects in alkali halides in the following way:

1. Early-stage coloration depends strongly on impurity type and concentration,[80,173–175] as well as on deformation,[176–178] radiation temperature,[173] intensity,[177,178] and other specific experimental conditions.[177,179] Because many of the experiments done on alkali halides deal with only the early stage and since it is so sensitive to many different parameters, we will devote a later subsection to a discussion of early-stage coloration.

2. Late-stage F-center coloration is characterized by a saturation concentration that is a function of radiation dose rate and temperature[180] as well as impurity.[181] In this stage, an increase in production efficiency occurs with an increase in dose rate and a decrease in efficiency is observed as the impurity concentration increases. This decrease can be eliminated by increasing the dose rate in samples that do not contain too many impurities, e.g., at high dose rates, specimens containing about 10^{17} impurities/cm^3 show the same type of growth curves as do pure crystals.

3. F-center aggregation, as indicated by the growth of F_2 and F_3 centers, can be caused by ionizing radiation[30,182–185] or by F-light bleaching.[186–189] It has also been demonstrated that radiation can destroy F-aggregate centers,[182–184,190] and indeed it is because radiation both creates and destroys aggregates that an equilibrium obtains between F and F-aggregate centers for given radiation conditions. The equilibrium between F and F_2 centers can be written

$$N_{F_2} = K(N_F)^2 \tag{10}$$

where N_{F_2} and N_F are the defect concentrations and K depends strongly on radiation temperature and intensity and on the impurity concentration.

First, we consider the late-stage defect saturation. At very high intensities, the F-center concentration saturates[84] at approximately 2×10^{19} cm^{-3}. For lower dose rates,[180] saturation occurs in most alkali halides* at much lower concentrations. The level depends upon temperature and dose rate, indicating that a thermally activated process is involved and that the forward (F-center production) and back (F-center destruction or inhibition) reactions have a dissimilar dependence on dose rate. One explanation that accounts for the observed behavior is as follows: F centers are

* Lithium fluoride might be an exception; 10^{19} F centers/cm^3 can be produced at temperatures up to 150°C (see Refs. 168, 169).

produced at a rate proportional to the absorbed energy and without any large temperature dependence of the production mechanism itself. A small, steady-state concentration of ionized *F* centers (F^+ centers) is present as long as the sample is being irradiated. These empty vacancies diffuse thermally, with activation energy about 0.5 eV in KCl, and annihilate at interstitial clusters. Since diffusion appears to be the major rate-controlling step in this reverse process, it is not surprising that the decrease in the saturation level with temperature is governed by a reaction energy whose magnitude[180] is the same as the activation energy for vacancy motion in this temperature range.[151]

The above explanation also accounts for the impurity and intensity dependences. The number of mobile vacancies (F^+ centers) is greater when the number of electron traps (impurities) is greater. Thus a larger impurity concentration will tend to enhance the back reaction, causing a lower net *F*-center production rate. Recent observations have shown that, in fact, moderate doping of KCl with lead causes the *F*-center concentration at saturation to be less than it is for the same temperature and dose rate in pure material. Such a result is shown in Fig. 10.

As the dose rate is increased, the concentration of F^+ centers does not increase proportionately, but saturates at a concentration comparable to that of the electron traps. Thus the black reaction becomes independent of dose rate at high radiation intensities, while the *F*-production reaction

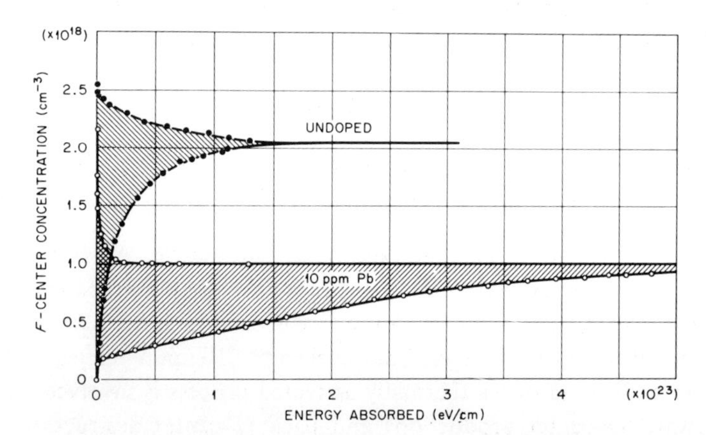

Fig. 10. *F*-center production and destruction curves in commercial undoped (Harshaw) KCl and deliberately doped material. The rising curves starting from the origin are normal coloring curves obtained at 20°C in a gamma source; the descending curves were obtained in the same gamma source after excess *F* centers had been produced by high-intensity electron irradiation.[180]

increases with the dose rate. It follows that higher radiation intensities should produce higher production rates for defects, as is observed.

The applicability of the above explanation can also be tested by optically stimulating the $F \rightarrow F^+$ reaction and seeing whether that decreases the net F-center production. This has been done by bleaching KCl in the F band while it was being irradiated at a moderate dose rate. A decrease in production rate was observed.[191]

Another explanation for the radiation behavior at room temperature exists. Instead of favoring the annihilation reaction, impurities can inhibit the production reaction, thus also depressing the saturation level. In fact, the production curve of the doped sample in Fig. 10 clearly exhibits a lower slope over the whole late-stage range of the curve. If the impurity were simply enhancing the back reaction, the initial slope would be invariant. If, however, the F-center production rate were affected by the presence of impurity while the back reaction remained unchanged, the slope of the production curve would be affected. Pooley[170] has analyzed the possibility of electron–hole recombination at impurity sites short-circuiting the excitonic production process. His calculations predict that the inhibition of the production mechanism depends upon a parameter,

$$S \approx (1/60)\dot{n}_\gamma N_0/\nu_h N_I^2 \tag{11}$$

where $\dot{n}_\gamma$ is the electron–hole pair production rate, N_0 the concentration of halogen sites, N_I the concentration of recombination centers, and ν_h the rate-limiting hole jump rate for electron–hole recombination. For $S \gg 1$, the short-circuiting is saturated and there is no visible effect on the colorations, but for $S \ll 1$, most of the electron–hole pairs produced by the radiation flux $\dot{n}_\gamma$ recombine at impurity recombination centers N_I and late-stage coloration is suppressed. There appears to be excellent agreement between these predictions and measurements made with lead-doped KCl. Moreover, Dawson and Pooley[150] have measured the net production rates in doped KI and KCl as a function of temperature between 125 and 300°K and have found that the decrease in production rate with temperature scales with the hole jump energy. This is what would be expected for electron–hole recombination at impurity sites since hole diffusion is the rate-limiting step. Similar experiments with NaCl also support this mechanism.[192]

Thus it appears that inhibition or "short-circuiting" of the production mechanism has a great influence on measured F-center production rates in the vicinity of room temperature. It might be argued that even the temperature- and dose-rate-dependent saturation is due simply to inhibition of the production mechanism. However, in the measurements of F-center saturation in KCl, it was shown[180] that large numbers of F centers (and at the same time, F-aggregate centers) that had been produced and were stable after irradiation could be destroyed by changing the irradiation temperature or intensity. These results cannot be caused solely by inhibition of the production mechanism and it must be concluded that back reactions as well as inhibition processes operate at room temperature.

Now let us discuss F-center aggregation and the F-center–M-center equilibrium. As mentioned, the equilibrium constant K in the expression $n_{F_2} = K(n_F)^2$ depends upon experimental conditions.[30] In fact, it depends upon the same parameters as does the F-center production rate, namely impurity content, dose rate, and temperature. Figures 11 and 12 show the dependence of K on radiation intensity and sample purity when the radiation is at room temperature.[183] At temperatures below about 200°K in KCl, K is determined by the statistical probability of forming one F center next to an existing F center by radiation[86] and is very small. The similar dependence of K and the F-center production process on impurity, intensity, and temperature suggests that the same processes enter the aggregation equilibrium as enter the room-temperature F-center production. As a matter of fact, the explanations given above for the saturation of the F-center

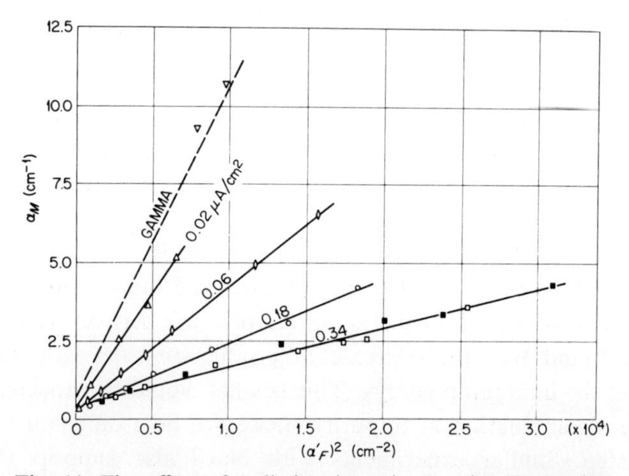

Fig. 11. The effect of radiation intensity on the F-center/M-center relation in Harshaw KCl. The dashed curve is for a sample gamma-irradiated in a 4.5×10^6 R source. The solid lines are for samples electron-irradiated with 1.5-MeV electrons and current densities shown.[183]

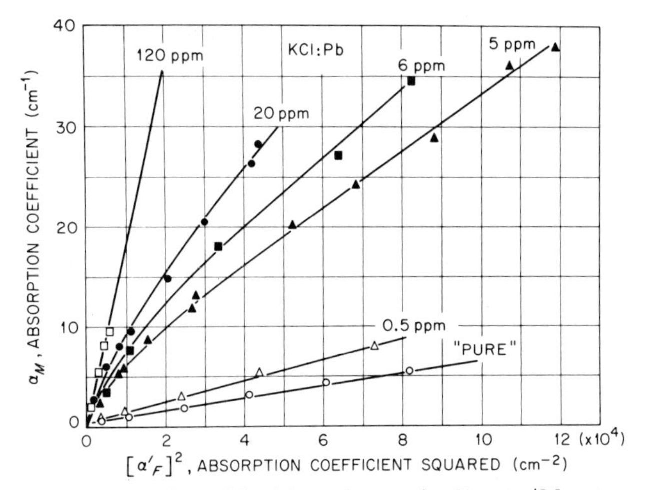

Fig. 12. The effect of lead impurity on the F-center/M-center relation in KCl electron-irradiated with a current density of 0.2 μA/cm². The curvature apparent at high M-center concentrations is due to formation of larger aggregates, F_3 and F_4 centers as discussed in Ref. 193.

production curves appear to be able to account admirably well for the aggregation behavior. If it is assumed that during irradiation some of the F centers present are ionized so that they become mobile until they bond with other F centers to form F_2^+ and later F_2 centers, then the aggregation process should go as the square of the F-center concentration and would increase with temperature according to the jump energy of the F^+ centers. Meanwhile, since fresh F centers and interstitials are continuously produced, a number of F_2 centers will be destroyed by interstitials.[193] This reaction would be proportional to dose rate and F_2-center concentration and would have no large temperature variation. The reader can conclude from this that K will decrease with dose rate and increase with temperature approximately as does the F^+-center mobility. Horn and Peisl[194] have measured K in KCl and have indeed found a reaction rate energy of 0.5 eV, which compares with the F^+-center mobility energy of 0.6 eV and the experimentally determined reaction rate energy for F-center saturation of 0.4 eV. Unfortunately, the hole mobility energy in KCl is of the same order of magnitude, 0.5 eV, and it is possible to devise models for the F/F_2 equilibrium in which hole motion is the rate-determining step.[195]

The effect of impurities can be explained in the same way as for F-center production. They can act as electron traps and can thereby cause a greater fraction of existing F centers to be mobile and to aggregate. At high dose rates, electron trapping would become less important as the traps become saturated. The change of K with impurity can also be affected by the impurity-caused short-circuiting of the production mechanism discussed before, since a decrease of Frenkel-pair production will cause a proportionate decrease in the F_2-center destruction, thus increasing K.

As the above discussion shows, the aggregation equilibrium can be accounted for qualitatively in terms of processes involving mobile interstitials and vacancies. Moreover, the processes may be complicated by electronic effects, particularly those involving electron trapping and electron–hole recombination at impurities. Recent careful measurements[195] of the equilibrium constant K as a function of temperature between 200 and 350°K for KCl suggest that the simple exponential dependence giving a 0.5-eV reaction rate energy may be only a first approximation to the behavior, and that the slope of log K plotted versus inverse temperature is not constant throughout this range but changes with temperature, impurity content, and dose rate.

(e) *Early-Stage Coloration at Room Temperature.* To this point, we have ignored what occurs at room temperature after light irradiation when samples are not very pure and when dose rates are not very large (conditions for which complete or partial suppression of the late stage occurs).

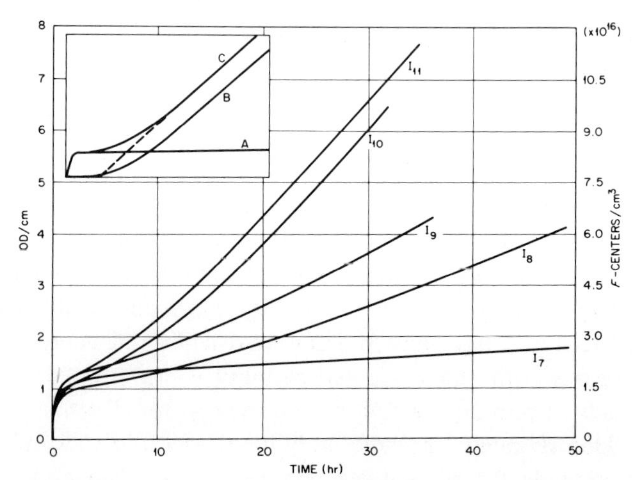

Fig. 13. Growth of F centers during the early stages of X-ray irradiation of KCl. The inset indicates how curves of this shape can be analyzed in terms of a saturating first stage and a delayed linear second stage.[178]

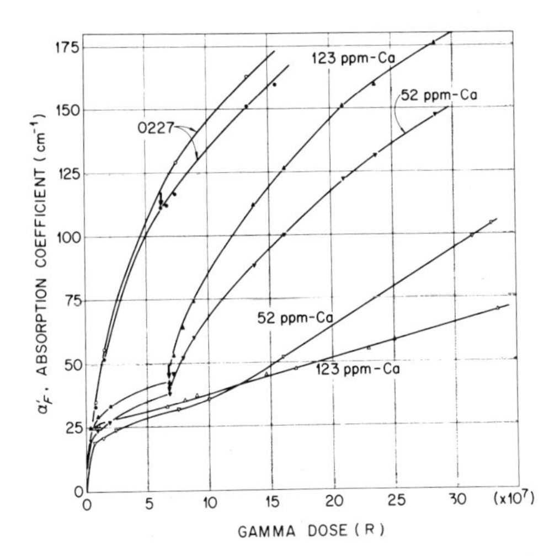

Fig. 14. *F*-center coloring curves for pure and doped KCl, showing the effect of a few per cent deformation. The arrows and the sudden change in slope for the doped samples indicate at what radiation dose the samples were deformed. Complete curves for the pure and doped samples are included to facilitate comparison.[176]

Historically, these are the conditions under which most of the early experimental work was done and, ironically, it is this early-stage defect production that is perhaps the least well understood of these processes. Figure 13 shows classical *F*-center production curves[178] in which the early stage saturates and the late stage comes in linearly (the dose is too small to show any of the late-stage saturation effects discussed in the previous subsection) but with a delayed start. It has now been shown that the height of the first stage in material moderately doped with a divalent cation increases as the square root of the impurity concentration.[80,173,174] On the other hand, the first-stage height is also very sensitive to deformation of the sample. It is possible to separate these two causes of early-stage enhancement by deforming samples after they have been irradiated sufficiently long that the impurity stage is saturated. In this way, it can be demonstrated that deformation causes production of excess *F* centers, as is illustrated in Fig. 14. The experiments[176] with KCl showed that this deformation stage is minimized in high-purity crystals and is a maximum for moderate doping. It now appears that there is no single early stage, but that saturating processes occur, depending on the type and number of defects present in the alkali halide sample. This concept of a number of saturating early-stage processes is supported by recent curve-fitting procedures applied to data from NaCl[172,196] and KCl[172] irradiated with rather low dose rates in a gamma source. A number of saturating exponentials were necessary to fit the data; there was an indication that the various exponential terms depended consistently upon experimental conditions, as, for example, temperature, impurity content, and strain.

Two classes of explanations are being considered for these early-stage processes. The first class involves secondary reactions and the second a modification of the initial production process.

An explanation of first-stage coloration in terms of secondary reaction requires either more efficient stabilization of the radiation defects or suppression of back reactions. We have mentioned that early stages are particularly apparent at room temperature in doped samples for which late-stage coloring is wholly or partly suppressed. This indicates that back reactions that are responsible for the late-stage suppression are definitely active. Thus, more efficient stabilization of radiation defects appears to be the more probable cause of the appearance of the early stage. At liquid nitrogen temperature, most defects and impurities are thought to be possible interstitial traps. At room temperature, the height of the impurity-caused early stage is appreciably smaller than the total impurity concentration and for divalent impurities is proportional to the square root of the concentration. These results suggest, as we will explain, that positive-ion vacancies are the efficient interstitial traps necessary to give rise to an early stage, as proposed by Ikeya *et al.*[173]

The addition of divalent impurity cations to an alkali halide requires that an equal number of cation vacancies be introduced to prevent charge buildup. Since these two defects are oppositely charged in the lattice, they will attract, and consistent with thermodynamics, an equilibrium between associated vacancy–impurity pairs and unassociated vacancies and impurity ions will exist. The mole fraction of cation vacancies, N_+/N_0, will be related to the mole fraction of impurities, N_I/N_0, according to the equation

$$N_+^2/N_0(N_I - N_+) = e^{-F/kT} \qquad (12)$$

where F is the free energy of association. For an association energy of a few tenths of an electron-volt, most of the impurity ions and vacancies will be bound, so that the vacancy concentration is proportional to the square root of the impurity concentration. Not only does the square-root relation indicate that the impurity-induced early stage is related to the free positive-ion vacancy, but the absolute number of F centers produced is also consistent. Moreover, it has recently been shown from measurements of positron annihilation[197] in NaCl that positive-ion vacancies in divalent cation-doped material do indeed decrease as a result of room-temperature X-ray irradiation. These experiments make use of the fact that positive-ion vacancies, being sites of net negative charge in an alkali halide, attract and trap positrons, giving rise to a resolvable annihilation peak that is proportional to the positive-ion vacancy concentration.

The explanation of the deformation-induced stage is on less firm ground since it is not even known what defect resulting from the deformation causes the saturating *F*-center production; i.e., whether it is dislocations or the debris left behind by dislocation interactions. The results of Sibley *et al.*[176,198] suggest that the debris produced by moving dislocations is very important and perhaps acts to stabilize interstitials.

The second class of explanations of the early-stage defect production characteristics can be devised in terms of modifications in the production mechanism. Crawford and Nelson[199] suggested some time ago that hole capture by a halide ion adjacent to a positive-ion vacancy could produce, with high efficiency, a negative-ion vacancy plus a halide atom trapped as a Cl_3^- molecular ion in what had been the positive-ion vacancy. If this mechanism operated only at *isolated* positive-ion vacancies, then this proposal would also be in agreement with the observed square-root impurity dependence.

An unsolved problem connected with explanations of the early stages is that of accounting for the details of the curve shapes, in particular the positive curvature that can be observed in many samples between the time the early stage saturates and the late stage attains its maximum slope. To explain that effect, one has normally resorted to additional, *ad hoc* assumptions. In the early work concerned with fitting these curves,[178] an annealing term was added in the rate equations. Recently, the importance of post-irradiation processes occurring before the *F*-band height is measured has been realized[172] and it has been shown that the shape of the curve depends strongly on the time interval between the cessation of radiation and measurement of the *F* band. Apparently, thermally-activated as well as radiation-triggered back reactions[200] are operative in the first stage.

(f) *Changes of Production Efficiency with Temperature.* From the foregoing discussions, it should be obvious that the temperature at which alkali halides are irradiated is perhaps the most important experimental parameter. On one hand, the temperature determines which of the secondary reactions can operate; on the other hand, the defect production

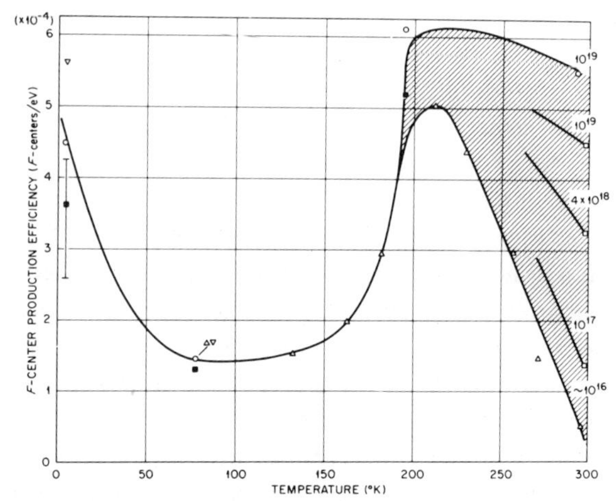

Fig. 15. Radiation production efficiency for F centers in KCl at different temperatures. Circled points obtained by Hughes *et al.*;[103] open and closed squares obtained by Sonder and Sibley.[160,183] The error bar shown at 5°K spans the results for different irradiation type and energy obtained by Ritz.[201] The triangles, Comins and Wedepohl,[203] and the inverted triangles, Behr *et al.*,[155] were not absolute determinations; they have been normalized to Pooley's value at liquid nitrogen temperature. The numbers above the room-temperature points give the dose rate in eV absorbed per cm^3 per sec.

mechanism itself is most likely temperature-dependent, particularly if it can be inhibited by energy loss due to the fundamental luminescence or by electron–hole recombination at non-defect-producing recombination sites. There are a number of difficulties inherent in obtaining significant temperature dependence data, as the following discussion will show. Figure 15 is a plot of F-center production efficiency versus temperature which includes data for KCl obtained by a number of investigators. Two things are immediately clear from the figure. The shape of the curve is complex; the production efficiency goes through a minimum between 20 and 150°K. Moreover, the efficiency above 200°K varies by more than a factor of 10, depending upon experimental conditions. Before we discuss these characteristics, let us point out in addition that when one compares data from different laboratories for which absolute values have been obtained, one often finds differences of a factor of two or three, which appear to depend upon uncontrolled experimental conditions. Nevertheless, the qualitative temperature dependence depicted in Fig. 15 is consistently observed.

The sharp decrease in production efficiency observed between liquid helium and liquid nitrogen temperature is, as we have discussed before, due to the instability of correlated Frenkel pairs and the mobility of interstitial ions and atoms that becomes significant in that temperature range. Let us point out that not only is there a drop of the order of a factor of five shown for *F* centers, but a much greater decrease of the empty anion vacancies that are produced efficiently at liquid helium temperatures.[148]

The rise in efficiency between liquid nitrogen temperature and 200°K is not yet well understood. It cannot be explained away by including other charge states of vacancies; the *F* centers are the majority charge state in this temperature range. Stability considerations would produce a decrease with temperature rather than the observed increase. Pooley has suggested that the increase stems from the fact that the recombination luminescence inhibits the production mechanism at 80°K, but not at 200°K. An anticorrelation in the temperature dependence of the recombination luminescence and defect production has indeed been observed for a number of alkali halides, particularly iodides. However, as the graph shows, there is no significant rise of the production efficiency with temperature for KCl below approximately 100°K, even though the recombination luminescence is three orders of magnitude lower at 80°K than at 4°K.[153] Anticorrelation between recombination luminescence and defect production also does not occur in KBr, as pointed out by Comins.[156] On the other hand, it has been observed[202,203] that the rise in production efficiency with temperature occurs at temperatures at which the self-trapped hole becomes mobile. This suggests that *F*-center production may be more efficient when electrons and holes recombine at special sites than when they recombine at a normal lattice site. However, to date, no generally accepted explanation for the rise in production efficiency exhibited in Fig. 15 exists.

The decrease in production efficiency with temperature above 200°K in KCl that has been referred to repeatedly[202–204] is to a large extent connected with the dose-rate dependence. We have included in Fig. 15 a number of determinations of production efficiency at room temperature obtained for differing dose rates. It is evident that for high dose rates, the large decrease is less pronounced. Unfortunately, there have not yet appeared any data that would show how high the production efficiency can get if high dose rates are used. Vilu and Elango[202] have compared the temperature dependence of the production efficiency in a number of alkali halides. Their results show that the decrease shown in Fig. 15 for KCl also occurs in NaCl, KBr, and KI. Moreover, if one orders the four halides according to the temperature range in which the decrease occurs, namely

KI, KBr, KCl and NaCl (with increasing temperature), one finds that the F^+-center mobility (see Table IX) scales similarly, suggesting that vacancy mobility may have something to do with the observed decrease in production efficiency.

4.3.4. *Summary*

From the large mass of data available on defect production, bleaching, and annealing, it has been possible to devise a somewhat consistent picture of processes that occur in alkali halide crystals under ionizing irradiation. The most important facets of this picture are as follows:

1. Frenkel defects in the halide sublattice can be produced with such a high degree of efficiency that a large fraction of ionization (electron–hole pair production) events must lead to the production of defect pairs. Recent experiments indicate that even excitons or nonionizing excitations of electron–hole pairs can produce Frenkel defects.

2. With few exceptions, the temperature, dose-rate, and impurity dependences of defect production can be accounted for without the necessity of invoking additional *F*-center production processes. The evidence is strong that behavior of the coloration at temperatures higher than liquid helium temperature depends upon secondary reactions and on inhibition of the primary production process.

3. The secondary reactions (namely recombination of primary defects, trapping, and nucleation and growth of defect clusters) necessary to account for the data are similar to those that have been proposed in connection with radiation damage in other materials. These reactions depend not only on radiation temperature and defect configuration, but also are affected by the radiation-induced changes in the charge state of certain defects, thus making them mobile.

4. The inhibition mechanism proposed to explain the decrease of damage efficiency near room temperature in doped alkali halides involves the annihilation of electrons and holes at impurity recombination centers with no defect production. Since holes are self-trapped at low temperatures, this inhibition becomes inoperative at low temperatures.

5. DEFECT CREATION IN OXIDE CRYSTALS

As has already been discussed, radiation damage results from a number of processes, ranging from the production of defects to their interaction and annihilation. These processes are affected in different ways by the presence of impurities and when the impurity content of a crystal is high enough (10 ppm or greater), it is difficult to untangle them. For example, when optical properties of impure irradiated crystals are measured, it is sometimes impossible to distinguish between changes due to electronic processes, such as valence changes of the impurities, and those due to radiation-produced interstitials or vacancies. In recent years, the growth, by a variety of techniques, of moderate quantities of different oxide single crystals containing only a few parts per million of unknown impurities has greatly enhanced research interest in these materials and the resultant experimental data have led to a significant increase in our understanding of radiation effects. Considerable progress has come from research on the alkaline earth oxides, which are the divalent structural analogs of the alkali halides, and much of our discussion will be concerned with these materials, although radiation effects on other oxide crystals, such as ZnO and Al_2O_3, will also be discussed. Radiolysis, which is most sensitive to impurities, does not appear to play a major role in radiation damage of oxides,[8,205,206] so we will not consider this case. Since impurity valence changes are so effective in altering many properties of crystals, electronic processes due to ionizing radiation will be discussed in detail. After this, the results of elastic collisions between high-energy electrons or neutrons and oxide lattice ions will be reviewed.

5.1. Electronic Processes

Even though oxide crystals which are considerably purer than those grown several years ago are now available, these crystals still have an appreciable amount of impurity. Table X shows the impurity concentration for some MgO crystals grown at several different laboratories and for a few 3M Company ZnO samples. The valence states of many of these impurities can be altered by radiation or heat treatment. Figure 16 illustrates such effects for four different impurities.[9] The data used are from spin resonance measurements of the concentration of doubly charged Mn and V ions, triply charged Fe and Cr ions, and singly charged positive-ion vacancies. As shown by the dashed portions of the curves, room-temperature gamma irradiation produces V^- centers and causes the $[V^{2+}]$ concentration to increase and the other species shown to decrease. Heating above 500°C, as shown, restores the preirradiation charge distribution. The figure also shows that postirradiation warming of MgO containing Fe or Cr causes the triply charged species to grow rapidly below 100°C, in the same temperature range in which the spin resonance of the V^- center decreases. This illustrates

Table X. Impurity Analysis[a]

Element	Impurity, μg/g					
	MgO				ZnO^{c}	
	Ingot #8	Ingot #6	MgO:Ni[b]	Kanto powder	Pure	ZnO:Li
Ag	—	—	—	<1	<0.5	<0.5
Al	42	41	100	50	≤5	≤5–10
As	<0.3	<0.4	—	<3	—	—
B	—	—	—	3	≤1	≤1
Ba	<0.7	<0.6	—	<2	<3	<3
Be	—	—	—	<1	<0.5	<0.5
Bi	—	—	—	<3	<2	<2
Ca	47	61	190	29	<1	—
Cr	<3	<5	20	—	<5	<5
Cu	—	—	—	<1	2	10
Fe	3	3	70	11	<1	—
Ge	—	—	—	—	<1	<1
K	—	—	—	<5	<3	—
Li	—	—	—	—	<1	20–190
Mg	—	—	—	—	3	1–10
Mn	0.2	0.3	<10	1.5	<1	<1
Mo	—	—	—	<1	<3	<3
N	9.2	9.3	—	<1	—	—
Na	0.3	0.3	7	25	1	1
Ni	<5	<5	1400	—	<5	<5
P	2	1	—	8.5	—	—
Pb	<0.5	<0.5	—	<0.8	<3	<3
S	<2	<2	—	140	—	—
Sb	—	—	—	<5	<5	<5
Si	27	19	49	76	<5	—
Sn	—	—	—	<1	<3	≤3
Ti	3	3	—	<2	<10	<10
V	—	—	—	<1	<5	<5
Zn	7	6	—	<50	—	—
Zr	<3	<6	3	<1	—	—

[a] The crystals were analyzed by means of wet chemistry, mass spectroscopy, neutron activation, and flame photometry.
[b] Crystal obtained from Muscle Shoals Electrochemical Corporation.
[c] Crystals obtained from the 3M Company.

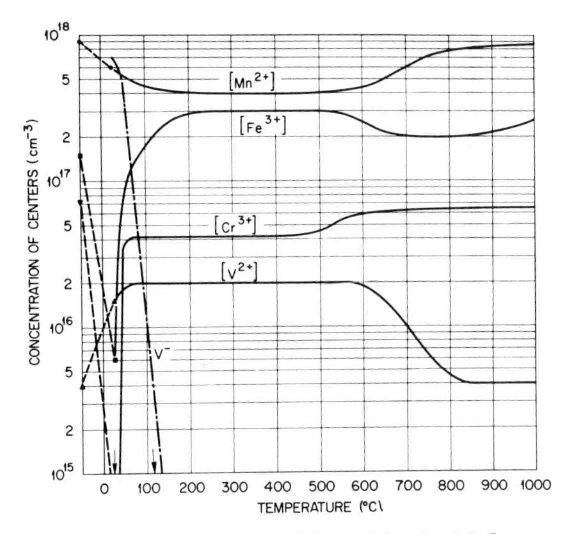

Fig. 16. Valence changes of impurities in MgO as a function of radiation and heat treatment. The dashed lines show the effect of gamma irradiation on the impurities; the decay of V^- centers with temperature is also indicated.[9]

even more clearly that charge is being transferred between the Fe or Cr and the positive-ion vacancies.

Such electronic processes are very common. Some of these effects have been discussed in a recent review of alkaline earth oxides by Henderson and Wertz.[7] They point out, for example, that irradiation of as-grown samples with electrons, gamma rays, or X-rays induces the following changes in the impurity states:

$$Ti^{4+} \rightarrow Ti^{3+}$$

$$V^{3+} \rightarrow V^{2+}$$

$$Cr^{3+} \rightarrow Cr^{2+}$$

$$Fe^{3+} \leftarrow Fe^{2+} \rightarrow Fe^{+}$$

$$Co^{2+} \rightarrow Co^{+}$$

$$Ni^{3+} \leftarrow Ni^{2+}$$

This is in general agreement with what is shown in Fig. 16.

In many instances, electronic processes are accompanied by luminescence, and Johnson[207] has reviewed the effects of valence changes on the luminescence of several oxygen-dominated lattices. Thermoluminescence, in particular, has been used to study the return of the impurities to their original charge states after irradiation. Wertz and co-workers[208] have found that the thermal destruction of V^- centers at 90°C leads to luminescence from the reaction

$$Cr^{2+} + e^+ \rightarrow (Cr^{3+})^* \rightarrow Cr^{3+} + h\nu_1$$

where e^+ is a hole and $h\nu_1$ is the energy of the emitted photon. Searle and Glass[209] measured the thermal decay of V^- centers and found that the hole is liberated from the vacancy with an activation energy of about 1.13 ± 0.05 eV. They also proposed that the released hole recombined with an electron at an iron impurity ion such that

$$Fe^{2+} + e^+ \rightarrow (Fe^{3+})^* \rightarrow Fe^{3+} + h\nu_2$$

An example of thermoluminescence emitted by irradiated MgO is shown in Fig. 17.[9] The blue luminescence shown in the inset is attributed to the reac-

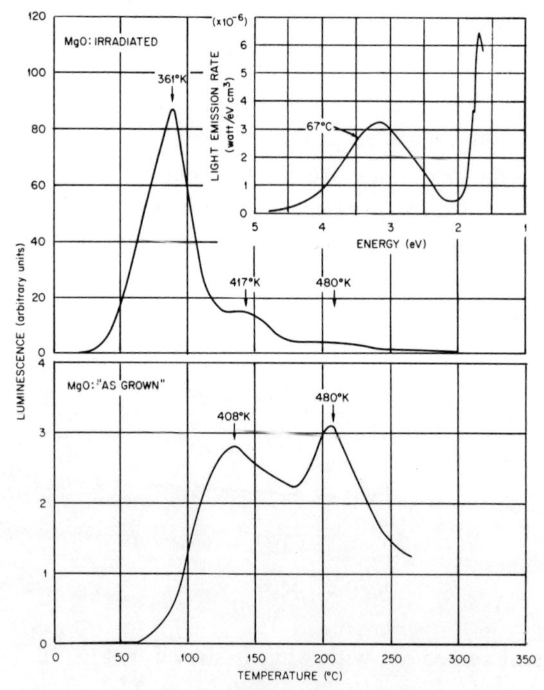

Fig. 17. Thermoluminescence from irradiated and unirradiated MgO single crystals.[9]

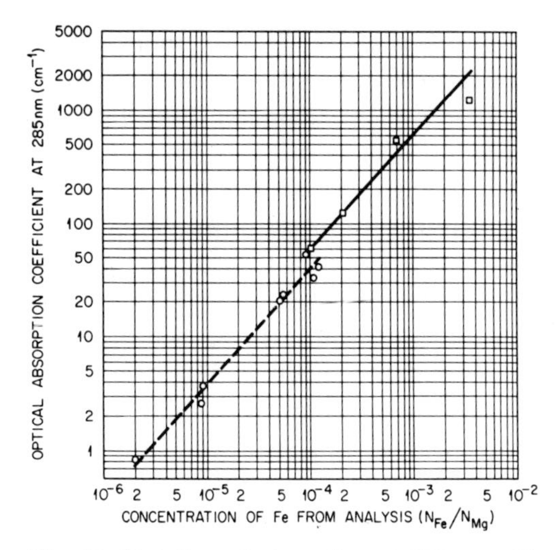

Fig. 18. Variation of absorption at 285 nm with iron content. Circles: saturation level on gamma irradiation.[8] Squares: saturation level on heating in air at 1400°C.[6]

tion proposed by Searle and Glass[209] and the red luminescence to that of Wertz *et al.*[208] It should be pointed out that it is necessary to obtain not only the glow curves (total light intensity versus sample temperature), but also the spectra of the light emitted in each glow peak before it is possible to determine which impurity is responsible for a particular glow peak. It is now standard practice to measure both the temperature dependence and spectral distribution of the emission.

Radiation-induced valence changes have also been observed through optical absorption studies. Since each impurity valence state has its own characteristic absorption spectrum, a detailed study of radiation- or temperature-induced changes in valence states is possible. One instance of this is the studies of [Fe^{3+}] in MgO. It has been proposed that there are at least two [Fe^{3+}] absorption bands occurring at 5.7 eV (210 nm) and 4.3 eV (285 nm).[210] Studies of the intensity of the 4.3-eV band as a function of Fe concentration, as illustrated in Fig. 18, have verified that this band is due to iron.[6,8] Therefore, using this band as a measure of Fe^{3+} concentration, it is possible to nondestructively monitor changes in this concentration in any particular crystal as a function of treatment. Wong *et al.*[211] have investigated charge transfer due to radiation of vanadium in Al_2O_3 by observing absorption in the far-infrared. They found that [V^{3+}] was transformed to [V^{2+}] and [V^{4+}] by X-ray irradiation.

In all polar materials, the condition of electrical neutrality must hold on a macroscopic scale. This means that when an impurity such as [Fe^{3+}] takes the place of a host divalent cation like [Mg^{2+}] in MgO, another defect with an effective negative charge must be incorporated to give a charge balance. In MgO containing no large impurity concentrations besides iron, the complementary defect is the positive-ion, [Mg^{2+}], vacancy. When the vacancy contains no electrons, it is doubly charged with respect to the perfect crystal and only one positive-ion vacancy needs to be formed for every two [Fe^{3+}] centers; when the vacancy is associated with a hole (V^- center) the vacancy concentration will be equal to that of the [Fe^{3+}]. The presence of such cation vacancy defects has been well documented in a number of other alkaline earth oxides[7] and may have been observed in irradiated Al_2O_3.[212]

The charge states of many electronic defects can be changed by means other than radiation. For example, when an MgO sample containing Fe is heat-treated in a reducing atmosphere, many of the [Fe^{3+}] centers trap electrons to become [Fe^{2+}]. Whether or not the complementary defect change involves annihilation of cation vacancies or some other means of charge compensation is not yet known. Extensive studies of cation vacancy

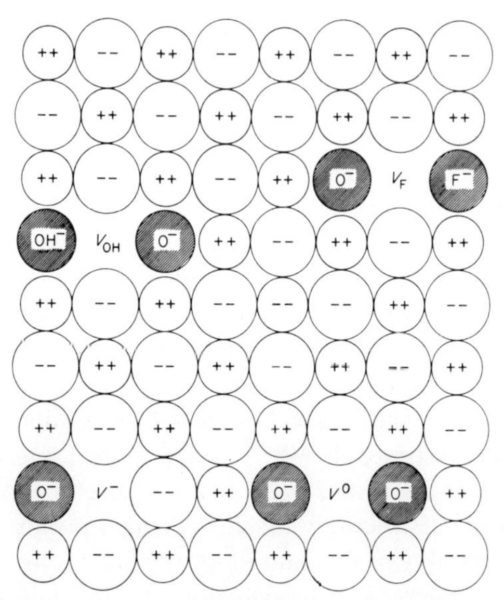

Fig. 19. Modified V centers produced in alkaline earth oxides: F and H designate fluorine and hydrogen impurities, respectively.

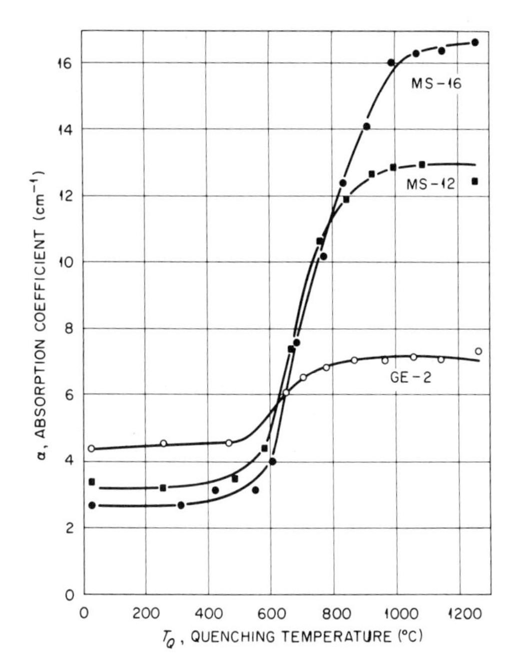

Fig. 20. Absorption coefficient of the 2.3-eV band (due to V^- centers) after a 20-min gamma radiation versus quenching temperature.[8]

centers (V centers) and valence changes have been made by Wertz and his collaborators[16,213–216] using ESR and optical techniques. They have shown that V^- centers (previously called V_1 centers) form in MgO due to ionizing radiation and that this center absorbs light at 540 nm and has an oscillator strength of about 0.1.[7–9,208] Modified V^- centers are also produced and the structure of some of these is shown in Fig. 19.

Vacancies associated with two holes, or V° centers, do not produce spin resonance signals in the ground state. Thus, it is impossible to study in detail all of the charge-transfer processes and to determine whether ionic defects are produced or destroyed when a polar crystal is annealed in an oxidizing or reducing atmosphere.

It should be possible to enhance the concentration of positive-ion vacancies by doping crystals with Me^{3+}-type impurities, by quenching in Schottky pairs from high temperature, or by heat treatment followed by rapid quenching to dissociate vacancy clusters and prevent reassociation. Figure 20 illustrates that quenching a sample from above 1000°C is effective in increasing the number of V^- centers that can be obtained by subsequent ionizing radiation.[8] Tohver and co-workers[217,218] have investigated V^- centers in irradiated CaO. Moreover, Henderson and Tomlinson[219] have shown that heat treatment of CaO yields V^- centers without subsequent irradiation. In Fig. 20, a definite saturation of induced concentration with temperature can be seen which is different for each of the crystals investigated. This indicates that the thermal formation of Schottky pairs is not important and that most likely it is the dissociation of vacancies from clusters or impurities that leads to the observed enhancement. Crawford[220] has analyzed the data shown in Fig. 20 on the assumption that V^- enhancement is due to the breakup of impurity–vacancy complexes, and he deduces a binding energy of about 1.4 eV. If positive ion vacancies are indeed mainly due to impurities, then it is puzzling why large concentrations of V^- centers can be produced by ionizing radiation even in the purest crystals. The possibility that these vacancies arise from thermal deformation during the growth process has, we believe, been eliminated by deformation studies.[9] It could be that impurities other than those shown in Table X are present in significant amounts and generate positive-ion vacancies.

5.2. Elastic Collision Processes

In Section 3 of this chapter, it was stated that the determination of displacement thresholds and cross sections is one of the goals of radiation damage research and that the ability to detect and count specific radiation-produced defects in polar crystals makes these materials especially attractive for such studies. In the oxide crystals studied to date, it appears possible to obtain displacement threshold and cross section data.[205,206] However, oxide systems have not been studied nearly as thoroughly as the alkali halides; so even though progress has been made in identifying certain defects produced by radiation and in using these defects to study elastic collisions in oxides, there is very little information available on mobilities, clustering, and annihilation. Therefore, in this section, we are limited to a discussion of the progress that has been made in various oxide materials and we cannot give as complete a treatment as in Section 4.

5.2.1. *Alkaline Earth Oxides*

In MgO, the F^+ center has been identified by both ESR and ENDOR measurements.[221,222] When these measurements are done with great care, it is possible to determine the concentration of centers responsible for the resonance. Optical absorption measurements are also related to the concentration of defects responsible for a particular band through the relation[25]

$$Nf = 0.87 \times 10^{17}[n/(n^2 + 2)^2] W_{1/2}\alpha_m \tag{13}$$

where N is the concentration of defects, f is the oscillator strength, n is the index of refraction of the crystal at the wavelength of the absorption band, $W_{1/2}$ is the width in electron-volts at half-maximum of the band, and α_m is the absorption coefficient at the peak position of the band. An accurate determination of f can be made by a combined optical and spin resonance experiment if the two types of data are for the same center. Kemp and his co-workers[223,224] have exploited a technique called Faraday rotation–ESR double resonance which has directly identified the optical absorption band at 5.0 eV in irradiated MgO as being due to F^+ centers. This method uses Faraday rotation and ESR and consists in mounting an oxide sample in a microwave cavity provided with windows. The sample is then placed at the center of a superconducting magnet in such a way that ESR and optical Faraday rotation measurements may be made simultaneously. Optical Faraday rotation due to paramagnetic centers in crystals is known to be decomposable into a diamagnetic or orbital part and a paramagnetic part as has been discussed by Mort *et al.*[225] and Brown.[226] Figure 21 illustrates this effect for the case of the F center in KCl. The temperature-independent

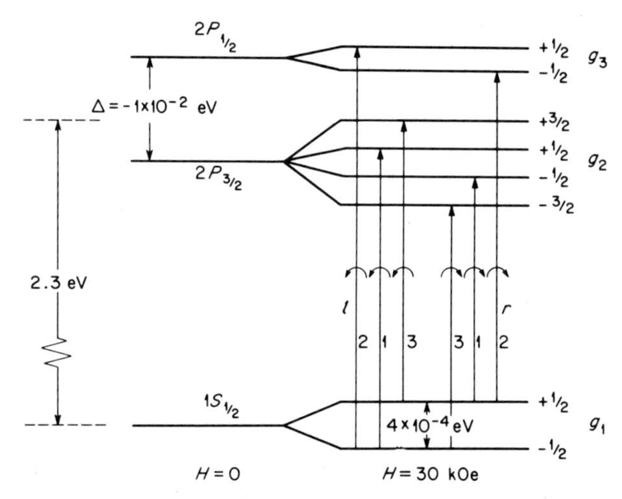

Fig. 21. The origin of Faraday rotation as illustrated by the energy level diagram of the F center in KCl which shows the approximate effect of a magnetic field.

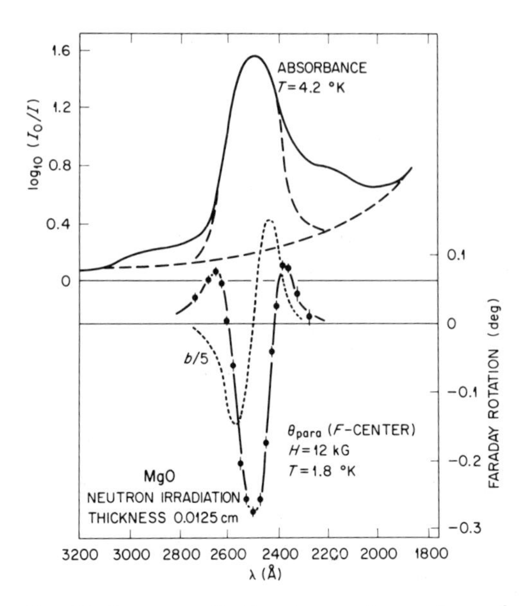

Fig. 22. Paramagnetic Faraday rotation pattern of the MgO F center, obtained by the Faraday rotation–ESR method. The dotted curve is the circular dichroism b, divided by 5, obtained by Kramers–Kronig inversion of θ; the same ordinate scale applies to θ and $\frac{1}{5}b$. The sample absorbance, taken on a spectrophotometer, is shown with a hand-sketched partial resolution of the absorption bands.[212]

diamagnetic term arises from the excited P states, whereas the paramagnetic component, which predominates at low temperatures, comes from the ground-state splitting. Thus, when the ground-state populations are changed by ESR excitation, an effect is observed on the Faraday rotation.

An example of the results of Kemp *et al.*[227] on F^+ centers in MgO is shown in Fig. 22. Kemp and his collaborators[223,224,227] and Bessent *et al.*[228] have identified the optical absorption bands due to F^+ centers in several other oxide systems.

Prior to Kemp's Faraday rotation–ESR measurements on MgO, Henderson and King[229] set out to see if the 5.0-eV absorption band in MgO was the result of F^+-center absorption and, if so, to determine accurately the oscillator strength of this center. Their experiments were performed on neutron-irradiated specimens and they obtained an oscillator strength of $f = 0.8$ for the F^+ center. Some complications became apparent when the same experiments were performed after electron irradiation rather than neutron irradiation.[206] After electron irradiation, only a small spin resonance signal attributable to F^+ centers could be observed, even though the optical absorption at 5.0 eV was quite strong. Moreover, it was found that in additively colored crystals, the 5.0-eV band was clearly present but little, if any, F^+-center ESR signal was observed.[230] Kemp and Neeley[231] had predicted from a theoretical calculation that F and F^+ centers could absorb at about the same photon energy in MgO, but it took a strong cooperative effort among Hensley, Kemp, Henderson, Wertz, and co-workers and the Oak Ridge group before it could be demonstrated that this was indeed the case. Finally, it was possible by a series of ultraviolet light bleaching and thermal annealing experiments, which release charge but do not produce ionic defects, to show, as can be seen from Fig. 23, that F centers in MgO absorb with a peak at 5.01 eV and F^+ centers at 4.92 eV.[19,232] It has also been shown that F^+-center luminescence occurs with a peak at 3.13 ± 0.03 eV. Kappers *et al.*[232] have observed F-center luminescence at 2.4 eV in MgO. Henderson *et al.*[233] and Evans *et al.*[234] have studied F- and F^+-center emission in CaO. Recently, Evans and Kemp[235] have found

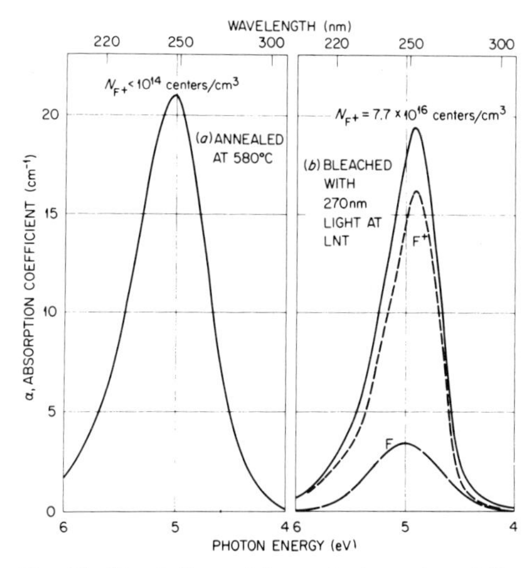

Fig. 23. Resolution of the optical spectra of the F^+ and the F bands of an additively colored MgO sample measured at room temperature. Experimental curve (b) was decomposed using a symmetric Gaussian with a half-width of 0.77 eV for the F^+ band.[19]

Table XI. *F*-Center Absorption and Emission in Oxides

Lattice	F^+ Center		F Center	
	Absorption, eV	Emission, eV	Absorption, eV	Emission, eV
MgO	4.92	3.13	5.01	2.4
CaO	3.70	3.3	3.10	2.0
SrO	3.10	2.42	2.49	—
BaO	2.00	—	—	—

the F-center luminescence in SrO. A summary of the absorption and emission peak energies for F centers in a number of oxides is given in Table XI. In additively colored or electron-irradiated crystals, F centers are more numerous than F^+ centers, but in neutron-irradiated samples, where many different types of cluster centers exist, some of which can trap electrons, F^+ centers are far more prevalent than F centers. This is why Henderson and King obtained such a reasonable oscillator strength for the F^+-center absorption.

As mentioned in connection with electronic properties, a positive-ion vacancy defect has also been well identified in MgO. This is the V^- center, a Mg vacancy with a nearby hole.

Cluster centers involving both positive- and negative-ion vacancies are formed in MgO when a sample is irradiated with heavy particles or is thermally heated.[7] Many of these complex defects give rise to optical absorption bands. An example of the bands that appear in neutron-irradiated and heat-treated material is shown in Fig. 24.[236] Some of the bands have been tentatively linked to F-aggregate centers of various charge states; one vacancy aggregate center has been clearly identified by a combination of optical and spin resonance studies.[216,237] It is a vacancy pair containing one electron which has been reported by King and Henderson to absorb light at 3.6 eV.

As Fig. 24 indicates, there are several very narrow absorption lines which can be seen in neutron-irradiated MgO when optical measurements are made at low temperature; and, in fact, the number of these lines increases upon annealing the samples.[236] The narrowest of these lines has been interpreted as being due to optical transitions in which no energy is transferred to lattice phonons, the so-called zero-phonon line; and the succeeding

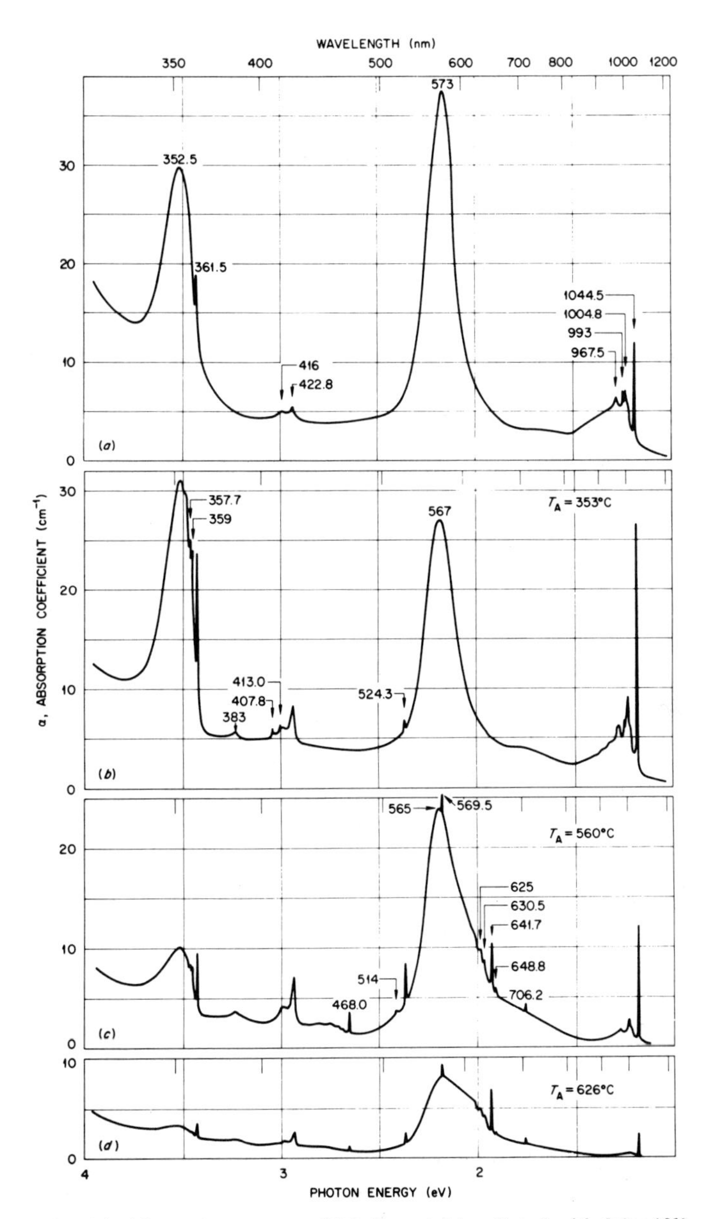

Fig. 24. Absorption spectra of MgO crystal irradiated with 2.2×10^{18} neutrons/cm² and annealed for 10 min at the temperatures indicated. Measurements were made at 77°K.[236]

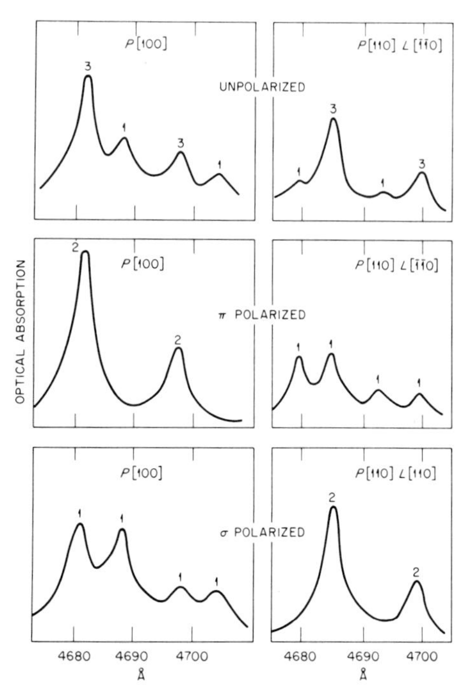

Fig. 25. Stress-splitting of the 468.5- and 470-nm zero-phonon lines in neutron-irradiated MgO. The direction of stress *P* and the propagation direction of the light *L* are indicated. π-polarized indicates that the electric vector of the incident light is parallel to the stress direction and σ-polarized means it is perpendicular to the stress direction. The numbers shown above the peaks are relative intensity indicators. Notice that the sum of the intensities for the polarized light measurements is equal to that for unpolarized light.[243]

peaks may correspond to electron–phonon transitions.[238] The theory of these transitions and of stress effects has been reviewed by several authors[239–241] and used by various workers[7,237,242,243] to determine the symmetry and electric dipole orientation of several defects in MgO. Figure 25 illustrates the effect of uniaxial stress in the [100] and [110] directions on the zero-phonon lines at 468.5 and 470 nm.[243] The number and intensity ratios of the components which the line splits into determine the orientation of the transition dipole moment at the center.[244,245] Thus in Fig. 25, we see the splitting characteristic of defects having $\langle 110 \rangle$ orientations and orthorhombic symmetry with linear electric dipole oscillators along a ⟨110⟩ direction. There is some evidence that the intensity of the band at 468.5 nm is influenced by impurities and therefore it has been attributed to F_2-type centers in the vicinity of vacancies or impurities.[7] From other, similar studies, it has been proposed that the F_2 center in MgO absorbs light at 362 nm and 1045 nm, the F_2^+ absorbs at 470 nm, the F_3 at 642 nm, the F_3^+ at 525 nm and the F_3^- at 649 nm.[7,246] As an aside, it should be mentioned that because these oxides are divalent in nature, various charge states exist for each type of structural defect, e.g., F_2^+, F_2, and F_2^-. This means that electronic processes play a strong role in observation of these defects. Appropriate optical bleaching can change F centers to F^+ centers[19,232] and F_2 centers into F_2^+ centers.[7] Moreover, the charge on the various defects obviously depends on competition between the radiation-induced defects and the impurities. This may be why in neutron-irradiated specimens, F^+ centers are formed almost to the exclusion of F centers, whereas in additively colored or electron-irradiated specimens, mostly F centers are created. Recently, Bessent[247] has observed an ESR signal in neutron-irradiated CaO which he ascribes to an F_4-type center with an unspecified charge state.

Identification of the absorption bands of simple defects as described above greatly simplifies studies of the radiation damage processes. The production of vacancy defects in MgO by 1.7-MeV electrons at room temperature has been studied. The experimental cross section $\sigma(E)$, as determined from Eq. (6) in Section 3 for the production of F centers, is $\sigma(1.7 \text{ MeV}) = 1.7$ barns.[206] Preliminary studies[206,248] of the energy dependence of the cross section show that only very few F centers are produced by 0.33-MeV electrons. From this observation and Eq. (4), a displacement threshold energy of about 60 eV for the oxygen ion can be computed. The theoretical displacement cross section for oxygen ions, assuming a displacement threshold of 60 eV, is 11.6 barns.[249] This is much greater than that obtained from experiment. Moreover, calculations of the electron-irradiation displacement cross sections for O and Mg ions suggest that positive- and negative-ion vacancies should be produced in approximately equal concentrations; no indication of any radiation-produced V centers could be found after room-temperature electron irradiation. These two results suggest that the F-center production observed at room temperature is the result not simply of elastic-collision production processes, but rather such processes followed by recombination and stabilization processes analogous to those described in Section 4 for alkali halides. When the irradiation temperature is sufficiently high that interstitials are mobile, then as they

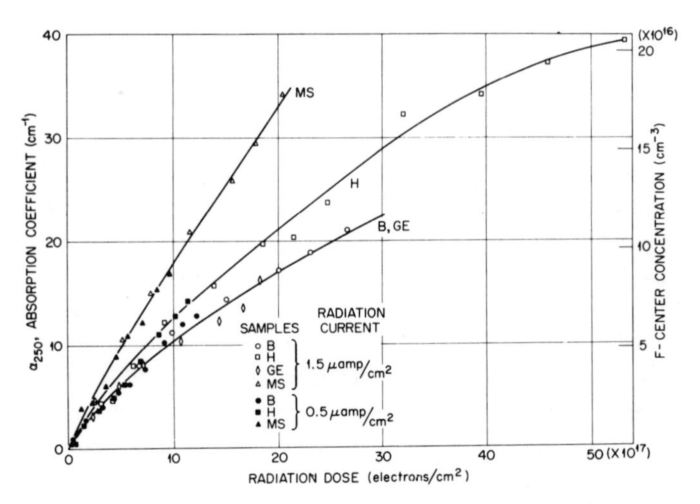

Fig. 26. A plot of the absorption coefficient at the peak of the *F* band versus electron dose for MgO crystals. The ordinate on the right-hand side of the figure shows the *F*-center concentration as calculated from Eq. (13). Samples *H, B, GE* contain less than 10 ppm Fe, whereas the crystals labeled *MS* contain 60 ppm Fe.[206]

move through the lattice after their creation, they will annihilate by recombining with vacancies or will become trapped at impurities or at other interstitials. In this regard, when the irradiations and measurements can be made at liquid helium temperature, we expect to see interstitial vacancy recombination steps much like those observed by Itoh *et al.*[119] in the alkali halides and those in metals.[46,49] Further evidence for the presence of secondary reactions in MgO irradiated at room temperature is presented in Fig. 26, which is a plot of *F*-center concentration versus electron radiation dose for a series of MgO crystals. The fact that the impurity-doped sample shows a greater coloration than the pure one is due, we believe, to a trapping of anion interstitials by impurities and thus an inhibition of the recombination process. The fact that "back reactions" or recombination processes occur at these temperatures is also substantiated by the saturation nature of the growth curves shown. At higher radiation doses, the saturation is even more evident.[206]

Defect mobility is, as stated in Section 4, very important in "back reactions" and an appreciable mobility for positive-ion interstitials may well explain the puzzling result that there is as yet no concrete evidence for positive-ion vacancy formation due to either electron or neutron irradiation of MgO crystals. As we noted previously, copious numbers of positive-ion vacancies are already present in these materials and one charge state of these, the V^- centers, can be enhanced through electronic processes. Recent measurements suggest that at liquid nitrogen temperature, the V^--center concentration can be increased by radiation.[250] If this result is not simply due to electronic processes, it would suggest that cation vacancies are produced, but that at room temperature, they are annihilated by their complementary interstitials.

An idea about the mobility of anion (O) interstitials and vacancies can be obtained by comparing the annealing of irradiated and additively colored crystals. In the latter case, only anion vacancies are present (no interstitials), whereas in irradiated crystals, both vacancies and interstitials are present. Figure 27 shows that F centers in irradiated crystals anneal out long before those in the additively colored ones.[206,251] This suggests that interstitial defects in the irradiated material break up or dissociate at around 400°C (the dissociation temperature is slightly different for neutron and electron irradiations) and annihilate the anion vacancies. In fact, the evidence is that both F and F^+ centers are destroyed at the same time. As can be seen from the figure, in additively colored crystals, vacancies do not become mobile below 900°C, and they probably do not play a very important role in radiation annealing processes.

We can summarize our present, incomplete view of the defect creation processes in the case of electron irradiation of MgO in the following manner: High-energy electrons ($E > 0.33$ MeV) are capable of producing widely separated anion and cation Frenkel pairs in a crystal by means of elastic collisions. At temperatures greater than 77°K (no measurements have yet

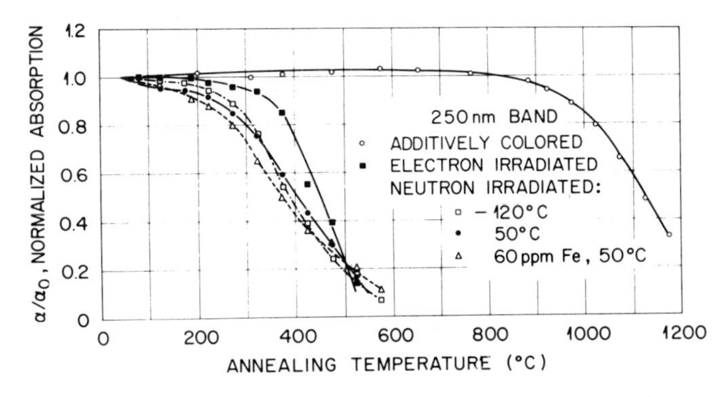

Fig. 27. Normalized absorption coefficient at the peak of the F band (5.0 eV) versus annealing temperature (10 min at temperature) for electron-irradiated, neutron-irradiated, and additively colored MgO crystals.[251]

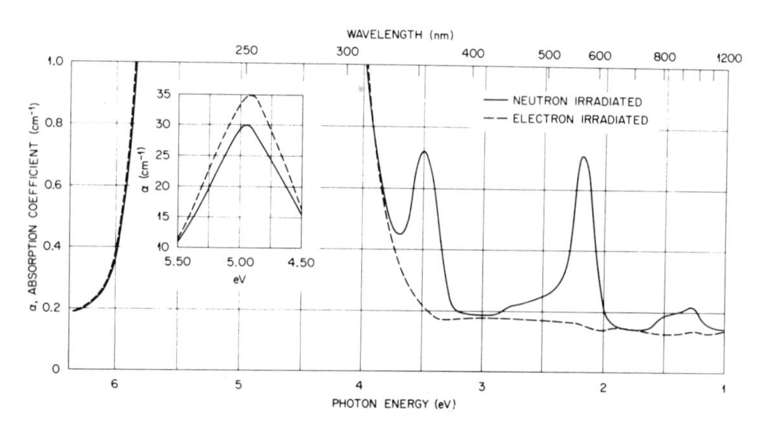

Fig. 28. Comparison of the spectra of MgO crystals irradiated with electrons and neutrons.

been made below this temperature), interstitial partners of the Frenkel pairs are mobile and wander through the lattice. They can be trapped at impurities, other interstitials, or dislocation lines; and when this occurs, stable damage remains. Apparently, cation interstitials and vacancies are highly mobile during electron irradiation and recombine with a high probability, whereas anion interstitials can be more easily stabilized. It is not unreasonable to assume that radiation damage processes for the other alkaline earth oxides are similar, with the primary differences being in defect mobility.

The defect structure resulting from electron and neutron irradiation is not the same. This is evidenced most clearly by the differences in the optical absorption spectra for the two cases. Figure 28 illustrates spectra for MgO that has been irradiated with 2-MeV electrons or neutrons such that about the same number of *F*-type centers are present. The absorption bands which appear in the neutron-irradiated samples have been attributed to cluster-type defects.[7] As one would expect, these bands also appear when 30-MeV electrons are used.[250] The massive neutrons and high-energy electrons impart considerable energy through elastic collisions to the indigenous lattice ions (as can be seen from Table III), so that according to Eq. (5) in Section 3, defect cascades are created. In the instance of neutron or 30-MeV electron irradiation, the probability for cluster formation should exceed that for isolated *F* centers; and, as a matter of fact, studies of the effect of irradiation on the flow stress of MgO suggest that at least twice as many small defect-cluster centers as isolated Frenkel pairs are produced by high-energy neutrons.[252,253] This is essentially in agreement with the work of Henderson and Bowen.[254] Their data on lattice-parameter and crystal-density changes with neutron irradiation suggest that below about 3×10^{19} nvt, Frenkel defects or very small cluster defects are formed. At higher radiation doses, saturation of the damage occurs which is due to the *spontaneous* recombination of vacancies and interstitials when these are separated by less than 18 nearest-neighbor distances. Preirradiation annealing in hydrogen, which produces large lenticular voids in the crystals, results in a large decrease in the amount of neutron damage observable in the crystals.[255] This is presumably due to vacancies migrating to the voids during irradiation since the data show an excess of interstitial defects with respect to vacancies.[255]

Wilks[256] has reviewed the formation of dislocation loops, macroscopic growth, and lattice parameter changes that occur in neutron-irradiated MgO and points out that numerous cluster defects and cavities are produced. Briggs and Bowen[257] report that when MgO crystals are irradiated to neutron doses of greater than 10^{20} nvt and then annealed at high temperature (1800°C), cavities are formed that contain the gaseous transmutation products Ne and He at pressures not exceeding 10 atm. They report, however, that diffusion of appreciable numbers of gas atoms into the cavities occurred only when the annealing period was sufficiently long to remove the majority of radiation-induced dislocations.

In summarizing the effect of neutrons on the creation of defects in the alkaline earth oxides, it can be said that the primary creation mechanism is elastic collisions between the incident neutrons and lattice ions followed by collisions between displaced lattice ions and other lattice ions. The latter effect creates large regions of displaced ions (displacement spikes) and favors the formation of cluster-type centers. Spontaneous interstitial-vacancy recombination should also be enhanced in these regions.

5.2.2. BeO *and* ZnO

Both BeO and ZnO have a hexagonal wurtzite structure. The F^+ center has been reported in both materials. DuVarney *et al.*[258] observed the ESR spectrum and made double-resonance measurements of F^+ centers in neutron-irradiated BeO. They found the g value of this defect to be 2.0030 with H parallel to the crystal c axis. Wilks[256] has reviewed the effects of large neutron doses on BeO and reports that numerous cluster centers are produced which give rise to macroscopic growth, decreased thermal conductivity, and bubble formation. There is some evidence for helium and tritium bubbles in heavily irradiated samples (10^{20} nvt) which are irradiated

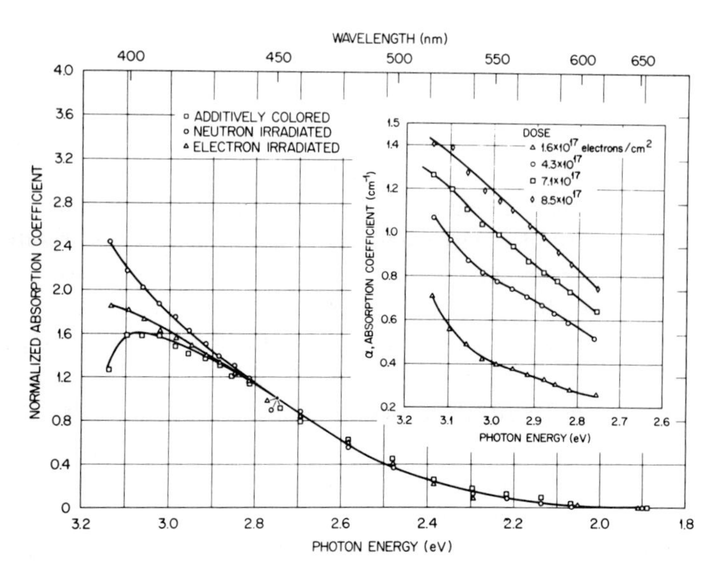

Fig. 29. A plot of the normalized absorption coefficient versus photon energy which shows the induced-absorption spectra for ZnO samples colored in different ways. The data are normalized to the absorption coefficient at 450 nm (2.755 eV). The inset shows the change in character of the absorption with dose for an electron-irradiated specimen.[260]

or heat-treated at high temperature, but most of the helium formed escapes during irradiation.

In ZnO, Smith and Vehse[259] report that F^+ centers are present after electron irradiation and optical bleaching. The g value for this material is $g_\parallel = 1.9948$ and $g_\perp = 1.9963$. The optical absorption of neutron- and electron-irradiated ZnO has also been determined.[259] In this compound, the radiation produces only a shoulder on the band edge,[260] as shown in Fig. 29. However, by observing the height of the shoulder as a function of radiation dose, it was possible to show that damage was produced only when the electron energy exceeded 0.55 MeV and that, moreover, fast neutrons (>1 MeV) produce about 30 times as many defects as do energetic electrons. These results indicate that damage is produced primarily by elastic collisions.

The radiation damage processes in these materials, even though they have a different crystal structure, seem to be very similar to those in the alkaline earth oxides. There has always been some question as to whether zinc interstitials or oxygen vacancies cause the coloration of ZnO[261] and further effort in this area should delineate which is the case.

5.2.3. Al_2O_3

The study of this material has not progressed as far as that of the alkaline earth oxides. There has been little attempt to correlate radiation-produced absorption bands with spin resonance spectra, nor have there been made any clear identifications of simple radiation defects. Gamble *et al.*[212] have found some ESR signals that appear in gamma-ray- or neutron-irradiated samples. It is, however, not clear whether the defects giving rise to these spectra are due to radiation production of defects or simply electronic processes. They did make suggestions about the identity of two of the defects giving rise to observed spectra. They attributed the ESR spectrum, consisting of 12 equally intense lines, the symmetry axis of which was perpendicular to the crystal *c* axis and nearly parallel to rows of [O^{2-}] ions, to an [AlO^{3-}] molecular ion in a pair of oxygen sites (i.e., an Al that has found its way into an oxygen vacancy). However, Henderson[262] has suggested that this spectrum is also consistent with a pair of holes trapped at adjacent O^{2-} ions, a defect that can be simply produced by charge redistribution. Gamble *et al.*[212] have attributed another of the observed spectra to an interstitial defect. They suggest that a three-line spectrum, which appears only when the sample is irradiated and held at liquid nitrogen temperature, is due to an interstitial [O^+]. The fact that the concentration saturates quickly with gamma-ray irradiation but continues to grow under neutron irradiation is explained by attributing interstitial oxygen-atom production to the heavy-particle irradiation and charge redistribution, producing O^+ interstitials from uncharged oxygens, to the gamma rays.

An optical absorption band with a peak at 6.0 eV (205 nm) has been observed both by Levy[263,264] and by Compton and Arnold[205] after electron and neutron irradiation of corundum crystals. They find that the band height increases with radiation dose. An energy dependence study with electrons indicated that electrons with energy less than 0.35 MeV did not generate the defects responsible for this band.[205] From this value, a displacement threshold can be estimated. If the band is due to the displacement of aluminum ions, the displacement threshold T_d is about 40 eV; whereas if is due to the displacement of oxygen, then T_d is around 70 eV.

A review of neutron irradiation damage, especially at high doses, in Al_2O_3 has been written by Wilks.[256,265] There is strong evidence for the production of dislocation loops and tangles by the radiation. Heat treatment tends to increase the size of the loops and decrease their density just as in the other oxide materials studied. Recovery of much of the damage occurred at around 800°C in this material.

5.3. Future Research Areas

5.3.1. *Identification of Defects*

The foregoing indicates how limited our information is about the identity of radiation-produced defects in the various oxide materials. The problems are especially difficult where radiation defects, being produced by elastic collisions, can appear both in the anion and cation sublattices. One way to decide whether a defect is in the anion or cation sublattice is to observe the effect of additive coloration. Several studies of this nature have already been completed by Hensley and his co-workers for materials such as MgO, CaO, and SrO, but much more effort is needed. This is especially true when cluster defects are to be investigated. When crystals are irradiated with high-energy electrons, protons, or neutrons, numerous cluster-type defects are produced and many times it is almost impossible to determine if these defects consist of both cation and anion vacancies or just *F*-aggregate centers. Another research area dealing with the identification of defects is one that has received much attention lately. This is the use of the Stark effect, Faraday rotation, and stress splitting studies to determine the configuration of centers responsible for absorption lines or bands. Such techniques are especially attractive since numerous zero-phonon lines and narrow phonon-assisted lines occur in many of these materials.

5.3.2. *Defect Production and Annealing*

Very little is known about the production and annealing kinetics of specific defects in the oxides. For example, a careful study of whether V^- centers can be produced by radiation in MgO single crystals has not been made. It has not been determined whether Mg Frenkel pairs are produced by the radiation and recombine rapidly due to high mobility of the interstitial or if they do not form. A complete study of the energy dependence of defect production in materials like ZnO and BeO would be helpful in determining whether predominantly cation or anion defects are formed. Moreover, this type of investigation would lead to considerable knowledge about the binding or packing of the ions in various materials.

5.3.3. *Electronic Processes*

It appears to us that one of the most interesting and profitable areas yet to be actively pursued in these materials is that of studying valence changes and charge transfer. It would be very useful to combine ESR and optical measurements with photoconductivity and delineate the electronic processes that take place. These measurements are of particular importance in device research and could have many far-reaching implications.

REFERENCES

1. I. Schneider, *Appl. Opt.* **6**, 2197 (1967).
2. W. Hayes and J. W. Twidwell, *J. Chem. Phys.* **35**, 1521 (1961).
3. D. S. McClure and Z. J. Kiss, *J. Chem. Phys.* **39**, 3251 (1963).
4. E. S. Sabisky, *Phys. Rev.* **141**, 352 (1966).
5. J. L. Mertz and P. S. Persham, *Phys. Rev.* **162**, 217 (1967).
6. R. W. Davidge, *J. Mat. Sci.* **2**, 339 (1967).
7. B. Henderson and J. E. Wertz, *Advan. Phys.* **17**, 749 (1968).
8. Y. Chen and W. A. Sibley, *Phys. Rev.* **154**, 842 (1967).
9. W. A. Sibley, J. L. Kolopus, and W. C. Mallard, *Phys. Stat. Sol.* **31**, 223 (1969).
10. W. Hayes and D. W. Jones, *Proc. Phys. Soc.* **71**, 503 (1958).
11. C. J. Delbecq, W. Hayes, M. C. M. O'Brien, and P. H. Yuster, *Proc. Roy. Soc.* **271A**, 243 (1963).
12. C. J. Delbecq, A. K. Ghosh, and P. H. Yuster, *Phys. Rev.* **151**, 599 (1966).
13. J. W. Wilkins and J. R. Gabriel, *Phys. Rev.* **132**, 1950 (1963).
14. E. Boesman and D. Schoemaker, *J. Chem. Phys.* **37**, 671 (1962).
15. F. Lüty, *Halbleiterprobleme*, Vol. VI (Springer Verlag, 1959), p. 238.
16. J. E. Wertz, P. Auzins, J. H. E. Griffiths, and J. W. Orton, *Disc. Faraday Soc.* **28**, 136 (1959).
17. A. J. Shuskus, *J. Chem. Phys.* **39**, 849 (1963); **40**, 1602 (1964).
18. J. C. Kemp, W. M. Ziniker, and E. B. Hensley, *Phys. Rev. Letters* **25A**, 43 (1967).
19. Y. Chen, J. L. Kolopus, and W. A. Sibley, *Phys. Rev.* **186**, 865 (1969).
20. T. G. Castner and W. Känzig, *J. Phys. Chem. Solids* **3**, 178 (1957).
21. T. O. Woodruff and W. Känzig, *J. Phys. Chem. Solids* **5**, 268 (1958).
22. C. J. Delbecq, W. Hayes, and P. H. Yuster, *Phys. Rev.* **121**, 1043 (1961).
23. G. D. Jones, *Phys. Rev.* **150**, 539 (1966).
24. R. B. Murray and F. J. Keller, *Phys. Rev.* **153**, 993 (1967).
25. J. H. Schulman and W. D. Compton, *Color Centers in Solids* (Macmillan, New York, 1962).
26. F. Lüty, in *Physics of Color Centers*, Ed. by W. Beall Fowler, (Academic Press, New York, 1968), p. 182.
27. F. Seitz, *Phys. Rev.* **79**, 529 (1950); *Rev. Mod. Phys.* **26**, 7 (1954).
28. N. Itoh and H. Saidoh, *Phys. Stat. Sol.* **33**, 649 (1969); N. Itoh and M. Ikeya, *J. Phys. Soc. Japan* **22**, 1170 (1967).
29. J. H. Crawford, *Advan. Phys.* **17**, 93 (1968).
30. W. D. Compton and H. Rabin, in *Solid State Physics*, Vol. 16, Ed. by F. Seitz and D. Turnbull (Academic Press, New York, 1964), p. 121.
31. F. Seitz, *Rev. Mod. Phys.* **18**, 384 (1946).
32. H. Pick, *Z. Phys.* **159**, 69 (1960).
33. C. Z. Van Doorn, *Phys. Rev. Letters* **4**, 236 (1960); *Philips Res. Rep.* **11**, 479 (1956); **12**, 309 (1957).

34. T. J. Turner, N. N. Isenhower, and P. K. Tse, *Solid State Comm.* **7**, 1661 (1969).
35. C. J. Delbecq, E. Hutchinson, D. Schoemaker, E. L. Yasaitis, and P. H. Yuster, *Phys. Rev.* **187**, 1103 (1969).
36. F. W. Patten and F. J. Keller, *Phys. Rev.* **187**, 1120 (1969).
37. G. W. Groves and A. Kelley, *Phil. Mag.* **8**, 1437 (1963).
38. T. Hibi and K. Yada, *J. Electronmicroscopy* **10**, 164 (1961); Y. Kawamata and T. Hibi, *J. Phys. Soc. Japan* **20**, 242 (1965).
39. J. Z. Damm and M. Suszynska, *J. Physique* **9**, 168 (1967).
40. D. Schoemaker, *Phys. Rev.* **149**, 693 (1966).
41. H. A. Bethe, *Ann. Physik* **5**, 325 (1930).
42. H. A. Bethe and J. Ashkin, in *Experimental Nuclear Physics I*, Part II, Ed. by E. Segré (John Wiley and Sons, New York, 1953).
43. *American Institute of Physics Handbook* (McGraw-Hill, New York, 1957).
44. L. Katz and A. S. Penfold, *Rev. Mod. Phys.* **24**, 28 (1952).
45. R. L. Heath, *Scintillation Spectrometry Gamma-Ray Spectrum Catalog*, AEC Res. and Dev. Report IDO 16880-1, 1964; G. W. Grodstein, *NBS Circular* 583 (1956).
46. D. K. Holmes, in *The Interaction of Radiation with Solids*, Ed. by R. Strumane, J. Nihond, R. Gevers, and S. Amelinckx (North-Holland Publishing Co., Amsterdam, 1964).
47. J. W. Corbett, *Electron Radiation Damage in Semiconductors and Metals* (Academic Press, New York, 1966).
48. F. Seitz and J. S. Koehler, *Solid State Physics*, Vol. II, Ed. by F. Seitz and D. Turnbull (Academic Press, New York, 1955), p. 305.
49. D. S. Billington and J. H. Crawford, Jr., *Radiation Damage in Solids* (Princeton University Press, 1961).
50. G. H. Kinchin and R. S. Pease, *Rept. Progr. Phys.* **18**, 1 (1955).
51. F. Eisen, *Phys. Rev.* **135A**, 1394 (1964).
52. Y. Farge, M. Lambert, and A. Guinier, *J. Phys. Chem. Solids* **27**, 499 (1966); Y. Farge, presented at Int. Symp. on Color Centers in Alkali Halides, Rome, 1968, Abstract 57, p. 89 (unpublished).
53. S. Datz, C. Erginsoy, G. Leibfried, and H. O. Lutz, *Ann. Rev. Nuclear Sci.* **17**, 129 (1967).
54. R. S. Knox and K. J. Teagarden, in *Physics of Color Centers*, Ed. by W. Beall Fowler (Academic Press, New York, 1968).
55. C. B. Lushchik, I. K. Vitol, and M. Elango, *Soviet Phys.—Solid State* **10**, 2166 (1969).
56. W. L. Wagner, *Z. Physik* **181**, 143 (1964).
57. C. J. Delbecq, A. K. Ghosh, and P. H. Yuster, *Phys. Rev.* **154**, 797 (1967).
58. D. Schoemaker and J. L. Kolopus, *Solid State Comm.* **8**, 435 (1970).
59. R. G. Kaufman and W. B. Hadley, *J. Chem. Phys.* **47**, 264 (1967).
60. H. W. Etzel and J. H. Schulman, *J. Chem. Phys.* **22**, 1549 (1954).
61. T. L. Gilbert, Lecture Notes for the NATO Summer School in Solid State Physics, Ghent, Belgium, 1966 (unpublished).
62. A. N. Jette, T. L. Gilbert, and T. P. Das, *Phys. Rev.* **184**, 884 (1969).
63. W. Känzig, *J. Phys. Chem. Solids* **17**, 88 (1960).
64. J. Wilkens, Thesis, Clarendon Laboratory, Oxford, 1961.
65. I. L. Bass and R. Miher, *Phys. Rev. Letters* **15**, 25 (1965).
66. L. A. Pung and Yu. Yu. Khaldre, *Bull. Acad. Sci. USSR* **30**, 1509 (1966).

67. R. B. Murray, *AEC Progress Report* NYO-3842-5, 1969 (private communication); R. B. Murray, H. B. Dietrich, and F. J. Keller, *J. Phys. Chem. Solids* **31**, 1185 (1970).
68. C. D. Delbecq, B. Smaller, and P. H. Yuster, *Phys. Rev.* **111**, 1235 (1958).
69. T. J. Neubert and T. A. Reffner, *J. Chem. Phys.* **36**, 2780 (1962).
70. F. J. Keller, R. B. Murray, M. M. Abraham, and R. A. Weeks, *Phys. Rev.* **154**, 812 (1967).
71. F. J. Keller and R. B. Murray, *Phys. Rev. Letters* **15**, 198 (1965); *Phys. Rev.* **150**, 670 (1966).
72. W. Känzig and T. O. Woodruff, *Phys. Rev.* **109**, 220 (1958); *J. Phys. Chem. Solids* **9**, 70 (1958).
73. B. S. H. Royce, *Progr. Solid State Chem.* **4**, 93 (1968).
74. H. Seidel and H. C. Wolf, in *Physics of Color Centers*, Ed. by W. Beall Fowler (Academic Press, New York, 1968), p. 594.
75. W. D. Compton and C. C. Klick, *Phys. Rev.* **110**, 349 (1958).
76. C. J. Delbecq, J. L. Kolopus, E. L. Yasaitis, and P. H. Yuster, *Phys. Rev.* **154**, 866 (1967).
77. C. J. Delbecq, J. L. Kolopus, E. L. Yasaitis, and P. H. Yuster (unpublished results).
78. J. S. Nadeau, *J. Appl. Phys.* **33**, 3480 (1962); **34**, 2248 (1963); **35**, 1248 (1964).
79. W. A. Sibley and E. Sonder, *J. Appl. Phys.* **34**, 2366 (1963).
80. W. A. Sibley and J. R. Russell, *J. Appl. Phys.* **36**, 810 (1965).
81. R. L. Fleischer, *Acta Met.* **9**, 835 (1962).
82. R. Balzer, H. Peisl, and W. Waidelich, *Phys. Letters* **27A**, 31 (1968); *Phys. Stat. Sol.* **15**, 495 (1966); **28**, 207 (1968); *Phys. Rev. Letters* **17**, 1129 (1966); *Z. Physik* **204**, 405 (1967).
83. A. Behr and W. Waidelich, *Physics Letters* **23**, 620 (1966).
84. D. Pooley, *Brit. J. Appl. Phys.* **17**, 1 (1966).
85. B. J. Faraday and W. D. Compton, *Phys. Rev.* **138A**, 893 (1965).
86. B. Faraday, H. Rabin, and W. D. Compton, *Phys. Rev. Letters* **7**, 57 (1961).
87. J. H. O. Varley, *Nature* **174**, 86 (1954); *J. Nucl. Eng.* **1**, 130 (1954).
88. V. H. Ritz, *Phys. Rev.* **142**, 505 (1966).
89. B. S. H. Royce and D. J. Treacy, presented at Int. Symp. on Color Centers in Alkali Halides, Rome, 1968, Abstract 155.
90. J. H. O. Varley, *J. Phys. Chem. Solids* **23**, 985 (1962).
91. N. Itoh, *Phys. Stat. Sol.* **30**, 199 (1968).
92. D. L. Dexter, *Phys. Rev.* **118**, 934 (1960).
93. A. N. Jette, T. L. Gilbert, and T. P. Das, *Phys. Rev.* **184**, 884 (1969).
94. D. Pooley and W. A. Runciman, *Solid State Comm.* **4**, 351 (1966).
95. A. Smakula, *Z. Phys.* **63**, 762 (1930).
96. C. B. Lushchik, G. G. Liidya, and I. V. Taek, in *Proc. Int. Conf. Semiconductors, Prague, 1960.*
97. J. H. Parker, Jr., *Phys. Rev.* **124**, 703 (1961).
98. T. P. P. Hall, D. Pooley, W. A. Runciman, and P. T. Wedepohl, *Proc. Phys. Soc.* **84**, 719 (1964).
99. C. B. Lushchik, G. K. Vale, E. R. Ilmas, N. S. Rooze, A. A. Elango, and M. A. Elango, *Optics and Spectr.* **21**, 377 (1966).
100. H. N. Hersh, *Phys. Rev.* **148**, 928 (1966); J. D. Konitzer and H. N. Hersh, *J. Phys. Chem. Solids* **27**, 771 (1966).
101. D. Pooley, *Proc. Phys. Soc.* **87**, 245 (1966).

102. R. B. Murray and F. J. Keller, *Phys. Rev.* **137**, 942 (1965).
103. A. E. Hughes, D. Pooley, H. V. Rohman, and W. A. Runciman, AERE R 5604 (1967).
104. J. F. Keller and F. W. Patten, *Solid State Comm.* **7**, 1603 (1969).
105. C. B. Lushchik, G. G. Liidya, and M. A. Elango, *Soviet Phys.—Solid State* **6**, 1789 (1965).
106. K. J. Teegarden and R. A. Weeks, *J. Phys. Chem. Sol.* **10**, 211 (1959).
107. D. Pooley, *Proc. Phys. Soc.* **87**, 257 (1966).
108. I. McC. Torrens and L. T. Chadderton, *Phys. Rev.* **159**, 671 (1967).
109. W. A. Sibley and O. E. Facey, *Phys. Rev.* **174**, 1076 (1968).
110. P. B. Still and D. Pooley, *Phys. Stat. Sol.* **32K**, 147 (1969).
111. M. Ueta, Y. Kondo, M. Hirai, and T. Yoshinary, *J. Phys. Soc. Japan* **26**, 1000 (1969).
112. R. K. Swank and F. C. Brown, *Phys. Rev.* **130**, 34 (1963).
113. M. L. Dakss and R. L. Mieher, *Phys. Rev.* **187**, 1053 (1969).
114. Y. H. Chu and R. L. Mieher, *Phys. Rev. Letters* **20**, 1289 (1968); *Phys. Rev.* **188**, 1311 (1969).
115. A. Behr, H. Peisl, and W. Waidelich, *Phys. Letters* **24A**, 379 (1967).
116. R. Balzer, H. Peisl, and W. Waidelich, *Phys. Stat. Sol.* **31**, 29K (1969).
117. W. Fuchs and A. Taylor, *Phys. Rev. B* **2**, 3393 (1970).
118. G. Giuliani, A. Perinati, E. Reguzzoni, and G. Chiarotti, *Solid State Comm.* **3**, 161 (1965).
119. N. Itoh, B. S. H. Royce, and R. Smoluchowski, *Phys. Rev.* **137A**, 1010 (1965).
120. H. Rüchardt, *Phys. Rev.* **103**, 873 (1956).
121. G. Kurtz and W. Gebhardt, *Phys. Stat. Sol.* **7**, 351 (1964).
122. W. Hertz, H. Peisl, and W. Waidelich, *Phys. Letters* **25A**, 403 (1967).
123. R. Balzer, H. Peisl, and W. Waidelich, *Phys. Stat. Sol.* **27K**, 165 (1968).
124. W. Gebhardt, *J. Phys. Chem. Solids* **23**, 1123 (1962).
125. K. Tharmalingham, *J. Phys. Chem. Solids* **25**, 255 (1964).
126. P. Suptitz and J. Teltow, *Phys. Stat. Sol.* **23**, 9 (1967).
127. E. Agathonikou-Rokohyllou, A. Costikas, and C. Manos, *J. Phys. Chem. Solids* **28**, 367 (1967).
128. I. Schneider and H. Rabin, *Phys. Rev.* **140A**, 1893 (1965).
129. J. Nahum and D. A. Wiegand, *Phys. Rev.* **154**, 817 (1967).
130. A. Chandra and D. F. Holcomb, *J. Chem. Phys.* **51**, 1509 (1969).
131. W. A. Sibley and J. R. Russell, in *Lattice Defects and Their Interactions*, Ed. by R. R. Hasiguti (Gordon and Breach, London, 1967), p. 837.
132. Yu. L. Lukantsever, F. N. Zaitov, and V. I. Sidlyarenko, *Bull. Acad. Sci. USSR (Physical Series)* **30**, 1544 (1966).
133. J. M. Bunch and E. Pearlstein, *Phys. Rev.* **181**, 1290 (1969).
134. K. Konrad and T. S. Neubert, *J. Chem. Phys.* **47**, 4946 (1967).
135. B. Gudden and R. W. Pohl, *Z. Physik* **31**, 651 (1925).
136. H. Pick, *Ann. Physik* **31**, 365 (1938).
137. H. Rabin, *Phys. Rev.* **129**, 129 (1963).
138. E. Sonder, W. A. Sibley, and W. C. Mallard, *Phys. Rev.* **159**, 755 (1967).
139. M. Hirai and A. B. Scott, *J. Chem. Phys.* **44**, 1753 (1966).
140. D. Dutton and R. Maurer, *Phys. Rev.* **90**, 126 (1953).
141. R. Fieschi and C. Paracchini, *Phys. Rev.* **182**, 935 (1969).
142. G. Kaufman (unpublished results).

143. Y. Farge, M. Lambert, and R. Smoluchowski, *Solid State Comm.* **4**, 333 (1966).
144. J. Nahum, *Phys. Rev.* **174**, 1000 (1968).
145. R. Onaka, I. Fujita, and A. Fukada, *J. Phys. Soc. Japan* **18** (Suppl. 2), 263 (1963).
146. R. Allen, W. T. Angel, and E. Sonder (unpublished data).
147. G. Chiarotti, G. Giuliani, and D. W. Lynch, *Nuovo Cimento* **17**, 989 (1960).
148. G. Giuliani and F. Reguzzoni, *Phys. Stat. Sol.* **25**, 457 (1968).
149. T. Matsuyama and M. Hirai, *J. Phys. Soc. Japan* **27**, 1526 (1969).
150. D. K. Dawson and D. Pooley, *Solid State Comm.* **7**, 1001 (1969).
151. H. Härtel and F. Lüty, *Z. Physik* **177**, 369 (1964); **182**, 111 (1964).
152. E. Link and F. Lüty, presented at Int. Symp. on Color Centers in Alkali Halides, Urbana, 1965, Abstract 120.
153. G. Giuliani, *J. Physique* (Suppl. C4), 175 (1967).
154. M. N. Kabler, *Phys. Rev.* **136A**, 1296 (1964).
155. A. Behr, H. Peisl, and W. Waidelich, *Phys. Stat. Sol.* **21K**, 9 (1967).
156. J. D. Comins, *Phys. Stat. Sol.* **33**, 445 (1969).
157. E. Sonder, *Phys. Stat. Sol.* **35**, 523 (1969).
158. J. L. Alvares-Rivas and P. W. Levy, presented at Int. Symp. on Color Centers in Alkali Halides, Rome, 1968, Abstract 3.
159. E. Sonder, G. Bassignani, and P. Camagni, *Phys. Rev.* **180**, 882 (1969).
160. E. Sonder and W. A. Sibley, *Phys. Rev.* **140A**, 539 (1965).
161. G. Giuliani, *Phys. Rev. B* **2**, 464 (1970).
162. N. Itoh, B. S. H. Royce, and R. Smoluchowski, *Phys. Rev.* **138**, A1766 (1965).
163. M. Ikeya, N. Itoh, and T. Suita, *J. Phys. Soc. Japan* **23**, 455 (1967); **26**, 291 (1969).
164. E. Sonder, W. A. Sibley, J. E. Rowe, and C. M. Nelson, *Phys. Rev.* **153**, 1000 (1967).
165. J. D. Comins, *Phys. Stat. Sol.* **42**, 101, 113 (1970).
166. E. Sonder and D. Walton, *Phys. Letters* **25A**, 222 (1967).
167. R. A. Guenther and H. Weinstock (to be published).
168. P. Durand, Y. Farge, and M. Lambert, *J. Phys. Chem. Solids* **30**, 1353 (1969).
169. Y. Farge, *J. Phys. Chem. Solids* **30**, 1375 (1969).
170. D. Pooley, *Proc. Phys. Soc.* **89**, 723 (1966).
171. H. G. Dahnke and K. Thommen, *Z. Physik* **184**, 367 (1965).
172. P. W. Levy, P. L. Mattern, and K. Lengweiler, *Phys. Rev. Letters* **24**, 13 (1970).
173. M. Ikeya, N. Itoh, T. Okada, and T. Suita, *J. Phys. Soc. Japan* **21**, 1304 (1966).
174. J. H. Crawford, *J. Phys. Soc. Japan* (Suppl. III) **18**, 329 (1964).
175. F. Fröhlich and S. Altrichter, *Ann. Physik* **17**, 143 (1966).
176. W. A. Sibley, C. M. Nelson, and J. H. Crawford, Jr., *Phys. Rev.* **139A**, 1328 (1965).
177. J. L. Alvares-Rivas and P. W. Levy, *Phys. Rev.* **162**, 816 (1967).
178. P. V. Mitchell, D. A. Wiegand, and R. Smoluchowski, *Phys. Rev.* **121**, 484 (1961).
179. P. V. Sastry and K. A. McCarthy, *Phys. Stat. Sol.* **10**, 585 (1965).
180. E. Sonder and L. C. Templeton, *Phys. Rev.* **164**, 1106 (1967).
181. W. A. Sibley, E. Sonder, and C. T. Butler, *Phys. Rev.* **136A**, 537 (1964).
182. K. Thommen, *Phys. Letters* **2**, 189 (1962).
183. E. Sonder and W. A. Sibley, *Phys. Rev.* **129**, 1578 (1963).
184. P. G. Harrison, *Phys. Rev.* **131**, 2505 (1963).
185. H. Rabin, *J. Phys. Soc. Japan* (Suppl. III) **18**, 334 (1963).
186. K. Asai and A. Okuda, *J. Phys. Soc. Japan* **21**, 2197 (1966).
187. C. J. Delbecq, *Z. Physik* **171**, 560 (1963).
188. N. Itoh and T. Suita, *J. Phys. Soc. Japan* **17**, 348 (1962); **15**, 2364 (1960).

189. W. E. Bron, *Phys. Rev.* **119**, 1853 (1960).
190. G. Baldini, L. Dalla Croce, and R. Fieschi, *Nuovo Cimento* **20**, 806 (1961).
191. J. W. Mathews, W. C. Mallard, and W. A. Sibley, *Phys. Rev.* **146**, 611 (1966).
192. E. Sonder, *Phys. Rev. B* **2**, 4189 (1970).
193. S. Schnatterly and W. D. Compton, *Phys. Rev.* **135A**, 227 (1964).
194. G. Horn and H. Peisl, *Z. Physik* **194**, 219 (1966).
195. C. H. Seager, D. O. Welch, and B. S. H. Royce, Abstract BD8, *Bull. Am. Phys. Soc. Series II*, **14**, 324 (1969); Int. Symp. on Color Centers in Alkali Halides, Rome, 1968, Abstract 162.
196. S. Kalbitzer and P. W. Levy, *Phys. Rev.* (to be published).
197. W. Brandt, H. F. Waung, and P. W. Levy, presented at Int. Symp. on Color Centers in Alkali Halides, Rome, 1968, Abstract 29.
198. W. A. Sibley and J. R. Russell, *Phys. Rev.* **154**, 831 (1967).
199. J. H. Crawford and C. M. Nelson, *Phys. Rev. Letters* **5**, 314 (1960).
200. C. Sanchez and F. Agullo Lopez, *Phys. Stat. Sol.* **29**, 217 (1968).
201. V. H. Ritz, *Phys. Rev.* **133A**, 1452 (1964).
202. R. O. Vilu and M. A. Elango, *Soviet Phys.—Solid State* **7**, 2967 (1966).
203. J. D. Comins and P. T. Wedepohl, *Solid State Comm.* **4**, 537 (1966).
204. C. L. Bauer and R. B. Gordon, *Phys. Rev.* **126**, 73 (1962).
205. W. D. Compton and G. W. Arnold, *Disc. Faraday Soc.* **31**, 130 (1961).
206. W. A. Sibley and Y. Chen, *Phys. Rev.* **160**, 712 (1967).
207. P. D. Johnson, in *Luminescence of Inorganic Solids*, Ed. by P. Goldberg (Academic Press, New York, 1966), p. 287.
208. J. E. Wertz, L. C. Hall, J. Hegelson, C. C. Chao, and W. S. Dykoski, in *Interaction of Radiation with Solids*, Ed. by A. Bishay (Plenum Press, New York, 1967), p. 617.
209. T. M. Searle and A. M. Glass, *J. Phys. Chem. Solids* **29**, 609 (1968).
210. R. W. Soshea, A. J. Dekker, and J. P. Sturtz, *J. Phys. Chem. Solids* **5**, 23 (1958).
211. J. Y. Wong, M. J. Berggren, and A. L. Schawlow, in *Optical Properties of Ions in Crystals*, Ed. by H. M. Crosswhite and H. W. Moos (Interscience Publishers, New York, 1967), p. 383.
212. F. T. Gamble, R. H. Bartram, C. G. Young, O. R. Gilliam, and P. W. Levy, *Phys. Rev.* **134A**, 589 (1964); **138A**, 577 (1965).
213. J. E. Wertz and P. V. Auzins, *Phys. Rev.* **139A**, 1645 (1965).
214. P. W. Kirklin, P. Auzins, and J. E. Wertz, *J. Phys. Chem. Solids* **26**, 1067 (1965).
215. W. C. O'Mara, J. J. Davies, and J. E. Wertz, *Phys. Rev.* **179**, 816 (1969).
216. J. E. Wertz, J. W. Orton, and P. Auzins, *Disc. Faraday Soc.* **31**, 140 (1961).
217. Y. Chen, B. Henderson, M. M. Abraham, and H. T. Tohver, *Bull. A.P.S. II* **16**, 421 (1971).
218. H. T. Tohver, B. Henderson, Y. Chen, and M. M. Abraham, *Phys. Rev.* (to be published).
219. B. Henderson and A. C. Tomlinson, *J. Phys. Chem. Solids* **30**, 1801 (1969).
220. J. H. Crawford, *Crystal Lattice Defects* **1**, 357 (1970).
221. J. E. Wertz, P. Auzins, R. A. Weeks, and R. H. Silsbee, *Phys. Rev.* **107**, 1535 (1957).
222. W. P. Unruh and J. W. Culvahouse, *Phys. Rev.* **154**, 861 (1967).
223. J. C. Kemp, W. M. Ziniker, and J. A. Glaze, *Phys. Letters* **22**, 37 (1966); *Proc. British Ceram. Soc.* **9**, 109 (1967).
224. J. C. Kemp, W. M. Ziniker, J. A. Glaze, and J. C. Chung, *Phys. Rev.* **171**, 1024 (1968).

225. J. Mort, F. Lüty, and F. C. Brown, *Phys. Rev.* **137A**, 566 (1965).
226. F. C. Brown, *The Physics of Solids* (W. A. Benjamin, New York, 1967).
227. J. C. Kemp, J. C. Cheng, E. H. Izen, and F. A. Modine, *Phys. Rev.* **179**, 818 (1969).
228. R. G. Bessent, B. C. Cavenett, and I. C. Hunter, *J. Phys. Chem. Solids* **29**, 1523 (1968).
229. B. Henderson and R. D. King, *Phil. Mag.* **13**, 1149 (1966).
230. Y. Chen, W. A. Sibley, F. D. Srygley, R. A. Weeks, E. B. Hensley, and R. L. Kroes, *J. Phys. Chem. Solids* **29**, 863 (1968).
231. J. C. Kemp and V. I. Neeley, *Phys. Rev.* **132**, 215 (1963).
232. L. A. Kappers, R. L. Kroes, and E. B. Hensley, *Phys. Rev. B* **1**, 4151 (1970).
233. B. Henderson, S. E. Stokowski, and T. C. Ensign, *Phys. Rev.* **183**, 826 (1968).
234. B. D. Evans, J. Cheng, and J. C. Kemp, *Phys. Letters* **27A**, 506 (1969).
235. B. D. Evans and J. C. Kemp, *Phys. Rev.* (to be published).
236. Y. Chen and W. A. Sibley, *Phil. Mag.* **20**, 217 (1969).
237. R. D. King and B. Henderson, *Proc. British Ceram. Soc.* **9**, 63 (1967).
238. D. B. Fitchen, R. H. Silsbee, T. A. Fulton, and E. L. Wolf, *Phys. Rev. Letters* **11**, 275 (1963).
239. F. Lanzl, W. von der Osten, U. Röder, and W. Waidelich, *Localized Excitations in Solids* (Plenum Press, New York, 1968), p. 575.
240. A. E. Hughes, *J. Physique* **28** (Suppl. C4), 55 (1967).
241. D. B. Fitchen, in *The Physics of Color Centers*, Ed. by W. B. Fowler (Academic Press, New York, 1968), p. 293.
242. I. K. Ludlow, *Proc. Phys. Soc.* **88**, 763 (1966).
243. R. D. King and B. Henderson, *J. Physique* **28** (Suppl. C4), 75 (1967); *Proc. Phys. Soc.* **89**, 153 (1966).
244. A. A. Kaplianskii, *Opt. Spectry.* (*USSR*) **6**, 267 (1959); **16**, 329, 557 (1964).
245. W. A. Runciman, *Proc. Phys. Soc.* **86**, 625 (1965).
246. I. K. Ludlow and W. A. Runciman, *J. Phys. C* **1**, 1194 (1968).
247. R. G. Bessent, *J. Phys. Chem.* **2**, 1101 (1969).
248. J. E. Sickles, Ph. D. Thesis, Northwestern University, 1969 (private communication).
249. O. S. Oen, Oak Ridge National Laboratory Report No. ORNL 3813 (1965) (unpublished).
250. Y. Chen, D. L. Trueblood, O. E. Schow, and H. T. Tohver, *J. Phys. C* **3**, 2501 (1970).
251. Y. Chen, R. T. Williams, and W. A. Sibley, *Phys. Rev.* **182**, 960 (1969).
252. W. C. McGowan and W. A. Sibley, *Phil. Mag.* **19**, 967 (1969).
253. R. W. Davidge, *J. Nucl. Mat.* **25**, 75 (1968).
254. B. Henderson and D. H. Bowen, *J. Phys. C* **4**, 1487 (1971).
255. B. Henderson, D. H. Bowen, A. Briggs, and R. D. King, *J. Phys. C* **4**, 1496 (1971).
256. R. S. Wilks, *J. Nucl. Mat.* **26**, 137 (1968).
257. A. Briggs and D. H. Bowen, in *Symposium on Mass Transport in Oxides*, NBS Publication 296 (1968), p. 103.
258. R. C. DuVarney, A. K. Garrison, and R. H. Thorland, *Phys. Rev.* **188**, 657 (1969).
259. J. M. Smith and W. E. Vehse, *Phys. Letters* **31A**, 147 (1970).
260. W. E. Vehse, W. A. Sibley, F. J. Keller, and Y. Chen, *Phys. Rev.* **167**, 828 (1968).
261. G. Heiland, E. Mollwo, and F. Stöckmann, *Advan. Solid State Phys.* **8**, 191 (1959).

262. B. Henderson, in *Symposium on Mass Transport in Oxides*, NBS Publication 296 (1968), p. 41.
263. P. W. Levy, *Disc. Faraday Soc.* **31**, 118 (1961).
264. P. W. Levy, *Phys. Rev.* **123**, 1226 (1961).
265. R. S. Wilks, J. A. Desport, and R. Bradley, *Proc. British Ceram. Soc.* **7**, 403 (1967).

Chapter 5

PROPERTIES OF ELECTRON CENTERS

Clifford C. Klick
Naval Research Laboratory
Washington, D. C.

1. SINGLE-VACANCY CENTERS

1.1. *F* Centers

1.1.1. *Absorption Spectra of the F Center*

In this chapter, as in the previous one, there will be a great deal of emphasis on color centers in alkali halides and especially on the *F* center.* It is surely the best understood and most widely investigated center in ionic crystals. Since it consists of an electron localized at a halide ion vacancy, it is a very simple center. Also, many of the other, more complex centers consist of clusters of two or more *F* centers. As a result, the *F* center holds a position in the area of defects in solids which is similar to that of the hydrogen atom in atomic physics. It is the simple system on which new models, theories, and techniques are usually first tried before being extended to more complex problems.

The *F* center is shown in Fig. 1; methods for producing it have been described in the previous chapter. The electron is held in the vicinity of the halide ion vacancy by the electrostatic forces of the remainder of the crystal. Chemically, it is a strange entity; one could describe it as a valence electron without a nucleus. This center has a number of electronic states, and optical

* A number of books have been published which discuss the *F* center in detail.[1]

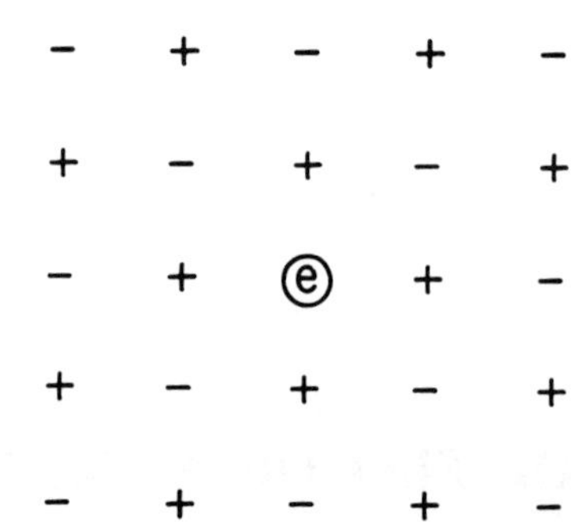

Fig. 1. The F center in alkali halides. Positive alkali ions are shown as $+$; negative halide ions as $-$. The electron e is held in the halide ion vacancy by the electric field of the surrounding ions.

absorption produces a transition of the electron from the ground state to its first excited state. There is a characteristic absorption band for the F center in each alkali halide. Often, this falls in the visible part of the spectrum, so that the crystals appear colored. Once in the excited state, the F center may return to the ground state by the emission of light or by radiationless transitions.

Figure 2 shows measurements of the broad absorption and emission spectra of F centers in KBr at various temperatures.[2] Several facts are worth noting about these spectra. First of all, the spectra are not lines at all as would be the case for the spectra of atoms in a gas. Typically, the spectra

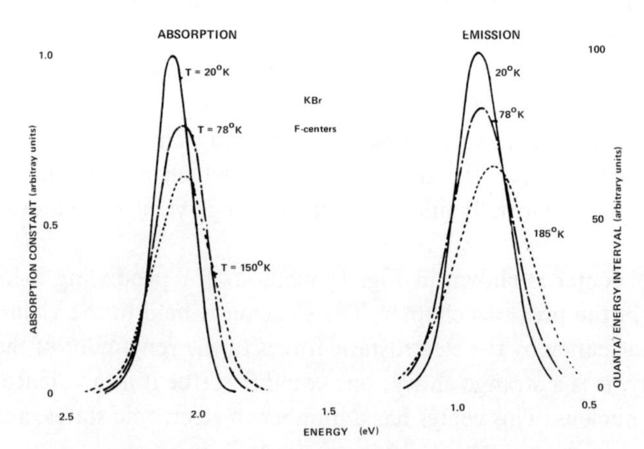

Fig. 2. Absorption and emission spectra of the F center in KBr as a function of temperature (Gebhardt and Kühnert[2]).

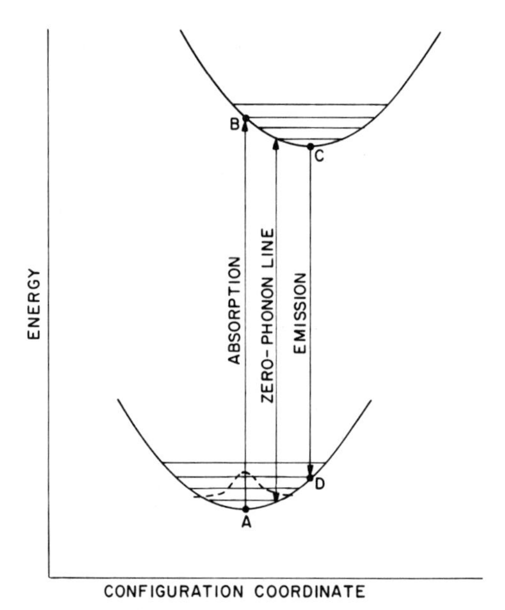

Fig. 3. Schematic of the configuration coordinate model.

may be several tenths of an electron-volt wide at points half-way down from the peak. As the temperature is decreased, the bands become narrower, but they do not become "lines" even at the lowest temperatures. Another major difference between these spectra and those in gases is the very large shift between the energy of the absorption and emission bands. In Fig. 2, this is more than a factor of two. This is called the Stokes shift and is characteristic of optical centers in ionic crystals.

A simple model exists which can be used to understand why the spectra of centers in solids differ so markedly from those in a gas. This is called the configuration coordinate model* and is illustrated in Fig. 3. In the illustration, the lowest-energy state of a center corresponds to the lower parabola. The energy of the state varies as the configuration coordinate changes. In the most general sense, this coordinate represents the positions of all the ions in the solid that interact with the center. For simplicity, we might think of the coordinate as representing the distance from the center to its nearest neighbors. As this distance varies, there is a lowest energy for the state at point A, which would be the equilibrium position for that state. If the forces acting on the system are proportional to the displacement from equilibrium, then the energy curve will be a parabola and such a system will have the

* A detailed description of the configuration coordinate model is given in Ref. 3.

properties of a simple harmonic oscillator. Because of thermal vibrations of the lattice, there will be appreciable deviations from A and these will increase as the temperature is raised.

If the system absorbs light, an electronic transition will occur that raises the system to an excited state indicated by the upper parabola. The electronic properties of the center will be different in the excited state and therefore the interaction of the center with its neighbors will also be different. As a result, the equilibrium displacement of the center in the excited state is at C and this corresponds to a different value of the configuration coordinate or a different distance of the nearest neighbors in our simplified description.

When a transition occurs, such as from A to B, it will be vertical on this diagram to a good approximation. A vertical transition means that the positions of the nuclei in the lattice do not change appreciably during the nearly instantaneous electronic transition. This approximation is called the Franck–Condon principle. If the center is displaced from the equilibrium point A by lattice vibrations, transitions will be to points displaced from B on the excited-state curve. However, the curve at B varies rapidly in energy for displacement from B. Therefore the lattice vibrations at A lead to a spread in energies due to the steepness of the curve near B. This leads to a broad absorption band with a considerable spread of energies as is seen experimentally.

Once the center has reached the excited state near B, it will proceed to its new equilibrium at C. This can be accomplished by having the neighbor ions change their position and having the center give off its excess energy to these lattice vibrations. At C, the center will again experience thermal vibrations. After some time, the center may make another transition to the ground state, with the emission of light. This transition is shown going from C to D. Here, too, D is on a sharply varying part of the energy curve, so that variations in the configuration coordinate around C lead to a variation in energy which appears as a broad emission band.

At D, the center will again relax to the starting point A by giving off the excess energy as heat. The energy absorbed was A to B and the energy emitted was only C to D. As a result, the emission energy will be at lower energy and longer wavelengths than the absorption, thus explaining the origin of the Stokes shift. The difference in energy is B to C and D to A, which appears in the lattice as heat during this cyclic process.

Much of what we know about the optical properties of centers at high temperatures can be understood using this simple classical model. The model would predict that the absorption and emission bands should narrow to lines as the temperature approaches absolute zero. This is not the case, as we see in Fig. 2. It is possible to describe the low-temperature case by introducing a quantum mechanical treatment of the center. It was pointed out earlier that the center undergoes a simple harmonic vibration if the energy curve is a parabola. The quantum mechanical treatment of the simple harmonic vibrator shows that it has a series of equally spaced energy levels with separation $h\nu$, where h is Planck's constant and ν is the frequency of the vibrator. The lowest level is $h\nu/2$ above the minimum of the classical parabola. Such levels are shown in Fig. 3.

At high temperatures, many of these vibrational levels are populated and the quantum mechanical treatment approaches the classical one. However, at low temperatures, the lowest energy level alone is populated. This does not mean that the system is at rest. It has a "zero-point vibration" and the probability of appearing at various values of the configuration coordinate is given by the dashed curve of Fig. 3. This is a Gauss curve; as the temperature is raised and higher levels begin to be populated, the probability distribution continues to be a Gauss curve with a half-width that increases with temperature. The broad distribution in the configuration coordinate even at the lowest temperatures leads again to a broad absorption band since the upper curve varies rapidly near *B*. The illustration shows a transition to a point of the upper curve that is only a few vibrational levels above the lowest one. In an actual case, the transition is often to levels that are 20 or 30 above the minimum. In this case, the quantum mechanical treatment of the upper state may be replaced by the classical curve.

Similarly, in emission, the zero-point vibration in the lowest level of the upper curve leads to a broadband emission. Here again, the state from which the transition begins must be treated quantum mechanically for the low-temperature case, while the state at which the transition ends may be treated classically if the transition is to levels far removed from the lowest level.

In some cases, the curves are close enough so that a transition from the lowest energy level of one state can be made to the lowest energy level of the other state. This results in an electronic transition without the production of heat. The absorption and emission spectra both have a sharp line, called a zero-phonon line, at the same wavelength. Such a case will be described later in the spectra of more complex centers. Figure 3 illustrates such a zero-phonon transition.

For the usual broad absorption and emission bands, the width of the band at some temperature T, called $H(T)$, can be related to the width at absolute zero, $H(0)$, by the equation

$$H^2(T)/H^2(0) = \coth(h\nu/2kT) \tag{1}$$

$H(0)$ is simple to measure since in practice it is a constant up to 30 or 40°K, where temperature broadening begins. Figure 4 shows data for the F-center absorption of various alkali halides plotted according to Eq. (1).[4] From the slope of the curves in Fig. 4, there can be obtained the effective vibrational frequencies which interact with the F center in various solids. A similar treatment can be made for the emission spectra.

This treatment leads to a single value of the vibrational frequency interacting with the center, but this should be viewed as some average frequency since there is a large distribution of lattice vibrational modes that can interact with the center. It is possible that, even so, there are single mode frequencies that largely determine the average. In particular, there are two which have been proposed as being dominant. If one assumes that the F center interacts primarily with lattice modes at a distance from the center, then the optical transition of the F center would be expected to couple with the longitudinal optical vibrational modes of the lattice. If, on the other hand, one assumes that the F-center interaction is primarily local, then an interaction with the nearest neighbors would dominate. Since these ions are next to a vacancy, their vibrational frequency would be expected to be considerably less than the typical lattice vibrational frequencies.

The experimental absorption data[4] for F centers in KCl yield a frequency of 2.96×10^{12} sec^{-1}; the longitudinal optical frequency is 6.40×10^{12}.

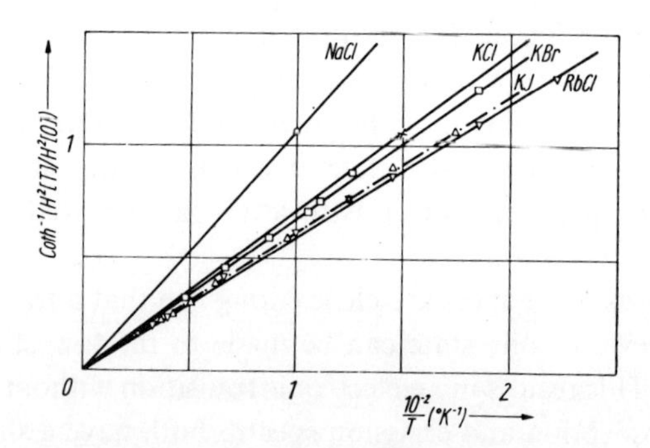

Fig. 4. The temperature dependence of the half-width of the F-center absorption bands plotted according to Eq. (1) in the text. The slope of these curves yields an effective vibrational frequency interacting with the center in its ground state. (Gebhardt and Kühnert.[4])

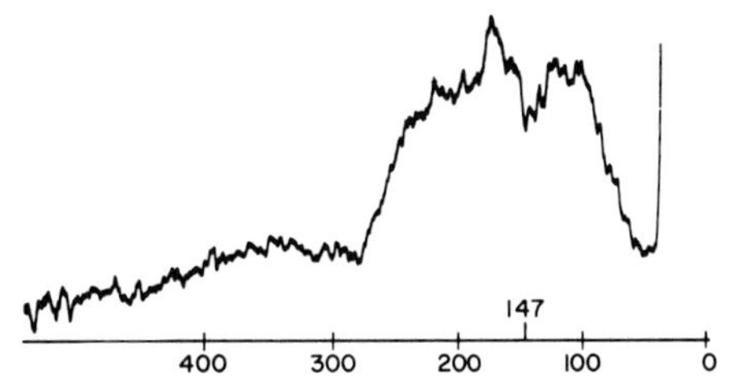

Fig. 5. The Raman spectrum of F centers in NaCl using unpolarized detectors and exciting with a 5145-Å He–Ne laser (Worlock and Porto[5]).

For NaCl, the absorption data[4] yield 4.40×10^{12} and the longitudinal optical frequency is 8.1×10^{12} sec^{-1}. These results indicate that there is appreciable interaction of the F center with near neighbors and that the distribution of modes of vibration are at a lower frequency near the F center than in the pure lattice.

1.1.2. *Raman Scattering of the F Center*

With the advent of laser light sources, it has become possible to measure the Raman spectra of F centers in alkali halides.[5] This is especially significant because it has been shown[6] that in this case, the Raman spectrum must be identical with the spectrum of lattice vibrations that broaden the absorption bands. Thus a detailed evaluation of the interacting lattice modes is possible. The data for NaCl are shown in Fig. 5. Also marked on the figure is the value of the vibrational frequency (147 cm^{-1}) obtained from the temperature variation of absorption. It is apparent that it is a good average of the main part of the scattering spectrum and that this lattice vibration spectrum is characteristic of the lattice near the F center rather than that of the pure lattice.

1.1.3. *Electron Spin Resonance of F Centers**

It is generally agreed that electron spin resonance has been the single most important tool in unraveling the nature of color centers. Some of the simpler centers, such as the F center, were already well understood before the advent of electron spin resonance, but even in these cases, the resonance data made the assignment of models absolutely certain. The use of these

* A recent extensive review of color center resonance is given by Seidel and Wolf.[7]

techniques has now given us a wide range of centers that are understood, and this has permitted elaborate studies of what might be considered as defect chemistry: the production, annihilation, and motion of defects in crystal lattices. This will be especially apparent in the later discussions of more complex centers. Since the subject of electron spin resonance is treated in detail elsewhere, we will be content here with showing how the method has been used with *F* centers.

A center is paramagnetic if it has unpaired electron spins. In the case of the *F* center, there is a single electron which gives rise to paramagnetism. This paramagnetism has been observed using sensitive magnetic balances; but in most cases, a paramagnetic center such as the *F* center can be detected with much greater sensitivity using electron spin resonance. For an *F*-center electron in a magnetic field, there are two energy levels, corresponding to the orientation of the electron spin parallel or antiparallel to the applied field. A transition between these states can be made by coupling the electron to an ac magnetic field of the proper frequency and oriented at right angles to the dc field. The resonance condition is

$$h\nu = \Delta E = g\mu H$$

where ν is the frequency of the ac magnetic field, usually in the microwave region, g is a factor usually near 2, μ is the Bohr magneton, and H is the dc magnetic field.

In addition to the externally applied dc field, the *F*-center electron is influenced by the magnetic field generated by the nuclei of its near neighbors. This change in the total magnetic field seen by the electron leads to a corresponding change of the resonance frequency and is called the hyperfine structure. In the case of the *F* center in KCl, the nearest-neighbor K^+ ions have a nuclear magnetic moment of 3/2. There are six of these equivalent ions and this total spin can vary from $+9$ to -9 in unit steps, leading to 19 lines. Furthermore, the statistical probability of each line can be computed and compared with the strength of the observed lines. Ions at greater distances also interact with the electron and, if this interaction is large enough, the effect may be to cause a smearing out of the details of the spectrum so that only a structureless broad absorption is seen. This is what is seen in the electron spin resonance of the *F* center in most of the alkali halides. A few, however (LiF, NaF, RbCl, CsCl, NaH), do show resolved hyperfine structure. The CsCl results[8] are shown in Fig. 6. In this spectrum, 35 out of 57 lines have been resolved. A further complication that frequently occurs is that more than one isotope of a particular ion exists and the nuclear

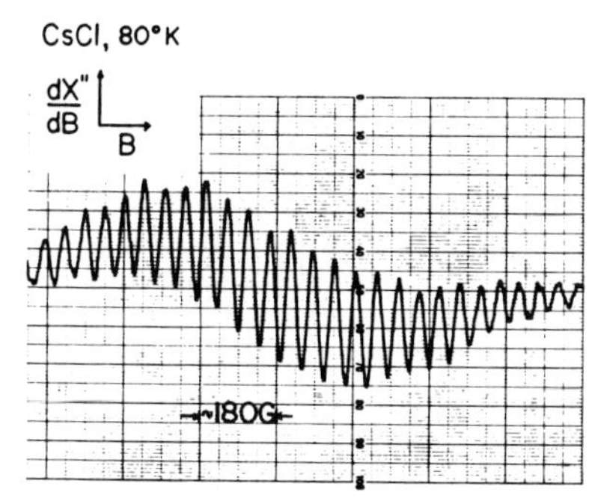

Fig. 6. Hyperfine structure seen in the absorption derivative spectrum of the electron spin resonance of the *F* center in CsCl (Hughes and Allärd[8]).

magnetic properties of the isotope can vary. This can make the unraveling of a spectrum quite difficult.

In addition to determining the number of nearest-neighbor ions surrounding an *F* center and identifying them from the hyperfine structure and from isotope substitutions, other information can be obtained from electron spin resonance. The highly isotropic nature of the *F* center is reflected in the constancy of the resonance spectrum as the crystal is rotated in the magnetic field. For more complex centers, which can sometimes be oriented, this is not the case and the variation of the spectra with the direction of the field is an important clue for determining the nature of an unknown center. Electron spin resonance has also been used to determine the number of centers in a sample. From this determination and a measurement of the magnitude of the optical absorption band, it is possible to determine the optical oscillator strength of an electronic transition in the center.

When electron spin resonance measurements are made on the *F* center at low temperatures and high powers, it is possible to saturate the microwave absorption. This happens when there are approximately equal numbers of centers in the magnetic ground and excited states and equal numbers of absorption and emission transitions are induced. The rate at which these excited centers relax to the ground state is called the relaxation time and it has been measured as a function of temperature and the concentration of centers. The physical processes involved in these relaxations are complex and in general it has not been possible to interpret the results in much detail. In the ENDOR (electron nuclear double resonance) measurement first used by Feher,[9] advantage is taken of the saturation properties of the electron spin resonance to obtain a great deal of information about the hyperfine structure. This technique depends on the development of a saturated electron spin resonance signal. At the same time, a radiofrequency signal is applied, capable of exciting nuclear resonances of nuclei coupled to the color center electron. The frequency of this interaction is a function of the color center electron density in the vicinity of the nucleus. This leads to a series of resonances for different nuclei at different distances from the color center. Detection of the nuclear resonance is made by observing changes in the saturated *F*-center electron spin resonance that occur as the nuclei go through resonance. An ENDOR spectrum for *F* centers in KBr is shown in Fig. 7, taken from work of Seidel.[10] The ions, isotopes, and shells of the various lines are identified.

The ENDOR method is able to pick out specific nuclei and it can give a measurement of interactions as far away as the eighth shell from the color center. In some instances, even more remote interactions have been noted. Since the ENDOR interaction frequency of a particular nucleus is a function of the color center electron density at that nucleus, the electron density can be measured at large distances from the color center, thus giving a detailed picture of its fall off with distance.

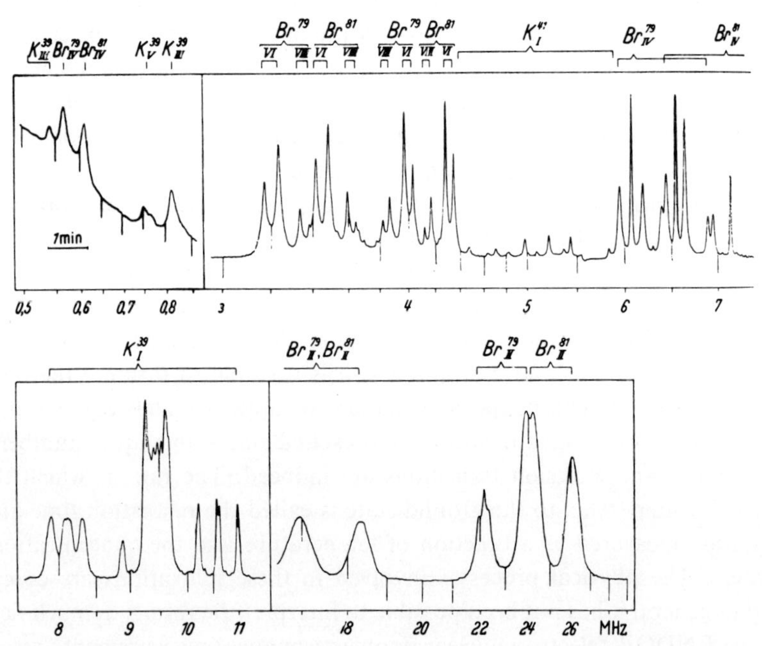

Fig. 7. ENDOR spectrum of *F* centers in KBr shown as a function of frequency. There are two isotopes for both K and Br. Assignment of shells I–VIII is shown (Seidel[10]).

The ENDOR measurement has been an especially difficult one for the theory of the *F* center to account for quantitatively. Reasonably simple approximations do lead to satisfactory values of the absorption energy of the *F* center and even properties such as the oscillator strength. One of the best known of these approximations is the point-ion-lattice approximation,[11] which uses a point-ion-lattice potential and ignores distortion of the ions and the necessity for orthogonalization of the wave functions of the color center electrons to the valence and core states of the neighboring ions. The model gives the density of the *F*-center electron quite well in the first few shells, but the density at distant shells can be an order of magnitude larger than predicted by theory. It appears that these approximations will need to be eliminated for theory to account adequately for these measurements. Attempts to work out an elaborate and complete theoretical treatment of the *F* center are currently underway.[12]

1.1.4. *Luminescence of the F Center*

The luminescence of the *F* center in alkali halides has shown a number of surprising properties and a complete understanding of them is still not in hand at this time. The absorption spectrum of the *F* center has a single Gaussian-like shape and an oscillator strength near unity, indicating a single allowed transition between isolated levels. The ground state of the center is known to be isotropic from electron spin resonance. All this suggests that the *F* center resembles the hydrogen atom, with the absorption being a $1s \rightarrow 2p$ transition. Luminescence would be expected to be the reverse of this, $2p \rightarrow 1s$, with some shift to longer wavelengths that could be understood on the basis of the configuration coordinate model of Fig. 1.

The measured luminescence for the *F* center in KBr is shown in Fig. 2 Its shape is much like that of the absorption band, but it is shifted very far into the red. In many cases, the ratio of the peak energy of the absorption and emission bands is more than a factor of two. A much smaller Stokes shift is usually found for luminescent ions such as Tl^{+}, Ag^{+}, and Mn^{2+}.

From the strong allowed transition of the *F* center in absorption, it would be expected that the emission would be similarly allowed, with a lifetime of between 10^{-7} and 10^{-8} sec and a quantum efficiency near unity. The quantum efficiency for luminescence of the *F* center at low temperatures has been found to be nearly unity as expected. However, the lifetime for luminescence has been found to be much longer than expected. Measurements by Swank and Brown,[13] shown in Fig. 8, lead to a value of nearly 10^{-6} sec for the *F*-center luminescence lifetime in KCl at the lowest tem-

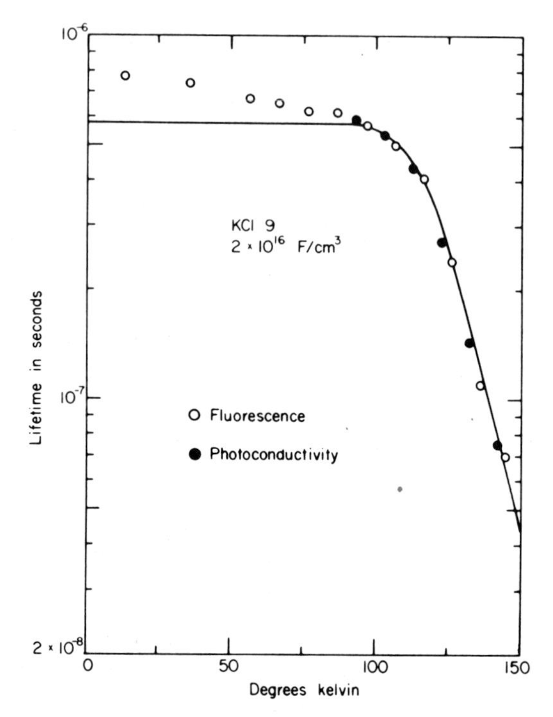

Fig. 8. Lifetime of the excited state of the *F* center in KCl as measured by luminescence decay (Swank and Brown[13]).

peratures. At higher temperatures, both the lifetime and the quantum efficiency fall off. The curve shapes can be reproduced by assuming that there is an activation energy to a photoconducting level which does not lead to luminescence. In KCl, the activation energy for this process is 0.16 eV.

It has not been easy to understand how these long lifetimes may arise. At present, two explanations of the effect have been proposed. It is possible that both make an important contribution to the increased lifetime. One possibility, studied by Fowler,[14] is that the excited-state wave function is very diffuse in comparison with the ground-state wave function. This would lead to a reduced transition probability. The computation by Fowler has two important facets. He assumed that in the excited state of the *F* center, the nearest-neighbor ions moved outward by 10%. This results from the lack of negative charge in the vacancy as the wave function of the excited-state electron becomes diffuse. A shift outward by the nearest-neighbor positive ions would be expected and has been calculated. The large Stokes shift in the energy of emission arises naturally from this shift. The second important feature of the calculation is the incorporation of an electronic polarization in the evaluation of the effective dielectric constant for the excited-state electron. For NaCl, the dielectric constant becomes quite large, 4.2, and the wave function was found to spread out over about four nearest-neighbor distances. Using these wave functions, the transition probabilities calculated gave values in satisfactory agreement with experiment.

Another quite different contribution to the long lifetime of the excited state of the *F* center may arise from metastable states of the center. While it is agreed that absorption in the *F* center arises from atomiclike $1s \rightarrow 2p$ transitions, the possible role of a $2s$ excited-state level in the emission process had been obscure. Calculations of the energy of such a state had not indicated clearly whether it was higher or lower in energy than the $2p$ level. Recent experiments on the Stark effect in absorption[15] and luminescence[16] of the *F* center have clarified this situation. The absorption measurements of Chiarotti *et al.*[15] give information about the *F* center in its unrelaxed configuration with the ions in the positions characteristic of the ground state of the *F* center. In this case, the $2s$ state lies 0.11 eV above the $2p$ state for the *F* center in KCl. However, the configuration involved in emission is the relaxed one which the *F* center comes to after the ions near it have had an opportunity to adjust to the excited electronic state. Stark-effect measurements have been made by Bogan and Fitchen[16] of the emission from the *F* center in KCl and other alkali halides. They conclude that the two states are extremely close in KCl and are mixed by local electric fields in the material. The lowest of the mixed states has about 60% *s* character and 40% *p* character. Lying above it by only 0.017 eV is the other mixed state. At low temperatures, the emission transition is from the lower state and the predominance of *s*-character wave function in that state will reduce the transition probability to the ground *s* state.

Both of these explanations of the long lifetime of the excited state of the *F* center seem plausible and in KCl, their contribution may be nearly equal in magnitude. Hopefully, one or the other mechanism might predominate in other alkali halides so that the importance of each effect could be studied separately.

The effects of a magnetic field on the optical properties of the *F* center have been studied intensively. If light passes through the sample parallel to the magnetic field, and if the light is circularly polarized, a difference can be seen in the absorption spectrum between that measured for right-handed and left-handed circularly polarized light. This is illustrated in Fig. 9, which shows the *F* absorption band in KBr.[17] Similar effects are seen

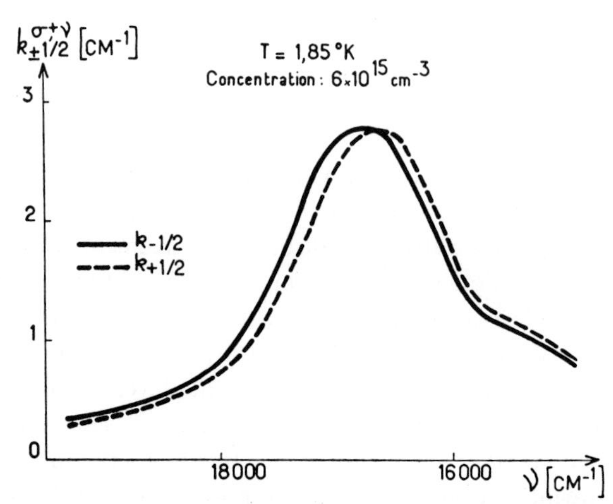

Fig. 9. The absorption spectrum of F centers in KBr measured by left-handed circularly polarized light passing through the sample parallel to an applied magnetic field. The solid curve is for the spin of the F center aligned parallel to the direction of propagation of the light; the dashed curve is for an antiparallel orientation of the spin (Karlov *et al.*[17]).

in measurements of Faraday rotation of light in the F band.[18] If the light is modulated by rapidly altering its polarization, ac detection techniques can be used and the effects can be measured with high precision.

A first approximation to the F-center case can be made by studying the situation in the alkali metals. An energy level diagram illustrating this case is shown in Fig. 10.[19] The S state is the ground state, with levels split by the magnetic field. The two groups of P states represent the excited states with a separation due to the spin–orbit coupling. In this case, the $P_{3/2}$ states lie higher than the $P_{1/2}$ states. Shown are the allowed transitions, their circular polarization, and their weight. The g values are the atomic Landé factors.

Two principal properties are measured in experiments on the F center. One is the dependence of the absorption on magnetic field and the other is its temperature dependence. From the data, it is found that the $P_{3/2}$ levels lie below the $P_{1/2}$ levels for the F center. This reversal in sign and the magnitude of the effects have been explained[20] by using wave functions carefully orthogonalized with respect to the nearest neighbors. The interaction of the F-center wave function with these ions makes an important contribution to the observed effects.

Recently, attempts have been made to see similar magnetooptic effects in the F-center emission in KCl, but without success.[21] From an analysis of the sensitivity of the apparatus, it is possible to say that the spin–orbit coupling for the relaxed excited state is less than that for the unrelaxed state by a factor of 30 or more. This difference is not presently understood.

An interesting property of the *F* center in a magnetic field concerns its spin memory after absorption of light. If an *F*-center electron with a particular spin direction, chosen by circularly polarized light, is excited and then returns by luminescence to its ground state, will its spin direction be altered during the relaxation processes attending these various processes? It has been shown by measuring the population in the split ground states[22] that the spin memory in this system is nearly complete. Furthermore, electron spin resonance has been detected in the excited electronic state by applying a simultaneous microwave field in this experiment. At resonance, some centers in the excited state will have their electron spins reversed and this will be seen as a change in the ground-state populations after those centers decay by luminescence. In this case, the electron spin resonance condition in the excited state is detected with great sensitivity by an optical measurement. Attempts to extend this technique to ENDOR measurements in the excited state are being made. Such measurements would be especially valuable since they should give a measure of the diffuseness of the excited-state wave function, which was discussed earlier as one explanation of the long *F*-center luminescence decay.

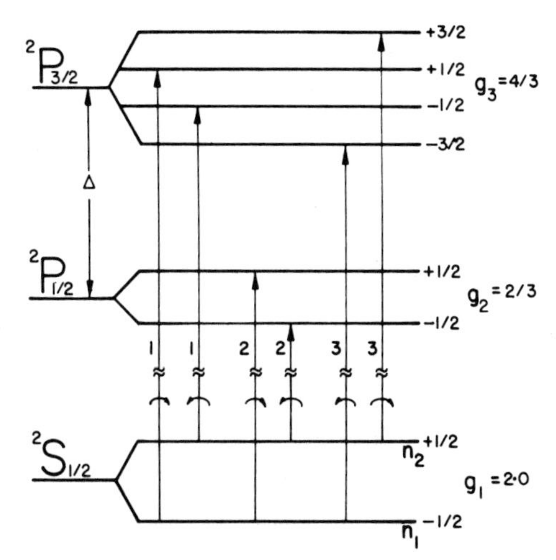

Fig. 10. An energy level diagram for an alkali metal used to illustrate the absorption of circularly polarized light in the *F* center. See the discussion in the text. (Lüty and Mort[19]).

1.1.5. *Higher Excited States of the F Center*

In addition to the transitions from the ground state to the first excited states of the *F* center, there are transitions of the *F* center to still higher states. These fall into two groups: levels that are below the conduction band edge, and levels that occur well above the edge. The lower energy levels will be discussed first.

It might be expected that the *F* center would resemble the hydrogen atom and have an infinite series of levels merging into the conduction band. There is an absorption band on the high-energy side of the *F* band, called the *K* band, that is believed to include transitions to these higher states. It is shown in Fig. 11 for the *F* center in RbCl. The *K* band is the broad, asymmetric band at 2.4 eV. In contrast to the *F* band, it does not sharpen as the temperature is reduced. Its absorption strength is roughly a tenth that of the *F* center. Calculations of the *F* band and transitions to higher states were made by Smith and Spinolo[23] for RbCl using a consistent set of

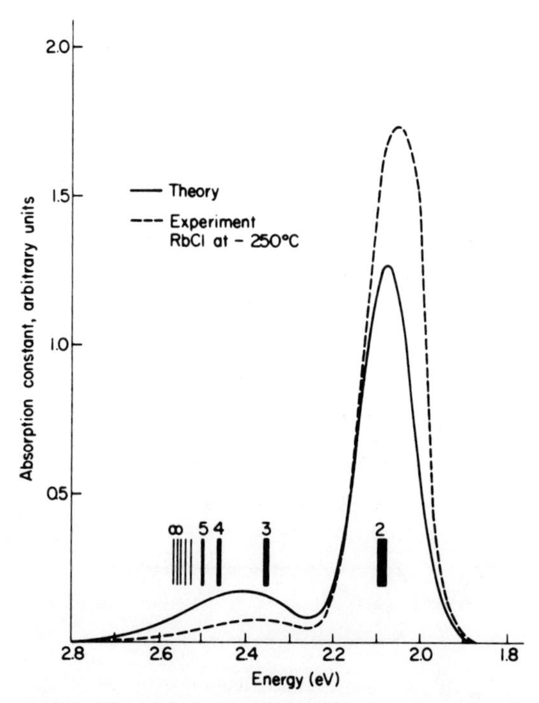

Fig. 11. Theoretical and experimental *F* and *K* bands for RbCl. The details of the figure are given in the text (Smith and Spinolo[23]).

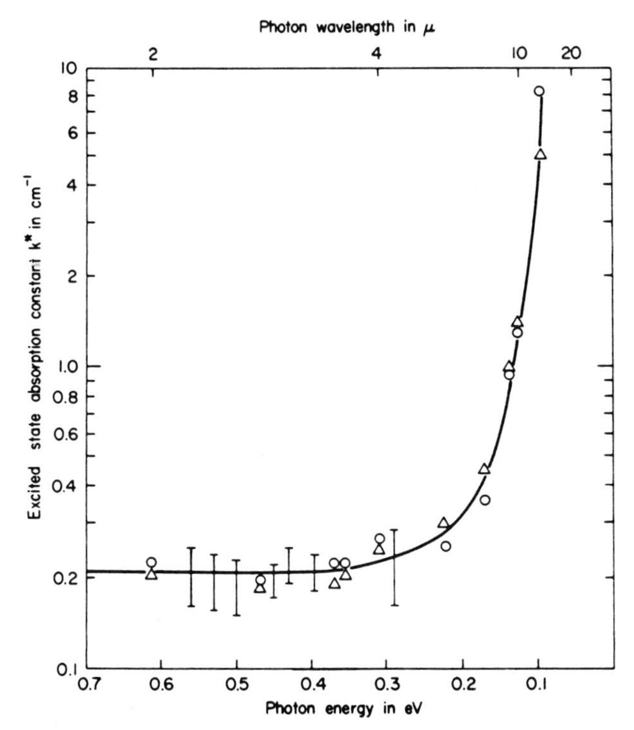

Fig. 12. Absorption from the relaxed excited F center in KI to still higher levels measured at 77°K (Park and Faust[24]).

wavefunction parameters. Their results are shown in Fig. 11. The peak position of the transitions is shown by the heavy lines, where, for instance, "3" refers to the $3p$ level. The width of the lines is proportional to the transition probability from the ground state to that line. If the lines are considered to be broadened by lattice vibrations and if the transitions are summed, the solid curve of Fig. 11 is obtained. Considering the difficulties previously mentioned with wave functions for the F center, the agreement with experiment suggests that this model is a good one.

Because the lifetime of the excited state of the F center is relatively long, it has also been shown to be possible to excite the F center from its relaxed luminescent state into still higher states such as $3p$ or $3d$. A ruby laser was used by Park and Faust[24] to excite the F centers in KI to the luminescent level and absorption measurements were made of the centers in that state. The results are shown in Fig. 12. The sharp rise in absorption near 0.1 eV is interpreted as resulting from transitions to still higher states. The thermal activation energy for ionization from the first excited state is

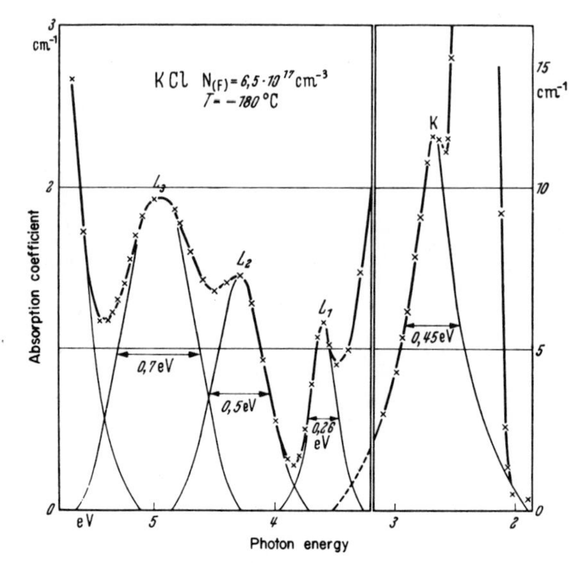

Fig. 13. The optical absorption of the F center bands in KCl. Note the different scales for the L bands and the stronger F and K bands. (Lüty[25]).

only 0.16 eV, as was mentioned earlier. As the data of Fig. 12 show, no additional structure has been seen which might indicate that there are other strong and well-resolved transitions to still higher levels or to the conduction band. These results are generally similar to those derived from studies of the K band.

In contrast to the K band, which was expected from even very qualitative considerations, the L bands were most unexpected and still pose a difficult problem in interpretation. They were first discovered by Lüty[25] in carefully prepared, additively colored samples. The spectra of the L_1, L_2, and L_3 bands in KCl are shown in Fig. 13. Their strength is quite low, being about 1/100 that of the F band. Because of this and their unexpectedly large transition energy, it was especially important to make sure that these transitions arose from the F center and were not some spurious effect. An especially convincing experiment was performed by Chiarotti and Grassano,[26] who modulated the ground-state population of the F center by shining chopped light into the F band and observed a corresponding frequency of the correct phase for the K- and L-band absorptions. This is strong evidence that all these transitions occur from the same ground state.

If the conduction band edge occurs for energies in the *K*-band region when exciting in the *F* center, then the *L* bands correspond to transitions well up into the conduction band. As would be expected, excitation into these bands gives rise to strong photoconductivity in the crystals[27] and pronounced exoelectron emission from the crystal surfaces.[28]

There is no general agreement on a model for the origin of the *L* bands. It may be that these transitions are connected with singularities in the density of states at the Brillouin zone boundaries. Another suggestion is that these transitions are very local in nature and are essentially independent of the band characteristics of the solid. One proposal[29] of this sort suggests that the *F* band be thought of as a transition of the *F*-center electron to a neighboring alkali ion in its ground state. The *L* bands would correspond to the same transition with the alkali ions raised to excited states. A moderately successful correlation of *L*-band positions with this model was found, but no detailed calculations have been attempted. Further guidance might be found by looking for similar effects with a variety of centers as well as in other materials.

1.1.6. F_A *Centers*

The F_A center consists of an *F* center with one of the neighboring alkali ions replaced by a foreign ion as shown in Fig. 14. In KCl, for example, small concentrations of Na^+ or Li^+ may be introduced and *F* centers formed with these ions nearby have unique properties. Because the symmetry of this center is lower than that of the *F* center, it is possible to observe optical polarization in this center and to measure very slow diffusion effects such as reorientation of the F_A center which result in the loss of this polarization. A large fraction of the work on these centers has been done by Lüty and

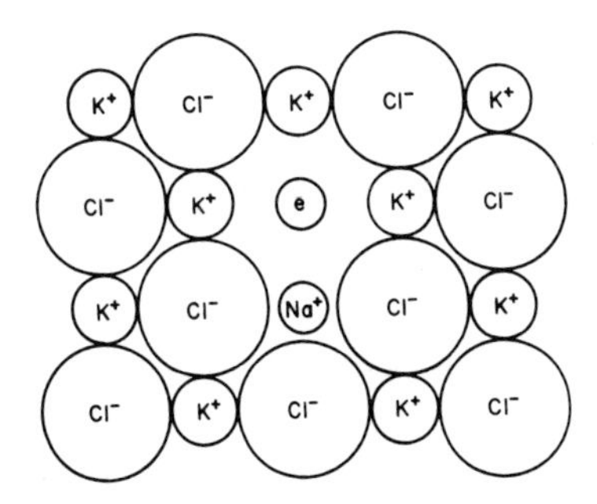

Fig. 14. The F_A(I) center n KCl with an impurity Na^+ ion next to the *F* center.

his associates. A comprehensive review of this work by Lüty[30] gives detailed references to these studies.

F_A centers in the alkali halides fall into two groups. The $F_A(I)$ have emission and ionization processes very similar to those of F centers; the F_A(II) have some interesting and novel properties. KCl:Na leads to F_A(I) centers and it will be described first.

The absorption spectrum for the F_A(I) centers consists of two absorption bands slightly shifted from the original F-center absorption. In KCl:Na, these bands occur at 2.35 and 2.12 eV; the F-center absorption is at 2.31 eV. The lower-energy band is seen for light polarized so that its electric vector is in the direction of the alkali impurity. The higher-energy transition is for the electric vector in the other two $\langle 100 \rangle$ directions.

In emission, only a single band is observed for the F_A(I) center quite independent of which of the two absorption bands is excited. In KCl:Na, the emission peaks at 1.12 eV and has a decay time of 53×10^{-8} sec; the pure KCl has an F-band emission at 1.24 eV and a decay time of 58×10^{-8} sec at liquid helium temperatures.

The F_A(II) centers may be illustrated by KCl:Li. Again, there are two absorption bands as in the F_A(I) center with the same polarizations. For KCl:Li, these bands occur at 2.25 and 1.98 eV. However, the emission properties of the F_A(II) centers are radically different from those of the F center. In KCl:Li, the F_A-center emission is a single sharp band at 0.46 eV, with a decay time of 8.5×10^{-8} sec.

The F_A center has higher excited K- and L-band transitions similar to those for the F center, with relatively small differences in peak positions of the bands and their absorption strengths.

Using the F_A centers as tools, it is possible to look for a variety of activation energies of motion of the ions in the alkali halides. One such measurement involves the production of F_A centers from F centers. This is done by shining light in the absorption band of the F center at temperatures which, for KCl, range from -20 to -70°C. At these temperatures, the F center loses its electron, which may be captured by some other F center. The F center with an additional electron is called the F' center; it will be discussed in a later section. The ionized F center is simply a negative-ion vacancy. At these temperatures, the vacancy diffuses through the lattice to the impurity alkali ion. On capturing a free electron, it appears as an F_A center. An analysis of the temperature dependence of this process yields the activation energy for the jump of halide ions into a neighboring vacancy. It is found to be 0.60 eV. Using radioactive tracer techniques at temperatures of about 600°C, an activation energy of 0.95 eV was found for the same process.[31] These varying results suggest that the activation energy is itself slightly temperature-dependent.

Once the F_A center is formed, it can be dissociated by heating and watching the change in spectra. The activation energy of this process is 0.99 eV for KCl:Li and 1.35 eV for KCl:Na. A value of 1.6 eV is found for the F center in KCl from electron spin resonance measurements. The F_A centers always consist of an impurity alkali ion that is smaller than that of the host lattice. This results in less tight ionic packing for the F_A center and a correspondingly lower activation energy for diffusion.

If the F_A(I) center is excited optically, some of its excited-state properties can be compared with those of the F center. The thermal ionization energy for the excited KCl:Na center is 0.138 eV, compared to the F-center energy of 0.160. Ionization also occurs in the two centers when a strong electric field is applied. This allows ionization of the F_A(I) centers to be produced at very low temperatures where the ionic motions are frozen in, and oriented F_A(I) centers with zero, one, or two electrons can be studied. With the F_A(I) centers, it is also possible to measure the reorientation activation energy of the center in its excited state by observing the optical polarization. For KCl:Na, this is 0.09 eV, which shows a remarkable decrease from the ground-state value of 0.60 eV. The reorientation activation energy for the ionized F_A(I) center, that is, the empty vacancy, is found to be 0.45 eV.

A very different situation exists for F_A(II) centers such as KCl:Li. First, it has already been noted that the emission band is shifted to lower energies by a factor of two and that the decay time of the luminescence is faster by nearly a factor of ten when compared with the F center. Even more surprising is the failure to find either thermal or electric field ionization of the F_A(II) center in its excited state.

These differences between the properties of the F_A(I) and F_A(II) in the excited state have led Lüty to propose that the F_A(II) excited state has a two-well configuration as shown in Fig. 15. In this case, a halide ion assumes an interstitial position, with the F_A(II) center electron occupying the wells on either side of the ion. This position is made possible by the small size of the Li^+ ion in KCl:Li with the resulting small repulsive energy for the interstitial ion. It is argued that the total sum of electronic and ionic energies of the system leads to this position as the one with minimum energy. Even with the F_A(I) center, this position may not be very different in energy from that of the vacancy-centered position. This may account for the low activation energy for reorientation of F_A(I) centers in the excited state which was given earlier for KCl:Na as being only 0.09 eV.

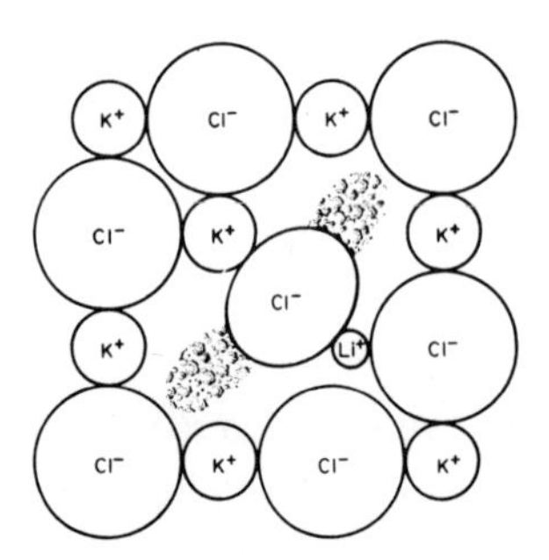

Fig. 15. The F_A(II) center in KCl with an impurity Li^+ ion. The center is shown in its excited state according to a model proposed by Lüty.[30] The central Cl^- ion has moved into an interstitial position and the *F*-center electron cloud density occurs in the two areas on either side of this ion.

The differences between F_A(I) and F_A(II) behavior should depend primarily on the relative ion sizes of the host lattice and of the alkali ion impurity. There is excellent correlation between this criterion and the many F_A systems studied in the alkali halides.

1.1.7. *Z Centers*

Another class of centers may be formed in alkali halides by the introduction of divalent alkali earth ions such as Ca^{2+} or Sr^{2+}. These ions replace normal alkali ions in the lattice, but they would give rise to a charge imbalance in the crystal if they were not compensated. From ionic conductivity measurements, it has been established that the compensation occurs by the introduction of an alkali ion vacancy for each alkaline earth ion. These divalent ions and the vacancies have a Coulomb attraction for each other and there can be association depending on conditions such as the temperature, the concentrations, and the degree of quenching from high temperatures. If *F* centers are introduced into such a crystal, a variety of complex aggregates is possible. The nature of most of these is uncertain, but the Z_1 center in KCl has been unraveled by ENDOR measurements[32] to give the two different arrangements shown in Fig. 16, both of which contribute to the same optical absorption spectrum. In these cases, one might think of the center as an *F* center strongly perturbed by a neighboring positive-ion vacancy and almost independent of the divalent positive ion.

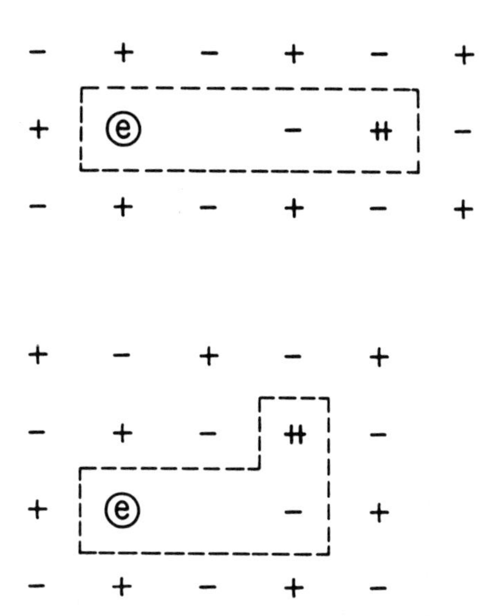

Fig. 16. Two possible configurations for the Z_1 centers according to Bushnell.[32]

1.2. F^- Centers

The F^- center (also referred to as the F' center) in alkali halides is closely related to the F center. It consists of a negative-ion vacancy that has trapped two electrons. The identification of the center goes back many years and it is one of that small group of centers that was not dependent on spin resonance for its unraveling. In fact, it does not show any resonance spectrum.

The F^- center is formed by taking an additively colored crystal and cooling it to about 100°K. Irradiation with light in the F band bleaches that band partially and the F^- band grows. Figure 17 shows the F^- band for KBr.[33] Irradiation with light in the F band ionizes the F center and creates a free electron. The electron drifts through the crystal until it is trapped by another F center to become the F^- center. As might be expected, the F^- center holds its second electron with a relatively low binding energy compared to the single electron in the F center and this is why the experiments are done at low temperatures.

It is apparent from the figure that the F^- band is extraordinarily wide and extends to both higher and lower energies than the F center. The shape of the band suggests that there is not a single well-defined excited state but that the situation is more complex. Photoconductivity measure-

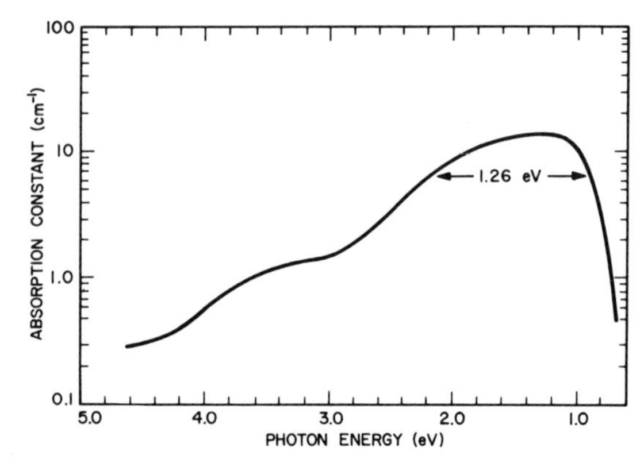

Fig. 17. The absorption spectrum of the F^- center in KBr at low temperatures (Crandall and Mikkor[33]).

ments on the F^- band show that radiation anywhere in the F^- band leads to the production of conduction electrons even at the lowest temperatures. It is believed that irradiation in the F^- center excites an electron directly into the conduction band.

Because the F and F^- absorption bands overlap, irradiation in the F band always leads to bleaching of the F^- centers by ionization. These electrons may be captured by vacancies to recreate F centers. As a result, it is not possible to change F centers completely into F^- centers. However, by irradiating at wavelengths longer than the F band, it is possible to eliminate the F^- band completely. An interesting property of the conversion between F and F^- centers is that the quantum efficiency for the processes

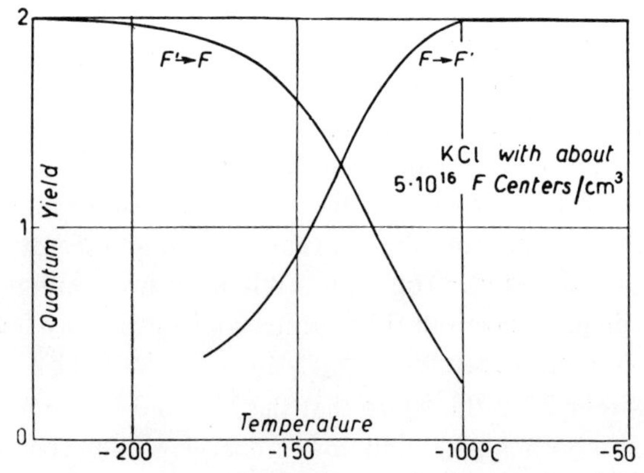

Fig. 18. The quantum efficiency of $F \rightarrow F^-$ and $F^- \rightarrow F$ conversion in KCl as a function of temperature (Pick[34]).

can be as high as two. When an F center is bleached and its electron is trapped at a second F center to form the F^- center, two F centers disappear to form a single F^- center and an empty vacancy. Bleaching the F^- center simply reverses the process. Figure 18 shows measurements by Pick[34] which followed the conversion between F and F^- centers in KCl as a function of temperature. The quantum yield has a maximum value of two in both cases and this has been very strong evidence for the correctness of the model for the F^- center.

1.3. F^+ Centers

The negative-ion vacancy with one and two electrons has already been discussed; we now turn to the vacancy with no electrons. Such a center could not have an absorption band in the same way that we see absorption for the F and F^- centers. However, it has been shown clearly that an ultraviolet absorption band called the α band arises from the vacancy.[35] In KI, the band was found at 2380 Å, which is close to the first exciton peak at 2130 Å. This has led to the interpretation of the band as arising from the formation of an exciton in the vicinity of a vacancy.

Calculations have been made of a simple model for this process[36] which is illustrated in Fig. 19. The transfer of an electron from a negative ion to a positive ion in the pure lattice is shown in the top part of the figure. This is taken to be the model for the exciton. The next illustration shows the exciton formed near an F center and the last is an exciton formed near a vacancy. Bassani and Inchauspé were able to show reasonable agreement between calculations based on this model and experiment. If the exciton is formed farther from the vacancy, it will be perturbed less and will therefore be closer to the fundamental exciton energy. This has been proposed as the explanation for one of the ultraviolet bands seen in KBr.[37]

The existence of the α band is very useful in unraveling the motion and capture of electrons in alkali halides. Starting with F centers in an additively colored crystal at low temperatures, irradiation in the F band will produce some F^- centers and some F^+ centers by transferring an electron from one F center to another. Similarly, other processes involving impurities, diffusion of ions, or electron transfer may sometimes be worked out. A very useful clue frequently is the presence and size of the α band, measuring the number of negative-ion vacancies.

Excitation in the α band also leads to luminescent emission. In KBr, the α absorption is at 2010 Å and the emission is at 5000 Å, showing an extremely large Stokes shift.[38] By measuring the temperature variation

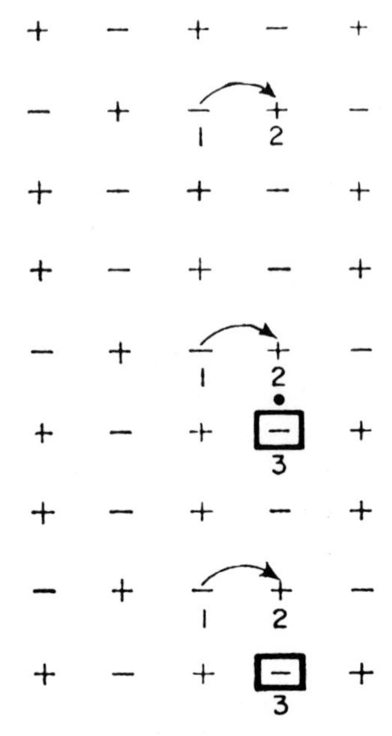

Fig. 19. A model for creation of an exciton in which the arrow represents the transfer of an electron. The first figure represents the pure crystal; the second shows an exciton near an F center; the last corresponds to the F^+ transition. (Bassani and Inchauspé.[36])

of bandwidths, Timusk was able to determine the vibrational frequencies of the F^+ center and compare them with those of the F center. In the α absorption, the electron starts far from the vacancy and the center vibrational frequency is that of the empty vacancy. This corresponds nicely to that found for the excited F center with its diffuse character. Similarly, the emission of the excited F^+ center starts from a state with the electron close to the vacancy and has a vibrational frequency near that of the F center in absorption.

An exciton formed near an F center gives rise to an ultraviolet absorption called the β band. In energy it is between the α band and the first exciton peak. Figure 19 illustrates how this may occur. The β band may be used to measure the presence of F centers, but it is usually more convenient to measure the F band itself.

It seems likely that all centers have perturbed excitons formed near them. In cases other than the α and β bands, these have not been extensively studied since other properties of the centers are usually easier to observe. The α band is uniquely valuable because it is the only optical property of the vacancy (F^+ center) that can easily be measured.

2. COMPLEX CENTERS

2.1. F_2 Centers

In the coloration of alkali halides, it is very common for several absorption bands to be seen in addition to the F band. X-ray coloration at room temperature, for instance, will give rise to a spectrum similar to that shown in Fig. 20. A similar spectrum can be produced by additively coloring a crystal to produce F centers and then shining light in the F band. The result of this process is a bleaching of the F band and a growth of the F_2 and F_3 bands (also referred to as M and R bands, respectively). It is now known that the F_2 band arises from a pair of F centers and the F_3 band from the group of three F centers.

The formation of the F_2 center from the F center is similar in many ways to the formation of the F_A center. Irradiation in the F band ionizes the electron from an F center, leaving an F^+ center or vacancy. The electron

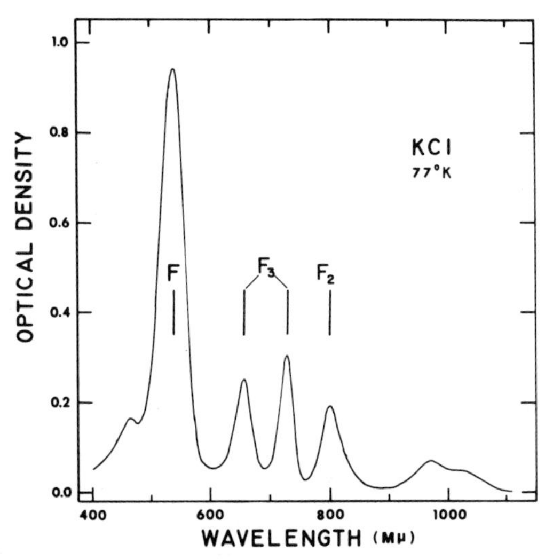

Fig. 20. A typical spectrum of KCl colored by X-rays at room temperature. The measurements were made at liquid nitrogen temperature.

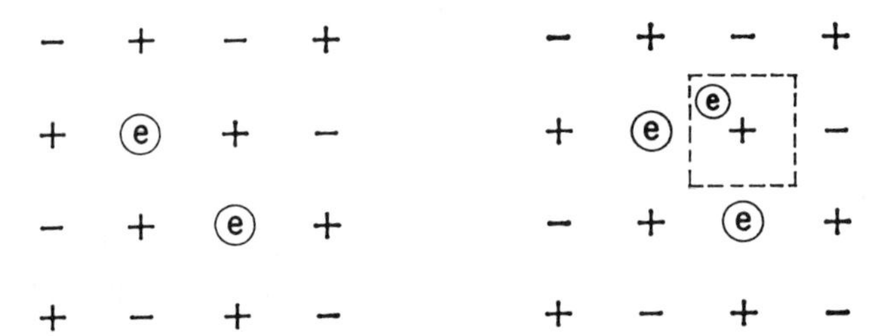

Fig. 21. Configurations of the F_2 and F_3 centers. The F_2 center consists of two F centers. The F_3 center contains three F centers that form an equilateral triangle. In the illustration, two of the F centers are in the plane of the paper; the third would be, for instance, in the atomic plane above the paper and situated directly over the positive ion surrounded by the dashed square.

will be captured temporarily by another F center to form the F^- center. The vacancy diffuses through the lattice until it reaches an F center or an F^- center. The vacancy has a net positive charge and the F^- center a net negative charge, so that there is a Coulomb attraction between them. The two vacancies eventually join and when there is an electron in each vacancy, the F_2 center is complete.

It is apparent from Fig. 21 that the F_2 center has a principal axis along a $\langle 110 \rangle$ direction. It is also found that light with its electric vector parallel to this direction is absorbed differently than light polarized perpendicular to this axis. There are two principal absorption areas. As shown in Fig. 20, the F_2 has an absorption peak to the long-wavelength side of the F band. This absorption is for light polarized parallel to the center. The other absorption area is under the F band itself and consists of several overlapping bands with polarization perpendicular to the F_2 axis.

Under normal conditions, the F_2 centers are distributed among all six $\langle 110 \rangle$ directions and the polarization properties of the center cannot be observed. It is possible, however, to orient the centers by irradiating with a specific [110] polarization at a wavelength corresponding to the higher-energy transitions under the F band. This treatment tends to excite all but the centers already oriented in a [110] direction; those already oriented have high-energy absorptions in the $[1\bar{1}0]$ direction. The excited centers tend to change orientations until they happen to be in the [110] direction, after which they no longer absorb the exciting polarized light. In this way, a large population of centers oriented in a particular direction can be created and the polarization properties of the centers in absorption and emission can then be observed.

In recent work,[39] it has been possible to unravel the process by which the F_2 center orients. At very low temperatures, the absorption of a photon leads to the ionization of the F_2 center by ejecting an electron. The ionized center, which we shall call an F_2^+ center, absorbs another photon and rotates to a new orientation in its excited state before returning to the F_2^+ ground state. Capture of an electron converts the reoriented F_2^+ center into a reoriented F_2 center. At temperatures above about 100°K in KCl, the absorption of a second photon in unnecessary since the F_2^+ center reorients by thermal excitation alone.

If the F_2 center is excited in its lowest-energy absorption band with light polarized parallel to the axis of the center, luminescence may be observed from the center. It is also polarized parallel to the center. In KCl, the F_2 emission is at 1.2 eV, which is close to that of the F center. The decay time of the F_2-center luminescence is faster than that of the F center by more than an order of magnitude and in general is found to fall in the range of 10^{-8}–10^{-7} sec. Another difference between the emissions of the F and F_2 centers is found in the temperature dependence. The F-center emission is seen only at low temperatures, dropping off above about 100°K in KCl. By contrast, the F_2-center emission is strong at room temperature. These facts lead to the conclusion that the F_2-center emission occurs between the same levels involved in the F_2-center absorption, and that both the ground and excited states are well localized.

Absorption of light in the higher excited states of the F_2 center produces the same emission band. There appear to be radiationless transitions to the lowest excited state, from which the luminescent transition then occurs.

The ground state of the F_2 center is a singlet state with the electrons on the two vacancies having their spins antiparallel. As a result, it is not possible to observe this state in spin resonance. It is possible, however, to create a triplet state with parallel spins of the F_2-center electrons by irradiating at a variety of wavelengths. This triplet state has a lifetime in the range of minutes, so that a sizeable population of triplets can be produced and both spin resonance and ENDOR may be observed.[40] These measurements were additional evidence in support of the model of the F_2 center. Optical absorption can be seen in the triplet F_2 center with energies slightly higher than in the singlet F_2. No luminescence has been reported for the triplet center.

An intersting problem is connected with the production of the F_2 triplet state by light. It has been shown[41] that excitation in the F band will produce triplet F_2 centers from the usual singlet F_2 centers. This is a forbidden optical transition and the problem is how the transition is made to occur. It is especially puzzling since excitation in the F band does not excite the F_2 singlet center, which could be detected by luminescence. The F_2 singlet transition is allowed, but is not seen; the F_2 triplet is forbidden, but is seen. It may be that the F_2 triplet is not produced directly by optical processes, but indirectly by thermal processes arising from excitation in the F center. This problem is still unresolved.

In addition to the singlet and triplet F_2 centers, a number of other F_2-type centers can be formed. If an F_2 center becomes localized at an alkali impurity, it becomes an F_{2A} center by analogy with the F_A center. The absorption and emission properties of the F_{2A} center are not very different from those of the F_2 center. However, it tends to reorient with greater efficiency than the F_2 center when excited[42] and this has useful practical applications. A high-density information storage system has been proposed using F_{2A} centers.[43] In this system, the F_{2A} centers are oriented by polarized exciting light in the higher-energy bands and read with polarized light in the lower-energy band. The ability to reorient the center easily is an important element in the system and the F_{2A} center is therefore promising for this application.

If an extra electron is added to the F_2 center, a three-electron F_2^- center is formed. Such centers have been produced and have been seen in optical absorption.

It has also been possible to produce and study F_2 centers that have lost an electron and are designated as F_2^+ centers.[44] The process used to produce this center is to form F_2 centers in an additively colored crystal at room temperature. The crystal is cooled to liquid helium temperatures and X-rayed to produce self-trapped holes. These are then excited optically and caused to diffuse through the lattice. When a hole combines with an F_2 center, the F_2^+ center is formed.

2.2. F_3 Centers

A cluster of three F centers at the points of an equilateral triangle is called an F_3 center and is illustrated in Fig. 21. It gives rise to two absorption bands as shown in Fig. 20. There is also an emission from the F_3 center which is near that of the F and F_2 centers at about 1.0 eV.

The F_3 center has three electrons and it might be expected that spin resonance of the center could be observed. This was not, however, the case. An analysis of the states of such a center by Silsbee,[45] however, showed that the ground state of the center was doubly degenerate. By applying

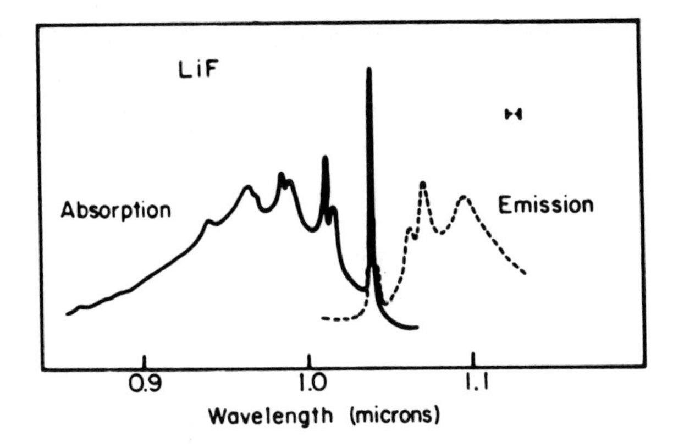

Fig. 22. Absorption and emission spectra of the F_2^- center in LiF. The zero-phonon peaks in the two spectra fall at identical energies. (Fitchen *et al.*[49]).

stress to the center, this degeneracy could be removed and spin resonance could then indeed be observed.

In addition to the ground state with spin 1/2, metastable states of spin 3/2 have also been seen in resonance using techniques very similar to those used in studying the triplet state of the F_2 center.[46] Here again, the detailed mechanism by which the spin of an electron is flipped is not understood.

2.3. Larger Aggregates

Still larger concentrations of F centers can be produced but their detailed analysis in terms of models has not yet been achieved. If the process of aggregation is permitted to continue by heating an additively colored crystal at temperatures of 400°C for a period of 15 min, a new absorption band appears in the visible that is believed to arise from metallic colloids with less than 1000 atoms.[47] The details of the breakdown of a large cluster of F centers to form the metal colloid are not yet known.

2.4. Zero-Phonon Lines*

Many of the centers derived from F_2, F_3, or more complex centers show very sharp lines on top of the broadband absorption and emission spectra. Figure 22 shows measurements on the F_2^- center in LiF taken at

* A comprehensive review of zero-phonon transitions is given by Fitchen.[48]

4°K.[49] The longest-wavelength peak in absorption and the shortest-wavelength line in emission are called zero-phonon lines. Using the diagram of Fig. 3, they would correspond to transitions between the lowest vibrational level of each electronic state. As a result, there is no production of phonons and extremely sharp lines with width of an angstrom or less can be observed. Other, less sharp lines are observed away from the zero-phonon lines and these are one-phonon lines that reflect the electronic transition with the creation (or absorption) of a single phonon. Since there is a complex distribution of vibrational modes in a solid, this one-phonon structure is correspondingly complex.

Because of the sharpness of the zero-phonon line, small shifts in its position can easily be observed. Such shifts as a result of uniaxial stress have been used to determine symmetry properties of the complex centers.

2.5. Interstitial Centers

As described in the chapter on radiation damage, X-ray irradiation produces interstitial halide ions in alkali halides with high efficiency. However, interstitial alkali ions or atoms are not produced with equal ease. In LiF, there is evidence that an interstitial lithium atom may be created by neutron and high-energy electron irradiation.[50] It is thought to combine with a normal lithium ion to form an Li_2^+ center. The observed absorption is near the F_2 and F_3 absorption bands.

3. OTHER MATERIALS

The discussion of electron centers has so far concentrated entirely on alkali halides because they were the first materials to be studied and the information on them is most complete. A great deal of attention has recently been paid to other materials and many of these have color center systems that are quite well understood.

Alkaline earth oxides are the direct analogy of alkali halides with divalent ions replacing monovalent ions. There is the additional complication that a negative-ion vacancy can bind two electrons tightly rather than only one. This field is advanced enough so that a separate chapter is devoted to it in this volume.

Another class of compounds are the ammonium halides, in which the ammonium radical replaces the alkali ion. Extensive spin resonance and optical studies have been made of these materials[51] and a number of hole-type centers have been found. The strange result is that there is no evidence that electron centers are formed in the ammonium halides.

A great deal of knowledge has recently been obtained about electron centers in alkaline earth halides. One group of materials consists of CaF_2, SrF_2, and BaF_2. The difficulties in studying these materials may be illustrated by taking the case of CaF_2. When pure, it does not color under X-rays. If additively colored and quenched, several strong bands appear. The longer-wavelength band has been identified as an F_2 band.[52] It has higher excited states that strongly overlap the F-band region, so that the F band has not been seen alone. It has been identified by Faraday rotation measurements.[53] For CaF_2, SrF_2, and BaF_2, the F center absorption bands peak at 3760, 4350, and 6110 Å, respectively; the F_2 center absorption bands peak at 5210, 5950, and 7250 Å, respectively. Spin resonance experiments, luminescence, and polarization measurements have all been made on these centers.

Other alkaline earth fluorides with more complex crystal structures include MgF_2 and $KMgF_3$. For MgF_2, the F and F_2 centers have been seen along with spin resonance and luminescence. The F_2 centers have four different symmetries in this material because of the more complex crystal structure, but the properties of two of these have been worked out.[54] In $KMgF_3$, F, F_2, and F_3 centers have been seen and their luminescence and polarization and spin resonance properties have been studied. This substance is nearly unique among these materials in that it colors efficiently under ionizing radiation,[55] as do many of the alkali halides. The oxides, ammonium halides, and the alkaline earth halides all seem to color primarily as a result of relatively inefficient elastic collision processes.

Much of the current effort in color centers is addressed to extending this work to new materials of more complex composition, structure, and bonding. As can be seen in the steps already taken, each new material is related to the others in some ways, but also introduces its own novelties and surprises.

REFERENCES

1. J. H. Schulman and W. D. Compton, *Color Centers in Solids* (Macmillan, New York, 1962); J. J. Markham, *F-Centers in Alkali Halides* (Academic Press, New York, 1966); W. B. Fowler (Ed.), *Physics of Color Centers* (Academic Press, New York, 1968).
2. W. Gebhardt and H. Kühnert, *Phys. Letters* **11**, 15 (1964).
3. C. C. Klick and J. H. Schulman, in *Solid State Physics*, Vol. V, Ed. by F. Seitz and D. Turnbull (Academic Press, New York, 1957).

4. W. Gebhardt and H. Kühnert, *Phys. Stat. Sol.* **14**, 157 (1966).
5. J. M. Worlock and S. P. S. Porto, *Phys. Rev. Letters* **15**, 697 (1965).
6. C. H. Henry, *Phys. Rev.* **152**, 699 (1966).
7. H. Seidel and H. C. Wolf, in *Physics of Color Centers*, Ed. by W. B. Fowler (Academic Press, New York, 1968).
8. F. Hughes and J. G. Allard, *Phys. Rev.* **125**, 173 (1962).
9. G. Feher, *Phys. Rev.* **114**, 1219, 1245 (1959).
10. H. Seidel, *Z. Physik* **165**, 218 (1961).
11. B. S. Gourary and F. J. Adrian, *Phys. Rev.* **105**, 1180 (1957).
12. U. Opik and R. F. Wood, *Bull. Am. Phys. Soc., Ser. II*, **15**, 339 (1970).
13. R. K. Swank and F. C. Brown, *Phys. Rev.* **130**, 34 (1963).
14. W. B. Fowler, *Phys. Rev.* **135**, A1725 (1964).
15. G. Chiarotti, U. M. Grassano, G. Margaritondo and R. Rosei, *Nuovo Cimento X*, **64B**, 159 (1969).
16. L. D. Bogan and D. B. Fitchen, *Phys. Rev.* **B1**, 4122 (1970).
17. N. V. Karlov, J. Margerie, and Y. Merle, D'Aubigne *J. Physique* **24**, 717 (1963).
18. J. Mort, F. Lüty, and F. C. Brown, *Phys. Rev.* **137**, A566 (1965).
19. F. Lüty and J. Mort, *Phys. Rev. Letters* **12**, 45 (1964).
20. D. Y. Smith, *Phys. Rev.* **137**, A574 (1965).
21. M. P. Fontana, private communication.
22. L. F. Mollenauer, S. Pan, and S. Yngvesson, *Phys. Rev. Letters* **23**, 683 (1969).
23. D. Y. Smith and G. Spinolo, *Phys. Rev.* **140**, A2121 (1965).
24. K. Park and W. L. Faust, *Phys. Rev. Letters* **17**, 137 (1966).
25. F. Lüty, *Z. Physik* **160**, 1 (1960).
26. G. Chiarotti and U. M. Grassano, *Phys. Rev. Letters* **16**, 124 (1966).
27. N. Inchauspé, *Phys. Rev.* **106**, 898 (1957).
28. P. Petrescu, *Phys. Stat. Sol.* **29**, 333 (1968).
29. C. C. Klick and M. N. Kabler, *Phys. Rev.* **131**, 1075 (1963).
30. F. Lüty, in *Physics of Solids*, Ed. by W. Beall Fowler (Academic Press, New York, 1968).
31. R. G. Fuller, *Phys. Rev.* **142**, 524 (1966).
32. J. C. Bushnell, Thesis, Univ. of Illinois, unpublished (1964).
33. R. S. Crandall and M. Mikkor, *Phys. Rev.* **138**, A1247 (1965).
34. H. Pick, *Nuovo Cimento VII* (*Ser. X*), No. 2, 498 (1958).
35. C. J. Delbecq, P. Pringsheim, and P. H. Yuster, *J. Chem. Phys.* **19**, 574 (1951).
36. F. Bassani and N. Inchauspé, *Phys. Rev.* **105**, 819 (1957).
37. C. C. Klick and D. A. Patterson, *Phys. Rev.* **130**, 2169 (1963).
38. T. Timusk, *J. Phys. Chem. Solids* **26**, 849 (1965).
39. I. Schneider, *Phys. Rev. Letters* **24**, 1296 (1970).
40. H. Seidel, *Phys. Letters* **7**, 27 (1963).
41. I. Schneider, *Phys. Rev. Letters* **17**, 1009 (1966).
42. I. Schneider, to be published.
43. I. Schneider, M. Marrone, and M. N. Kabler, *Appl. Opt.* **9**, 1163 (1970).
44. I. Schneider and H. Rabin, *Phys. Rev.* **140**, A1983 (1965).
45. R. H. Silsbee, *Phys. Rev.* **138**, A180 (1965).
46. H. Seidel, M. Schwoerer, and D. Schmid, *Z. Physik* **182**, 398 (1965).
47. A. B. Scott and W. A. Smith, *Phys. Rev.* **83**, 982 (1951).

48. D. B. Fitchen, in *Physics of Color Centers*, Ed. by W. Beall Fowler (Academic Press, New York, 1968).
49. D. B. Fitchen, H. R. Fetterman, and C. B. Pierce, *Solid State Comm.* **4**, 205 (1966).
50. Y. Farge, *Phys. Rev.* **B1**, 4797 (1970).
51. F. Patten, *Phys. Rev.* **175**, 1216 (1968).
52. J. H. Beaumont and W. Hayes, *Proc. Roy. Soc.* **A309**, 41 (1969).
53. B. C. Cavenett, W. Hayes, I. C. Hunter, and A. M. Stoneham, *Proc. Roy. Soc.* **A309**, 53 (1969).
54. O. E. Facey, and W. A. Sibley, to be published.
55. C. R. Riley and W. A. Sibley, *Phys. Rev.* **B1**, 2789 (1970).

Chapter 6

HOLE CENTERS IN HALIDE LATTICES

M. N. Kabler
Solid State Division
Naval Research Laboratory
Washington, D.C.

1. INTRODUCTION

Irradiation of ionic crystals can produce a large variety of stable defects involving both charge separation and ionic displacement. Some of the resulting electronic energy levels fall within the band gap and are thus accessible to spectroscopic investigation wherever the crystal is transparent. It is convenient to classify a given defect or color center according to whether its character is primarily that of a trapped electron or a trapped hole. In the former case, the behavior of the center is conditioned primarily by the existence of an occupied level lying within the band gap, a donor level in semiconductor terminology. In the case of a hole center, the concern of the present chapter, the behavior is dominated by an unoccupied level in the gap, an acceptor level.

Any general discussion of the physics of color centers will invariably place strong emphasis on a single type of crystal, the alkali halides. Due both to their experimental and theoretical tractability, these crystals have been the most extensively investigated of all color-center-related materials. The physics of hole centers in particular has developed primarily within the alkali halide context. Our attention to other materials will be brief and confined to halides or halogen-containing crystals. Another area in which a significant amount of hole center information has recently accumulated, that comprising the simple oxide crystals, is covered in the chapter by Hughes and Henderson.

Electronic wave functions for hole centers in simple halides are generally strongly localized and largely confined to the anions; this contrasts with the case of shallow acceptor states in semiconductors, for example, where the hole is largely delocalized. Although the situation was not anticipated in the early days of hole center research, the diatomic halide molecular ion has turned out to be the essential element in the structure of nearly all hole centers which are presently identified. We shall deal almost exclusively with centers in this category.

The first tangible evidence for hole centers arose from the observation of a group of absorption bands in the near-ultraviolet in certain alkali halides which had either been heated in halogen vapor to produce a stoichiometric excess of halogen or irradiated with X-rays.[1,2] We shall not recount the earlier arguments and speculations concerning the natures of the centers responsible for these so-called V bands and shall thereby ignore the interesting historical development of this subject.[3,4] Instead, our discussion will be cast largely in terms of modern methods and current data. Emphasis will be placed on physical principles and representative examples, and no attempt will be made to catalog all pertinent investigations. Thus some significant research will unavoidably escape mention.

The electronic structures of intrinsic alkali halide lattices call for a brief comment here. These crystals are strongly ionic in character, the anions and cations having the rare-gas electronic configuration. The energies of the electrons in the outermost filled p shells on the halide ions are lowered substantially by the Madelung potential, on the order of 10 eV, but in general these energies remain at least several electron-volts above the energies of the outermost p electrons on the alkali ion. Thus the highest valence band, the band most relevant to hole center considerations, consists primarily of contributions from the outer p shells of the halide ions. Several band-structure calculations have been carried out, and these are in general qualitative agreement as to the broad features of the valence and conduction bands. For example, the conduction band minimum and valence band maximum both occur at $k = 0$.[5–8] The width of the highest valence band turns out to be roughly in the range 0.5–3 eV for most crystals. The wider valence bands are associated with the smaller cations, this being evidently and indication that the alkali ion is acting essentially as a spacer; smaller alkali ions give smaller halide ion separations, and therefore wider valence bands.

2. THE SELF-TRAPPED HOLE

This center is intrinsic to the simple halide lattice in the sense that no defects such as vacancies or interstitials are involved. The self-trapping is evidently spontaneous and results from localization of the hole in a covalent bond between two adjacent halide ions. The stable center thus formed may be characterized as a slightly perturbed X_2^- molecular ion frozen into specific crystallographic orientations within the lattice. We shall denote it as $[X_2^-]$ in order to emphasize its basic makeup, the square brackets implying the presence of the surrounding lattice. It is also commonly referred to as the V_K center.

By virtue of its unpaired spin, the $[X_2^-]$ center is paramagnetic. It was first identified through its microwave or EPR spectrum by Castner and Känzig in 1957.[9] It has since been observed in a number of other crystals, including ammonium halides,[10,11] alkaline earth halides,[12,13] PH_4I,[14] and $KMgF_3$.[15] Ionizing radiation will produce the $[X_2^-]$ center providing the temperature is sufficiently low to nullify the hopping mobility of the hole and providing the crystal contains extrinsic electron traps, impurities or other color centers, which can stabilize the holes against annihilation by recombination.

Because of its sensitivity to the symmetry of a center and to the hyperfine interactions of nuclei located within the extent of the unpaired spin, the EPR technique has generally proved the most powerful for identifying centers and for determining ground-state electronic properties. Thus the history of hole center discovery has become to a large extent a description of imaginative applications of EPR spectroscopy.[16] The present treatment of this method will, however, be brief since the original paper by Castner and Känzig[9] is straightforward and readable and the book by Slichter[17] gives a thorough and lucid review of the fundamental physics involved.

2.1. EPR Spectra

The crystal KCl provides a convenient example with which to begin. The reasons for choosing a chloride are that the EPR spectra of the $[Cl_2^-]$ center show distinctive isotope effects, but at the same time, second-order hyperfine and quadrupole effects are small enough to neglect in a qualitative description. A spectrum, which is actually the derivative of the microwave absorption for $[Cl_2^-]$ in KCl, is shown in Fig. 1.[9] The applied magnetic field, which for experimental convenience is made the independent variable, is parallel to a $\langle 100\rangle$ crystallographic axis. The seven prominent, equally

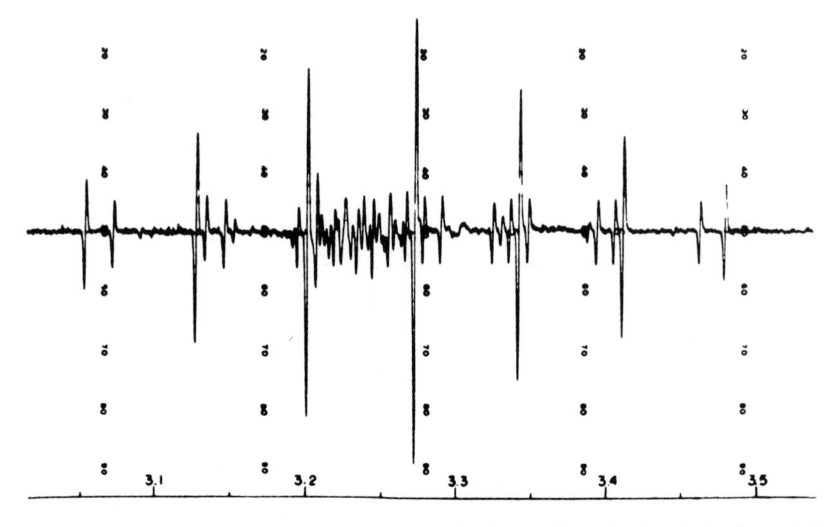

Fig. 1. EPR spectrum of $[Cl_2^-]$ centers in KCl. The ordinate is the derivative of the absorption strength, and the magnetic field is applied in a ⟨100⟩ direction. The microwave frequency is 9263 MHz (after Ref. 9).

spaced lines suggest interaction with a total nuclear spin of 3. The most abundant (75%) isotope of chlorine, ^{35}Cl, has a nuclear spin of 3/2. Thus one might at this stage speculate that the seven-line hyperfine pattern arises from interactions with two equivalent ^{35}Cl nuclei. We note that the spectra due to this center alone are essentially independent of impurities and do not change substantially with the identity of the alkali, as work on the alkali fluorides has shown.[18]

It is apparent that the electron Zeeman interaction, the energy at the approximate center of the spectrum, is large compared to the hyperfine interaction, measured roughly by the width of the overall pattern. These interactions combine to give energy levels which can be represented approximately by

$$E = g_{z'}\beta H m_S + A m_S m_I \tag{1}$$

where the applied field H is in the z' direction and the electron is presumed quantized along H. Here, β is the Bohr magneton, and $g_{z'}$ is the appropriate spectroscopic splitting factor or g factor. One expects the orbital angular momentum to be largely quenched in the crystal, and therefore $g_{z'} \approx 2$. The hyperfine constant A is, for the moment, assumed to be isotropic. The nuclear magnetic quantum number $m_I = m_1 + m_2$ ranges from -3 to 3 as m_1 and m_2 for the individual nuclei range from $-3/2$ to $+3/2$, giving seven equally spaced levels for each of the two values of m_S, $+1/2$ or $-1/2$. Figure 2 represents schematically these levels and the allowed magnetic dipole transitions, the relevant selection rules being $\Delta m_S = \pm 1$, $\Delta m_I = 0$. The 1:2:3:4:3:2:1 intensity distribution evident in Fig. 1 is a consequence of the equivalence of the two nuclei and reflects the statistical weights of the hyperfine levels. For example, there are three sets of m_1 and m_2 which give $m_I = 1$: $3/2, -1/2$; $1/2, 1/2$; $-1/2, 3/2$.

In the chlorides, there are two isotopes, ^{35}Cl and ^{37}Cl, which have slightly different magnetic moments, the ratio being $\mu_{37}/\mu_{35} = 0.83$. Furthermore, the abundance of the scarcer isotope is 25%, hardly negligible. For mixed nuclear pairs, each combination of m_1 and m_2 now has a distinct magnetic moment, and thus one would expect a *multiplet* distribution similar to the *intensity* distribution noted for ^{35}Cl pairs. This expectation is borne out in Fig. 1: The 7 groups of lines, 16 lines in all, each with an intensity 2/3 that of the outermost line due to ^{35}Cl pairs, is of the correct distribution and intensity for ^{35}Cl–^{37}Cl pairs. This is a strong indication indeed that a diatomic halogen configuration is responsible.

The anisotropy of this molecular ion and the orientational constraints imposed upon it by the lattice now enter the picture, and a more complete

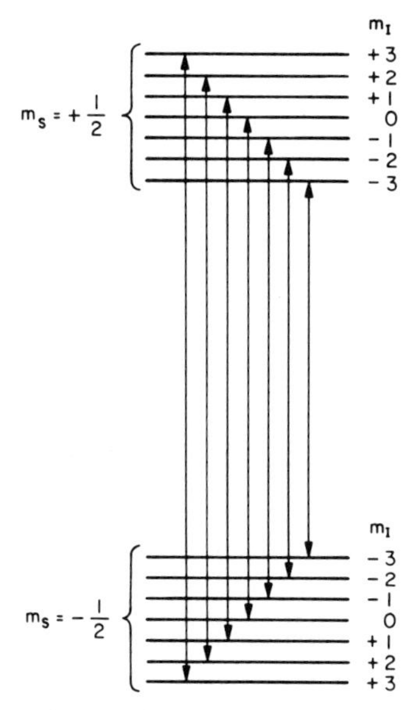

Fig. 2. Schematic level diagram showing principal EPR transitions for a center with nuclear spin $I = 3$ ($= \frac{3}{2} + \frac{3}{2}$ for Cl_2^-).

description of the interactions of the electron spin with the ionic environment and with the applied magnetic field is required. These phenomena can be accommodated to a good approximation by a more general spin Hamiltonian which again comprises an electron Zeeman term and a hyperfine term,

$$H = \beta \mathbf{H} \cdot \mathbf{g} \cdot \mathbf{S} + \mathbf{I} \cdot \mathbf{A} \cdot \mathbf{S} \tag{2}$$

H is the applied field, **S** and **I** are the electron and nuclear spins for the system, and **A** and **g** are symmetric tensors, which are diagonal in the principal axis system. The nuclear Zeeman interaction is neglected because it is small and has no important bearing on the present discussion.

Let us concentrate, for the moment, on the hyperfine interaction. It is advantageous to express this term in an axially symmetric form,

$$H_{\text{hf}} = g_0 \beta \sum_{i=1}^{2} (a_i \mathbf{I}_i \cdot \mathbf{S} + b_i I_{i,z} S_z) \tag{3}$$

where the z axis is along the line connecting the two X^- nuclei. The nuclei may be different isotopes, and so a sum of contributions from each nucleus i is given. As will be seen, the departures from axial symmetry imposed by the crystal lattice are small and cause no particular difficulty. In first order, a_i and b_i are

$$a_i = a_{\text{F}i} - \tfrac{1}{3} b_i \tag{4}$$

$$b_i = (6/5)(\mu_i/I_i)\langle 1/r^3 \rangle_i \tag{5}$$

where

$$a_{\text{F}i} = (8\pi/3)(\mu_i/I_i) \mid \psi_i(0) \mid^2 \tag{6}$$

I_i and μ_i are the spin and magnetic moments, respectively, of nucleus i. The Fermi contact interaction a_{F} depends on the electronic wave function only in terms of its amplitude at either nucleus $\psi_i(0)$; it is therefore a measure of the extent to which s atomic orbitals contribute. The anisotropic part b_i represents the dipole–dipole interaction between the electron and nucleus i at finite separation, and the average $\langle 1/r^3 \rangle_i$ is to be taken over only the p-like part of the orbital. In using Eq. (3) to analyze the hyperfine data, account must be taken of the fact that for an arbitrary direction of the applied field, the electron and nuclear moments will in general be quantized along different directions. Treatment of this situation is straightforward and, as our primary interest is the $[X_2^-]$ center itself, we shall refer the reader to the relevant literature for details.[17,19]

Equation (3) has been found to fit the hyperfine splittings observed in fluorides and chlorides quite well when the angle θ between **H** and the axis of the X_2^- is not too close to 90°. The z axis turns out to be along the nearest-like-neighbor directions, i.e., $\langle 110 \rangle$ in NaCl-type lattices and $\langle 100 \rangle$ in CsCl-type lattices. In general, the EPR spectra are complicated by contributions from several nonequivalently oriented centers. In Fig. 1, for instance, the transitions we have discussed actually arise from those centers for which $\theta = 45°$; these comprise two-thirds of the total centers. The remaining one-third make an angle of 90° with the applied field, and these contribute lines crowded near the center of the pattern. The strong anisotropy is a consequence of the fact that the electron wave function is concentrated along the z axis, and a large contribution from p_z orbitals is thus indicated. In the present approximation, the magnitude of the splitting is proportional to $a_i + b_i$ for $\theta = 0°$ and to $|a_i|$ for $\theta = 90°$.

The simple picture sketched so far gives good qualitative agreement with the data and is of considerable heuristic value. However, general quantitative agreement with experimental results can be obtained only by taking into account various high-order hyperfine interactions and nuclear quadrupole effects which are significant in any case and which become quite important for the heavier halogens.[9,19–21] These hyperfine interactions include splittings of states with the same m_I value but different I values due to interactions between the two nuclei, and corrections to the values of a_i and b_i arising from the combined effects of spin–orbit interaction with the nuclear-electron orbital interaction and with the nuclear-electron spin dipolar interaction. For details, the reader is again referred to the literature. When adequate care is taken, the agreement between the appropriate spin Hamiltonian and the data is generally good.

The premise that the $[X_2^-]$ center is intrinsic, that is, comprises no other lattice defect, is supported by a considerable body of circumstantial evidence derived from analysis of EPR and optical data. The basic factors involved are that no other defect is required to account for any of the data and that deliberate incorporation of a variety of other defects has no effect on the measured properties of the $[X_2^-]$ center. For LiF and NaF, the intrinsic nature of the $[F_2^-]$ center has been confirmed directly by means of electron-nuclear double resonance, or ENDOR experiments.[22,23] ENDOR is rendered feasible by the fact that the EPR lines are broadened inhomogeneously through hyperfine interactions with many of the lattice nuclei which fall within the volume occupied by the unpaired-electron wave function. In this situation, the ENDOR technique in effect allows the investigator to make high-sensitivity NMR measurements on these nuclei.

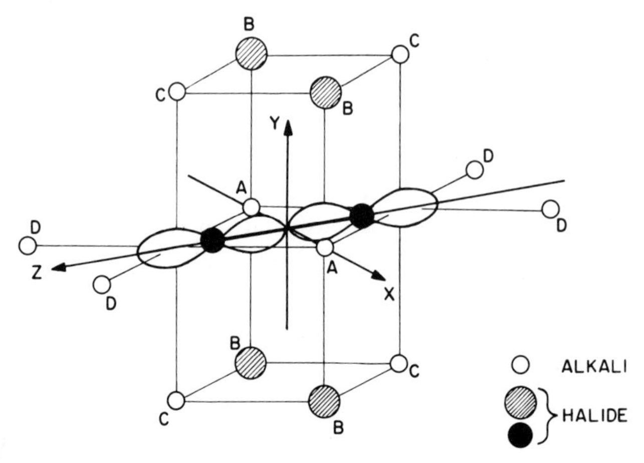

Fig. 3. The $[X_2^-]$ center in an NaCl-type lattice.

Working with LiF, Daly and Mieher were able to observe three nonequivalent Li^+ ions and three nonequivalent nearby F^- ions, in addition to the two principal F^- ions.[22] By using an appropriate electronic wave function, they were able to obtain approximate values for the displacements of these ions from their normal lattice sites due to the influence of the self-trapped-hole. Figure 3 illustrates schematically the ionic environment of the X_2^- in an NaCl-type lattice. The two lattice ions which share the hole are drawn toward each other somewhat from their normal lattice positions, and the localized positive charge causes small but significant outward displacements of the nearby Li^+ ions. The structure of the center in CsCl-type lattices is similar; two nearest-neighbor halide ions again share the hole, but the axis of the $[X_2^-]$ lies along $\langle 100 \rangle$.

2.2. An Atomic Model of Electronic Structure

In order to provide grounds for a quantitative description of the $[X_2^-]$-center electronic properties, it is convenient at this stage to construct a simple molecular orbital (MO) model. These orbitals will comprise linear combinations of atomic orbitals (LCAO's) and will thereby facilitate comparison of the data with known properties of halogen atoms. This type of model is appropriate for the $[X_2^-]$ center both because of the weak interaction of the X_2^- molecular ion with the surrounding lattice and the moderate interaction between the two halide ions sharing the hole.

Consider a set of one-electron LCAO-MO's made up of valence orbitals, $2p$ for fluorine, $3p$ for chlorine, etc. Electron spin and the crystal field will be ignored for the moment, thereby allowing orbital angular momentum parallel to the molecular axis to be conserved. Thus one obtains the $\sigma_g np$, $\pi_u np$, $\pi_g np$, and $\sigma_u np$ MO's shown in the schematic orbital energy diagram, Fig. 4. For information on how this sequence of levels comes about, the reader is referred to the book by Herzberg[24] or to other standard texts on molecular spectroscopy. If the hole were not present and the system were therefore isoelectronic with a diatomic rare gas molecule, then the available electrons would just fill the molecular orbitals, two electrons occupying each σ orbital and four each π orbital. This situation would give no net molecular bonding, since each bonding electron, σ_g or π_u, would be approximately counterbalanced by an antibonding electron, σ_u or π_g. However, when the system traps a hole in its ground state, that is, in the σ_u orbital, an antibonding electron is lost and the Coulomb repulsion between the two negative ions is eliminated. This gives a bound state, as confirmed by the observed stability of the X_2^- molecular ion in free space as well as in the crystal. It is apparent that the ground state must be $^2\Sigma_u^+$ with electron configuration $(\sigma_g np)^2(\pi_u np)^4(\pi_g np)^4(\sigma_u np)$. The $\sigma_u np$ hole orbital is represented schematically in Fig. 3. Transfer of the hole to the three lower orbitals will yield $^2\Pi_g$, $^2\Pi_u$, and $^2\Sigma_g^+$ excited states. Figure 5 displays these states along with other features to be discussed presently.

The spin of the hole (or unpaired σ_u electron) must now be brought into the picture. Spin–orbit coupling, a relatively strong factor in the halogens, will split the $^2\Pi$ states into components characterized by the total angular momentum parallel to the internuclear axis, the corresponding

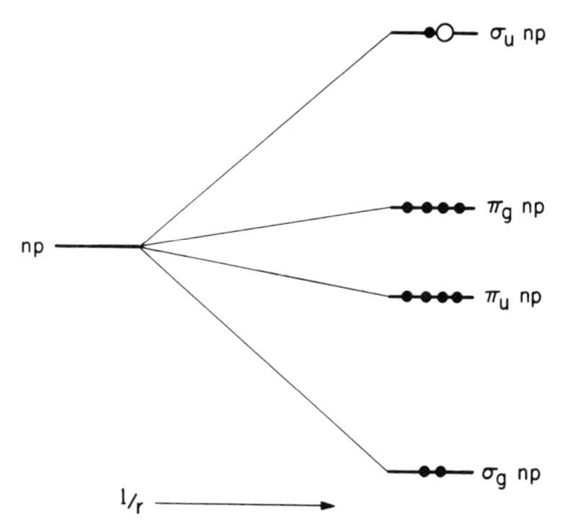

Fig. 4. Orbital energy diagram for MO's arising from *np* AO's on a pair of identical atoms.

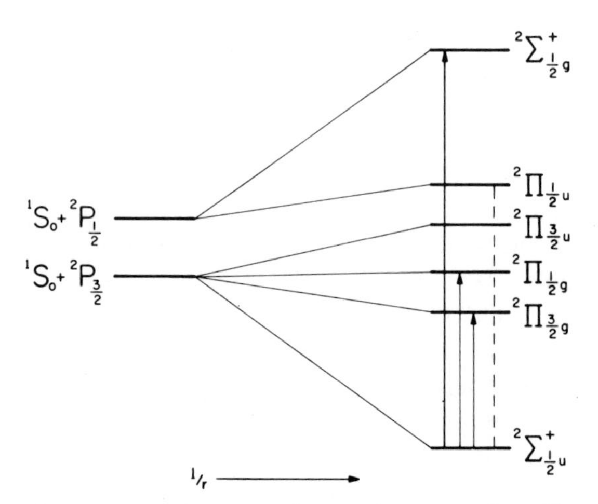

Fig. 5. Schematic energy diagram for states of an X_2^- molecular ion. Observed optical transitions are shown by solid lines; the states which mix to produce the g shift are connected by the dashed line. The two lowest-energy optical transitions are not resolved for the lighter halides.

quantum number being Ω. The splitting of the $\Omega = 1/2$ and $\Omega = 3/2$ states is analogous to the splitting of the lowest $^2P_{1/2}$ and $^2P_{3/2}$ states of the halogen atom. In the atom, the magnitude of this splitting is $3\lambda/2$, where λ is the spin–orbit coupling constant. The splitting will be approximately λ in the molecule. This difference arises because the spin–orbit interaction mixes p_x, p_y, and p_z orbitals in the atom but only orbitals comprising p_x and p_y in the molecule. Overlap effects and configuration interaction are being neglected. Since the sign of λ is negative for the nearly closed-shell configuration, the $\Omega = 3/2$ states lie lower in energy than the $\Omega = 1/2$ states. At large internuclear separation, the states of the X_2^- become those of a free halide ion in its $^1S_0$ ground state and a halogen atom in its lowest $^2P_{1/2}$ or $^2P_{3/2}$ state. Adiabatic correlations among these states are sketched in Fig. 5.

One furthere refinement of the model is useful before comparing it with the data. The MO's made from atomic p orbitals alone will produce no isotropic hyperfine interaction and, as we have already implied, a_{Fi} in Eq. (6) is in fact observed to be nonzero. Actually, in constructing the σ_u and σ_g orbitals, configuration interaction makes it essential in principle to include contributions from other atomic orbitals. Both $\sigma_u ns$ and $\sigma_u(n+1)s$ contributions to the primary $\sigma_u np$ MO might be expected. The $\sigma_u ns$ contribution is evidently the more important,[20] and thus for the present purpose, we shall consider the MO

$$\sigma_u = \sum_{i=1}^{2} [\alpha_i(ns)_i + \beta_i(np)_i] \tag{7}$$

2.3. Hyperfine Structure

Through Eqs. (4)–(6), it is possible to extract from the hyperfine structure data values for the coefficients α_i and β_i in Eq. (7). Since $\langle 1/r^3 \rangle_i$ is to be averaged over the $(np)_i$ component of σ_u, the right-hand side of Eq. (5) must be multiplied by an additional factor β_i^2; similarly, $\psi_i(0)$ pertains to the $(ns)_i$ component and the right-hand side of Eq. (6) must include a factor α_i^2. Values of $\langle 1/r^3 \rangle_i$ for the *np* functions of halogen atoms are available either from calculations or from atomic beam experiments, and calculated values of $|\psi(0)|^2$ for *ns* functions are also available. Values for α_i^2 and β_i^2 obtained in this way are included in Table I, which also contains hyperfine data and *g* factors for a representative sample of $[X_2^-]$ centers. From the measured parameters $a_i + b_i$ and $|a_i|$, values for a_i' and b_i' have been recalculated using higher-order hyperfine corrections in the form given by Jette.[21] The effects of these corrections are quite apparent, especially for the bromides. Table I adopts Schoemaker's choices for the magnitudes of $\langle 1/r^3 \rangle_i$ and $|\psi_i(0)|^2$ and for the signs of the measured $|a_i|$ (negative for fluorides and positive for all others).[20] The degree of $(ns)_i$

Table I. Measured Hyperfine Constants and *g* Factors and Derived Contributions α_i and β_i of *ns* and *np* AO's to the Unpaired σ_u MO of the $[X_2^-]$ center[a]

	Measured		Corrected		α_i^2	β_i^2	$g_\parallel$	$g_\perp$	λ, eV	Ref.
	a_i+b_i	$\lvert a_i \rvert$	a_i'	b_i'						
LiF	887	59	−79	976	0.014	0.60	2.0031	2.0230	−0.0334	19
LiCl	95.6	8.4	4.5	93.0	0.021	0.59	—	—	−0.0729	28
KCl	101.3	12.5	8.4	94.8	0.024	0.60	2.0012	2.0426	−0.0729	20
NH_4Cl	97.6	10	6	93.5	0.022	0.59	2.0010	2.0437	−0.0729	29
KBr	450.0	79	−12	505	0.018	0.69	1.9833	2.169	−0.305	20
NH_4Br	435	80	6	463	0.019	0.63	1.9826	2.150	−0.305	29

[a] Here, a_i and b_i are measured in gauss; the corrected values a_i' and b_i' take into account higher-order hyperfine interactions. The λ are measurements for halogen atoms.

mixture is seen to be quite small, consistent with the relatively large energy separations (several electron-volts) between the $(ns)_i$ and $(np)_i$ orbitals in the atoms. The normalization condition arising from Eq. (7) is

$$\sum_{i=1}^{2} (\alpha_i^2 + \beta_i^2) = 1 \tag{8}$$

Inspection of Table I shows only rough compliance with this condition. Most of the discrepancy can, however, be attributed to the fact that the overlap of the atomic orbitals on different atoms has been ignored. Interpreted in this way, the data imply a value of roughly -0.3 for the overalp integral, which is of reasonable sign and magnitude for an antibonding $\sigma_u np$ orbital of this type. Thus the measured hyperfine interactions are quite consistent with the general behavior one might expect for an X_2^- molecular ion. It should be kept in mind, however, that this atomic model represents a considerable simplification, and therefore the exact magnitudes of α_i^2 and β_i^2 should not be given too much significance.

We shall consider next the g factor and the optical transitions, both of which depend strongly on the excited electronic states as well as the ground state of the $[X_2^-]$ center.

2.4. Electron Zeeman Interaction

Through its spin magnetic moment, the unpaired electron experiences magnetic forces due not only to the applied field **H** directly, but also to the slight net orbital motion induced by **H**. The effects of this orbital motion can be described in terms of a deviation of the g-factor components from the constant free-electron value.[9,17] For the centers under discussion, the orbital angular momentum in the ground state is quenched by molecular electric fields, and thus the g shifts arise entirely from coupling by means of the spin–orbit interaction to excited states with nonzero orbital angular momenta. As is the case for the hyperfine tensor **A**, the anisotropy of **g** reflects the symmetry of the center. For relatively symmetric centers, such as the $[X_2^-]$, the principle axes of **A** and **g** coincide. This is not, however, a general requirement, as we shall find for the V_1 or H_A center.

Detailed discussions of the g-shift problem are available in the previously mentioned literature. For a system with several force centers or nuclei, it is customary to treat the spin–orbit interaction as a sum of atomlike contributions from each of the nuclei. This is practical because the spin–orbit interaction falls off very rapidly with distance from the force center and is therefore relatively insensitive to the overlap of orbitals on different atoms. The effect of the spin–orbit interaction, which is usually expressed in terms of orbital and spin angular momenta **l** and **s** as $\lambda\mathbf{l}\cdot\mathbf{s}$, is to mix the nondegenerate p orbitals on a given atom. For the X_2^-, this mixes the $^2\Pi_{1/2,u}$ state into the $^2\Sigma_u^+$ ground state, thereby supplying the orbital components which sustain a net circulation in response to the field **H**. If we continue to ignore the crystal field, there are two independent components of **g**, parallel and perpendicular to the axis of the X_2^-. The g shifts can be shown to be given to second order in $\lambda/E_{\Delta g}$ by

$$\Delta g_{\parallel} = g_{\parallel} - g_0 = -\beta^2\lambda^2/E_{\Delta g}^2 \tag{9}$$

$$\Delta g_{\perp} = g_{\perp} - g_0 = -(2\beta^2\lambda/E_{\Delta g}) - (2\beta^2\lambda^2/E_{\Delta g}^2) \tag{10}$$

The free-electron spectroscopic splitting factor $g_0 = 2.0023$ has been approximated as $g_0 = 2$ where it enters on the right-hand sides of Eqs. (9) and (10). For simplicity, the distinction between the two nuclei has been dropped in Eqs. (9) and (10), and β represents the total p amplitude on both nuclei. $E_{\Delta g}$ is the energy difference between the $^2\Sigma_u^+$ and $^2\Pi_u$ states and is indicated by the dashed line in Fig. 5. Data on the g shifts are included in Table I, along with λ values for halogen atoms. Note that $\lambda < 0$ because of the nearly filled p shell, and this in turn gives a positive $\Delta g_{\perp}$. The appearance of a positive g shift is often used as a preliminary criterion for ascribing a given spectrum to a hole center, although this procedure is not reliable in general.

An obvious method to obtain the location of the $^2\Pi_u$ excited state is to substitute the measured values of $g_{\parallel}$ and $g_{\perp}$ into Eqs. (9) and (10), along with values for β^2 obtained from the hyperfine data. This results in values of $E_{\Delta g}$ generally in the range 1.5–2 eV, which are quite consistent with the optical absorption data, as we shall see. There are several ways in which this rough estimate of the g shift could be improved, but so far, more accurate calculations have not been performed.

2.5. Optical Absorption

For convenience in making optical measurements, it is desirable to work with a well-defined system having isolated absorption bands. However, additional centers are almost invariably produced by the standard coloration methods, and a major part of the task is to separate overlapping absorption bands due to other known or unknown centers. The problem is particularly acute for hole centers, since different species often display similar absorptions. Fortunately, the anisotropy of these centers provides a powerful

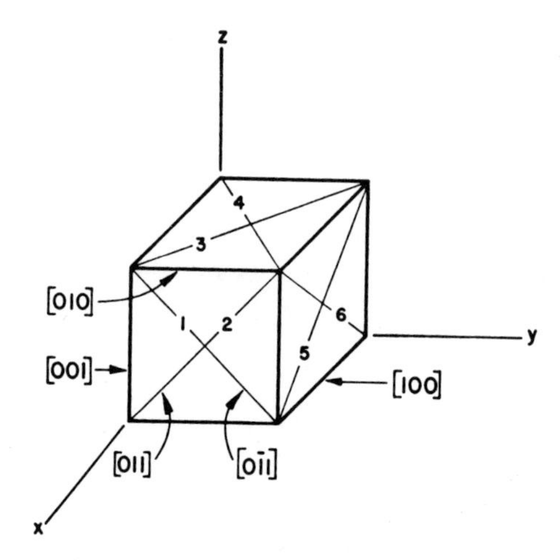

Fig. 6. Crystallographic representation of $[X_2^-]$-center internuclear axes (after Ref. 33).

means of identifying and measuring bands through their polarization or dichroic absorption.

Consider an alkali halide crystal doped with a suitable electron-trapping impurity, which has been irradiated at low temperature to produce $[X_2^-]$ centers. As formed, the $[X_2^-]$ centers will be distributed uniformly among the six $\langle 110 \rangle$ directions. With reference to Fig. 6, let the crystal be illuminated with a beam of light propagating along the [100] direction and with electric vector parallel to the $[0\bar{1}1]$ direction. If the wavelength of this light is tuned to a transition of the $[X_2^-]$ center, then dichroism will be produced in regions of $[X_2^-]$ absorption. This is caused by actual reorientation of the centers during excitation by the light; when a center finally settles into a favorable orientation, due to its anisotropy it no longer absorbs the $[0\bar{1}1]$-polarized light strongly. In the present case, with near-ultraviolet excitation, the [011] direction will be preferentially populated. By subtracting the optical absorption spectra for light polarized in the [011] direction and in the $[0\bar{1}1]$ direction, a dichroic absorption spectrum is now obtained which consists only of the anisotropic bands of the $[X_2^-]$ center. Bands due to centers which have not been reoriented or which have symmetries different from that of the $[X_2^-]$ will have been subtracted out.

This method was first used for the $[X_2^-]$ center by Delbecq, Yuster, and co-workers[25,26] at Argonne National Laboratory, a group which has played a major role in hole center research over the years. They were able to locate the principal optical transitions of the $[X_2^-]$ in a number of alkali halides. Furthermore, by orienting centers optically and noting the corresponding changes in the relative intensities of EPR spectra for the various directions, a correlation was established between the optical and EPR spectra of the [X_2^-], and the directions of the optical dipole moments for the [X_2^-] transitions relative to the internuclear axis were determined.

A [Br_2^-]-center dichroic spectrum for KBr is shown in Fig. 7.[26] This illustration is typical in that the principal transitions invariably comprise a symmetrically shaped band in the 3–4-eV range and a weaker, narrower band in the 1.5–1.7-eV range. Table II contains optical data for [X_2^-] centers in a number of halide crystals. A variety of experimental procedures were used to obtain the tabulated parameters, and the original papers should be consulted for details.[26–30] A cursory inspection of Table II shows that a given parameter depends primarily on the identity of the halogen and less on the cation; thus one has another indication that the X_2^- molecular ion in the self-trapped hole configuration is only slightly perturbed by the lattice.

Returning now to Fig. 5, the electronic transitions to which these two main absorption bands correspond are immediately apparent. Those indicated by the solid arrows fall in the proper energy range and are allowed according to the usual selection rules for electric dipole transitions. The ${}^2\Sigma_g^+ \leftarrow {}^2\Sigma_u^+$ transition will be strong since it connects corresponding bonding and antibonding orbitals and there is considerable charge transfer along the internuclear axis. It will be a σ transition, that is, its transition

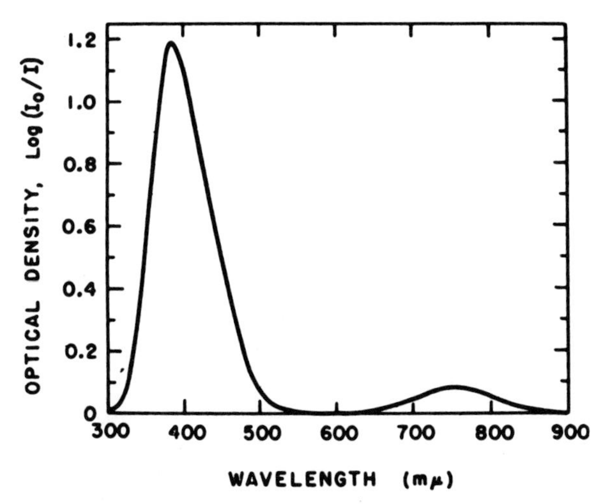

Fig. 7. Dichroic absorption spectrum for the [Br_2^-] center in KBr (after Ref. 26).

Table II. $[X_2^-]$-Center Optical Absorption and Reorientation. Peak Energies E, Widths at Half-Maximum $W_{1/2}$; and Relative Strengths I_r Are Given Along with Disorientation (or Translation[a]) Temperatures T_d

	E, eV	$W_{1/2}$, eV	I_r	T_d, °K	Ref.
LiF	3.65	1.20	>200	110	26
	1.65	—	1		
NaF	3.38	0.66	—	155	27
CaF_2	3.88	1.0	>200	130[a]	13
	~1.65	—	1		
SrF_2	3.80	1.0	>200	120	13
	~1.65	—	1		
BaF_2	3.69	1.0	>200	100	13
	~1.65	—	1		
LiCl	3.15	1.47	—	100	28
NaCl	3.28	1.11	—	130	28
KCl	3.40	0.81	100	170	26
	1.65	0.37	1		
RbCl	3.40	0.76	—	190	28
NH_4Cl	3.31	0.83–1.0	100	130[a]	11, 29
	1.57	0.37	1		
KBr	3.22	0.73	445	140	26
	1.65	0.26	9.5		
	1.38		1		
NH_4Br	3.10	0.80	40	—	11
	1.55	0.27	1		
NaI	2.87	—	~1	50	30
	1.41	—	1		
	~2	—	—		
KI	3.10	0.55	340	90	26
	1.55	0.22	56		
	2.12	0.36	—		
	1.08	0.19	1		
RbI	3.06	—	~2	100	30
	1.56	—	1		
	2.2	—	—		

[a] Temperature at onset of 0° jumps; T_d is higher.

moment will be parallel to the internuclear axis. The experimental data confirm this polarization in all cases. An approximately symmetric splitting of bonding and antibonding states with respect to the associated atom-ion states, as indicated in Fig. 5, is another consequence of the simple LCAO-MO model. In view of this fact, the measured optical transition energies, E_{uv}, and the $E_{\Delta g}$ values obtained in the preceding section may be judged to bear a reasonable relationship to each other, the former being roughly twice the latter.

Interpretation of the infrared transition is somewhat more involved. Although its energy, E_{ir}, is consistent with the expected location of the ${}^2\Pi_g$ state, its oscillator strength is low, particularly for the lighter halogens. Also, instead of being π-polarized as a ${}^2\Pi_g \leftarrow {}^2\Sigma_u{}^+$ transition should be, it generally shows a net σ polarization which increases in magnitude with the atomic number of the halogen. The root of these irregularities lies in the fact that the σ_u hole orbital is derived primarily from p_z atomic orbitals, while the π_g hole orbitals are derived from p_x and p_y orbitals. The transition is thus dipole-allowed only to the extent that interactions between the two halogens bring about mixing of other orbitals through configuration interaction or spin–orbit coupling. In the case of the $[X_2{}^-]$ center, the spin–orbit coupling is evidently the primary agent; the alternative, configuration mixing of s atomic orbitals into the $\sigma_u np$ orbitals, gives a π transition moment which can be shown to be roughly two orders of magnitude smaller than that of the ultraviolet transition. The appropriate σ polarization can be produced most reasonably by mixing of the ${}^2\Sigma^+_{1/2g}$ state into the ${}^2\Pi_{1/2g}$ state through the spin–orbit coupling, in the same way that the g shift is produced by mixing of ${}^2\Pi_{1/2u}$ into ${}^2\Sigma^+_{1/2u}$. The oscillator strength of this σ-polarized component should be lower than that of the ultraviolet transition by a factor of roughly $(\lambda/E_{\text{ir}})^2$, a conclusion in adequate agreement in the experiment. For example, the 1.65-eV band of the $[F_2{}^-]$ center in LiF is at least a factor of 200 lower in oscillator strength than the ultraviolet band; the π- and σ-polarized contributions are evidently comparable in magnitude and yield an essentially unpolarized transition. At the other extreme, in NaI, the large σ-polarized component in the 1.4-eV transition makes this band nearly as strong as the ultraviolet transition.[30]

It should be pointed out that dichroic spectra such as that of Fig. 7 cannot be relied upon to give the relative strengths of transitions which differ in their anisotropies. The subtraction which produces a dichroic spectrum reveals only the excess of transition moment in one direction relative to another, thus giving undue weight to the more anisotropic transitions.

A complete specification of the optical anisotropy of a given transition cannot be obtained from measurements with light propagating in only one direction. For example, the procedure previously described in connection with Fig. 6 will not reveal transition moments in the [100] direction parallel to the direction of propagation of the light. An additional measurement of dichroism with light propagating along, say, the [010] direction is needed to reveal any [100] oscillator strength. A sample of data of this type, which is evidently unique in the literature, is shown in Fig. 8.[26] The spectrum pertains to the ${}^2\Pi_{1/2g} \leftarrow {}^2\Sigma^+_{1/2u}$ transition in KCl. These are not difference spectra, but are simply the optical densities obtained upon measuring with light polarized in the [011], [0$\bar{1}$1], and [100] directions, the [Cl_2^-] centers being roughly 90% aligned along [011]. The relative displacements of these three spectra are an effect of the crystal field on the center. The coupling to the crystal field evidently exceeds the spin–orbit coupling in KCl, giving rise to the approximately 0.14 eV splitting in peak energy between the [0$\bar{1}$1] and [100] spectra. The data of Fig. 8 force recognition of the fact that the D_{2h} site symmetry admits no orbital degeneracy. Instead of a ${}^2\Pi_g$ state split only by spin-orbit coupling into $J = 1/2$ and $J = 3/2$ states, we have instead an upper state, ${}^2B_{3g}$, for which the hole orbital consists mainly of p_x atomic contributions and a lower state, ${}^2B_{2g}$, for which the hole orbital consists mainly of p_y atomic orbitals. The two interact through spin–orbit coupling, and each remains doubly degenerate in the absence of an applied field. A similar situation obtains for the ${}^2\Pi_u$ states and gives rise to slightly

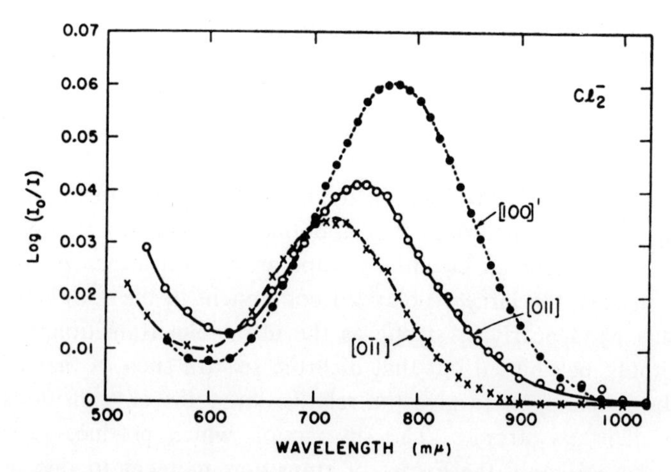

Fig. 8. Spectra corresponding to the [011], [0$\bar{1}$1], and [100] dipole components of the infrared transition for [011]-oriented [Cl_2^-] centers (after Ref. 26).

differing values of $g_\perp$ depending upon whether the magnetic field is applied in the x or y direction.[9] Note that for the CsCl structure, the x and y directions remain equivalent and there is no crystal field splitting in principle.

For $[Br_2^-]$ and $[I_2^-]$ the spin–orbit coupling is larger, and presumably dichroic splittings such as that of Fig. 8 are not readily observable. There appears, nevertheless, an additional very weak optical absorption band at the low-energy side of the 1.6-eV band; the splitting is roughly 0.3 eV for KBr and 0.5 eV for KI, for example.[26] The anisotropies of these weak infrared transitions are substantially less than those of the other transitions. The magnitudes of the splittings of the infrared band are close to the respective λ values which is just what one would expect for the $^2\Pi_{1/2g}$ $-{}^2\Pi_{3/2g}$ separation. Since spin–orbit coupling mixes $^2\Sigma^+_{1/2g}$ into $^2\Pi_{1/2g}$ but not $^2\Pi_{3/2g}$, the lower-energy transition should be weaker and less σ-polarized, as is observed. This transition is also marked in Fig. 5. To summarize, both spin–orbit coupling and crystal field effects (in NaCl-type lattices) split the $^2\Pi$ states. For the bromides and iodides, spin–orbit coupling is evidently the predominant effect, while in chlorides (and probably fluorides), the crystal fields are more important. In any case, these splittings are small compared to the separations caused by the X_2^- molecular field. There appears to be even a fourth optical transition for $[I_2^-]$ in KI,[26] NaI,[30] and RbI,[30] located between the ultraviolet and infrared bands. It is also weak and may be due to a $^2\Pi_u \leftarrow {}^2\Sigma_u^+$ transition, the forbiddenness being partially broken by vibrational distortions around the center.

The optical absorption discussed so far has been ascribed exclusively to transitions in which the hole is excited into the filled πnp and σnp orbitals. Transitions in which the unpaired $\sigma_u np$ *electron* is excited into higher Rydberg-like orbitals have not as yet been reported for the $[X_2^-]$ center. The probable reason is that these transitions begin at an energy near the fundamental edge, where they tend to be obscured by the absorptions arising from electron-trapping impurities. A very rough indication that the energy should indeed be in this range may be obtained by considering the following hypothetical cycle of operations, which transforms the $[X_2^-]$ into an F^+ center and back again: (1) Dissociate the $[X_2^-]$ into a normal substitutional atom and ion at lattice sites, expending roughly 1 eV in the process. (2) Remove the atom to infinity, thereby creating an F^+ center and expending a small energy ε. (3) Excite the F^+ optically at the α-band energy. (4) Return the X atom to the vacancy at a gain of roughly ε. (5) Restore the lattice configuration of the original $[X_2^-]$. The difference between the energy gained in step 5 and that lost in step 1 is probably small. Since all ions have been returned to their original positions, the cycle should represent an optical transition which adds up to an energy roughly that of the α band and thus would occur near the fundamental absorption edge of the crystal. A further search for this transition of the $[X_2^-]$ center would appear to be worthwhile.

No luminescence due to the self-trapped hole *alone* has been reported. With reference to Fig. 5, a plausible explanation of why this is the case is that the X_2^- spontaneously dissociates when placed in its excited states. This is consistent with the fact that reorientations can be brought about by optical excitation in any of the transitions. The $[X_2^-]$ center does play a prime role in luminescent processes involving intrinsic electron–hole recombination. These processes will be discussed briefly in a subsequent section.

2.6. Thermally Induced Reorientation

The general fact that dichroism can be induced in the absorption bands of the $[X_2^-]$ center by exciting with polarized light has already been mentioned as a crucial element in the analysis of optical spectra. During this excitation, not only does the absorption for one direction of polarization decrease, but also absorption for the opposite direction of polarization increases. These changes can, of course, be monitored as well by EPR measurements. Clearly, an actual reorientation is taking place, and not merely a selective bleaching. Thermally activated reorientation without optical excitation is also observed in most crystals, the exceptions being among those having the CsCl structure. That is, if a crystal containing oriented $[X_2^-]$ centers is warmed slowly in the dark, in a certain temperature range the $[X_2^-]$ bands begin to loose their polarization before their overall intensity is diminished significantly.

Reference to the morphology of the $[X_2^-]$ center in either the NaCl or CsCl lattice leads one to conclude that a change in location of the $[X_2^-]$ is a necessary result of reorientation (though not vice versa). Illumination in the absorption bands of the $[Br_2^-]$ center in KBr at low temperature has been found to produce measurable photoconductivity.[31] Translation and reorientation are thus aspects of the same process, except for cases such as CaF_2 [13] and NH_4Cl,[29] in which the $[X_2^-]$ moves parallel to its axis and undergoes translation but no reorientation. Of the two, reorientation is the more convenient for study, since a single jump drastically alters the optical moments but produces very little charge transport. The usual procedure has been to prepare an oriented array of $[X_2^-]$ centers and then to observe the rates of loss of dichroism or EPR anisotropy upon optical excitation or annealing.

Thermally activated processes can be studied by means of isochronal or isothermal annealing experiments, from which frequency factors and activation energies can be extracted. The initial work on $[X_2^-]$-center reorientation in alkali halides employed isochronal annealing.[25,26] The results were expressed in terms of a disorientation temperature T_d, at which the slope of a concentration-versus-temperature plot is greatest. T_d values are included in Table II; they have in some cases been estimated from related data.

There are actually two distinct minimum jumps which can produce reorientation of an $[X_2^-]$ center in the NaCl lattice. One involves a rotation of axis through an angle of 60°, the other through an angle of 90°. With respect to an $[X_2^-]$ in a given orientation, there are four alternative $\langle 110 \rangle$ directions at a 60° angle, but only one at 90°. A simple experiment which specifies the relative likelihood of these two jumps has been carried out by Keller and Murray.[32] Consider again a beam of $[0\bar{1}1]$-polarized light resonant with one of the $[X_2^-]$ transitions and propagating along the [100] axis in Fig. 6. This light will, as we have seen, align the centers in the [011] direction. Now, warm the crystal in the dark to a selected temperature, and record the resulting isothermal annealing curve for the dichroism. Both 60° and 90° jumps will contribute to the reorientation if they are active, and the rate constant will be a simple weighted sum of the two jump probabilities P_{60} and P_{90}. Now repeat the experiment, this time aligning with [010]-polarized light and thereby producing preferential (and equal) populations in the [101] and $[\bar{1}01]$ directions which lie in the x, z plane perpendicular to the polarization vector of the light. Consideration of Fig. 6 will show that now only 60° jumps will be effective in destroying the dichroism, since 90° jumps will not alter the net components of the transition moment. These two measurements thus allow the evaluation of both P_{90} and P_{60}. Results have been reported for KCl[33] and KI,[32] P_{90} having been found to be effectively zero for both materials. The data fit well the formula $P_{60} = \nu \exp(-E_{\text{reor}}/kT)$, with $E_{\text{reor}} = 0.54$ eV and $\nu = 10^{13}$ sec^{-1} for KCl, and $E_{\text{reor}} = 0.27$ eV and $\nu = 10^{13}$ sec^{-1} for KI. These activation energies agree within experimental error with activation energies measured for hole diffusion. They are small compared to the energies of the excited electronic states, indicating that thermally induced reorientation takes place entirely within the ground electronic state.

Referring again to Fig. 3, it may be seen that a 60° jump entails replacement by a B ion of only one of the two X^- ions sharing the hole, whereas in a 90° jump, both of the original X^- ions must be replaced. One might therefore expect a 60° jump to involve minimum ionic displacements and thus a lower activation energy. An additional consideration relating to the atomic reorientation mechanism is the rough correlation, evident in the alkali halide data of Table II, between low T_d values and low cation–anion size ratios. This correlation is consistent with the straightforward premise that the smaller the cation, the smaller will be the ionic displacements accompanying reorientation and, consequently, the smaller the activation energy.[34]

In CsCl-type lattices, $[X_2^-]$-center reorientation can occur only through 90° jumps. However, it is found that linear motion parallel to the X_2^- axis, that is, by 0° jumps, is an important competitive process. In CaF_2, only 0° jumps are observed; translation effectively destroys the holes before they can reorient.[13] In SrF_2 and BaF_2, both 0° and 90° jumps are observed. The activation energies for 0° jumps in CaF_2, SrF_2, and BaF_2 are 0.31, 0.21, and 0.30 eV, respectively; the activation energy for 90° jumps in both SrF_2 and BaF_2 is 0.30 eV.

There is evidence that 0° jumps are predominant also in NH_4Cl. The activation energies for both 0° and 90° jumps fall again in the 0.3-eV range.[29]

The process of optically stimulated reorientation differs in principle from the thermally activated process just discussed. In the case of ultraviolet excitation, the $[X_2^-]$ center passes through an antibonding excited state from which the molecular ion dissociates. Experiments already described imply that the probability of the center reforming in a different orientation is appreciable. Additional work dealing with the ultraviolet-stimulated reorientation in KCl at 78°K has recently been reported.[35] By methods similar to those outlined above, it has been shown that 60°, 90°, and 0° jumps occur with equal probability, that is, the absorption of a single photon causes the $[Cl_2^-]$ to assume a new orientation having no correlation with initial orientation. The distance the "free" hole migrates following dissociation is undetermined; it might be several lattice spaces, or a single ion might be common to both orientations.

The fact that in many cases, at least with the alkali halides, it is possible to reorient $[X_2^-]$ centers at temperatures down to that of liquid helium with light in any of the absorption bands suggests that optical excitation energies are generally high enough to bring about dissociation from any of the excited states. This conclusion is consistent with the theoretical potential curves to be considered in the next section.

2.7. A Theory for the $[X_2^-]$ Center

As it stands, the simple atomic model discussed thus far can provide no quantitative predictions of parameters which depend upon the bonding of the molecular ion or its interaction with the surrounding lattice. It is clear at the outset that the theoretical problems relating to realistic models will be quite complex. Relatively little theoretical work has actually been attempted. There does exist, however, a significant beginning in this direction in the form of a theory initiated by Das *et al.*[36] and developed by Jette *et al.*[37] This theory is worthy of attention not only because it gives fair numerical results for $[F_2^-]$ and $[Cl_2^-]$ centers, but also because the approach can serve as a reasonable model for future calculations on the $[X_2^-]$ and more complicated hole centers. The few other published theoretical efforts, which are of more limited scope, will not be dealt with explicitly.

As a conceptual preliminary, it is useful to trace a general line of reasoning, due originally to Gilbert,[38] concerning the occurrence and basic energetics of self-trapping. The conversion of a free carrier in an otherwise perfect crystal to a self-trapped carrier may be artificially regarded as two separate processes, one entailing localization of the charge at a point in the lattice, the other the relaxation of the nearby ions into an optimum configuration which in the present case includes formation of the molecular ion. Upon localization, polarization energy E_{pol} is gained as the ions relax around the charge. However, an energy E_{loc} must simultaneously be expended to bring about breakoff of a localized state from the band. To localize a state in real space requires delocalizing it in k space; other considerations aside, this produces a state whose energy is intermediate within the particular band. E_{loc} may thus be presumed to be roughly one-half the width of the band. The final contribution is the energy E_{bond} gained in the formation of the molecular unit. Adding these together, we arrive at the net energy of self-trapping,

$$E_{\text{s-t}} = E_{\text{pol}} + E_{\text{loc}} + E_{\text{bond}} \tag{11}$$

Taking hole self-trapping in KCl as an example, Gilbert has estimated on the basis of a simple classical model that $E_{\text{pol}} \approx -0.5$ eV. The appropriate valence band appears to have a width of the order of 1.0 eV.[7,8] This gives $E_{\text{loc}} \approx +0.5$ eV, and thus E_{pol} and E_{loc} approximately cancel each other. Typical energies of single bonds between halogens are of the order of -1 eV, and thus, according to Eq. (11), $E_{\text{s-t}} \approx -1$ eV, clearly in favor of hole self-trapping.

Although $E_{\text{s-t}}$ itself has not been measured, the estimate can nevertheless be tested to some extent through comparison with the measured value of E_{reor} given in the last section. Since it is reasonable to assume that the hole remains localized, E_{pol} and E_{loc} should fluctuate only slightly during a 60° reorientation. One can argue also that E_{reor} should be somewhat less than E_{bond}, because in the saddlepoint configuration, the three participating halide ions will remain partially bonded to each other.[34] The net self-trapping energy might then be expected to be of the order of, but somewhat greater than, E_{reor}. Since for KCl, $E_{reor} = 0.54$ eV, this criterion is met satisfactorily in the chosen example. Unfortunately, theoretical and experimental data are insufficient to provide another distinct test case.

It is reasonable to ask how these considerations relate to the hypothetical event of electron self-trapping in a diatomic alkali molecular ion. The conduction band width is considerably greater than the valence band width; for an electron in KCl, E_{loc} would probably be in the vicinity of 2 eV. E_{pol} should be roughly the same as for the hole, but E_{bond} for K_2^+ is estimated to be only about -0.3 eV.[38] Thus electron self-trapping would appear to require a net expenditure of energy of the order of an electron-volt; consistent with this conclusion, electron self-trapping is not observed experimentally.

To formulate the self-trapping problem more explicitly, one must in principle look for that path in configuration space along which the system proceeds, beginning with a free hole and all ions in the perfect lattice configuration and ending with a self-trapped hole and ions displaced around it. This forbidding task has not been attempted, and only the final self-trapped state is actually analyzed. Even for this state, some sort of approximation is clearly demanded, and reliance is placed on the experimental evidence that the self-trapped hole behaves very much as if it were a free X_2^- molecular ion. This approach leads to the requirement for calculated parameters for the electronic states of the free X_2^- as a function of the internuclear separation R.[37] Let us designate the resulting energy of the electronic state n of the free molecular ion as $E_b{}^n(R)$. Note that $E_b{}^\circ(R_e) \approx E_{bond}$, where R_e is the equilibrium separation. The surrounding lattice is taken into account through another term in the energy, $E_l(Q)$, where the configuration coordinate Q comprises the positions of all ions in the crystal and includes the coordinate R. The term $E_l(Q)$ is assumed independent of the state of the molecular ion, a severe though necessary approximation. The complex matter of the localization energy is simply ignored, although one thereby foregoes knowledge of the relative locations of the band states as well as the dynamics of the self-trapping process. Thus for configurations Q not too near the perfect lattice configuration and for states in which the hole remains well localized on the X_2^-, the net energy of the self-trapped configuration can be written

$$E_{s\text{-}t}^n(Q) = E_b{}^n(R) + E_l(Q) \tag{12}$$

Jette *et al.*[37] obtain values for $E_b^n(R)$ from molecular orbital calculations of Wahl, Gilbert, and others,[39] based upon a self-consistent field (SCF) method. These SCF-MO data are available only for F_2^- and Cl_2^- and include wave functions and energies for the states ${}^2\Sigma_u^+$, ${}^2\Pi_g$, ${}^2\Pi_u$, and ${}^2\Sigma_g^+$. To evaluate $E_l(Q)$, the electrostatic monopole energies, the short-range repulsive energies, and the electronic and ionic polarization energies are computed as functions of R and of the positions of the six nearest-neighbor alkali ions in the (100) plane containing the X_2^-. For each value of R, the sum of these contributions is minimized with respect to displacements of the six alkalis, in effect converting $E_l(Q)$ into a term $E_l(R)$ depending only on R. All ions, including the two comprising the X_2^-, are described in terms of standard Born–Mayer ionic potentials. This is another important approximation, both because the ionic displacements are not small and because the necessary parameters for the molecular ion itself are not known. The resulting $E_l(R)$ term contributes significant repulsion in the vicinity of R_e because of interactions with the two adjacent nearest-neighbor alkali ions, labeled A in Fig. 3. The effect, relative to the calculated potential curves for free X_2^-, is to increase R_e somewhat and to lessen the depth of the ground-state potential curve. In KCl, for example, R_e in the crystal is calculated to be 2.90 Å; this is to be compared with the value of 2.64 Å from the SCF-MO calculation for free Cl_2^- and with a perfect-lattice spacing $R_0 = 4.45$ Å.

Calculations were performed for eight crystals, the fluorides and chlorides of lithium, sodium, potassium, and rubidium. Sample potential curves are reproduced in Fig. 9. The SCF-MO energies for free Cl_2^- are shown in Fig. 9(a), and in Fig. 9(b) is shown the result for the [Cl_2^-] center in KCl. The internuclear separation R is given in atomic units, that is, in Bohr radii. Note that in Fig. 9(b), the energy scale doubles for $E < 0$. The effect of the term $E_l(Q)$ on the ${}^2\Sigma_u^+$ state is apparent.

For a detailed comparison of theory and experiment, the reader is referred to the original paper.[37] In reviewing the situation, we begin with the optical absorption data. The ultraviolet spectra have been measured for all the relevant crystals except KF and RbF, and the transition energies agree with theory usually to within about 30% or better. For the chlorides, experimental values exceed theoretical values, but by a margin which decreases with decreasing size of alkali ion. The agreement is probably fortuitous to some extent, in view of the approximations employed; this matter will be discussed presently.

The calculations specify displacements for ions A and D in Fig. 3, as well as for the two halide ions sharing the hole. As mentioned in Section

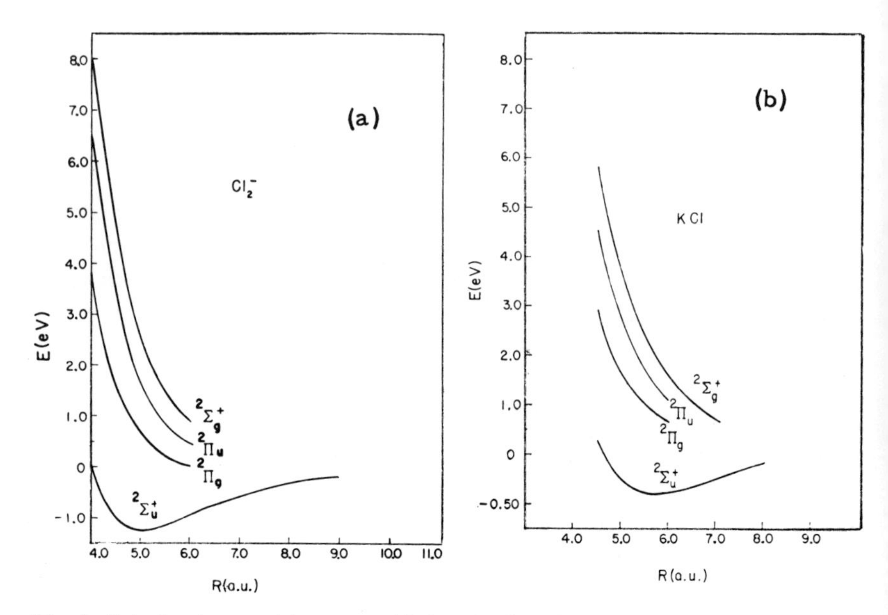

Fig. 9. Calculated potential curves: (a) for the free Cl_2^- molecular ion; (b) for the $[Cl_2^-]$ center in KCl (after Refs. 37 and 39).

2.1, ionic displacements in and around the $[F_2^-]$ in LiF and NaF have also been determined from ENDOR measurements.[22,37] These displacements were derived on the basis of the same electronic wave function as used by Jette *et al.*, but the two determinations are otherwise independent. They agree to within about 10% for the $[F_2^-]$ internuclear separation and to within about 20% for the outward displacements of the two A ions (Fig. 3) in LiF. Agreement is poorer for NaF and for the alkali ions D in LiF. The displacement of the A ions is radially outward and in LiF amounts to roughly 10% of the nearest-like-neighbor distance in the perfect lattice. These results do not appear to be unreasonable.

The most significant comparison with the experiment probably lies in the hyperfine constants a_i and b_i defined in Eqs. (4)–(6), since they are a direct test of the accuracy of the ground-state wave function. A comparison which depends primarily on the contact term a_{Fi} is in fact too stringent a test, since this parameter can be affected strongly by minor variations in the wave function and by unaccounted phenomena such as spin polarization. The dipolar hyperfine constant b_i, which depends on an integral over the entire unpaired σ_u orbital rather than its amplitude at a single point, is a more reasonable parameter to use for evaluating the present theory. Table III lists some of the theoretical hyperfine data and the cal-

Table III. Comparison of Experimental Hyperfine Constants (units of gauss) for [F_2^-] and [Cl_2^-] Centers with the Theory of Jette *et al.*[a] Data Are Shown Also for the Free Molecular Ions[b]

	LiF	KF	F_2^-	LiCl	KCl	Cl_2^-
b_i (experiment)	945	936	—	87	89	—
b_i (theory)	836	828	861	79	78	80
a_i (experiment)	−58	−28	—	8.5	12.5	—
a_i (theory)	−158	−179	−61	−11	−12	0.3
R_e (a.u.) (theory)	4.00	4.15	3.60	5.50	5.60	5.00

[a] Ref. 37; experimental data of Schoemaker as quoted in Ref. 37.
[b] Ref. 39.

culated R_e, along with comparable experimental results. The a_i and b_i values refer directly to Eq. (3); in this case, it is the theory to which the correction for higher-order hyperfine interactions has been applied. This accounts for the apparent discrepancies between data common to both Tables I and III. The choice of sign for the *experimental* a_i is as before; $a_i > 0$ for chlorides and $a_i < 0$ for fluorides.[20]

On the whole, the theory is sufficiently consistent with the data to give one confidence in the theoretical approach as well as the [X_2^-] model. Several plausible sources for the discrepancies can be singled out among the various approximations in the theory. In this regard, let us first note that observable parameters derived from theory all depend on the computed value of R_e. In fact, the theory takes account of the lattice only through its influence on R_e. It is apparent from the data in Table III that if one actually regards R_e as an adjustable parameter and arbitrarily sets it near R_e for the free X_2^-, significantly better agreement with experiment is obtained for b_i and even for a_i. This change is also generally in the proper direction to improve agreement with the optical transition energies. The problem with the calculated R_e evidently lies in the approximation whereby the repulsive interactions of the X_2^- with the lattice are expressed in terms of X^--ion potentials without taking account of the presence of the hole; the resulting overestimate of the size of the X_2^- has increased the repulsion between it and the two closest alkali ions and produced an erroneously large R_e. The derivation and use of Born–Mayer parameters appropriate to X_2^- is an obvious though formidable refinement for future theoretical work.

Another approximation of prime importance is the assumption of complete localization of the hole on the $[X_2^-]$, even in the excited states. The $^2\Sigma_g^+$ state in particular appears to lie close to the valence band, and thus significant interaction with these band states and spreading of the hole wave function is to be expected. This interaction is difficult to treat theoretically, but it would almost certainly have the effect of lowering somewhat the $^2\Sigma_g^+$ state. This would in turn move the optical transitions to lower energy, a trend opposite to that which would result from the R_e correction. Thus the interplay of these effects might well be anticipated to produce much better quantitative agreement with the experiment.

Although the SCF-MO calculations for free F_2^- and Cl_2^- are a strong point in the present theory, they are inevitably subject to improvement through the inclusion of configuration interaction, electron correlation, and other related factors. For example, to narrow the uncertainty in a_i, it would be useful to take account of core polarization, which involves unpaired spin components induced in the core orbitals of the X_2^- by exchange interactions with the unpaired valence electron. It is interesting to note a related calculation for LiF by Ikenberry *et al.*[40] This calculation concerns not the molecular ion itself, but the so-called transferred hyperfine interaction between the unpaired σ_u electron in the $[F_2^-]$ and the nearest-neighbor F^- and Li^+ ions, *A* and *B* in Fig. 3. These ions lie in a nodal plane for the σ_u orbital, and thus there should be no direct overlap of the hole onto the *A* and *B* nuclei. ENDOR spectra nevertheless record a substantial negative isotropic hyperfine interaction for these ions.[23] Although spin–orbit interaction produces a π_u contribution which does have nonzero amplitude at these nuclei, in LiF this addition is too small by two orders of magnitude to explain the observed spin density. It was shown that the principal source of this transferred hyperfine interaction is direct exchange coupling between the unpaired σ_u electron and the *s* electrons in the cores of the Li^+ and F^- ions involved.[40] That is, due to the Pauli principle, the unpaired electron attracts nearby electrons with parallel spin, leaving net antiparallel spin at the *A* and *B* nuclei. The theory gives -5.03 MHz for the isotropic hyperfine interaction with Li^+ ion *A* and -6.60 MHz for F^- ion *B*, whereas the measured values are -4.12 MHz for *A* and -6.69 MHz for *B*, rather satisfactory agreement.

Employing a procedure very similar to that described,[37] Druger and Knox[41] have calculated energies and ionic relaxations pertaining to $[X_2^-]$-like hole self-trapping in rare-gas solids. They conclude that the R_2^+ molecular ion should be stable at low temperature in Ar, Kr, and Xe crystals, although there is as yet no direct experimental observation of such a center.

Let us mention finally an application by Song of small-polaron-migration theory to the problem of reorientation and diffusion of $[X_2^-]$ centers in alkali halides.[42] A somewhat simplified approach gives activation energies E_{reor} for 60° jumps in KCl and KI which are within a factor of two of the measured values. The calculated energies for 0° and 90° jumps turn out to be considerably higher, in general agreement with experiment.

2.8. Electron–Hole Recombination

The $[X_2^-]$ center is a very effective electron trap or recombination center. During prolonged high energy irradiation of a halide crystal, a substantial fraction of the electron–hole pairs produced will recombine and thus disappear immediately. The excitation energy released as a result of the recombination can be dissipated through any of four principal channels: (1) emission of luminescence, (2) phonon creation, (3) production of lattice defects such as vacancy–interstitial pairs, and (4) transfer of energy to some existing imperfection. The relative importance of these channels depends on temperature, on the particular material, and on its state of purity.

As discussed in a previous chapter, channel (3) can under certain circumstances consume a significant fraction of the available recombination energy. Evidently, the system is able to cross over into the ground electronic state after only a small fraction of the energy has been dissipated. The highly distorted pair of ground-state halide ions thus produced possesses several electron-volts of kinetic energy, which often is sufficient to thrust an ion or atom, not necessarily one of the two originally sharing the hole, into an interstitial position. In the present context, a novel aspect of this mechanism is that it can bring about conversion of a self-trapped hole into another similar hole center, the *H*. This transformation has been observed directly by following changes in the concentrations of the two centers during recombination produced by optical stimulation.[43]

The extrinsic processes of channel (4) occur only in special circumstances and are extraneous to the present topic. Channel (2) is active to a greater or lesser extent in all crystals at all temperatures and competes with or occurs in series with the other processes. However, no detailed investigations of phonon emission in this system have been carried out.

Knowledge of the electronic states of the intrinsic relaxed electron–hole pair has been derived almost exclusively from investigation of recombination luminescence. States which initiate the luminescent transitions can be populated by conduction band electrons, produced either with ionizing radiation or with light which frees electrons previously trapped at other imperfections. These states can also be populated with ultraviolet photons of an energy equal to or exceeding the lowest exciton absorption. The relaxed exciton states produced directly in this way without the isolated $[X_2^-]$ intermediate are, with a few possible exceptions, identical to the metastable states which arise from the recombination of a free electron with a preexisting $[X_2^-]$ center.

There exists a fairly extensive body of literature on intrinsic luminescence in halide lattices. In the present context, this phenomenon is to be regarded simply as one aspect of hole center behavior and thus will be accorded only a brief review, emphasizing those properties which reflect most strongly the nature of the self-trapped hole.

The most direct experimental correlation between the morphologies of the self-trapped hole and exciton is the observation of linearly polarized luminescence upon recombination of electrons with an oriented array of $[X_2^-]$ centers.[44,45] The oriented $[X_2^-]$ centers are prepared in the usual way; for example, by irradiating a Tl^+-doped alkali halide and then aligning the $[X_2^-]$ centers optically. By either additional high energy radiation or by optical excitation from traps, conduction electrons are then produced and allowed to recombine with the oriented $[X_2^-]$ centers. The resulting recombination luminescence exhibits partial plane polarization having either σ or π character with respect to the axis of the parent $[X_2^-]$ centers. Emission of polarized recombination luminescence from oriented $[X_2^-]$ centers is evidently a general characteristic of halide lattices; several ammonium halides[46] and alkaline earth fluorides,[13] in addition to a wide sampling of alkali halides, have been shown to exhibit the effect. Thus capture of an electron by the self-trapped hole does not cause reorientation or a change in site symmetry until the ground state is reached. However, direct excitation with polarized ultraviolet light in the exciton absorption bands in KI does not produce polarized recombination luminescence, presumably because the exciton loses memory of the polarization before it can relax.[47] We note in passing that when the electron–hole pair is created as a free and potentially mobile exciton, it is not clear that there are no metastable states whose symmetries are distinct from that of the self-trapped exciton. This question is of current interest and is pertinent, in principle, also to the matter of hole self-trapping.

Although the intrinsic luminescent efficiency is relatively high in most halides at sufficiently low temperatures, the radiative efficiencies and the lifetimes as well become strongly temperature dependent at higher temperatures, due to nonradiative recombination. The extreme examples are KCl and RbCl, both of which luminesce at 12°K with an efficiency roughly half that at 5°K.[48] For KI, on the other hand, the luminescent intensity begins to fall off strongly only above about 100°K.[45] There is no simple correlation between the decay temperatures of the recombination luminescence and the disorientation temperatures of the parent $[X_2^-]$ centers; except for the existence of the vacancy–interstitial production channel, little is known at present about the nonradiative decay processes.

Figure 10 illustrates emission spectra for several nominally pure alkali halides at liquid helium temperature under X-ray excitation.[44] The emission bands are quite broad for the same basic reason that the ultraviolet absorption bands of the comparable $[X_2^-]$ centers are broad, that is, because of strong coupling to the internal vibrational mode of the halide molecular ion. Also for this reason, the luminescent transitions display a large Stokes shift, meaning that the difference between the minimum energy necessary to create an electron–hole pair and the average emitted-photon energy is comparable to the energy of the photon itself.

Note that in Fig. 10 there appear two bands each for NaCl, KBr, and KI. The higher energy band is σ-polarized and exhibits a lifetime of the order of 10^{-8} sec, consistent with the transition being dipole-allowed.[49] This implies that the upper state is a singlet state in which the electron occupies a σ_g or s-like orbital about the self-trapped hole. The lower energy transition is partially π-polarized and shows a much longer and more variable lifetime, e.g., 4×10^{-6} sec in KI, 1.3×10^{-4} sec in KBr, and 5×10^{-3}

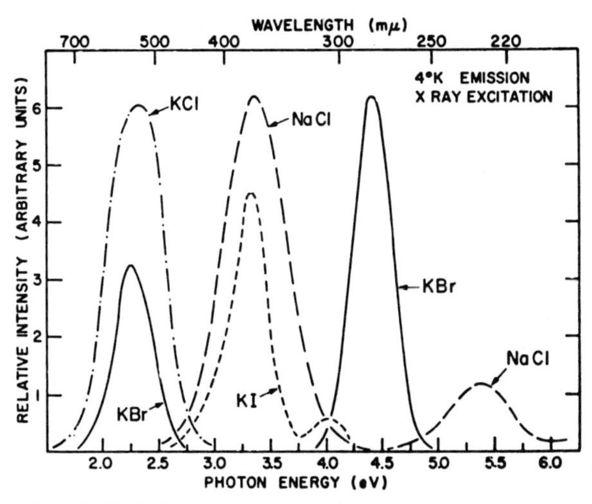

Fig. 10. Emission spectra for electron-$[X_2^-]$-center recombination under X-ray excitation at liquid helium temperature (after Ref. 44).

sec in KCl at 6°K. This transition is inferred to originate in a state which is predominately triplet in character.[49] The breaking of multiplicity forbiddenness has been ascribed to spin–orbit interaction, which mixes the $\sigma_u np$ and $\pi_u np$ orbitals, or ${}^3\Sigma_u{}^+$ and ${}^1\Pi_u$ states, of the molecular ions. This is analogous to the interaction which was shown to produce the g shifts in the EPR spectra of the $[X_2{}^-]$ centers themselves. Other factors being equal, the mixing of singlet into triplet states will increase in the sequence Cl, Br, I, thereby producing the observed decrease in lifetime along this sequence.

In some crystals, for example KBr,[50] KI,[47] or RbI,[30,51] there are indications of a second triplet–singlet transition, making a total of three distinct luminescent transitions in each. On the other hand, only one transition is active in KCl and RbCl. The source of these differences is only in the beginning stages of clarification. There is also little quantitative information available about trap depths and cross sections for capture of electrons at $[X_2{}^-]$ centers. Regarding the locations of the self-trapped exciton states relative to the conduction band, there do exist experimental data which may be interpreted to yield approximate optical binding energies for the luminescent triplet states as well as the locations of certain other states which do not originate luminescence. The experiment in question consists in measuring transient absorption spectra due to self-trapped excitons produced by pulsed electron irradiation at low temperature.[52] Absorption bands originating from the luminescent triplet states were identified primarily on the basis of coincidence between the lifetimes of the luminescence and the absorption. The spectra for three alkali chlorides are given by the solid curves of Fig. 11. Stable color centers created by the radiation pulse produce the absorption shown by the dashed curves; this has in each case been subtracted from the self-trapped exciton absorption. The absorption in the 1.6–2.8-eV range has been accounted for in terms of a Rydberg-like sequence of transitions involving excitation of the bound electron. On the basis of this premise and by analogy with other Rydberg-like spectra, one can argue that the conduction band edge should occur in the high-energy tail, between about 2.2 and 2.4 eV in RbCl and 2.5 and 2.9 eV in KCl. This is a rather large optical binding energy for the self-trapped electron–hole pair, larger by a factor of roughly three than those of unrelaxed excitons.[53] This is probably an indication that the electron is distributed primarily on the nearest-neighbor alkali ions.[52] Note also that the luminescent states are probably not the initial states into which the conduction electrons are trapped; short-lived, large-radius intermediate states may be expected to play an important role here.

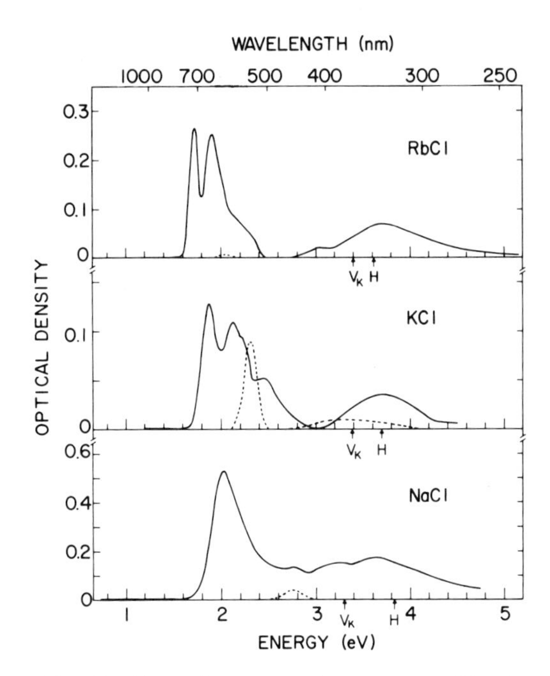

Fig. 11. Absorption spectra due to metastable electron-[X_2^-]-center pairs (solid lines). Absorption due to stable centers (dashed lines) produced by the pulsed electron excitation has been subtracted from the pair spectra. Peak energies of the stable [Cl_2^-]- and *H*-center absorptions are marked for reference; the contributions of these centers to the spectra shown are negligible (after Ref. 52).

The self-trapped exciton absorption in the 3–5-eV range in Fig. 11 has been attributed to a transition of the bound pair involving excitation of the hole rather than the electron. This absorption falls in the general region of the ultraviolet band due to the [X_2^-] center alone, and it is reasonable to presume that the self-trapped exciton transition is similar except for a small perturbation due to the presence of the trapped electron. For comparison, the peak energies of the *H*- and [X_2^-]-center absorption bands are also noted in Fig. 11.

At temperatures for which the [X_2^-] center is mobile, the migrating hole can seek out trapped electrons, and the resulting recombination luminescence is then generally characteristic of the electron-trapping impurity. The system comprising a thallium-activated alkali iodide is a good example.[54] The emission can be observed either as thermoluminescence upon warming an irradiated crystal or as luminescence produced directly by the irradiation at higher temperatures. Self-trapped hole migration appears in fact to be an essential element in the scintillation response of NaI:Tl.

3. $[X_2^-]$ CENTERS ASSOCIATED WITH NONINTERSTITIAL DEFECTS

There exist several varieties of hole center which EPR data show to be slightly perturbed $[X_2^-]$ centers. These perturbed centers can generally be produced in a suitably doped crystal by first irradiating at a low temperature to produce intrinsic $[X_2^-]$ centers and then annealing at a temperature at which the $[X_2^-]$ is mobile but becomes trapped stably upon encountering a lattice defect. The hole often can be shifted from one variety of defect to another by exploiting the inherent differences in trap depths. The most fundamental of these centers involves a single positive-ion vacancy at one of the A positions in Fig. 3. The present system of nomenclature designates this center simply as V, although it has in the past been designated V_F by virtue of its being the structural antimorph of the F center.

3.1. The V Center

This center has been identified in LiF,[55] KCl,[56] and NaCl.[56] Its EPR spectra are very similar to those for the $[X_2^-]$, and nearly axial symmetry of the g tensor with principal axis along a $\langle 110 \rangle$ direction is retained. However, the principal axes for the hyperfine interactions of the two halogen nuclei are canted slightly from $\langle 110 \rangle$ due to the presence of the vacancy; that is, the bond of the X_2^- molecular ion is bent. The actual symmetry is that of an isosceles triangle with base along a $\langle 110 \rangle$ direction and apex at the alkali-ion vacancy. The fact that the nuclei are equivalent only for the applied magnetic field aligned along one of the symmetry axes of the triangle is readily perceived from the EPR data. The bending of the bond is actually not very great; the angle between the principal axis of the hyperfine tensor of a halogen nucleus and the internuclear axis is 3° in LiF, 3.5° in NaCl, and 1.5° in KCl.

The hyperfine splittings are slightly larger for the V center in LiF than for the $[F_2^-]$ center. This implies a somewhat smaller separation between the two halogen nuclei in the V center, which is probably due to the diminished repulsion from ions A (Fig. 3) upon relaxation of the F_2^- toward the vacancy. The g-tensor components, and probably also the optical absorption bands, do not differ strongly from those of the $[X_2^-]$ center.

It is possible in principle for the hole to jump between equivalent positions around the alkali-ion vacancy, the motion ultimately being limited only by the Coulomb field of the vacancy. In LiF, it is known that the hole remains trapped in the vicinity of the vacancy at least to 230°K, and the EPR spectra show no motional broadening up to at least 130°K. Thus the frequency of jumping around the vacancy below 130°K is much less than the microwave frequency, and thus the system is clearly dominated by a static Jahn–Teller-like effect. Photodichroic experiments which might show whether or not the hole is frozen on a single halide pair at sufficiently low temperatures evidently have not been given an exhaustive trial.

3.2. The $[X_2^-]_A$ Center

A species of hole center for which one of the A ions in Fig. 3 is replaced by an alkali-ion impurity also has been observed in several alkali halides. This $[X_2^-]_A$ center differs from the self-trapped hole only slightly, and an enhanced thermal stability is its chief distinguishing characteristic. ENDOR experiments have positively identified the $[F_2^-]_{Li}$ in lithium-doped NaF.[57] In lithium-doped NaCl and KCl, $[Cl_2^-]_{Li}$ centers have been observed, as well as $[Cl_2^-]_{Na}$ centers in sodium-doped KCl.[56] Although qualitatively quite similar to V_F centers, these centers show almost negligible bond bending. This is probably related to the fact that the $[X_2^-]_A$ has a net positive charge whereas V_F does not.

3.3. Mixed Halogen Centers $[XY^-]$

In doped or mixed crystals containing two different halide ions in normal lattice positions, the heteronuclear diatomic center $[XY^-]$ can be produced by irradiation and thermal treatment. Let us take as an example a KCl crystal doped with roughly 0.1 mole % KBr and a smaller concentration of an electron-trapping impurity. This is, incidentally, an appropriate example from the experimental viewpoint, as bromine is a common impurity in KCl and special precautions must be taken during growth to eliminate it. Irradiation around 77°K creates primarily $[Cl_2^-]$ centers, since the radiation is absorbed uniformly and the hole is evidently self-trapped where formed. If the crystal is now warmed to the vicinity of 200°K, the hole migrates under thermal activation and has an appreciable probability of being trapped at one of the substitutional Br^- ions. This event produces the stable species

$[BrCl^-]$, whose internuclear axis is, as one would expect, along a $\langle 110 \rangle$ direction.[58]

This center can be observed and studied by means of EPR spectroscopy in the same way as can the homonuclear species. Equations (3)–(6) still describe the observed hyperfine interactions to a reasonable first approximation, the index $i = 1, 2$ now referring to the different ions. In the present example, all four isotopes ^{35}Cl, ^{37}Cl, ^{79}Br, and ^{81}Br each have nuclear spin 3/2; however, the nuclear magnetic moments of the bromine isotopes are substantially larger than the magnetic moments of the chlorine isotopes. Accordingly, the principal feature of the EPR spectrum for a given pair of isotopes will be 16 lines of approximately equal intensity divided into 4 groups of 4 lines each. The bromine hyperfine interaction is primarily responsible for the splitting into groups; the chlorine hyperfine interaction for the splitting within groups. In an actual measurement, each of the 4 possible isotopic combinations will contribute its own unique 16-line pattern, providing a complex and unmistakable overall spectrum. Higher-order hyperfine and quadrupole interactions alter the splittings and intensities significantly, as one would anticipate from the previous discussions.

The fact that the $[BrCl^-]$ center is a deeper hole trap than the $[Cl_2^-]$ can be attributed to the lower electronegativity of the bromine. The failure so far to observe a $[BrCl^-]$ center in KBr is consistent with this electronegativity difference, as are data indicating that the hyperfine interaction with the bromine nucleus is larger for $[BrCl^-]$ than for $[Br_2^-]$, while interaction with the chlorine nucleus is smaller for $[BrCl^-]$ than for $[Cl_2^-]$. Thus the $[BrCl^-]$ has a dipole moment, the hole tending to favor the bromine.

The electronic structure of the $[XY^-]$ center is very similar to that of the $[X_2^-]$ or $[Y_2^-]$ center. The $[BrCl^-]$ center exhibits the usual optical transitions, which fall at 382 and 760 nm in KCl and which are both σ-polarized.[58] The center can be reoriented by excitation in either of these transitions, although illumination at 382 nm will also readily convert $[BrCl^-]$ back into $[Cl_2^-]$. With regard to thermal reorientation, the $[BrCl^-]$ center has been reported to undergo 90° jumps with appreciable probability at temperatures around 88°K.

A center analogous to the previously described V center can also be produced in KCl–KBr mixed crystals containing an electron trap such as Pb^{2+}, which brings additional positive-ion vacancies into the lattice.[59] If, after creation of $[BrCl^-]$ centers, the crystal is heated further to about 280°K, the positive-ion vacancies become mobile and some are eventually trapped at one of the two nearest-neighbor sites to the $[BrCl^-]$ center. As the reader would by now perhaps expect, this new composite center retains the character of the parent [$BrCl^-$] center. However, the triangle in the (100) plane described by the constituents is no longer isosceles; the Br–Cl axis is in fact inclined at an angle of 12° to the [011] direction, the bromine end of the molecular ion being closer to the vacancy. The presence of the vacancy also causes the bond to bend, as for the *V* center. The optical transitions for the vacancy-associated [$BrCl^-$] center are σ-polarized and lie at 367 and 904 nm. It is reasonable to speculate again that these higher transition energies relative to the pure [$BrCl^-$] center arise from closer bonding of the molecular ion brought about by the absence of hard-core repulsion of one of the nearest-neighbor alkali ions. This interpretation is, however, complicated by the fact that the net charge of the center has vanished upon incorporation of the vacancy. Photodichroism can still be produced, evidently as a result of rotation about the fixed Br-vacancy axis, that is, by exchange of the chlorine constituent.

Other ⟨110⟩-oriented [XY^-] centers have been reported, including [IBr^-] in KBr:I[60] and [ICl^-] in KCl:I.[61] The latter is notable for the fact that its EPR disappears reversibly above 35°K. This is thought to arise from broadening due to the increased thermal motion of the hole among the 12 nearest-neighbor halide ions surrounding the iodine. Luminescence originating from the recombination of an electron with an [ICl^-] center has been observed.[62] The properties of this emission are generally quite similar to those of the intrinsic recombination luminescence.

It is interesting to note that fluorine-doped KCl and KBr apparently do not accommodate ⟨110⟩-oriented [XF^-] centers, at least at temperatures near or above 77°K.[20] Evidently, such centers are unstable under these circumstances because of the large electronegativity of the substitutional fluoride ion, which renders it a poor hole trap relative to the heavier halides. Further work at lower temperatures and in other alkali halides is needed to clarify the situation. The ⟨111⟩-oriented [XF^-] centers containing interstitial fluoride will be discussed in Section 4.5.

4. INTERSTITIAL-ASSOCIATED HOLE CENTERS

Although the [X_2^-] center is the simplest and most fundamental of the hole centers, its relative production efficiency under long-term irradiation is seldom high enough to enable it to predominate in pure halide crystals. Unless the crystal is deliberately doped with some electron-trapping impurity, electron–hole recombination severely limits the buildup of self-trapped holes. The principal hole centers arising from long-term irradiation involve interstitial halogen atoms. The production process is in some cases quite efficient and does not saturate rapidly with dose. The reader is referred to an earlier chapter for a description of the damage process itself. At low temperatures, the halogen atom is stable in a split-interstitial configuration known as the H center, which consists basically of an X_2^- molecular ion centered in a single halide-ion vacancy. At intermediate temperatures, generally somewhat below liquid nitrogen temperature, the interstitial becomes mobile and either annihilates at a vacancy or associates with another defect in the lattice. The prime example of such an association is the H_A or V_1 center. These centers evidently break up at higher temperatures, and the halogen interstitials combine to form more stable aggregates. Little is known about the nature of these complex hole centers, which appear to be generally nonparamagnetic. We shall not discuss this category except to mention, somewhat arbitrarily, the existence of work which indirectly associates an ultraviolet absorption in KCl and KBr, the H' band, with a diinterstitial configuration.[63]

4.1. The H Center

This center was first discovered through its optical absorption band in the near-ultraviolet.[64] Subsequent optical work on KCl and KBr showed it to be anisotropic, the ultraviolet transition having a $\langle 110 \rangle$ dipole moment.[65] The EPR investigations of Känzig and Woodruff provided additional information on the morphology of the center in three crystals, LiF, KCl, and KBr.[66] However, subsequent ENDOR work has shown the principal $\langle 110 \rangle$ interstitial in LiF to be associated with an alkali-ion impurity (specifically Na^+) and thus an extrinsic center.[67] The structure of the H center is illustrated in Fig. 12. The labeled ions are those identified from the LiF ENDOR data, the Na^+ impurity which is evidently necessary to stabilize this orientation in LiF being at B'. Although ENDOR measurements have not been made on KCl and KBr, a considerable body of indirect evidence suggests that the original intrinsic H-center model is correct for these crystals. We shall describe the EPR data on the two potassium salts only briefly, since the basic exposition is the same as that given in connection with the $[X_2^-]$ center.

The EPR spectra of the H center at 20°K are qualitatively similar to those of the $[X_2^-]$ except that each line is split into a group of lines which is itself a miniature reproduction of the $[X_2^-]$ spectrum.[66] Thus it is clear that although the hole is located mostly on two equivalent halide ions, it nevertheless spends a small fraction of its time on two other halide

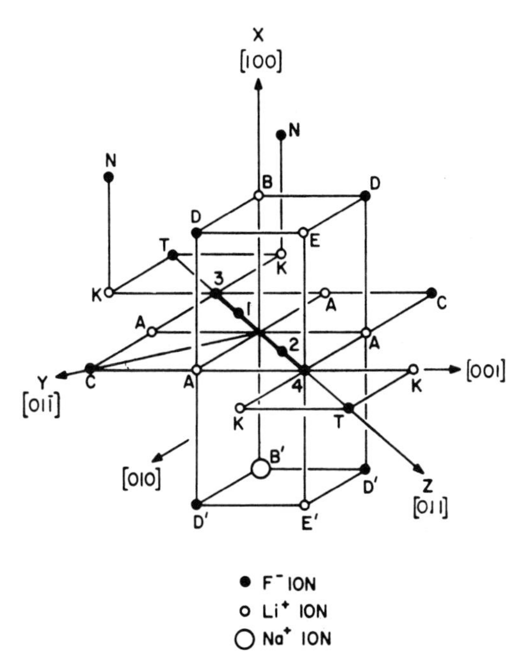

Fig. 12. The H (or H_A) center in an NaCl-type lattice (after Ref. 67).

ions which are equivalent to each other but distinct from the first two. The principal axes of the hyperfine tensors for the four constituent halogen nuclei coincide, and further quantitative analysis shows the four to be collinear. The overall symmetry is orthorhombic, as for the $[X_2^-]$ center. In the crystals investigated, the primary hyperfine splittings are slightly greater than those for the corresponding $[X_2^-]$ center (at least for $\theta < 70°$), indicating increased overlap of the central ions in the H center and thus a somewhat closer spacing than that for the $[X_2^-]$ center. The data are consistent with the contribution of atomic *ns* orbital on a principal halide ion being roughly 4% that of the *np* orbital on the same ion. This does not deviate markedly from the $[X_2^-]$ data in Table I. The shift of the spectroscopic splitting factor g is roughly one-half that for the $[X_2^-]$ center, again qualitatively consistent with a closer spacing. The hyperfine data imply a 4–10% probability of the hole being located on the outer two halide ions, the uncertainty arising in part from ambiguity in the sign of the relevant isotropic hyperfine interaction.

Since the perturbation arising from the third and fourth halide ions is small, the general features of the level scheme in Fig. 5 are adequate to characterize the optical absorption of the H as well as the $[X_2^-]$ center. Taking KCl as an example, the H center shows a strong band at 3.69 eV and a much weaker band at 2.38 eV.[68] Comparison with $[Cl_2^-]$-center energies in Table II suggests again a closer spacing of the molecular ion in the H center. Irradiation in either of these two bands at liquid helium temperature produces reorientation of the centers and dichroism in both bands. Simultaneous optical and EPR measurements have shown the higher energy transition to be strongly σ-polarized and the lower energy transition to be weakly σ-polarized. Evidence has recently been obtained for an additional transition further into the ultraviolet; in KBr, the peak occurs at 5.40 eV.[69] The preliminary identification of the source of this transition is based upon polarization, reorientation, and thermal conversion correlations. The electron configuration of this excited state is not known, but a Rydberg-like state comprising an electron bound weakly to a Br_2 molecule would appear to be a prime candidate. The possible existence of this type of state for the $[X_2^-]$ center has been mentioned in Section 2.5, and its probable observation in spectra of the $V_1(Li^+)$ center in KBr will be discussed in Section 4.3. For the H center, the ionization limit of this Rydberg-like sequence would consist of a free electron and a neutral X_2 molecule at the anion site. This X_2 center should be stable, and it was suggested that this is the center responsible for the V_1 band.[66,70] However, the V_1 center has since been identified, and no experimental observation of a substitutional halogen molecule appears to have been documented.

In contrast to the $[X_2^-]$ center, reorientation does not necessarily imply translation for the H center. This fact is apparent from the model (Fig. 12), and as an experimental example, the disorientation temperature for the H center in KCl is 11°K, whereas it does not begin to migrate below about 40°K.[68] Thus the center begins to rotate within its single vacancy well before the interstitial is able to escape to another site, probably by crowdion-like motion parallel to its axis. Furthermore, the lowness of the disorientation temperature (13°K in KBr)[70] gives general support to the view that the difference in energy between the $\langle 110 \rangle$ and $\langle 111 \rangle$ orientations is quite small. An elastically anisotropic defect with a low reorientation energy should produce a large paraelastic effect, which has in fact been observed for the H centers in KCl and KBr.[70] A manifestation of this effect is the appearance of dichroism in the H-center absorption bands when uniaxial stress is applied at a temperature slightly above the disorientation temperature. The static coupling to the lattice was measured and the reorientation kinetics were found to be of first order. The reorientation upon removal of the stress was found to obey an Arrhenius relation with a barrier energy of 3.1×10^{-2} eV in KCl and 3.7×10^{-2} in KBr. These numbers probably can be interpreted as a rough measure of the energy differences between the $\langle 110 \rangle$ and $\langle 111 \rangle$ orientations. Thermal activation was found to produce 60° reorientations, the situation being analogous to that described for the $[X_2^-]$ center.

4.2. *H*-Center Theory

Theoretical work on the *H* center has been concerned primarily with the stabilities of the several conceivable interstitial configurations. In particular, Dienes *et al.*[71] have investigated the relationship between the $\langle 110 \rangle$ and $\langle 111 \rangle$ orientations of a Cl_2^- molecular ion at a single anion site in KCl and NaCl. Their work predates but is actually quite similar to the work on the $[X_2^-]$ center described in Section 2.7. They, too, make use of the Gilbert–Wahl SCF-MO wave functions for free Cl_2^-, and again the model may be characterized as essentially a Cl_2^- molecular ion in a polarizable and deformable, but otherwise inert, ionic lattice.

The procedure followed was to compute the energies of the interstitial configurations as functions of the displacements of the 20 or so ions nearest the Cl_2^-, beginning with a simplified set of estimated displacements and minimizing the total energy with respect to each displacement in an iterative sequence. A numerical accuracy of approximately 0.01 eV was demanded. The halide-ion separation in the Cl_2^- was always included as one of the displacement parameters. The Coulomb and dipole–dipole energies were calculated in the usual way on the basis of a point-ion model; these energies are relatively small because of the overall neutrality of the center. The most important contributions arose from the repulsive interactions among the affected ions. Born–Mayer repulsive potentials were used, and the problem of repulsive parameters for the Cl_2^- itself was handled by assuming each constituent to be $Cl^{-1/2}$ and expeditiously choosing the parameters to be averages of those for the Cl^- ion and the Cl° atom. Although still somewhat arbitrary, this evaluation probably is superior to the previously mentioned approach of taking repulsive parameters for Cl_2^- to be the same as those for the Cl^- ion.[39] The assumptions made concerning the repulsive interactions are even more important for the *H* center because of the relative crowding, and for this reason, the absolute energies obtained are probably not very accurate. However, the investigators were interested primarily in the relative energies of the $\langle 110 \rangle$ and $\langle 111 \rangle$ orientations, and they found the differences between these to be relatively insensitive to the details of the potentials. The end result gave a lower energy for the $\langle 111 \rangle$ orientation by roughly 0.2 eV in both KCl and NaCl. This energy difference persisted under small arbitrary variations in the values assigned to parameters such as constants in the Born potentials or the polarizability of Cl^-.

In light of the previously mentioned experimental results which give indication that the ⟨111⟩ and ⟨110⟩ energies nearly coincide, this calculation would appear to be sound as far as it goes. With respect to the ⟨110⟩ center, however, the model could clearly be improved qualitatively by taking account of the spreading of the hole onto the third and fourth Cl^- ions, as observed in the EPR data. In principle, this spreading should be much less important for a ⟨111⟩ orientation. The net gain in energy from this additional weak covalent bonding is difficult to calculate accurately, but the 0.2 eV needed to give numerical agreement for KCl would appear to be within reasonable expectation. With respect to the observed ⟨111⟩ orientation of the intrinsic interstitial defect in LiF, one concludes that the added covalent bonding in the ⟨110⟩ crowdion fails to outweigh the ion-size effect. It may be noted that even for the ⟨111⟩ orientation, there is some indication in the hyperfine data that the hole is not completely localized on the F_2^-.[72]

A Cl_2^- internuclear separation of about 2.47 Å was obtained for the *H* center in NaCl and 2.55 Å in KCl. These values may be compared with the 2.64 Å calculated for the free Cl_2^- molecular ion[39] and the 2.9 Å calculated for the $[Cl_2^-]$ center.[37] Thus the relative crowding which experiment seems to indicate for the *H* center is a logical result of the theory at its present stage.

The excited electronic states of the *H* center have not been investigated theoretically. The general relevance of the level diagram of Fig. 5 is clear, and the fact that the ultraviolet absorption energy for the *H* center is larger than that for the $[Cl_2^-]$ center in KCl and also in NaCl is at least superficially consistent with the expected smaller internuclear separation for the molecular ion in the *H*. However, that the actual physical situation cannot be this simple is immediately apparent when KBr and KI are compared, since the *H* and $[Br_2^-]$ bands nearly coincide in KBr but the absorption thought to be the *H* band in KI appears at an energy roughly 10% *lower* than that of the $[I_2^-]$ center.[73]

Although the location of the *H*-center hole or acceptor level within the band gap has not been calculated in detail, it is possible to make a rough estimate based on a variation of the well-known Born–Haber cycle.[74] Consider an initial state consisting of an *H* center and an electron in the conduction band and carry through the following sequence of operations: (1) Remove the electron to infinity, expending an energy equal to the work function φ of the crystal; (2) remove the halogen atom to infinity, gaining the formation energy of an H center relative to the perfect lattice; (3) combine the electron and atom, gaining the electron affinity of the halogen; (4) finally, place the resulting halide ion in the crystal in an interstitial position. The net energy of this cycle is the acceptor energy relative to the conduction band. For NaCl, photoelectric yield measurements give $\varphi = 0.9$ eV for step (1). The calculation discussed above gives roughly 2 eV for step (2), and the energy of step (3) is known to be 3.6 eV. A calculation by Tharmalingam[76] must be used for step (4), and this is probably the least accurate term because of the larger displacements involved. Using Born–Mayer–Verwey repulsive potentials, which are considered more reliable for this purpose because of their increased hardness relative to Born–Mayer potentials, an energy of about -1.6 eV was obtained. The relaxed acceptor level is thus 3.1 eV below the conduction band in NaCl; a similar estimate gives 3.9 eV for KCl. Note that the accuracy of this estimate is aided by the fact that steps (2) and (4) tend to cancel each other, and thus to some extent errors in the potentials should affect both calculations in the same direction. Therefore, one can be reasonably confident that the H-center hole level is somewhere near the middle of the gap. This placement is consistent with the fact that the principal absorption band does not appear to be perturbed by interaction with the valence band, a situation which obtained also for the $[X_2^-]$ center, as one may recall. The trapping of electrons or the release of holes by H centers has not been observed directly, but the probabilities for these processes might be expected to be quite small in comparison with those for $[X_2^-]$ centers.

It should be noted that, because of the strong relaxation effects, optical transitions which excite or free an electron trapped by an H center probably occur at energies substantially higher than the trapping energy estimated above. The estimate is thus not inconsistent with the energies observed for absorption bands of interstitial halide ions, 5.4 eV in KBr[77] and 6.4 eV in KCl.[78] Application of the Born–Haber cycle to the H center appears to have been made originally by Dienes and Smoluchowski.[79]

4.3. Impurity-Associated Interstitials

Irradiation of a typical potassium halide, KCl or KBr, at a temperature near that of liquid nitrogen produces another distinct near-ultraviolet absorption known for many years as the V_1 band. Speculation concerning the nature of the center responsible for this band had given rise to a variety of alternative models prior to the recent firm identification. All early models presumed the V_1 center to be intrinsic, but this has not proved to be the case. The V_1 band is actually a transition due to an H center located adjacent to a Na^+ impurity; it is designated herein as an H_{Na}.[80,81] It is apparent in retrospect that essentially all potassium halide crystals have contained sufficient sodium to produce a substantial V_1 band in the proper temperature range. Only specially treated KCl has thus far proved sufficiently pure to give no evidence of this center. By selective doping, one would expect to be able to produce a wide variety of distinct H_A centers. As will be mentioned presently, an initial step in this direction has been taken with the production and identification of the H_{Li} center in both KCl and KBr. As with other hole centers, the general applicability of the model is presently circumscribed by the fact that only two or three of the most popular crystals have been studied.

The difficulty in identifying the H_A center arose primarily from the elusive character of its EPR absorption. For several years, it had been known that the V_1 absorption depended on the sodium content in KCl.[82] Coupled with the earlier observations that H and H_A centers can be transformed into each other by appropriate optical excitation and annealing treatments,[83] this gave preliminary indication that the center incorporated both an interstitial halogen atom and a sodium impurity. An interpretable EPR spectrum was finally found by searching through the temperature range between liquid hydrogen and liquid nitrogen. It was found that in KCl below about 25°K, the resonance is severely saturated and that above about 45°K, the characteristic hyperfine structure transforms into a broad, structureless singlet due to motional averaging of the hyperfine interactions as the interstitial hops around the Na^+ ion.[80,81] This broad singlet resonance had been observed at liquid nitrogen temperature, but had proved of little value in revealing the structure of the center.

The arguments which establish the nature of the H_{Na} center from the available EPR and optical data are similar to those described in other examples and need not be repeated here. We shall simply point out some of the more important results. Although H_{Na} is basically a perturbed H center, the position of the Na^+ ion is such as to destroy the overall symmetry of the center, leaving only reflection symmetry through the (100) plane containing the X_2^- and the impurity. Figure 13 gives a schematic representation of the H_{Na} center in KCl.[80] All four halide nuclei sharing the hole are nonequivalent, and the axes of each of the individual hyperfine tensors deviate measurably from the crystallographic axes. The angle indicated for the axis of the two principal chlorine nuclei, labeled 1 and 2, is derived from the EPR data, but the displacements of the ions are only plausible estimates. Relative to the H center, the parallel components of the hyperfine

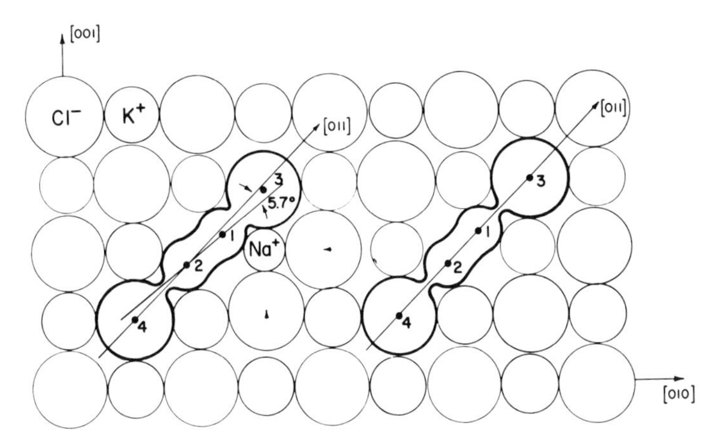

Fig. 13. The H_{Na} and H centers in KCl (after Ref. 84).

tensors (components approximately along the axis) for the H_{Na} are about 7% smaller for nucleus 2, 0.5% larger for nucleus 1, half as large for nucleus 4, and twice as large for nucleus 3. Thus there is a net shift of the hole toward, though not onto, the Na^+ ion. This shift very probably arises from the relatively reduced spacing of ions 1 and 3, brought about by the smaller size of the Na^+ ion. The linewidths for the H_{Na}-center EPR transitions are roughly twice as great as those for the H center, but, when considered along with comparable data for the H_{Li} center, this fact cannot be interpreted as indicating a significant interaction of the unpaired spin with the alkali impurity.[84]

The magnitudes of the hyperfine constants, g shifts, and optical transition energies all are consistent with a slightly larger spacing between ions 1 and 2 for the H_{Na} center than for the H center in KCl. The characteristic optical absorption bands for the Cl_2^- molecular ion are again observed: a relatively strong σ-polarized transition at 3.47 eV and a considerably weaker σ-polarized transition at 2.22 eV.[80] In addition to producing the usual optically induced reorientation effect, irradiation in the 3.47-eV band below about 40°K will free the interstitial halogen atom from the Na^+ ion and thereby convert the H_{Na} center into an H center.

A disorientation temperature of 16.8°K and a decay temperature of 113°K are characteristic of the H_{Na} center in KCl.[80] When compared with the H-center data, these parameters give evidence of the stabilizing effect of the sodium impurity. Two types of motion appear to contribute to the reorientation process, a crowdion-like jump of the Cl_2^- along [011], so that the Na^+ ion then lies between ions 2 and 4, and a rotation of the Cl_2^- about a single halide site among the four available H_{Na} orientations. Motional broadening of the EPR spectra with increasing temperature is quantitatively consistent with the measured reorientation rates.

The morphology of the H_{Li} center in KCl presents an interesting contrast to that of the H_{Na}, particularly in light of the previous discussions concerning the relative energies of $\langle 110 \rangle$ and $\langle 111 \rangle$ H-center orientations. The observed orientation of the Cl_2^- in H_{Li} is actually intermediate between [001] and [111], where the line connecting the halide-ion site and the Li^+ ion is taken to be [001]. Figure 14 shows the situation as determined from EPR data.[84] In this case, as well as with the H_{Na}, the stable arrangement would appear to be that which is most effective in relieving the compressive stress along the axis of the Cl_2^-. The EPR data give indication of a relatively slight spreading of the hole onto the halide ions 3 and 4 in Fig. 14, giving an overall nonplanar Y shape to the hole distribution rather than the distorted linear shape of the H_{Na} distribution.

The effects of different alkali-ion impurities on the stability of the H_A center are further illustrated by the optical measurements of Giuliani[85] on F-center production efficiencies in KBr irradiated at 80°K. At a nominal 1% doping level, Li^+ and Na^+ were found to increase the F-center production efficiencies relative to undoped KBr by factors of roughly three and two, respectively, an effect reasonably ascribed to interstitial stabilization. However, Rb^+-doped KBr showed no increase. Ion size considerations would lead one to expect a stable association with the larger cation only when

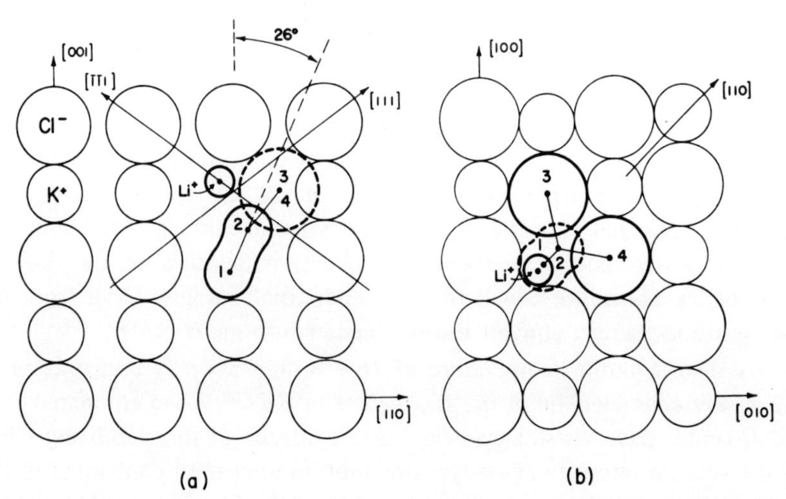

Fig. 14. The H_{Li} center in KCl. In (a), the plane of the diagram is (110); in (b), (001) (after Ref. 84).

Table IV. Optical, Paramagnetic, and Motional Parameters for Hole Centers Involving the Cl_2^- Molecular Ion in KCl

	E, eV	$W_{1/2}$, eV	Relative strength	$\langle A_{\parallel} \rangle_{1,2}$, gauss	T_d, °K	Decay temperature °K	Ref.
H	3.69	0.66	100	108.6	10.9	42	68, 80
	2.38	0.25	1				
H_{Na}	3.47	0.65	70	105.1	16.8	120	80
	2.22	0.32	1				
H_{Li}	4.23	0.7	9				
	3.50	0.70	35	101.5	23.5	240	84
	2.01	0.30	1				
$[Cl_2^-]$	3.40	0.81	100	101.3	170	—	26
	1.65	0.37	1				

located at the position B' in Fig. 12, the example being the previously mentioned H_{Na} center in LiF.[67] Evidently, a similar H_{Rb} center in KBr is simply not stable at 80°K.

It is of interest finally to compare properties of four well-characterized hole centers whose basic constituent is the Cl_2^- molecular ion. The necessary data are collected in Table IV. For convenience and brevity, the EPR data are represented by a parameter $\langle A_{\parallel} \rangle_{1,2}$, which is the arithmetic average over the two halide ions of the hyperfine tensor components most nearly parallel to the internuclear axis. These two components differ by roughly 10% for the H_{Na} and H_{Li} centers, but are identical by symmetry for the H and $[Cl_2^-]$ centers. The overall trends in the optical and hyperfine data are seen to be consistent with a small monotonic increase of internuclear separation from the H center at the near extreme to the $[Cl_2^-]$ center at the far extreme. The relatively small variation in the optical half-widths is consistent with the previous arguments whereby the broadening for the $[Cl_2^-]$ was attributed primarily to the internuclear vibrational mode of the Cl_2^- ion. The effectiveness of the alkali impurities in pinning the Cl_2^- is reflected in the variations of disorientation and decay temperatures.

Among the centers in Table IV, the H_{Li} is unique in displaying an additional higher optical absorption at 4.23 eV. This transition is σ-polarized, as are all transitions in this table, although not as strongly so as a near-ultraviolet transition. It has been suggested that this absorption is due to a transfer of the unpaired electron on the Cl_2^- to the Li^+ ion.[84] A transfer transition similar to this has already been mentioned in Section 4.1 as a possible source of the *H*-center absorption band in KBr at 5.40 eV. However, hole transfer from the Cl_2^- to the neighboring halide ions 3 and 4 in Fig. 14 would appear to be a likely alternative. On the basis of the rough considerations given previously, we can conclude only that the observed transition energy is probably not unreasonable for either of these possibilities.

4.4. Interstitial Halogen–Alkali-Vacancy Centers

At this point, it is of interest to consider the center comprising an interstitial halogen atom trapped by an alkali-ion vacancy. The form such a center might assume may be surmised with the help of Figs. 13 and 14 by simply visualizing the removal of the Na^+ or Li^+ ion. It would appear reasonable to expect the X_2^- to tip over into a [001] orientation (referring now to Fig. 14) with one end projecting into the vacancy. Repulsive interactions would be minimized in this way and, as we have seen, these are generally the interactions which influence most strongly the precise morphology of a given hole center.

The actual observation of this center, which we shall designate H_V, is a matter of question. EPR spectra due to $\langle 100 \rangle$-oriented X_2^- molecular ions have indeed been produced by irradiating KCl and KBr crystals containing alkaline earth impurities and compensating positive-ion vacancies.[86] In KCl, these centers are stable up to about 230°K.[87] The halogen nuclei were found to be nonequivalent, and new optical absorption was observed in the near-ultraviolet on the high-energy side of the *H* band. This center was interpreted in terms of a complex model involving both a divacancy and an alkaline earth ion. It was subsequently suggested that the data might be consistent with the simpler H_V model.[87] Note that the H_V center has a net negative charge, which would tend to complicate quantitative comparison with the corresponding *H* or H_A centers.

4.5. Heteronuclear Molecular Ions Involving Interstitials

In alkali chlorides and bromides containing small quantities of fluorine, a heteronuclear variety of *H* center can be produced by irradiation. This center has so far only been observed with fluorine as one of the constituents. It will be referred to here as interstitial FX^-. It was first observed in KCl containing about 0.5 mole % of KF,[88] and this FCl^- as well as interstitial FX^- species subsequently discovered occupy single halide-ion sites with their internuclear axes along $\langle 111 \rangle$ directions. EPR data on interstitial

FCl^-, FBr^-, and FI^- in various alkali halides have been thoroughly analyzed.[20] As might be anticipated, one of the results is that the hole density is greater on the less electronegative constituent; that is, for FX^-, the hole density on X increases monotonically through the sequence $X = $ F, Cl, Br, I. Those factors which are probably responsible for the observed $\langle 111 \rangle$ orientation have been mentioned in connection with the H center. As before, simple estimates are inadequate to predict whether $\langle 111 \rangle$ or $\langle 110 \rangle$ should be the stable orientation, and the necessary detailed calculations have not been carried out. The observed thermal stability of these mixed interstitial centers is generally substantially greater than that of the related homonuclear varieties. In KCl, the usual dominant optical absorption band appears for FCl^-, FBr^-, and FI^- at 4.14, 4.22, and 4.50 eV, respectively.

Unlike the homonuclear varieties, the FX^- molecular ion possesses a small dipole moment, and thus the internal vibrational mode should be infrared-active. Susman has identified the lowest infrared transition due to this mode for FCl^- in heavily irradiated KCl.[89] The transition is very weak; nevertheless, a net polarization parallel to the $\langle 111 \rangle$ directions, as it should be, was established. The isotope splitting due to ^{19}F–^{35}Cl and ^{19}F–^{37}Cl pairs was observed as a doublet with peaks at 403.7 and 399.8 cm^{-1}. The overall location of these peaks and their splitting are consistent with data for the neutral molecule in the gas phase, extrapolated on the basis that the presence of the antibonding electron reduces the binding energy by one-half. Considering the range of hole-center data covered thus far, one would not expect internal properties of an FX^- in and out of a crystalline matrix to differ markedly.

In this connection, the experimentally observed temperature dependence of the hyperfine components of various X_2^- and FX^- centers in alkali halides[90] deserve comment. This temperature dependence has been interpreted as arising from modulation of the electronic wave function by modes of vibration of the molecular ion. However, the dominant vibrational frequency for interstitial FCl^- in KCl measured by this method is an order of magnitude lower than the observed infrared frequency just mentioned. It would thus appear that in this case, the hyperfine data cannot be interpreted simply in terms of internuclear vibration of the molecular ion alone. Possibly other modes involving the lattice are effective in modulating the hyperfine components; however, low-frequency modes were searched for in the infrared absorption experiment but not found. On the positive side, the vibrational frequencies for $[Cl_2^-]$ and $[F_2^-]$ centers determined by the temperature variations of the hyperfine components do fall within a reasonable range of expectation.

Although infrared-inactive, the fundamental vibrational mode of $[X_2^-]$ should be Raman-active. A preliminary observation of a Raman peak at 235 cm^{-1} for the $[Cl_2^-]$ center in KCl has in fact been reported.[91] Furthermore, a Raman transition for a Cl_2^- molecular ion in a potassium borate glass matrix has been recorded at 265 cm^{-1}.[92] These are convincingly near the value of 210 cm^{-1} obtained from the hyperfine component variation[90] and the 260 cm^{-1} value for free Cl_2^- resulting from the SCF-MO calculations of Gilbert and Wahl.[39] For comparison, note that the vibrational frequency for the free Cl_2 molecule is 564.9 cm^{-1}.

5. HALIDE MOLECULAR IONS IN OTHER MATERIALS

Since extensive discussions of hole centers in other halide-containing materials would largely involve repeated application of physical principles already covered, we will at this point simply note the existence of some of these centers and refer the reader to the original literature for further data. Certain data are included in Tables I and II. No attempt will be made to cover all centers which have been studied, even though relatively little research has been carried out in this area.

As mentioned previously, in alkaline earth fluorides, the self-trapped hole has been observed through both its EPR and optical spectra.[13] In these crystals, the nearest-like-neighbor directions, and therefore the axes of the $[F_2^-]$ centers, are along $\langle 100 \rangle$ directions. The actual spectra are quite similar to those of alkali fluorides. Electron–hole recombination luminescence associated with the $[F_2^-]$ has also been observed.

Crystals of NH_4Cl and NH_4Br also support self-trapped holes under proper conditions.[10,11,29] The structure is that of CsCl (but with a slight tetragonal distortion in NH_4Br), and again the $[X_2^-]$ axes are along $\langle 100 \rangle$ directions. The similarities to the $[X_2^-]$ center in alkali halides with respect to both optical and EPR parameters are close indeed. An exception is the EPR linewidth, which is substantially greater in NH_4X due to interactions with the nearby protons. Electron–hole recombination luminescence has also been identified in NH_4X crystals.[46] A closely related, though structurally more complex crystal, PH_4I, has yielded EPR spectra of $[I_2^-]$ centers.[14]

Optical and EPR studies have identified the self-trapped hole in irradiated $KMgF_3$, a crystal having the perovskite structure.[15] This structure imposes a readily identifiable bent-bond configuration upon the F_2^- molecular ions. F_2^- has also been observed by EPR in the bifluorides KHF_2, $NaHF_2$, and NH_4HF_2.[93] The linear FHF^- molecular ion is evidently converted to F_2^- upon the ejection of the hydrogen atom by electron irradiation; gamma-ray irradiation is much less effective in producing the center. Again, the characteristics of the F_2^- seem little affected by the details of its environment.

Single crystals of hydroxylamine hydrochloride, NH_3OHCl, have been found to exhibit the characteristic [Cl_2^-]-center EPR spectra upon gamma-ray irradiation at 77°K.[94] These centers appear to be self-trapped holes rather than interstitials, but the complexity of the crystal structure renders this conclusion somewhat speculative.

The H center has been observed in BaF_2 and SrF_2 crystals through EPR measurements.[95] In this case, the fluoride interstitial enters the material as compensation for trivalent rare-earth dopants. Radiation then causes hole trapping at the interstitial ion, thereby producing the F_2^- molecular ion. Its axis is along a $\langle 111 \rangle$ direction, which is the most reasonable orientation in the fluorite lattice.

A halide center which is qualitatively distinct from the others we have discussed has been observed in NH_4Cl and NH_4Br crystals.[96,29] This is the NH_3X radical. Evidently, the radiation dislodges a hydrogen atom from an NH_4^+ ion, and a bond then forms between the NH_3^+ and one of its eight halide-ion neighbors. There is substantial amplitude of the hole wave function on both the nitrogen and the halogen. The electronic energy-level scheme is very similar to that of a heteronuclear XY^- molecular ion. In the ground state, the hole evidently occupies an antibonding orbital comprising mostly nitrogen $2p_z$ and chlorine $3p_z$ (or bromine $4p_z$) orbitals. Above 80°K, the EPR spectra undergo broadening and, ultimately, motional averaging due apparently to tumbling of the NH_3^+ ion from one corner of its halide-ion cage to another. A broad absorption band at 3.31 eV in NH_4Cl and 3.14 eV in NH_4Br has been assigned to a charge-transfer transition within the NH_4X radical, in close analogy with the strong ${}^2\Sigma_g{}^+ \leftarrow {}^2\Sigma_u{}^+$ transition of the X_2^-.

Let us finally note an observation of a Cl_2^- molecular ion in an amorphous matrix, that is, in alkali borate glasses doped with alkali halide.[97] Relative to a self-trapped hole in KCl, the halogen hyperfine interaction and the ultraviolet absorption energy are both slightly higher, indicative of a slightly decreased halogen spacing in the glass. Nothing is presently known about the site which the Cl_2^- occupies in the glass.

ACKNOWLEDGMENTS

The author wishes to thank Dr. F. W. Patten for informative discussions and for a critical reading of the manuscript.

REFERENCES

1. E. Mollwo, *Nachr. Akad. Wiss. Göttingen* **1**, 215 (1935); *Ann. Physik* **29**, 394 (1937).
2. J. P. Molnar, Thesis (unpublished), M.I.T. (1940).
3. F. Seitz, *Revs. Mod. Phys.* **26**, 7 (1954) and references therein.
4. J. H. Schulman and W. D. Compton, *Color Centers in Solids* (Pergamon Press, New York, 1962).
5. R. S. Knox and K. J. Teegarden, in *Physics of Color Centers*, Ed. by W. B. Fowler (Academic Press, New York, 1968), p. 20.
6. A. B. Kunz, *J. Phys. Chem. Solids* **31**, 265 (1970).
7. A. B. Kunz, *Phys. Rev.* **175**, 1147 (1968).
8. L. J. Page and E. H. Hygh, *Phys. Rev.* **B1**, 3472 (1970).
9. T. G. Castner and W. Känzig, *J. Phys. Chem. Solids* **3**, 178 (1957).
10. H. R. Zeller, L. Vannotti, and W. Känzig, *Phys. Kondens. Materie* **2**, 133 (1964).
11. F. W. Patten and M. J. Marrone, *Phys. Rev.* **142**, 513 (1966).
12. W. Hayes and J. W. Twidell, *Proc. Phys. Soc.* **79**, 1295 (1962).
13. J. H. Beaumont, W. Hayes, D. L. Kir, K., and G. P. Summers, *Proc. Roy. Soc. London* **A315**, 69 (1970).
14. C. L. Marquardt, *J. Chem. Phys.* **48**, 994 (1968).
15. T. P. P. Hall, *Brit. J. Appl. Phys.* **17**, 1011 (1966).
16. H. Seidel and H. C. Wolf, in *Physics of Color Centers*, Ed. by W. B. Fowler (Academic Press, New York, 1968), p. 537.
17. C. P. Slichter, *Principles of Magnetic Resonance* (Harper and Row, New York, 1963).
18. C. E. Bailey, *Phys. Rev.* **136**, A1311 (1964).
19. T. O. Woodruff and W. Känzig, *J. Phys. Chem. Solids* **5**, 268 (1958).
20. D. Schoemaker, *Phys. Rev.* **149**, 693 (1966).
21. A. N. Jette, *Phys. Rev.* **184**, 604 (1969).
22. D. F. Daly and R. L. Mieher, *Phys. Rev. Letters* **19**, 637 (1967); *Phys. Rev.* **175**, 412 (1968).
23. R. Gazzinelli and R. L. Mieher, *Phys. Rev.* **175**, 395 (1968).
24. G. Herzberg, *Molecular Spectra and Molecular Structure, I. Spectra of Diatomic Molecules* (D. Van Nostrand, New York, 1950), 2nd ed.
25. C. J. Delbecq, B. Smaller, and P. H. Yuster, *Phys. Rev.* **111**, 1235 (1958).
26. C. J. Delbecq, W. Hayes, and P. H. Yuster, *Phys. Rev.* **121**, 1043 (1961).
27. G. D. Jones, *Phys. Rev.* **150**, 539 (1966).
28. D. Schoemaker, C. J. Delbecq, and P. H. Yuster, unpublished.
29. L. Vanotti, H. R. Zeller, K. Bachmann, and W. Känzig, *Phys. Kondens. Materie* **6**, 51 (1967).
30. R. B. Murray and F. J. Keller, *Phys. Rev.* **153**, 993 (1967).
31. C. J. Delbecq, *J. Phys. Chem. Solids* **22**, 323 (1961).
32. F. J. Keller and R. B. Murray, *Phys. Rev. Letters* **15**, 198 (1965); *Phys. Rev.* **150**, 670 (1966).
33. F. J. Keller, R. B. Murray, M. M. Abraham, and R. A. Weeks, *Phys. Rev.* **154**, 812 (1967).
34. R. F. Wood, *Phys. Rev.* **151**, 629 (1966).
35. R. B. Murray and P. G. Bethers, *Phys. Rev.* **177**, 1269 (1969).
36. T. P. Das, A. N. Jette, and R. S. Knox, *Phys. Rev.* **134**, A1079 (1964).
37. A. N. Jette, T. L. Gilbert, and T. P. Das, *Phys. Rev.* **184**, 884 (1969).

38. T. L. Gilbert, Lectures Notes for the NATO Summer School in Solid State Physics, Ghent, Belgium, 1966 (unpublished).
39. A. C. Wahl, P. J. Bertoncini, G. Das, and T. L. Gilbert, *Int. J. Quant. Chem.* **15**, 123 (1967); T. L. Gilbert and A. C. Wahl, to be published.
40. D. Ikenberry, A. N. Jette, and T. P. Das, *Phys. Rev.* **B1**, 2785 (1970).
41. S. D. Druger and R. S. Knox, *J. Chem. Phys.* **50**, 3143 (1969).
42. R. S. Song, *J. Phys. Chem. Solids* **31**, 1389 (1970).
43. F. J. Keller and F. W. Patten, *Solid State Commun.* **7**, 1603 (1969).
44. M. N. Kabler, *Phys. Rev.* **136**, A1296 (1964).
45. R. B. Murray and F. J. Keller, *Phys. Rev.* **137**, A942 (1965).
46. M. J. Marrone and M. N. Kabler, *Phys. Rev.* **176**, 1070 (1968).
47. R. A. Kink and G. G. Liid'ya, *Fiz. Tverd. Tela* **11**, 1641 (1969) [English translation: *Soviet Phys.—Solid State* **11**, 1331 (1969)].
48. D. Pooley and W. A. Runciman, *J. Phys. C* **3**, 1815 (1970).
49. M. N. Kabler and D. A. Patterson, *Phys. Rev. Letters* **19**, 652 (1967).
50. M. N. Kabler and D. A. Patterson, unpublished.
51. D. Fröhlich, unpublished.
52. R. G. Fuller, R. T. Williams, and M. N. Kabler, *Phys. Rev. Letters* **25**, 446 (1970).
53. D. Fröhlich and B. Staginnus, *Phys. Rev. Letters* **19**, 496 (1967).
54. W. B. Hadley, S. Polick, R. G. Kaufman, and H. N. Hersh, *J. Chem. Phys.* **45**, 2040 (1966); C. J. Delbecq, A. K. Ghosh and P. H. Yuster, *Phys. Rev.* **151**, 599 (1966); R. G. Kaufman, W. B. Hadley, and H. N. Hersh, *IEEE, Trans. Nuc. Sci.* **NS-17**, No. 3, 82 (1970).
55. W. Känzig, *Phys. Rev. Letters* **4**, 117 (1960); *J. Phys. Chem. Solids* **17**, 80 (1960).
56. D. Schoemaker, Intern. Symp. Color Centers Alkali Halides, Urbana, Ill. (1965), unpublished.
57. I. L. Bass and R. L. Mieher, *Phys. Rev.* **175**, 421 (1968).
58. C. J. Delbecq, D. Schoemaker, and P. H. Yuster, *Phys. Rev.* **B3**, 473 (1971).
59. C. J. Delbecq, D. Schoemaker, and P. H. Yuster, Intern. Symp. Color Centers in Alkali Halides, Rome, Italy (1968), unpublished.
60. D. Schoemaker, *ibid.*
61. M. L. Meistrich and L. S. Goldberg, *Solid State Commun.* **4**, 469 (1966).
62. L. S. Goldberg, *Phys. Rev.* **168**, 989 (1968).
63. N. Itoh and M. Saidoh, *Phys. Stat. Sol.* **33**, 649 (1969).
64. W. H. Duerig and J. J. Markham, *Phys. Rev.* **88**, 1043 (1952).
65. W. D. Compton and C. C. Klick, *Phys. Rev.* **110**, 349 (1958).
66. W. Känzig and T. O. Woodruff, *J. Phys. Chem. Solids* **9**, 70 (1958).
67. M. L. Dakss and R. L. Mieher, *Phys. Rev.* **187**, 1053 (1969).
68. C. J. Delbecq, J. L. Kolopus, E. L. Yasaitis, and P. H. Yuster, *Phys. Rev.* **154**, 866 (1967).
69. D. Y. Lecorgne, H. Peisl, and W. D. Compton, unpublished.
70. K. Bachmann and W. Känzig, *Phys. Kondens. Materie* **7**, 284 (1968).
71. G. J. Dienes, R. D. Hatcher, and R. Smoluchowski, *Phys. Rev.* **157**, 692 (1967).
72. Y. H. Chu and R. L. Mieher, *Phys. Rev.* **188**, 1311 (1969).
73. J. D. Konitzer and H. N. Hersh, *J. Phys. Chem. Solids* **27**, 771 (1966).
74. M. Born and K. Huang, *Dynamical Theory of Crystal Lattices* (Oxford University Press, London, 1954).
75. P. H. Metzger, *J. Phys. Chem. Solids* **26**, 1879 (1965).

76. K. Tharmalingam, *J. Phys. Chem. Solids* **25**, 255 (1964).
77. G. Kurz and W. Gebhardt, *Phys. Stat. Sol.* **7**, 351 (1964).
78. A. Behr, H. Peisl, and W. Waidelich, *Phys. Letters* **24A**, 379 (1967).
79. J. G. Dienes and R. Smoluchowski, *J. Phys. Chem. Solids* **27**, 611 (1966).
80. C. J. Delbecq, E. Hutchinson, D. Schoemaker, E. L. Yasaitis, and P. H. Yuster, *Phys. Rev.* **187**, 1103 (1969).
81. F. W. Patten and F. J. Keller, *Phys. Rev.* **187**, 1120 (1969).
82. C. J. Delbecq, Intern. Symp. Color Centers in Alkali Halides, Urbana, Ill. (1965), unpublished.
83. J. A. Cape, *Phys. Rev.* **122**, 18 (1961).
84. D. Schoemaker and J. L. Kolopus, *Phys. Rev.* **B2**, 1148 (1970).
85. G. Giuliani, *Solid State Commun.* **7**, 79 (1969).
86. W. Hayes and G. M. Nichols, *Phys. Rev.* **117**, 993 (1960).
87. J. H. Crawford, Jr., and C. M. Nelson, *Phys. Rev. Letters* **5**, 314 (1960).
88. J. W. Wilkins and J. R. Gabriel, *Phys. Rev.* **132**, 1950 (1963).
89. S. Susman, *Phys. Stat. Sol.* **37**, 561 (1970).
90. W. Dreybrodt, *Phys. Stat. Sol.* **21**, 99 (1967).
91. A. N. Jette, J. Bohandy, J. C. Murphy, D. S. O'Shea, and C. M. Wilson, *Phys. Letters* **31A**, 449 (1970).
92. M. Hass and D. L. Griscom, *J. Chem. Phys.* **51**, 5185 (1969).
93. L. J. Vande Kieft and O. R. Gilliam, *Phys. Rev.* **B1**, 2015 (1970).
94. H. Ohigashi and Y. Kurita, *J. Phys. Soc. Japan* **24**, 654 (1968).
95. A. Tzalmona and P. S. Pershan, *Phys. Rev.* **182**, 906 (1969).
96. F. W. Patten, *Phys. Rev.* **175**, 1216 (1968); *Phys. Letters* **21**, 277 (1966).
97. D. L. Griscom, *J. Chem. Phys.* **51**, 5186 (1969); D. L. Griscom, P. C. Taylor, and P. J. Bray, *J. Chem. Phys.* **50**, 977 (1969).

Chapter 7

COLOR CENTERS IN SIMPLE OXIDES

A. E. Hughes

Materials Development Division
Atomic Energy Research Establishment—Harwell
Berkshire, England

and

B. Henderson*

Department of Physics
University of Keele
Staffordshire, England

1. INTRODUCTION

"Simple oxides" is a vague term, and the reader of this chapter may well be excused for remarking at the end of his labors that nothing in these oxides appears simple. Furthermore, even if the oxides are simple, the color centers surely are not. What we really mean by this expression in our title is that we intend to confine our attention to those oxygen-containing materials in which point defect phenomena are sufficiently well understood that we can present a substantial discussion. This means that nearly all of what we have to say will concern the alkaline earth oxides, simply because these materials have received most attention and are now in a position where the seeds of much hard work throughout the last twenty years are producing fruit in the form of well-categorized color centers.

* On leave during 1970 at Oak Ridge National Laboratory, Oak Ridge, Tennessee.

The field of defects in the alkaline earth oxides has recently been reviewed in detail by Henderson and Wertz.[1] In order to take a fresh approach in our present review, we have decided to concentrate on some specific features of the subject, in the hope that by doing so, we can augment rather than duplicate the previous work. We aim to concentrate on the electronic structure of color centers and their interactions with the host lattice and with themselves in the form of simple aggregates. In particular, we will, wherever possible, compare and contrast the situation in the oxides with that in the alkali halides, since these materials have given much guidance in understanding the properties of oxide color centers. We also hope to show how, now that more color centers are being identified in the oxides, certain interesting features not apparent in the alkali halides are contributing to our overall understanding of the physics of point defects. In order to remind ourselves that the alkaline earth oxides are not the only "simple oxides," comparative results on other materials will be presented wherever possible. In most cases, however, these sections will of necessity be short.

The alkaline earth oxides MgO, CaO, SrO, and BaO are the divalent cousins of the alkali halides. They have the face-centered cubic rocksalt structure (the remaining compound BeO, is hexagonal), and their binding is dominantly ionic. Their optical properties in perfect form are therefore typical of ionic crystals: They have electronic band gaps of several electron-volts, and show *reststrahl* absorption in the infrared. Relatively little work has been done on the band structure and lattice dynamics, but we give a survey of basic crystal data in Table I. It is certainly not unprecedented that defect studies are progressing in parallel or even in advance of those on

Table I. Crystal Data on FCC Alkaline Earth Oxides[a]

	MgO	CaO	SrO	BaO
Band gap, eV	7.3,[2] 8.7,[3] 7.8[5]	6–7,[4] 7.0,[11] 7.7[12]	6[4]	4.0,[2] 4.8[4]
$k = 0$ Optical phonon frequency, cm^{-1} TO	400,[6] 393[7]	295[9]	227[9]	?[b]
$k = 0$ Optical phonon frequency, cm^{-1} LO	730,[6] 707[8]	570[9]	480[9]	?[b]
ε_0	9.65,[4] 9.8[6]	11.8,[4] 11.1[9]	13.3,[4] 13.1[9]	34[4,4a,c]
ε_∞	2.95[4,6]	3.33[9]	3.46[9a]	3.83[10]
Melting temperature, °C	2800[4]	2600[4]	2415[4]	1928[4]

[a] Original references are given as superscripts.
[b] Parodi[9b] has reported infrared absorption at 189 cm^{-1} in BaO.
[c] A value of 14.4 for ε_0 is often quoted for BaO. This value is apparently without real foundation, being extrapolated from the values in the other oxides.[4a]

the perfect lattice, and we may cite the alkali halides as examples where this has occurred as a result of the natural desire of physicists to advance their knowledge on a broad front.

The alkaline earth oxides have lagged behind the alkali halides in fundamental research because they present more severe materials problems. Their melting points are between 1900 and 3000°C, and consequently they are difficult to grow as single crystals. Such high melting points rule out the commonly used techniques of crystal growth (e.g., Kyropoulos or Stockbarger growth) and purification (zone-refining), which have been so successful in the alkali halides and semiconductors. Crystal growth in the alkaline earth oxides is usually by the arc-fusion method. An arc is struck between carbon electrodes embedded in a mass of several kilograms of oxide or carbonate powder. The arc melts the powder and the slow cooling of the large mass when the arc is switched off produces single crystals of varying size. To prepare crystals of high purity, it is important to start with high-grade powders. Single crystals of several cubic centimeters are obtainable with MgO, and rather smaller ones of up to centimeter dimensions of CaO, SrO, and BaO. Small MgO crystals of millimeter dimensions can be grown from a lead molybdate or lead fluoride flux,[13] a technique that has applications for growing isotopically enriched crystals when only small amounts of material are available. Boules of CaO, SrO, and BaO have been grown by flame fusion (Verneuil method) using a plasma torch,[14] but the technique is a very difficult one and so far has not found widespread application. Crystals produced in this way tend to be rather badly strained due to the high cooling rate.

Through careful choice of starting materials, single crystals of oxides grown by arc-fusion are now available commercially with relatively low impurity levels. Some typical analyses are given by Henderson and Wertz[1] in their Table 1. The most common impurities are Fe, Si, Al, Zn, and foreign alkaline earths, all of which may be around the 10–50 ppm level. Other metal ions tend to be present at rather lower levels. It appears that transition metal ions tend to go into solution in either the 2+ or 3+ valence state, and in fact the oxides have long been used as hosts for studying the crystal field properties of these ions.[1] Small concentrations of impurities have profound effects on the types of color center found in the oxides, since the occurrence of tripositive cations or uninegative anions results in the presence of charge-compensating cation vacancies. These vacancies give rise to a wide variety of trapped-hole centers (V centers) which may be formed by ionizing radiation. Transition metal impurities themselves may also act as electron or hole traps, and therefore play an important role in determining the electronic structure of intrinsic point defects which may be present at comparable or lower concentrations. The valence state of transition metal impurities may be changed by heat treatments in oxidizing or reducing atmospheres, which has often been used to separate out impurity effects. Hydroxyl ions may also be present and can be detected through infrared absorption near 3 μm. Other types of impurities are often found to precipitate onto dislocation lines during cooling, and hence have less effect on color center properties. Large single crystals grown by arc-fusion often have a cloudy appearance, probably due to the presence of voids and precipitates.[15] Annealing in vacuum at high temperatures (1800–2000°C) usually renders thin, cleaved platelets perfectly transparent.

New defects or color centers may be introduced into the crystals by most of the traditional methods. As we have pointed out, ionizing radiation is useful in producing *V* centers and has some application in changing the charge state of electron-trapping centers (see Section 4, for example). However, in contrast to the alkali halides, low-energy ionizing radiation does not produce ionic displacement, so that to produce new vacancies (or interstitials), it is necessary to use particle irradiation. The damage is thus by knock-on collisions, and, insofar as detailed studies have been made, it shows all the characteristics of such processes[16]: an energy threshold, linear intensity dependence, and insensitivity to impurities. Neutron irradiation has been most widely used,[1] but recently, electron and proton irradiations have been successfully exploited.[16,17] They have the advantage of more controlled energy and dose measurements than reactor irradiations with neutrons. Additive coloration may be used to produce *F*-type centers by heating in metal vapor,[18] but so far, attempts to create cationic defects by heating in oxygen have not been successful; the principal result is the change of impurity valence states.[1] High temperatures are naturally required for additive coloration, which makes the technique more difficult than in the alkali halides. Arc-fusion growth itself sometimes produces "accidentally" colored portions of the melt. This seems to apply particularly to CaO (e.g., Section 4), where even in uncolored portions of the melt, it is sometimes possible to detect the presence of color centers by EPR or optical methods. This wide variety of methods for producing color centers is of great assistance in identifying different species.

The general types of color centers which have been identified are similar to those familiar from work in the alkali halides. However, the divalent nature of the oxide lattice does make possible a wider range of defect types than occurs in the alkali halides. A word on notation is consequently appropriate. Even in the alkali halides, a proliferation of different notations exists to describe the various defect types, such that much confusion is apparent in the literature. The need has clearly arisen for a rationalized system of describing defects: If such a system is applicable to insulating crystals in general, so much the better. Such a scheme has been proposed by Sonder and Sibley earlier in this volume, and we adhere to their notation. Although this may lead to some confusion, we have tried to eliminate this by giving the old notation in parentheses where the defect is first mentioned in detail. In any event, the rules of the new notation enable one to work out what is meant with a minimum of mental gymnastics. In the alkaline earth oxides, two species of simple anionic defects, the F^+ and F centers, have been widely studied. These are analogs of the F and F^- centers in the alkali halides. In addition, several other types of centers have been conclusively or speculatively identified. These include possible analogues of F_2 (M) and F_3 (R) centers, as well as other simple aggregates of F-type centers which as yet have not been found in the alkali halides. This latter class may be peculiar to the alkaline earth oxides because of the participation of cation vacancies. The V centers differ from those observed in the alkali halides; nevertheless, the notation scheme is easily adapted to include them. For a detailed portrayal of these defects, the reader is referred to Fig. 20 of this chapter and Table I of the chapter by Sonder and Sibley.

The plan of this chapter is as follows. Sections 2 and 3 discuss the magnetic and optical properties of the F^+ center: one electron trapped at an oxygen vacancy. Evidence for the observation of an oxygen vacancy with two trapped electrons, the F center, is presented in Section 4, along with a description of its optical properties. Centers identified principally by EPR as either perturbed F^+ centers or V-type centers are described in Sections 5 and 6. Section 7 covers the various defects studied primarily through uniaxial stress and magnetooptical experiments on zero-phonon lines, some of which have been assigned to particular types of aggregate centers. Finally, the magnetic properties of exchange-coupled pairs of defects (including impurities) are discussed in Section 8.

It is worth emphasizing at the outset that, as with most families of compounds, one (or a small group) of the compounds receives preferential treatment at first, and when ideas begin to develop, they are extended to the other members of the family. In the case of the alkaline earth oxides, the historical pecking order is MgO, CaO, SrO, and BaO, so that the amount of work done on each system varies considerably. There are, of course, fluctuations from this order; for example, BaO received some early attention because of its application in oxide-coated cathodes. However, by and large, the ordering holds good as far as color centers are concerned.

The choice, naturally, is not arbitrary, but is determined by the availability of good samples. In the future, we can expect that the less accessible systems will receive more attention, but the immediate effect on our presentation is that we shall have more to say about some compounds than others.

2. MAGNETIC PROPERTIES OF F^+ CENTERS

Unlike the situation in the alkali halides, where the *F* optical absorption band can be considered as the founder member of the color center field, the optical properties of the *F*-type centers in the alkaline earth oxides have only recently been studied successfully. The F^+ center was first identified as a clear species of defect using EPR techniques, not only in oxides, but also in sulfides, selenides, etc. The observation of clearly identifiable optical bands occurred much later. Consequently it is pertinent to open our discussion with a description of the magnetic resonance properties of F^+ centers, first pausing briefly to outline the principles used to analyze the observed spectra.

2.1. Principles of Magnetic Resonance Methods Applied to *F*-Center Problems

The ground state of an *F*-like center containing a single electron is orbitally nondegenerate with spin quantum number $s = \frac{1}{2}$. Consequently it is often referred to as a "1*s*" state, in analogy with the hydrogen atom. In the presence of a magnetic field **B**, there are two eigenstates for the electron, $s_z = \pm\frac{1}{2}$, between which electron paramagnetic resonance (EPR) transitions are induced by an alternating magnetic field of frequency $\nu = g\beta B/h$ applied perpendicular to **B**. (The constants are h, Planck's constant, β, the Bohr magneton, and g, the spectroscopic splitting factor.) The electronic wave function $\psi(1s)$, although symmetric about the vacancy center, is more diffuse than in atomic hydrogen and overlaps appreciably onto neighboring ions. Such wavefunction overlap causes fractional admixing of ion-core functions into the $|1s\rangle$ wave function. Spin–orbit coupling on the neighboring ions is then manifest in second-order perturbation theory as a shift in the g value for resonance absorption relative to that for free electrons. The electron also undergoes mutual magnetic interactions with neighboring nuclei which have nonzero nuclear spin; this results in the presence of hyperfine structure in the EPR spectrum, which affords the clearest identification of the structure of *F*-type centers.[19]

The EPR of F^+ centers is conveniently described by the spin-Hamiltonian

$$\mathcal{H} = \beta \mathbf{B} \cdot \mathsf{g} \cdot \mathbf{s} + \mathbf{I} \cdot \mathsf{A} \cdot \mathbf{s} - g_N \beta_N \mathbf{B} \cdot \mathbf{I} \tag{1}$$

in which β and β_N are electron and nuclear Bohr magnetons, $\mathbf{s}$ and $\mathbf{I}$ represent electron and nuclear spin operators, g_N is the nuclear g value for any neighboring isotope, and the electronic g factor is in general a tensor quantity. The term $\mathbf{I} \cdot \mathsf{A} \cdot \mathbf{s}$ splits each electronic level into $2I + 1$ hyperfine levels, EPR transitions between the levels being governed by the selection rule $\Delta m_s = \pm 1$, $\Delta m_I = 0$.

It is usual to write the hyperfine tensor A as the sum of a scalar a and a traceless tensor b. The scalar a is then referred to as the isotropic hyperfine interaction constant, and b describes the anisotropic hyperfine structure. Using first-order perturbation theory with the assumptions of high field ($g\beta B \gg a$) and small anisotropic interaction ($b_{ij} \ll a \pm g_N \beta_N B$), the eigenvalues of Eq. (1) are given by

$$E = m_s g \beta B - m_I g_N \beta_N B + m_s m_I (a + b_{z'z'}) \tag{2}$$

where z' is the field direction. For the simplest case of a nucleus on an axis (z) of axial symmetry, the elements of b referred to the principal axes x, y, z are $b_{xx} = b_{yy} = -b$, $b_{zz} = 2b$. Equation (2) may then be written

$$E = m_s g \beta B - m_I g_N \beta_N B + m_s m_I [a + b(3 \cos^2 \theta - 1)] \tag{3}$$

where θ is the angle between $\mathbf{B}$ (the z' axis) and the symmetry axis z. Resonant absorption for allowed EPR transitions then occurs at magnetic field values

$$h\nu / g\beta = B_0 = B + m_I [a + b(3 \cos^2 \theta - 1)] \tag{4}$$

where the constants a and b are now expressed in magnetic field units.

The magnitude of the hyperfine interaction is directly related to the extent of electronic overlap onto the near-neighbor ions. In the classical Fermi–Segrè form, the isotropic interaction is related to the square of the wave function at the nucleus, $|\psi(0)|^2$, by

$$a = (8/3) \pi g \beta g_N \beta_N |\psi(0)|^2 \tag{5}$$

In this expression, we use the CGS system of units such that $\beta = 0.92731 \times 10^{-20}$ erg/G, $\beta_N = 0.50504 \times 10^{-23}$ erg/G and $|\psi(0)|^2$ has units of cm^{-3} when a is measured in ergs. Most experimentalists quote the hyperfine interaction energies in units of frequency (MHz), wave number (cm^{-1}), or magnetic field (gauss G). The conversion factors into energy E expressed in ergs are: $E = 10^6 h \times (\text{MHz})$, $E = (hc) \times (\text{cm}^{-1})$ and $E = (g\beta) \times (\text{G})$. In the following discussion, the hyperfine interaction constants are quoted in MHz.

The form of the wave function $\psi(0)$ used in the calculation of a depends on the approximate description of the ground state of the F-type center. In a pseudopotential theory, such as the point-ion theory of Gourary and Adrian,[20] the wave function is written

$$\psi = N\Big\{\psi_F - \sum_i \psi_i \langle \psi_i^{*} \mid \psi_F \rangle \Big\} \tag{5a}$$

where ψ_F is a smoothly varying, vacancy-centered envelope function, and $\{\psi_i\}$ is the set of ion-core orbitals for the neighbors. N is a normalization factor, and ψ is orthogonalized to the ion-core wave functions ψ_i. If ψ is substituted in Eq. (5), Gourary and Adrian[20] have shown that the result is equivalent to replacing $\mid \psi(0) \mid^2$ by $A_R \mid \psi_F(R) \mid^2$, where A_R is an "amplification factor" for the nucleus at distance R from the vacancy center. Holton and Blum[21] have suggested that $A_R \propto Z^{3/2}$, Z being the atomic number of the nucleus concerned. The isotropic hyperfine interaction thus provides a measure of $\mid \psi_F(R) \mid^2$.

The anisotropic interaction is represented by a tensor b with zero trace, which depends upon the dipolar interaction between electronic and nuclear spins. For axial symmetry, the anisotropic constant b [see Eq. (3)], may be written as[20]:

$$b = \tfrac{1}{2} g\beta g_N \beta_N \langle \psi \mid [3\cos^2\alpha - 1]/\mid \mathbf{r} \mid^3 \mid \psi \rangle \tag{6}$$

where $\mathbf{r}$ is the electron–nuclear separation and α is the angle between $\mathbf{r}$ and the symmetry axis z. It is clear from the form of ψ in Eq. (5a) that several terms are involved in a proper calculation of b. Usually, the anisotropic interaction is determined mostly by the admixture of ion-core p orbitals into the ground-state wave function.[22] This part of b gives an approximate measure of $\mid \partial\psi_F/\partial R \mid^2$ at the nucleus.[20] The same matrix elements which determine this interaction are involved in the negative g shift of F centers relative to the free-electron g value[19] (see also Section 2.2).

Equation (4) describes, to first order only, the resonant fields when the hyperfine interaction involves only a single nucleus. However, around the anion site, there are 6 nearest-neighbor cations, 12 nearest neighbor anions, etc., all of which may be involved in the hyperfine interaction, depending upon the isotopic abundance of magnetic nuclei and the delocalization of the trapped electron. Thus, we rewrite Eq. (3) in the more general form

$$E = m_s g\beta B + \sum_n \{-M_I(n) g_N(n)\beta_N B + m_s M_I(n)[a(n) + b(n)(3\cos^2\theta_n - 1)]\} \quad (3a)$$

$\sum_n$ running over groups of equivalent nuclei. Thus Eq. (4) becomes

$$B_0 = B + \sum_n M_I(n)[a(n) + b(n)(3\cos^2\theta_n - 1)] \quad (4a)$$

where the total nuclear spin quantum number $M_I(n)$ of the nth shell of equivalent nuclei is given by

$$M_I(n) = \sum_{i=1}^{x} m_I(i) = xI, xI - 1, \ldots, -(xI - 1), -xI \quad (7)$$

x being the number of equivalent nuclei in the nth group. The intensity of each hyperfine component depends upon the number of ways of compounding the total nuclear quantum number, as well as the abundance of the isotopic species.

Since F-center wavefunctions overlap onto several shells of neighboring ions, the resulting hyperfine structure may be extremely complex, resulting in a broad, structureless line in unfavorable cases. In this situation, the ENDOR (electron–nuclear double-resonance) technique[19,23] becomes particularly useful, in view of its capability for resolving the hyperfine components of these inhomogeneously broadened lines. The ENDOR method detects the nuclear magnetic resonance (NMR) of the lattice nuclei which are near neighbors of the defect. It has become an extremely powerful tool in color center research, since it combines the inherent sensitivity of EPR with the narrow linewidths typical of NMR. The double-resonance effect is obtained via the desaturation of a partially saturated EPR transition.[19,23]

2.2. EPR Spectra of F^+ Centers

In the alkaline earth oxides, F^+ centers have particularly simple EPR spectra, since the near-neighbor ions have relatively low natural abundances of magnetic nuclides. For MgO, CaO, SrO, and BaO, although the cations have quite small abundances of magnetic nuclides, hyperfine splittings due to ^{25}Mg, ^{43}Ca, ^{87}Sr, ^{135}Ba, and ^{137}Ba have been observed in single-crystal and powder samples. The isotopic spin and natural abundances of these isotopes, which are needed to interpret the hyperfine structure, are given in Table II. The only naturally occurring oxygen isotope with $I > 0$,

Table II. Isotopic Abundances and Probabilities of Cation-Site Occupation around F^+ Centers in Oxides

Host	Nuclide	I	Natural abundance, %	P_0	P_1	P_2
MgO	^{25}Mg	5/2	10.05	0.52	0.356	0.10
CaO	^{43}Ca	7/2	0.13	0.9921	0.0077	0.00003
SrO	^{87}Sr	9/2	7.02	0.646	0.292	0.055
BaO	^{135}Ba	3/2	6.59	0.300	0.398	0.219
	^{137}Ba	3/2	11.32			

^{17}O ($I = 5/2$), is only 0.037% abundant. Thus, hyperfine interaction due to anion neighbors of the F^+ center may usually be neglected. The intensity of the hyperfine components in the EPR spectra is determined by the isotopic abundances and the probability that the near-neighbor sites are occupied by magnetic nuclei. The probabilities P_m for site occupation by a nucleus of fractional abundance x are given by the binomial expansion

$$P_m = \binom{n}{m} x^m (1 - x)^{n-m} \tag{8}$$

where n is the number of sites in a near-neighbor shell and m is the number of those sites occupied by magnetic nuclei. Values of P_0, P_1, and P_2 for MgO, CaO, SrO, and BaO are given in Table II; for BaO, we have not differentiated between ^{135}Ba and ^{137}Ba. It will be evident that in each case, the most intense line in the spectrum corresponds to that fraction of the F^+ centers with no magnetic nuclides. Also, Table II shows that the probability of observing hyperfine structure from F^+ centers in CaO is very low; nevertheless, such structure has been discerned even in the absence of isotopic enrichment.[24]

The first EPR identification of F^+ centers in the alkaline earth oxides was made by Wertz and his associates.[25] Since the spectra are essentially similar, we describe only one in detail, that resulting from F^+ centers in SrO. This spectrum is chosen because in addition to the normal features present in the spectrum, nuclear quadrupole interactions can be measured directly in EPR without resort to ENDOR.[26]

The EPR spectrum of F^+ centers in SrO for **B** directed along a [111] direction is shown in Fig. 1. The strong central line corresponds to $\Delta m_s = +\frac{1}{2} \leftrightarrow -\frac{1}{2}$ transitions from the 65% of F^+ centers that have only the even isotopes, ^{86}Sr and ^{88}Sr, as nearest cation neighbors. In this field orienta-

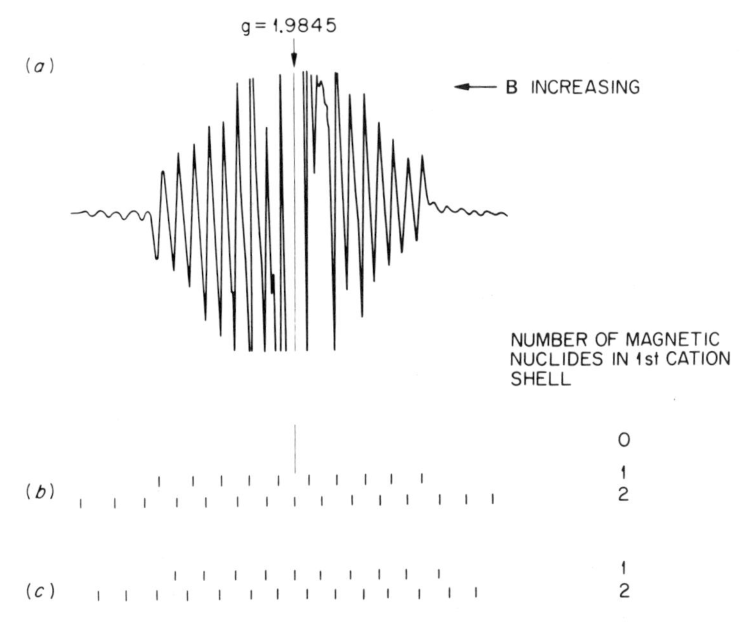

Fig. 1. (a) The EPR spectrum of F^+ centers in SrO for **B** ∥ [111]. The spectrum is reconstructed for F^+ centers with 0, 1, and 2 ^{87}Sr neighbors in the first shell, (b) referring to allowed transitions with $\Delta m_I = 0$ and (c) referring to forbidden transitions $\Delta m_I = \pm 1$ which become allowed by the strong quadrupole interaction. The separation between neighboring vertical lines in (b) and (c) is approximately 15 G. (After Culvahouse *et al.*[26])

tion, all nearest-neighbor ^{87}Sr nuclei in the nearest cation sites are equivalent, and 10 lines separated by a ($\approx$41 MHz, equivalent to 15 G), are observed in accord with the nuclear spin of $I = 9/2$. The total intensity of the indicated 10 lines is approximately 45% of that due to the single intense line, as expected for centers with one nearest cation site occupied by the 7% abundant ^{87}Sr isotope (Table II). A significant fraction, ~5%, of the F^+ centers will have two nearest-neighbor sites occupied by ^{87}Sr nuclides. For such centers, Eq. (7) leads to $M_I(n)$ values of 9, 8, 7, ..., 0, ..., −7, −8, −9. Consequently, there are 19 hyperfine lines from such centers, the splitting between neighboring lines being a. These lines have a total intensity of 8% of the central line, their individual intensities being in the ratio 1:2:3 ··· :10: ··· 3:2:1. Examination of Fig. 1 shows that outside the central region of the spectrum, there are lines with proper intensity and spacing for those centers having two ^{87}Sr nearest-neighbor nuclides. However, the lines which occur between the main hyperfine lines are too strong to be due to the remainder of such lines. These lines are identified as "forbidden" transitions in which m_I as well as m_s changes by one unit, in transitions between the states $| m_s - 1, m_I\rangle$ and $| m_s, m_I \pm 1\rangle$ of centers with one ^{87}Sr nearest neighbor.[26]

Orientation dependence studies[26] showed that the allowed lines are well accounted for by Eq. (4) with $g = 1.9845$, $a = -40.89 \pm 0.23$ MHz, and $b = -4.22 \pm 0.09$ MHz. However, the strong forbidden lines in the spectrum confirm that additional terms are needed in the spin-Hamiltonian. Culvahouse *et al.*[26] suggest that the only likely mechanism is an electric quadrupole interaction:

$$\mathscr{H}_Q = P[3(\mathbf{I} \cdot \mathbf{R}_1)(\mathbf{I} \cdot \mathbf{R}_1)R_1^{-2} - I(I+1)] \tag{9}$$

$\mathbf{R}_1$ being a radius vector directed from the vacancy center to the nearest-neighbor site, and $P = Qqe^2/4I(2I-1)$. This interaction introduces off-diagonal terms into the energies which couple states separated by a and $a/2$. It is reported that the intensities of the forbidden transitions shown in Fig. 1 for **B** parallel to [111] can be predicted with $Qqe^2/h \approx 60$ MHz. A precise measurement of P can be obtained from accurate measurements of the positions of allowed and forbidden lines; the best fit between the measured and predicted lines is obtained with $Qqe^2/h = 46 \pm 4$ MHz, $a = -40.70 \pm 0.23$ MHz, and $b = -4.30 \pm 0.09$ MHz.[26] These results are in close agreement with recent ENDOR measurements[27] (see Table III).

Table III. Spin-Hamiltonian Parameters for F^+ Centers

Host	g Value	Nuclide	a, MHz	b, MHz	P/h, MHz	Ref.
BeO	2.0030 (‖)	^{9}Be	−12.3[a]	−2.0[a]	0.092[a]	36
			−5.8[b]	−1.7[b]	0.125[b]	36
MgO	2.0023	^{25}Mg	−11.03	−1.33	+0.141	25, 28
CaO	2.0001	^{43}Ca	−25.66	−2.71	—	24
SrO	1.9845	^{87}Sr	−40.775	−4.279	+0.333	26, 27
BaO	1.9355	^{135}Ba	184.7	18.4	3.4	31
		^{137}Ba	206.9	20.3	4.8	31
ZnO[c]	1.9948 (‖)	^{67}Zn	55.6	—	—	37
	1.9963 (⊥)	^{67}Zn	77.8	—	—	37
BaS	1.9637	^{135}Ba	99.17	10.31	2.83	27, 33
		^{137}Ba	110.27	11.83	4.365	27, 33

[a] Values for basal plane Be nuclide nearest to F^+ center.
[b] Values for c-axis Be nuclide near F^+ center.
[c] Orientation dependence of hfs not specifically reported.

Table IV. The g Shifts and $|\psi_F(R_1)|^2 R_1^3$ for F^+ Centers

	b_i^{*},[a] MHz	$\Delta g \times 10^3$			$\lvert \psi_F(R_1) \rvert^2 R_1^3$	
		Calc. [Eq. (10)]	Calc. (Ref. 32)[b]	Expt.	Theory	Expt.
MgO	−0.81	−0.9	+2.9	0.0	1.0	1.0
CaO	−2.34	−5.1	+4.9	− 2.2	1.0	1.47
SrO	−4.07	−32.4	+5.4	−17.8	1.0	1.68
^{135}BaO	18.03	−138.2	+6.0	−66.8	1.0	2.38
^{137}BaO	19.88	−135.9				
^{135}BaS	—	—	—	—	1.0	1.89
^{137}BaS	—	—	—	—		

[a] b_i^{*} is that part of the anisotropic hyperfine constant due to overlap onto neighboring cation p orbitals (see text).
[b] Contribution from charge-transfer configurations only.

Quadrupole effects have also been observed in the ENDOR spectrum of F^+ centers in MgO.[27] In both oxides, the magnitude of the quadrupolar effects is consistent with an outward displacement of ions in the nearest cation shell of 0.05–0.08 of the interionic separation. This is consistent with the theoretical computations of Kemp and Neeley[29] (Section 3.2).

The EPR spectra of F^+ centers in MgO, CaO, SrO, and BaO single crystals have been reported, as have also the ENDOR spectra of MgO and SrO. In general, the hyperfine structure for the first shell of cations only has been investigated, although in the case of MgO, ENDOR transitions from the third and fifth shells of neighboring ions have been identified.[30] The results for the first shell are summarized in Tables III and IV. In each case, there is a negative g shift, compared with the free-electron g value, which increases through the series MgO to BaO. As noted previously, this g shift arises out of the admixture of orbital momentum into the $|1s\rangle$ ground state, consequent upon the electron overlap onto the nearest cation neighbors. The major contribution to this is from ion-core p orbitals. According to perturbation theory, the spin–orbit interaction in the field of the neighboring ions leads to a g shift proportional to

$$\sum_i \langle \psi \mid Z_i \cdot L_i^2/r^3 \mid \psi \rangle / E_F$$

where E_F is the optical F^+-band transition energy (Section 3), Z_i is the effective nuclear charge of the ith ion, r is the electron–nuclear separation, and L_i is the orbital angular momentum operator relative to the ith nucleus. Since good wave functions for the F^+ centers are not known, accurate calculations of the g shift are not possible. However, that part of the anisotropic hyperfine constant b resulting from p-function overlap, written b^*, is determined by the same matrix element as the p-orbital contribution to the g shift. Hence it may be shown that[19]

$$\Delta g = -\frac{20}{3} \frac{\beta}{\beta_N} \frac{1}{gE_F} \sum_i \frac{Z_i b_i^*}{g_{N_i}} \tag{10}$$

where g_{N_i} is the nuclear g value of the ith nucleus. In calculating b^* from the measured value of b, allowance should be made for contributions to b from terms in the matrix element of Eq. (6) other than those due to p-orbital overlap. The approach usually adopted is to subtract from b the interaction between the point dipole of an electron at the center of the vacancy and the nucleus at site i, the "classical" or "direct" dipolar term. This is only an approximate procedure, but has been followed here. The appropriate values of b_i^* are given in Table IV, where, by comparison with Table III, it can be seen that b^* is indeed the major term in b. Table IV also compares the Δg values calculated using Eq. (10) with the experimental Δg values. Evidently, the increasingly negative g shift in going to alkaline earths with higher atomic weight is qualitatively accounted for in terms of Eq. (10). The large discrepancies between experimental and calculated values of Δg for SrO and BaO have not really been satisfactorily explained. Configuration mixing of charge-transfer states from the next-nearest-neighbor anions may be important. Such admixture results in a positive contribution to the g shift[32,32a] (see Table IV). The magnitude of the positive contribution is of the right order for MgO and CaO, but in SrO and BaO, a large discrepancy still exists. Hagston[32a] has suggested ways of improving the situation within the framework of the charge-transfer model by allowing holes to have large g shifts, but this speculation is unconfirmed. An alternative explanation[31] is that contributions to Δg from higher excited states of the F^+ center effectively increase the value of E_F in Eq. (10), thus reducing the negative contribution to Δg. It appears that a more elaborate theory, which takes account of the structure of all the ions and states involved, is required before a good quantitative understanding of the g shifts is achieved. For F^+ centers in the alkaline earth selenides, tellurides, and sulfides, configurational mixing of charge-transfer states seems to be the dominant mechanism in the g shift, since in all cases examined, Δg is positive.[25]

Table III shows that the isotropic hyperfine term, which is related to the square of the wave function at the cation nucleus by Eq. (5), also increases on going from ^{25}Mg to ^{137}Ba. The derived values of the electron densities are given in Table V. The term $|\psi(0)|^2$ has been calculated directly from Eq. (5), and shows the same trends as the hyperfine constant a. In order to calculate $|\psi_F(R)|^2$ from $|\psi(0)|^2$, it is necessary to have values for the amplification factors A_R. These are not known precisely, but we have made an approximate choice by assuming the $Z^{3/2}$ variation (Section 2.1) and extrapolating from the values given for cations and anions in alkali halides.[19,20] The chosen values of A_R and the derived values of $|\psi_F(R)|^2$ are also given in Table V. Note that $|\psi_F(R)|^2$ does not vary greatly across the oxide series, in contrast to $|\psi(0)|^2$ and a. By using the point-ion lattice wave functions of Gourary and Adrian,[20] it can be shown that $|\psi_F(R_1)|^2$ is approximately proportional to R_1^{-3}, so that the quantity $|\psi_F(R_1)|^2 R_1{}^3$ should be nearly constant across the series.* The last column in Table IV shows the result of this analysis using the results of Table V, the data being normalized to unity in MgO. There is a clear trend upward from unity with increasing cation size. This may indicate some inadequacy of the point-ion model for the more covalently bonded oxides, although the value for BaS does not fit in well with this point of view. Perhaps not too much weight should be put on these figures, however, since the constancy of $|\psi_F(R_1)|^2 R_1{}^3$ and the choice of the amplification factors are both only approximate. Our results in Table IV differ slightly from those given by Tench and Nelson,[34] probably because their values of a and A_R are different from the ones we have chosen. Despite these uncertainties in the precise values of the various parameters, this analysis of the isotropic hyperfine interaction does bring out some of the expected features. There is room, however, for considerable improvement in the theoretical aspects of the problem.

Hyperfine structure from the more remote shells of nuclei has been observed for F^+ centers in CaO.[24] In this case, crystals taken from the melt were found to contain F^+ centers whose EPR linewidth is not more than 15 mG. With the increased sensitivity available, the possibility of observing hyperfine structure due to ^{43}Ca (0.13% abundant) and ^{17}O (0.037% abundant) is considerably enhanced. The single-crystal data for ^{43}Ca are given in Tables III–V. For ^{17}O, Henderson and Tomlinson[24] find $a = -2.435$ MHz; consequently $|\psi_F(R)|^2 = 0.077\times 10^{21}$ cm^{-3} (see Table V). If we compare this value of $|\psi_F(R)|^2$ with that of ^{19}F for F centers in KF,[35] where $|\psi_F(R)|^2 = 0.163\times 10^{21}$ cm^{-3}, it is evident that there is less overlap

* Reference 26 is in error on this point, as shown by Tench and Nelson.[34]

Table V. Comparison of F Centers in Alkali Halides with F^+ Centers in Alkaline Earth Oxides

Host	g-Value	First cation shell				First anion shell			
		a, MHz	$\lvert\psi(0)\rvert^2\times 10^{-24}$ cm^{-3}	A_R	$\lvert\psi_F(R)\rvert^2\times 10^{-21}$ cm^{-3}	a, MHz	$\lvert\psi(0)\rvert^2\times 10^{-24}$ cm^{-3}	A_R	$\lvert\psi_F(R)\rvert^2\times 10^{-21}$ cm^{-3}
^{7}LiF	2.0018	39.06	0.15	57	2.63	105.94	0.170	350	0.486
BeO[a]	2.0030	−12.3	0.13	65	2.00	—	—	—	—
NaF	2.0001	107.0	0.61	260	2.35	96.8	0.155	350	0.443
MgO	2.0023	−11.03	0.278	325	0.84	—	—	—	—
^{39}KF	1.9964	34.3	1.11	650	1.70	35.5	0.057	350	0.163
CaO	2.0001	−25.66	0.58	700	0.83	−2.43	0.027	350	0.077
^{85}RbCl	1.9804	98.0	1.53	1750	0.87	5.79	0.089	1500	0.059
SrO	1.9845	−40.775	1.42	1850	0.77	—	—	—	—
^{135}BaO	1.9355	184.7	2.90	3300	0.88	—	—	—	—
^{137}BaO	1.9355	206.9	2.90	3300	0.88	—	—	—	—
^{135}BaS	1.9637	99.17	1.53	3300	0.46	—	—	—	—
^{137}BaS	1.9637	110.27	1.53	3300	0.46	—	—	—	—

[a] Values quoted for basal plane Be nuclei.

of the wave function onto the nearest shell of anions in the oxides than in alkali halide crystals.

2.3. Comparison with *F* Centers in the Alkali Halides

It is of some interest to compare the magnetic resonance properties of the F^+ centers with the F centers in the alkali halides. An immediate and obvious point of disimilarity is the positive charge associated with the defect in the oxides. The charged nature of the defects in oxides will result in large distortions of the lattice around the F^+ center; thus the electric field gradients important in nuclear quadrupole effects will be enhanced. Consequently the nuclear quadrupole interaction is readily detected in EPR and ENDOR measurements in the oxides. This is not the case in the alkali halide F centers. The positive charge on the F^+ center in oxides localizes the electronic wave function inside the anion vacancy to a greater extent than for F centers in alkali halides. This is evident in Table V, where for a particular alkali halide (e.g., KF), the value of $|\psi_F(R)|^2$ is larger than for the corresponding oxide (CaO). This has also been demonstrated using a somewhat different analysis by Tench and Nelson,[34] who compare the isotropic hyperfine interaction for F^+ and F centers with the hyperfine splitting for the appropriate cations in the gas phase. The g shifts, being determined by overlap of the F-center wave function onto neighboring ion-core orbitals, are greater for a particular alkali halide than its counterpart in the oxides (Table V).

2.4. F^+ Centers in Other Oxides

There have been a number of recent reports of F-like centers in other oxide systems, including BeO,[36] ZnO,[37] ThO_2,[38] Al_2O_3,[39] GeO_2,[40] TiO_2,[41] and $BaTiO_3$.[42] In the last three mentioned materials (GeO_2, TiO_2, and $BaTiO_3$), it is questionable whether the term F center is applicable. An essential ingredient of the F-like centers in the alkali halides, as well as the alkaline earth halides and oxides, is the localization of the trapped electron inside the anion vacancy. In crystals where the oxygen is covalently bonded to a neighboring metal ion, electrons trapped at oxygen vacancies are likely to be localized in "broken" covalent bonds. Consequently these centers have greater similarity to the vacancy centers in silicon[43] and vitreous systems.[44]

Both BeO and ZnO have the wurtzite structure, in which there are two possible nonequivalent anion sites at which F^+ centers may be produced. In each site, the anion is surrounded by four cation nuclei at the corners of the tetrahedron. The tetrahedra are related by a 60° rotation about the

c axis (Figure 2). Thus a single electron trapped at an anion vacancy, an F^+ center, will have an EPR spectrum showing this symmetry and which is described by Eq. (1). The general properties of the spectrum in BeO are as follows[36]:

1. When **B** is parallel to [0001], the two sites will have the same hyperfine pattern.
2. Rotation of **B** in a $\{01\bar{1}0\}$ plane produces a different angular variation from each of the four nuclei at each site, both sites being equivalent.
3. In a $\{1\bar{2}10\}$ plane, the eight possible transitions are reduced to five, since the angular variation from the nucleus along the c axis is the same for each site and there are two equivalent nuclei in each site.

The EPR spectrum for F^+ centers in BeO is shown in Fig. 3.[36] There are 13 hyperfine lines spaced at intervals of $\sim$4 G for **B** in the [0001] direction, as expected from a center with four almost equivalent nuclei with $I = 3/2$. The identification is beautifully confirmed by ENDOR measurements;[36] it is evident from Tables III and V that the results are just those which are to be expected for F^+ centers in divalent host lattices. Table III also shows that the electron density at the Be nucleus is greater for basal plane nuclides than for c-axis beryllium. However, the electric field gradient,

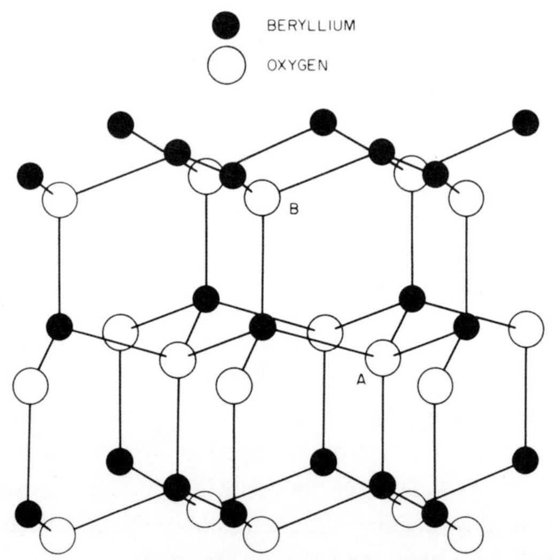

Fig. 2. The wurtzite structure of BeO and ZnO. Sites A and B are nonequivalent anion sites at which F^+ centers may be formed: they are related by a 60° rotation about the z axis.

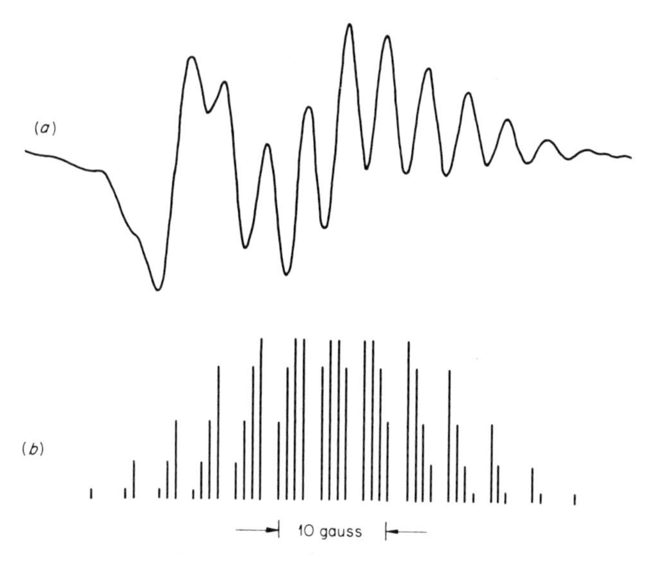

Fig. 3. (a) The EPR spectrum of F^+ centers in BeO for magnetic field **B** || [0001]. (b) Reconstruction of the EPR spectrum using hyperfine constants measured by ENDOR. (After DuVarney *et al.*[36])

as determined by the quadrupole interaction, is greater for the *c*-axis nuclide, presumably indicating that lattice distortion is greatest at this site. DuVarney and Garrison have reported hyperfine interaction with more distant Be nuclei.[36a] The EPR spectrum of F^+ centers in ZnO is somewhat simpler since only 4.5% of the centers experience hyperfine structure with ^{67}Zn ($I = 5/2$).[37] No definitive correlation of the EPR spectra with the optical spectra has been reported in either BeO or ZnO (see Section 4.4).

In ThO_2,[38] Neeley *et al.* have observed a broad line near $g = 2$ which they identify with F^+ centers; Kolopus *et al.*, working on a range of ThO_2 crystals under various conditions of impurity and radiation level, have identified this resonance with Pb^{3+} in axially symmetric sites. The situation is umpromising in Al_2O_3. Although resonant absorption near $g = 2$ is observed,[39] the lines are broad and lack the necessary hyperfine structure required to identify the detailed structure of the defect.

3. OPTICAL PROPERTIES OF F^+ CENTERS

As noted in Section 2, the identification of F^+ optical absorption bands came much later than detailed studies of the center by EPR methods. Nevertheless, the study of these optical bands has now developed into a flourishing activity, demonstrating some of the features of their counterparts in the alkali halides, as well as some very interesting developments specific to the oxides. The aim of this section will be to review the available data, with particular emphasis on the way the oxide situation can be guided by experience with EPR, theoretical calculations, and data from the alkali halides.

3.1. F^+ Centers in Absorption

In MgO, none of the absorption bands in the visible region of the spectrum can be correlated with the F^+-center EPR signal, and the near-UV spectrum is complicated by impurity bands, particularly those due to Fe^{3+} at 4.3 and 5.6 eV. However, Wertz *et al.*[45] found a peak near 5.0 eV which appeared to correlate with the F^+-center EPR signal during thermal annealing. This peak was quite close to the predicted transition energy of 4.7 eV calculated by Kemp and Neeley.[29] A subsequent detailed analysis by Henderson and King[46] of the growth of this band during neutron irradiation showed a clear correlation with the EPR signal. Combined with the annealing behavior, such a correlation argues strongly for the identification of the optical band with the F^+ center. At 4°K, the transition occurs at 4.95 eV, and an oscillator strength of 0.8 was derived for the band. Following this apparently good evidence that the 4.95-eV band was indeed the F^+ band, Chen *et al.*[47] observed a strong 4.95-eV band in additively colored and electron-irradiated MgO without any associated EPR signal. They suggested that the 4.95-eV band was therefore not wholly due to the F^+ center, if at all. This question has now been adequately resolved: The 4.95-eV band is actually a composite of the F^+ and F bands. Neutron irradiation produces predominantly F^+ centers, whereas additive coloration and electron irradiation also produce F centers. Some interconversion by bleaching and heat treatment is possible, and this will be described in more detail in Section 4.

The somewhat stormy passage of the F^+ optical absorption band in MgO illustrates the relative difficulty of optical work in the oxides compared with the alkali halides. Not only does one have to contend with high impurity levels, particularly of the optically and magnetically complex transition metals, but the divalent nature of the host allows more possible charge states for each defect type. Also, the lack of high-efficiency ion displacement by ionizing radiation requires the use of more violent methods, with the attendant increase in complexity (e.g., damage in the cation sublattice).

Once the likelihood that the 4.95-eV band in MgO was the F^+ band had been established, the assignment of F^+ bands in CaO, SrO, and BaO

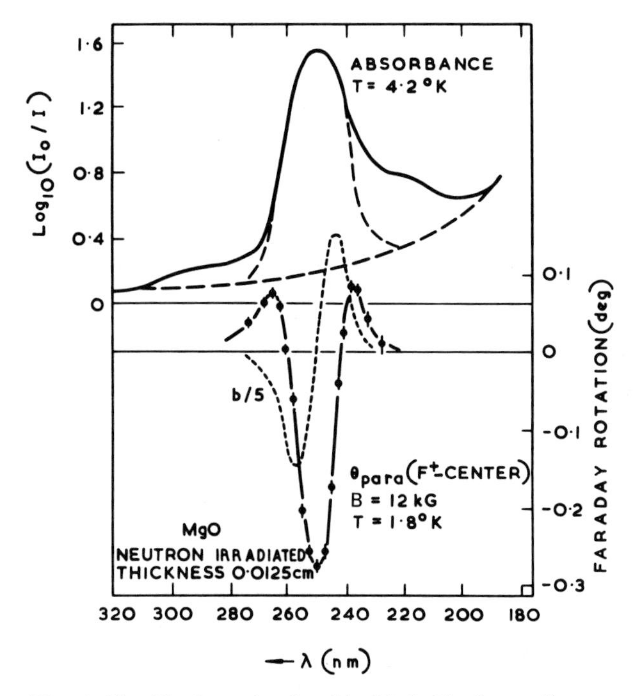

Fig. 4. The F^+ absorption band in MgO. The lower diagram shows the Faraday rotation pattern and the derived circular dichroism *b*. (After Kemp *et al.*[50])

followed fairly quickly. A band at 3.7 eV in neutron-irradiated and additively colored CaO is correlated with the F^+-center EPR signal, and Faraday rotation measurements (see Section 3.3) have now conclusively identified this as the F^+ band. In SrO and BaO, bands at 3.0 eV and 2.0 eV, respectively, have been identified as the F^+ bands, again primarily by Faraday rotation. The optical spectra of the F^+ bands in these four oxides are shown in Figs. 4 and 5, and a collection of their basic band parameters is given in Table VI.

In the alkali halides, the Mollwo–Ivey relation has been extremely useful in extrapolating F-band energies from one material to another. No such rule appears to hold in the four simple oxides, although it does roughly if BaO is excluded.[53,54] However, it is possible to construct an empirical relation between the logarithm of the F^+-band wavelength and the fourth power of the refractive index, which fits all four materials.[55] It remains to be shown whether this rule is more than a mathematical curiosity, and it is difficult to produce another material on which to test it.

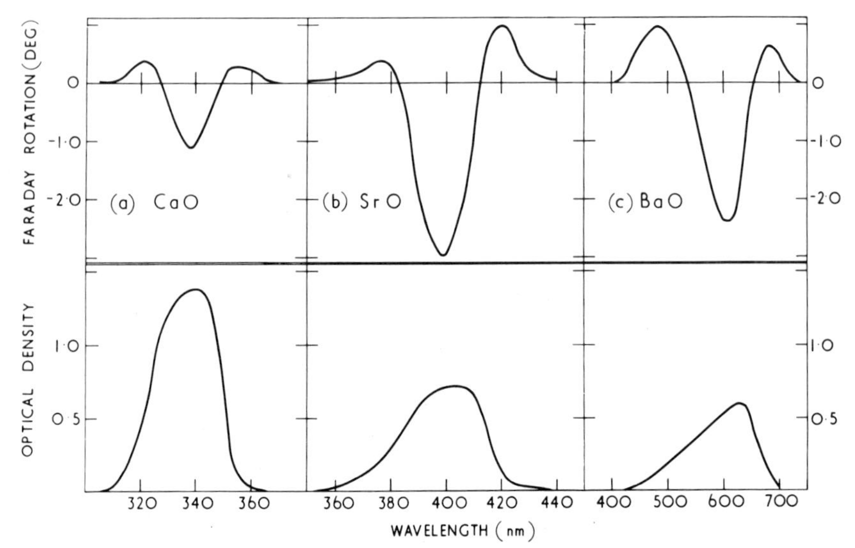

Fig. 5. The F^+ absorption bands in CaO, SrO, and BaO at 4.2°K. The upper diagram shows the Faraday rotation patterns. (After Bessent *et al.*[17])

3.2. Theoretical Position

Having outlined briefly the basis for the identification of the F^+ bands, we now give a short summary of the theoretical position before describing the optical properties of the F^+ centers in more detail. Calculations of the F^+-center transition energy have been made using the point-ion model by Kemp and Neeley.[29] As for the F center in the alkali halides, the ground state is an orbital singlet of symmetry A_{1g} in the local O_h symmetry, and the optically accessible excited state is an orbital triplet T_{1u}. These states

Table VI. Data on the F+ Absorption Bands in Alkaline Earth Oxides

	Band parameters at 4°K, eV		Calculated transition energy,[29] eV	Spin–orbit coupling λ', cm^{-1}	
	Peak energy	Half-width		Expt.	Calc.[17]
MgO	4.95[48]	0.48[48] 0.57[51] 0.73[52]	4.7	−12[50]	—
CaO	3.7[17,49]	0.30[17,76]	3.8	−60[49] −24[17]	−20
SrO	3.0[17]	0.28[17,76]	3.4	−185[17]	−80
BaO	2.0[17]	0.47[17]	3.0	−265[17]	−180

are frequently referred to as "1*s*"- and "2*p*"-like, respectively, in analogy to the hydrogen atom. Kemp and Neeley use a linear combination of 3*s* and 3*p* wave functions on the nearest-neighbor Mg ions in MgO, and also consider the possibility of admixture of next-nearest-neighbor cation 3*s* and 3*p* orbitals. In the actual calculation, only the nearest-neighbor wave functions are used, with the point-ion Hamiltonian of Gourary and Adrian.[20] Including polarization energy terms and allowing for nearest-neighbor displacements (which they found to be 5% outward in MgO), the transition energies $A_{1g} \rightarrow T_{1u}$ calculated by Kemp and Neeley are given in Table VI. The agreement with the experimental assignments is particularly good for MgO and CaO, and somewhat less striking for SrO and BaO. An improvement in the theoretical transition energies is achieved for all four oxides[56] using the wave functions discussed by Stoneham *et al.*[57] for *F* centers in the alkaline earth fluorides. An interesting feature of Kemp and Neeley's calculations in that they put the A_{1g} state below the top of the oxygen 2*p* valence band by several electron-volts. This was interpreted to mean that the "valence band" close to the defect is in fact depressed in energy (cf. the importance of charge-transfer configurations mentioned in Section 2.2). More recent calculations of the transition energy have been made by Bennett[58] and Wood and Öpik.[59] They obtain similar good agreement with experiment for the transition energies, but they also predict the coupling to the lattice in greater detail. This is discussed briefly in Section 3.4. The overall state of the art as regards such defect calculations in the oxides seems to be similar to that in the alkali halides: Agreement with experiment on the question of transition energies is on the whole good, but the reliability of the calculations in predicting effects of smaller energy, such as relaxations, seems less satisfactory. One success of the calculations is the prediction of the proximity of the F^+ and F bands in MgO, which, as we have pointed out already, seems well borne out by experiments (see also Section 4).

Bennett[58a] has estimated that the 2*s* excited state of the F^+ center in CaO is above the 2*p* level. Experimentally, it should be possible to obtain information about the 2*s* state through the Stark effect.[60] Measurements on the F^+ bands in CaO and SrO[60a] show that the effect is much smaller than in KCl.[60] The Stark effect of the zero-phonon line of the F^+ center in CaO (Section 3.4) has been interpreted by assuming that the 2*s* and 2*p* levels are nearly degenerate.[60b] Now that emission spectra have been identified (see Section 3.5), it is expected that more detailed information on the relaxed excited state will become available through lifetime measurements, etc. These will be extremely useful in building up data on the electronic states of the F^+ center.

3.3. Magnetooptical Experiments

We now turn to a closer examination of the F^+ absorption bands themselves, to present the information they yield about the electronic states of the F^+ center. Since a major tool in identifying the F^+ bands has been the technique of Faraday rotation, it is logical to start this more detailed account with this topic.

As is now well known, a beam of linearly polarized light with wavelength λ traversing a crystal of length l in a longitudinal magnetic field suffers a rotation θ of the plane of polarization given by

$$\theta = (2\pi l/\lambda)(n_+ - n_-) \quad (11)$$

where n_+ and n_- are the refractive indices for the two circularly polarized components of the incident radiation. Faraday rotation is, of course, closely related to the measurement of circular dichroism, where the difference in absorption coefficients rather than refractive indices is measured. The two are related through the Kramers–Kronig relations, provided n_+ and n_- are taken together as $(n_+ - n_-)$ and not separately (similarly for the imaginary components giving rise to the absorption).[61] The difference between n_+ and n_- in a magnetic field arises from the Zeeman splitting of the energy levels of the system and frequently Eq. (11) is rewritten as

$$\theta = (2\pi l/\lambda)(dn/dE)\, \Delta E \quad (12)$$

ΔE being the net energy shift between right and left circularly polarized components. This approximation is valid for spectral features (such as the F^+ band) which are broad compared to the Zeeman splittings represented by ΔE. The spectral shape of $\theta(\lambda)$ is then determined by dn/dE, and results in the usually observed dispersion-derivative shape.

Dexter[62] originally pointed out the application of Faraday rotation to color centers, and the first experimental observations were made on F centers in the alkali halides by Lüty *et al.*[63] Karlov *et al.*[64] simultaneously observed circular dichroism in the same systems. The great power of Faraday rotation for F-type centers stems from the magnetic splitting of the spin-doublet ground state combined with spin–orbit splitting of the excited state. The $^2T_{1u}$ excited state behaves very much like the ^2P excited states of alkali atoms. Spin–orbit coupling of the form* $\lambda \mathbf{L} \cdot \mathbf{S}$ splits the

* The spin–orbit coupling may be taken in more general form without involving **L** directly.[66] For our purposes, the simpler form is equivalent and convenient to use.

state into a doubly degenerate Γ_6 ($J = \frac{1}{2}$) and a quartet Γ_8 ($J = \frac{3}{2}$). The splitting between the quartet and doublet is given by

$$E(\Gamma_8) - E(\Gamma_6) = \Delta = \tfrac{3}{2}\lambda g_0 = \tfrac{3}{2}\lambda' \tag{13}$$

where g_0 is the orbital g factor for the $^2T_{1u}$ state (sometimes referred to[63] as m/m^*). Various different approaches have been used to relate the observed rotation through Eqs. (11) and (12) to the spin–orbit splitting Δ. The pioneering work on alkali halides[63] made use of the "alkali atom" model and the rigid-shift approximation, which respectively involve calculating ΔE in Eq. (12) using transition probabilities appropriate to the alkali atom case, and calculating dn/dE by making $n(E)$ consistent with a Gaussian absorption band. The alkali atom model is, in fact, entirely appropriate to the present case, since only one independent parameter governs all the transition probabilities, which are directly applicable to the $^2A_{1g} \leftrightarrow {}^2T_{1u}$ F^+-center transition. The rigid-shift approximation [essentially the use of Eq. (12)] is not so well-founded since it neglects any coupling to noncubic lattice vibrations. The essential result of this analysis[63] is that ΔE is given by

$$\Delta E = 2g_0\beta B + \tfrac{4}{3}\,\Delta\langle S_z\rangle \tag{14}$$

where

$$\langle S_z\rangle = -\tfrac{1}{2}\tanh(\beta B/kT) \tag{15}$$

The Faraday rotation thus has a diamagnetic term $2g_0\beta B$ and a paramagnetic term $\frac{4}{3}\,\Delta\langle S_z\rangle$. In practice, the paramagnetic term dominates at low temperatures and gives rise to rotations which are typically of the order of a few degrees for $T = 4.2°\text{K}$, $l \approx 1$ mm, $B \approx 20$ kG, and a reasonable concentration of about 10^{17} centers/cm^3.

The disadvantages of the rigid-shift model may be circumvented by using the method of moments developed by Henry *et al.*[65] The way this is usually applied is to calculate the circular dichroism from the Faraday rotation using the Kramers–Kronig relations. The first moment of the circular dichroism spectrum is then given by the same expression as in Eq. (14), although some care must be taken to avoid dropping factors of two, as various different definitions of circular dichroism are used.[49] The method of moments is preferable to the rigid-shift model, and has been used in most of the recent work on Faraday rotation in the oxides. The beauty of the moments method is that the formula for the first moment change ΔE is found to be independent of linear coupling to lattice vibrations (and, to lowest order, of coupling terms quadratic in the normal mode displacements[66]). Thus, despite this coupling, which is responsible for broadening the absorption band and obliterating the direct observation of the spin–orbit splitting Δ, the spin–orbit parameters may be extracted from the rotation (or circular dichroism) experiments on the broad band.

Faraday rotation has been used rather than circular dichroism in most experiments reported so far for F^+ bands in the oxides. Again, the proximity of the bands to the near-ultraviolet and the presence of other overlapping bands make measurements that much more difficult, particularly in MgO, where the band is at 4.95 eV (250 nm). On the side of experimental techniques, special mention should be made of the contributions of Kemp and collaborators,[67] who pioneered the use in the oxides of the double-resonance method of ESR-sensitive Faraday rotation (FR-ESR).[68] The idea here is that since the spin–lattice relaxation times for the $S = \frac{1}{2}$ ground state of the F^+ center are long (tens of seconds at low temperatures), it is easy to saturate the spin system by pumping in microwave power. This makes $\langle S_z \rangle = 0$ in Eq. (14), and wipes out the paramagnetic part of the rotation. Hence by turning the microwave power on and off, one can isolate that part of the Faraday rotation due to the paramagnetic species from other effects such as interband (or exciton) rotation. As with all "modulation" techniques, this effectively increases the sensitivity of the Faraday rotation experiment. This method was particularly valuable in the early studies of MgO,[67] where only the low-energy edge of the F^+ band was readily accessible for rotation experiments because of high optical density, weak light sources, etc. Another feature of the method is that it isolates the paramagnetic contribution, although in practice, this usually dominates anyway at low temperatures.

The philosophy of these magnetooptical experiments in the oxides has been slightly different from those performed in the alkali halides. There, the identity of the F band was well established, and the experiments have been used primarily (at least in the case of F centers) as probes into the spin–orbit splitting of the excited state. In the oxides, it was recognized that the observation of Faraday rotation provided a good means of actually identifying the F^+ band and (perhaps with some argument!) this may be reckoned as its prime achievment. However, the information on spin–orbit splitting and other features must not be belittled, since once the identity of the band has been established, these become the points of major interest.

The rotation patterns for the F^+ center in MgO and CaO have been measured by Kemp *et al.* using the FR–ESR technique,[49,50] and Bessent *et al.* have studied CaO, SrO, and BaO using conventional Faraday rotation techniques.[17] The best evidence for the positions of the F^+ bands in SrO and BaO in fact comes from the latter experiments, since as far as the authors are aware, no detailed quantitative optical–EPR correlations have been made in these materials. The optical and EPR signals do, however, show parallel behavior during irradiation and annealing.[17]

The observed rotation spectra are shown in Fig. 4 for MgO and in Fig. 5 for CaO, SrO, and BaO. The initial studies by Kemp *et al.*[67] suggested that in MgO and CaO, the rotation patterns were dispersion-like, but their later results[49,50] (as for MgO in Fig. 4), and the results of Bessent *et al.*[17] on more lightly colored samples, show that the broadband patterns have the expected dispersion-derivative shape, albeit somewhat asymmetric in some cases. Nevertheless, these difficulties in assessing the precise rotation shape, stemming from high optical densities, etc., introduce errors into the analyses in terms of the method of moments, so that considerable uncertainties exist in precise values of the splitting Δ. The values of the spin–orbit coupling constant λ' [see Eq. (13)] are given in Table VI. The kinds of errors that are possible are demonstrated by the two values given for CaO, the only case of overalp between different investigators. Undoubtedly this reflects the experimental difficulties and the problems associated with measuring moments of band shapes in the presence of large backgrounds, and possibly overlapping bands.

The values of λ' show clearly the expected trend toward higher values for the heavier ions, and λ' is *negative* as for the alkali halides. Thus the quartet Γ_8 lies *below* the doublet Γ_6, opposite to the case for an alkali atom. Bessent *et al.*[17] calculated λ' along the lines used by Smith[69] for the alkali halides, and obtained the values shown in Table VI. The agreement with experiment is only moderate, but good enough to give confidence that the correct trends are present and that F^+ centers are being observed. Bessent *et al.*[17] suggest that the wave functions they used (based on the point-ion calculations of Bartram *et al.*[70]) are too compact, an expansion of approximately one-third being required to improve the calculated values for SrO and BaO.

Bessent *et al.*[17] assume $g_0 = 1$ [see Eq. (13)] in their calculations of λ'. No measurements of the diamagnetic Faraday rotation have been made for the oxides, so there are no experimental values of g_0. In the alkali halides, g_0 appears to be rather larger than unity.*

* We are grateful to Professor F. C. Brown for correspondence on this point. There is a discrepancy between the values of $m/m^* = g_{\text{orb}}$ quoted in Refs. 63 and 66. Those in Ref. 63 are correct.

We have so far made no mention of the detailed features of the F^+ absorption bands and their Faraday rotation band shapes, since we have been concerned mainly with establishing the identity and basic parameters of the bands. In fact, the data available to date on these more detailed properties have raised some provocative speculations, with which we shall be concerned in the next section. Prominent among these is the observation that at low temperatures, the F^+ band in CaO shows vibronic structure, a situation unprecedented in the "usual" alkali halides (excepting the structure, spin–orbit and vibronic, seen on F bands in the cesium salts[71]).

Before discussing these features, it is just as well to recapitulate for a moment and consider briefly the reliability of the F^+-band assignments discussed so far. The Faraday rotation spectra demonstrate without doubt that the bands correspond to a paramagnetic species with negative spin–orbit coupling in the excited state. Doubts have been expressed (often in private) about the association of the well-documented F^+ EPR spectrum with the paramagnetic species, and it is here that most concern is evident. In particular, the FR-ESR experiments pump *everything* with g close to 2, which includes almost every kind of paramagnetic defect in the oxides. Nevertheless, Kemp and his collaborators[49] have shown fairly conclusively that if it is *not* the F^+ center which is responsible for the FR-ESR curves, it must be a species with a very large spin–orbit coupling of at least 1000 cm^{-1}. This seems unlikely. Similarly, in SrO and BaO, the only paramagnetic species observed with EPR which is a good candidate for being responsible for the Faraday rotation is the F^+ center.[17] These considerations, as well as the encouraging agreement between theoretical and experimental transition energies, provide good evidence that the correct identifications have been made. In the authors' opinion, the ball is now in the dissenters' court: The identifications must be regarded as correct until proved wrong, rather than vice versa. We should nevertheless not lose sight of the remote possibility that only a part of the "F^+" optical band in CaO, SrO, and BaO is due to the F^+ center. The situation in MgO, where F^+ and F bands nearly coincide, should keep us aware of some similar situation being possible in the other oxides. This does not seem to be the case in CaO, where the F band has itself been identified (see Section 4), but various other unknown species could be candidates instead, in CaO as well as in SrO and BaO. Only more experience in understanding the optical properties of color centers in these oxides will finally clear up these nagging doubts, but it is fair to point out that so far all the signs indicate no such effects.

3.4. Band Shapes and Vibronic Properties

Reference to the F^+-band spectra shown in Fig. 5 immediately draws attention to the asymmetric shape of some of the bands. This is in contrast to the usual alkali halide F bands, which have a closely Gaussian bandshape. Some of the asymmetry in the oxides could be due to the problem of overlapping bands, but experimental evidence so far suggests that the situation is not quite so simple, and reflects rather more basic properties of the F^+ centers themselves. Indeed, subjection to various annealing treatments has always indicated that a single band is involved.

Detailed measurements which yield information about the vibronic properties of the F^+ center have been made only on MgO and CaO. Preliminary measurements on SrO indicate that it fits in with the MgO and CaO pattern. In BaO, the exceedingly long high-energy tail on the F^+ band (see Fig. 5) has been interpreted as implying that the excited state is close to or in the conduction band.[17] Photoconductivity[72] in the band at 2.0 eV supports this view, but the scheme is highly inconsistent with the theoretical calculations of the absolute position of the excited energy levels, which, as in the other oxides, are predicted to be well below the bottom of the conduction band. It is not yet clear whether the photoconductivity and the F^+ band are really correlated, since photoconductivity measurements have not been made on samples similar to those used for the Faraday rotation experiments. Dash[72] observed the 2.0-eV band in additively colored samples, and reported the photoconductivity results in X-irradiated samples. The photoconductivity was enhanced by simultaneous ultraviolet irradiation, possibly because this releases electrons from the valence band into the trap responsible for the 2.0-eV photoconductivity. The shape of the photoconductivity spectrum is very similar to that shown for the F^+ band in Fig. 5. If the excited state of the F^+ center in BaO really is close to the conduction band, it would have interesting consequences for the theoretical calculations. For instance, it might imply that polarization terms are very important where the large Ba^{2+} ion is involved, since these would have to be responsible for shifting the absolute energies. Also, it may mean that the F center in BaO has no bound excited state (as for the F^- center in the alkali halides), which would be different from the situation apparently found in MgO and CaO (Section 4).

By measuring the half-widths of the F^+ absorption bands as a function of temperature, it is possible to derive some of the important parameters governing the coupling of the electronic states to the lattice vibrations. The half-widths of the F^+ bands at 4°K have been given in Table VI. The differences in half-widths reported in MgO require some explanation. The measurements of Chen *et al.*[51] may well represent the most reasonable value, since their experiments were made on additively colored samples in which F centers (see Section 4) had been converted to F^+ centers by ultraviolet radiation. In this case, one does not encounter problems due to F aggregate centers (Section 7) or Fe^{3+} absorption in the F^+-band region. The smallest value reported, $H = 0.48$ eV, was obtained using relatively impure, neutron-irradiated crystals by computer decomposition of the optical spectrum in the F^+ band region into a series of Gaussian bands.[48] The object of this analysis was to obviate difficulties caused by impurity bands which occur in this region. However, in purer crystals, it is evident that the F^+ band is asymmetric in shape[51,52] (e.g., see Fig. 13, Section 4). Thus it is apparent that the half-width determined by Henderson *et al.*[48] may be an underestimate. The larger half-width obtained by Kappers *et al.*[52] is more difficult to explain. However, their samples were irradiated in the UMC Research Reactor, which has no facilities for reducing γ-heating inherent in pile experiments. Experience in other reactor systems suggests that the specimen temperatures may have been as high as 350–500°C, with the consequent formation of F-aggregate or small cluster defects (Section 7). These may have optical transitions underlying the F^+ band, as in the alkali halides, thus contributing an additional apparent width to the true F^+ band.

Measurements of the temperature dependence of band half-widths and their interpretation have been extensively pursued in the alkali halides. In discussing the results, it is customary to confine oneself to simple models of the "vibronic" problem, since the kind of data obtained are not sufficiently detailed to warrant more complex models. Also, the simple treatment gives considerable insight into the processes involved. The simplest picture imagines the defect coupled to a single vibrational mode Q of angular frequency ω. This mode is usually assumed to be a "breathing mode" of the defect nearest neighbors, although, in principle, it could be more complex. In the ground state, we imagine that the equilibrium position of the ions corresponds to $Q = 0$, but in the excited state, there is a displacement to $Q = Q_0$. This situation can be represented by the familiar configuration coordinate diagram given in Fig. 6. Both the ground- and excited-state parabolas have the same curvature if the coupling to mode Q is linear, i.e., involves an energy AQ, where A is a coupling constant. The absorption transition at $Q = 0$ represents the peak of the band (The Franck–Condon principle), and semiclassically, the zero-point motion in the ground state results in a Gaussian-shaped band. Quantum mechanically, the probability

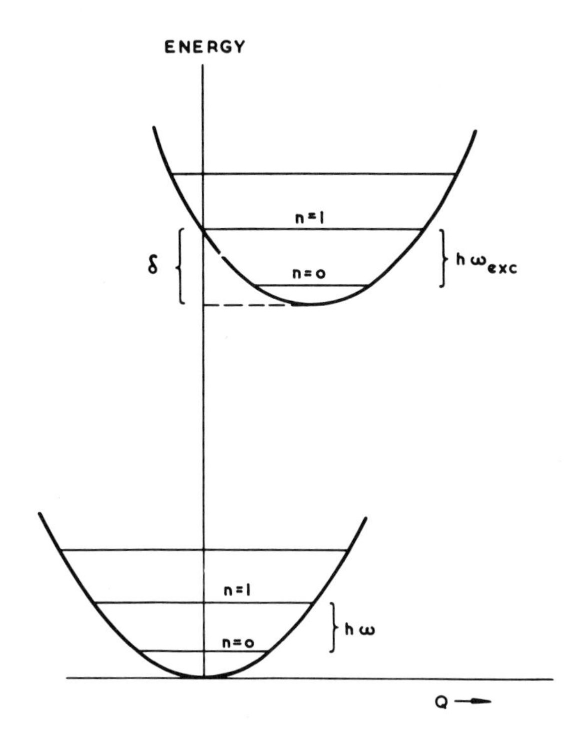

Fig. 6. Configuration coordinate diagram.

of a transition from the $n = 0$ vibrational level of the ground state to the $n = m$ level of the excited state is given at $T = 0°K$ by the Poisson distribution[73]

$$P_{0m} = e^{-S}S^m/m! \tag{16}$$

where $S = A^2/2M\hbar\omega^3$, the Huang–Rhys factor, is a convenient dimensionless measure of the linear coupling. M is an effective mass for mode Q. For large values of S, P_{0m} approximates to a Gaussian shape, consistent with the semiclassical picture. One assumption made in deriving Eq. (16) is the so-called Condon approximation that the electronic transition matrix element connecting the two states is independent of Q. For further details of these models, the reader is referred to the many good reviews of this field.[74]

A convenient property of a Gaussian curve (or P_{0m} for large S) is that its half-width H is related to its second moment M_2 by

$$H^2 = (8 \ln 2)M_2 \tag{17}$$

The moments of the band shape can be calculated rather easily following a method developed by Lax,[75] with the result that $M_2 = S\hbar^2\omega^2 \coth(\hbar\omega/2kT)$. Thus we arrive at the oft-quoted result for Gaussian bands

$$(H/\hbar\omega)^2 = (8 \ln 2)S \coth(\hbar\omega/2kT) \tag{18}$$

By measuring H as a function of T, it is then possible to derive the effective frequency ω and the Huang–Rhys factor S.

Henderson *et al.*[48] and Kappers *et al.*[52] have made measurements of this sort on the F^+ band in MgO, and Bessent[76] has performed an analysis of CaO and, in less detail, SrO. The results are given in Table VII, and the temperature dependence of the bandshape in CaO, about which we shall have more to say later, is shown in Fig. 7. Note the disagreement between the sets of values for MgO, stemming from the disparate values of H in Table VI. Except for the rather high value of ω obtained by Kappers *et al.*[52] for MgO, the values of the effective frequencies ω are well in line with the trends found in the alkali halides: ω tends to lie in the acoustic branches of the phonon spectrum. The phonon dispersion curves have been experimentally investigated by neutron scattering in MgO[7,8,77] and CaO,[77a] and breathing shell model calculations for MgO, CaO, and SrO have been made by Mon.[78] The $k = 0$ optical frequencies have been given in Table I. Since ω is in practice only an effective average frequency (the excited electronic state of the F^+ center interacts with a whole spectrum of normal modes of the defect lattice), no deep significance can be put on its measurement other than the satisfaction of having a consistent trend and the indication that acoustic modes are important. The value of S for MgO is also in line with alkali halide F centers, where "strong coupling" (large S, $\gg 10$) is always apparent. The values of S for CaO and SrO are rather lower: this point will be developed further on.

Table VII. Vibronic Properties of F^+ Centers

	Huang–Rhys factor S from half-width	Effective frequency ω, cm^{-1}	Excited-state frequency ω_{exc}, cm^{-1}	Ref.
MgO	39	260	247	48
	29	467	—	52
CaO	13–16	200–280	—	76, 91
SrO	~24	130–190	—	76

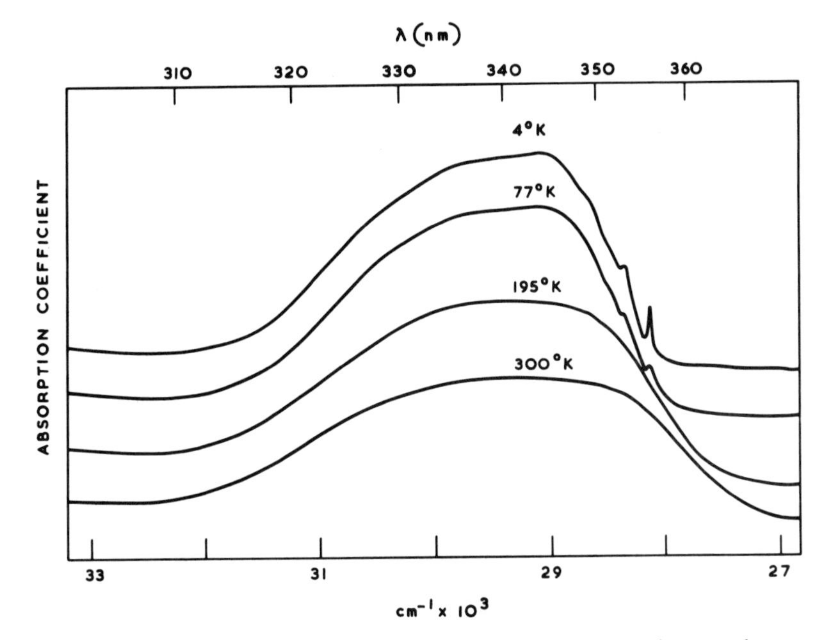

Fig. 7. The F^+ band in CaO at four different temperatures. Note the zero-phonon line and asymmetric shape of the band. The zero-phonon line is not completely resolved in this figure. (After Hall and Bessent.[91])

The peak of the F^+ band shifts to lower energies with increasing temperature. In MgO, the peak energies are 4.95 eV at 4°K and 4.85 eV at 400°K. Two effects contribute to this shift: thermal expansion of the lattice, and higher-order coupling terms in the simple model we have discussed.[74] The effect of the latter is most easily found through the expression for the first moment (i.e., centroid) of the broad band.[75] To a first approximation, the higher-order coupling terms can be represented by introducing a different frequency ω_{exc} for the excited-state parabola in Fig. 6. The centroid position is then given by[74]

$$M_1 = E_2 - E_1 + \tfrac{1}{4}\hbar\omega[(\omega_{\text{exc}}^2/\omega^2) - 1] \coth(\hbar\omega/2kT) \qquad (19)$$

where $E_2 - E_1$ is the energy difference at $Q = 0$ in Fig. 6. In practice, it is difficult to estimate ω_{exc} accurately, because the contribution to the shift from thermal expansion is not known exactly. In principle, it can be determined from the known lattice expansion and the measurement of the band shift under external pressure. In the oxides, the latter measurement is only available for CaO, and then indirectly through stress experiments on the zero-phonon line (see later). Henderson *et al.*[48] have used an Ivey–Mollwo relation to estimate the expansion effect in MgO. The estimated value of ω_{exc} is listed in Table VII. For MgO, $\omega_{\text{exc}} < \omega$, which in general is also the case in alkali halides when ω_{exc} is measured in this way.[79] Experience there, however, where ω_{exc} can also be estimated from the temperature dependence of the half-width of the emission band,[80] suggests that not too much reliability should be placed on the physical interpretation of the parameter ω_{exc} as an "effective frequency" in the excited state. The analysis in terms of only one vibrational mode undoubtedly has a bearing on this point, since only even-parity modes couple linearly and therefore enter into the expression (18) for the half-width, whereas both odd and even modes can couple quadratically to give a contribution to M_1. Thus different types of measurement do not necessarily measure the same modes, and the averaging process implied by the single-mode picture is somewhat arbitrary.

Henderson *et al.*[48] and Kappers *et al.*[52] use their values of S and ω to predict the position of the F^+ emission band in MgO. The Stokes shift (see Fig. 6) is given by $2\delta = 2S\hbar\omega$ and they obtain $E_{\text{emission}} = 2.4$ eV,[48] or 1.65 eV,[52] about half the absorption energy. A similar analysis in CaO would yield an emission energy of 2.5 eV. Both these values underestimate the observed emission energy (see Section 3.5).

We now discuss in more detail the F^+ band in CaO, which shows (Fig. 7), rather unexpectedly, a clearly resolved zero-phonon line at 355.7 nm and associated vibronic structure. Zero-phonon transitions occur on many color center optical bands in the alkali halides,[81] but so far, none has been detected on any F band. This is simply because the values of S are usually too large. From Eq. (16), we see that at 0°K, the intensity of the zero-phonon transition (P_{00}) is only a fraction e^{-S} of the whole band, and is usually undetectable unless $S \lesssim 6$. At higher temperatures, the fraction becomes smaller as $P_{00}(T) \approx \exp[-S \coth(\hbar\omega/2kT)]$. Thus CaO stands alone among simple F-type centers in showing such structure. Since the Huang–Rhys factor S is crucial in determining the appearance of the zero-phonon line, it is worthwhile considering its value in CaO more carefully. As well as the derivation from measuring the band half-width, S may be determined in two other ways from the simple model if a zero-phonon line is observed. From the fractional intensity e^{-S}, S is found to be in the range 4.5–6.0 for the F^+ center in CaO. The separation of the band centroid and the zero-phonon line should be $\delta = S\hbar\omega$, which, with the value of ω from Table VII, yields $S \sim 4$–5. These values of S are shown in Table VIII. Both these values are much smaller than the values of S derived

Table VIII. Measured Values of S for the CaO F^+ Center[b]

Method	Absorption	Emission[a]
Zero-phonon line fraction P_{00}	6,[49] 4.5,[94] 5.6[91a]	4.9[94]
Zero-phonon-line centroid separation, δ	5,[49] 3.9,[94] 5.8[91a]	3.5[94]
Band half-width $H^2 = (8 \ln 2) S\hbar^2\omega^2$	13–16[76]	4[95]

[a] See Section 3.5.
[b] For more recent estimates see Refs. 89a, 91, 91a, 95a, 95b.

from the half-width. A discrepancy as large as this is unlikely to be a result of the single-mode model, since extending the model to many modes produces the generalization[74]

$$P_{00} = e^{-S}, \qquad \delta = S\hbar\langle\omega\rangle, \qquad M_2 = S\hbar^2\langle\omega^2\rangle; \qquad T = 0°\text{K} \tag{20}$$

where $S = \sum_i S_i$ in an obvious notation. It seems doubtful that differences between $\langle\omega\rangle^2$ and $\langle\omega^2\rangle$ could explain the anomaly.

Faraday rotation and uniaxial stress experiments on the 355.7-nm zero-phonon line have now led to an explanation of these features, albeit after raising some controversial issues. The Faraday rotation results of Kemp *et al.*[49] on the zero-phonon line are shown in Fig. 8. Most noticeable is the dispersion-like shape of the rotation pattern, and the fact that the magnitude of the rotation observed is similar to that observed for the broad F^+ band (the peak to peak rotation for the broad F^+ band was about 140 min under the same conditions as Fig. 8; see Fig. 1 of Ref. 49). Intuitively, one might have expected much larger peak rotations for the zero-phonon line, since from Eq. (12), dn/dE is much larger for the sharp line than the broad band. (Very roughly, dn/dE is proportional to $\alpha_{\max}/H$, where $\alpha_{\max}$ is the absorption coefficient at the peak of a band or line, and H is its halfwidth. For the F^+ band in CaO, $\alpha_{\max}$ is comparable for the broad band and zero-phonon line.) Kemp *et al.*[49] explain both the shape and size of the zero-phonon line rotation in the following way. A dispersion-like rotation implies that the circular dichroism spectrum has the same shape as the absorption line itself. Such a situation can arise for transitions between states which are effectively orbitally nondegenerate.[82] In such a case, application of a magnetic field can only in lowest order affect the transition probabilities for left and right circularly polarized light, without producing energy shifts, so that the circular dichroism spectrum is absorption-like. When the ground state is a spin doublet (as for the F^+ center), the transition probability difference is proportional to $\langle S_z \rangle$. The paramagnetic nature of the zero-phonon line rotation was confirmed by Kemp *et al.* using the FR-ESR technique. The apparent orbital nondegeneracy of the $^2T_{1u}$ excited state of the F^+ center is explained as a consequence of the Jahn–Teller effect. An orbitally degenerate state is unstable with respect to a symmetry-lowering distortion of its surroundings, so that in general such a distortion occurs and removes the degeneracy.[83] Schematically, the situation can be represented by a configuration-coordinate diagram as in Fig. 9. Linear coupling to a Jahn–Teller mode Q (that is, a distortion mode which lowers the symmetry) splits the potential energy curve into two branches, lowering each by an energy E_{JT}, the Jahn–Teller energy. Since a zero-phonon transition by definition takes the system to one of the energy minima, it "feels" the Jahn–Teller effect. By the Franck–Condon principle, the peak of the broad band corresponds to absorption at $Q = 0$, and does not "feel" the distortion. This insensitivity of the broad band to vibronic coupling is the physical reason for the power of the method of moments.[65,66]

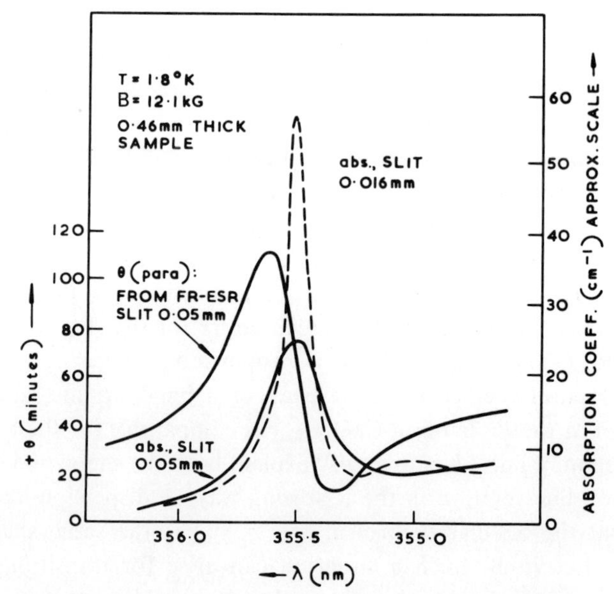

Fig. 8. Faraday rotation pattern and absorption coefficient for the 355.7 nm F^+-center zero-phonon line in CaO (after Kemp *et al.*[49]).

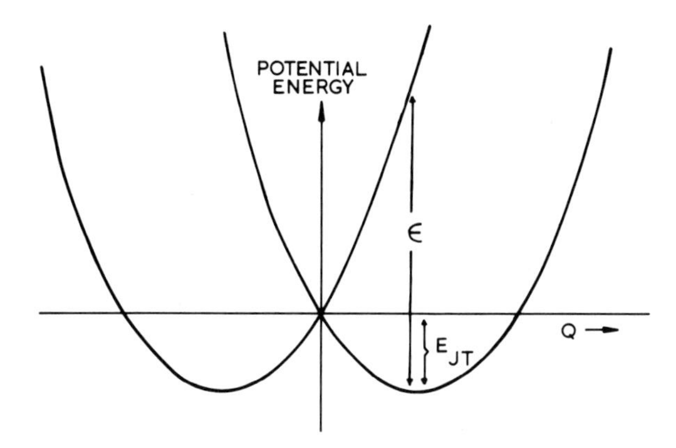

Fig. 9. Simple illustration of the Jahn–Teller effect using a configuration coordinate diagram. The separation of the two branches at a value of Q corresponding to a minimum on the lower branch is the Jahn–Teller "gap" ε. The depression of the lower branch defines the Jahn–Teller energy E_{JT}.

The Jahn–Teller effect of a T_{1u} orbital state is much more complicated than the situation depicted by Fig. 9, but the latter has the merit of explaining Kemp *et al.*'s interpretation in a simple way (they in fact use a similar model). At each minimum, the orbital state is nondegenerate, as required by the dispersion-like rotation. The transition probability difference is caused by spin–orbit coupling across the Jahn–Teller "gap" ε, which is equal to $4E_{JT}$ in this two-dimensional model. This gives rise to a peak-to-peak paramagnetic rotation[49]

$$\theta_{pp}/l = \alpha_0(\lambda'/\varepsilon)\langle S_z\rangle \tag{21}$$

where α_0 is the peak absorption coefficient of the zero-phonon line and l is the sample thickness. Taking λ' as $-60\ \text{cm}^{-1}$ from their broadband Faraday rotation measurement (see Table VI), and making some corrections for instrumental broadening and degeneration, Kemp *et al.* deduce $\varepsilon \approx 1000\ \text{cm}^{-1}$. The negative λ' also predicts the correct sign for the rotation. Remembering that E_{JT} is an energy analogous to the quantity $S\hbar\omega$ discussed earlier in this section, the deduced value of ε seems reasonable.

The interpretation as we have described it clearly explains both the shape and magnitude of the zero-phonon line rotation. It is worth noting that for the F^+ broad band,

$$\theta_{pp}/l \approx \alpha_{max}(\lambda'/H)\langle S_z\rangle \tag{22}$$

so that in the zero-phonon line rotation, Eq. (21), the Jahn–Teller energy is playing the role of the band half-width. This explains why the zero-phonon line and band rotations are similar in magnitude.

The Jahn–Teller effect for a T_{1u} orbital state comes about because of coupling to vibrational modes of E_g and T_{2g} symmetry[84] (a state described by an irreducible representation Γ couples linearly to modes of symmetry contained in the symmetric square $[\Gamma \times \Gamma]$; for T_{1u}, this is $A_{1g} + E_g + T_{2g}$, A_{1g} being a "breathing" mode). In the limit of a large Jahn–Teller effect ($E_{JT} \gg \hbar\omega$; $\hbar\omega$ = frequency of E_g or T_{2g} mode), an effectively static distortion occurs which is either tetragonal or trigonal according to whether the E_g mode coupling is larger or less than the T_{2g} coupling.[84] In this limit, the argument we have presented to explain the zero-phonon line rotation follows through roughly as with the simple model above. When the static limit is not reached, the Jahn–Teller effect must be considered dynamically, as a problem involving coupled electronic and vibrational motion (vibronic states).[85] The lowest vibronic state, which is the terminal state for the zero-phonon transition, always turns out to have the same degeneracy and symmetry as the original electronic level, i.e., T_{1u} in our case. This is simply a consequence of group-theoretical compatibility with the uncoupled situation. The threefold degeneracy for the T_{1u} state corresponds, in the static limit, to the three possible directions for a tetragonal distortion if the E_g coupling is dominant. The T_{2g} coupling case is more complicated because "accidental" degeneracy occurs.[86] Spin–orbit effects within the $^2T_{1u}$ vibronic state can be described in two parts: first, coupling to higher vibronic states in second- or higher-order perturbation theory. This corresponds to the coupling across the Jahn–Teller gap which we described before, and gives a similar contribution to the rotation[91a] ($\varepsilon = 3E_{JT}$ for the Jahn–Teller effect of a T_{1u} state). Second, there is some residual first-order splitting of $^2T_{1u}$ into Γ_6 and Γ_8, but with a reduced spin–orbit coupling parameter $\lambda'' = \lambda' e^{-x}$. The parameter x is given by $3E_{JT}/2\hbar\omega$ for E_g mode coupling, and approximately $9E_{JT}/4\hbar\omega$ for T_{2g} mode coupling.[85] This first-order effect does not appear in the static model we discussed, but should give rise to a dispersion-derivative-shaped rotation (like that of the broad band) with peak-to-peak rotation given by Eq. (22), where H is now the zero-phonon line half-width and λ' is to be replaced by λ''.

In principle, both types of rotation occur together. Kemp *et al.*[49] discuss the reasons why they only see the dispersion-like component. They attribute this to the fact that $x \approx 2$, and that their imperfect resolution (about 40 cm^{-1}, whereas the half-width of the line at 4°K can be as small as 4 cm^{-1}) discriminates against the more rapidly varying derivative signal. Despite these uncertainties, it is clear that the zero-phonon line is due to a paramagnetic center, and that any first-order spin–orbit splitting is severely quenched from the value appropriate to the broad component of the F^+ band.

The lack of spin–orbit splitting is further evidenced by the observation that the zero-phonon line is a single, narrow line, with no trace of a doublet character.[87] This implies that the spin–orbit splitting is much less than the half-width, 4 cm^{-1}, so that, remembering λ' is at least 25 cm^{-1} (Table VI), e^{-x} must be of the order of 0.1 or less. This is, of course, consistent with $x \approx 2$ quoted above.

Before the publication of the Faraday rotation results, Hughes and Runciman[87] investigated the splitting of the 355.7-nm line under an applied uniaxial stress. They found that the line split very clearly into two polarized components under [001] and [111] stress, and into three components under [110] stress. The splittings were linear with applied stress, as shown in Fig. 10. This splitting pattern can only be explained as being due to an electric dipole transition between an "A" symmetry ground state and a "T" symmetry excited state in a center of cubic symmetry.[87] At first sight, this confirms beautifully that the line is indeed due to the F^+ center, until one remembers the implications of our discussion of the Jahn–Teller effect. A relatively strong effect is required to quench the spin–orbit splitting, and we have stated that in the static limit, this implies a tetragonal or trigonal distortion. If this were the case, the stress splitting patterns would reflect this lower symmetry. For instance, for a tetragonal distortion, there would be no [111] splitting, and for a trigonal distortion, no [001] splitting, whereas both are observed. In the language of the dynamic Jahn–Teller effect,[85] for coupling to E_g modes, the [111] stress splitting (described by a coupling coefficient C') is quenched, whereas for coupling to T_{2g}, the [001] splitting (coupling coefficient B') is quenched. The observed values of B' and C' (and the "hydrostatic" coefficient A') shown in Table IX are the order of magnitude one expects for defects ($\sim 10^4$ cm^{-1}/unit strain) and any violent quenching seems out of the question. This difficulty, taken together with the disparity in Huang–Rhys factors S derived for the CaO F^+ band already described, led for a time to a controversy over whether the zero-phonon line really was due to the F^+ center.[49,87]

The situation may be resolved by taking into account dynamic Jahn–Teller coupling to E_g and T_{2g} modes simultaneously. At the time of the controversy, this problem had not been solved, but recently, O'Brien[88] has produced a model solution which provides the basis for understanding the difficulty. When the coupling to both types of mode is more or less

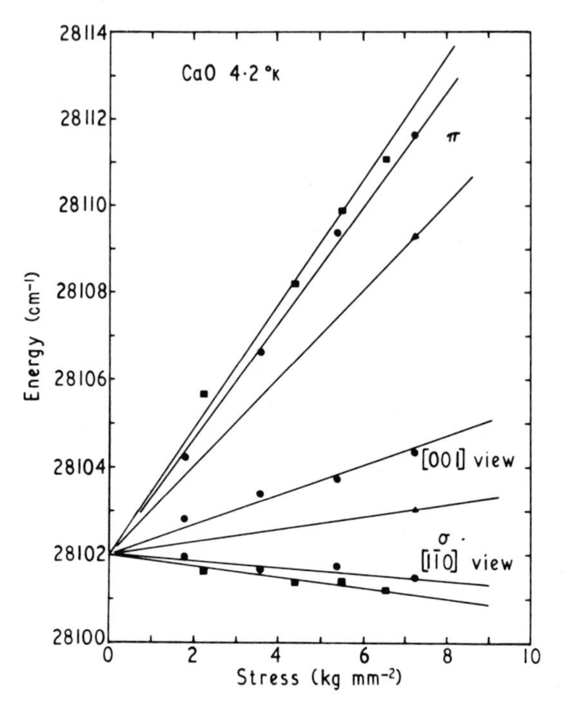

Fig. 10. Uniaxial stress splitting of the 355.7-nm line in CaO at 4.2°K: (■) stress ‖ [001]; (▲) stress ‖ [111]; (●) stress ‖ [110]. π and σ refer to lines polarized with the electric vector parallel and perpendicular to the stress direction, respectively. (After Hughes and Runciman.[87])

equal, there can be a large quenching of the spin–orbit splitting (as required by the Faraday rotation results and the unsplit zero-phonon line) without producing a static tetragonal or trigonal distortion. In other words, no drastic quenching of the strain coupling parameters need occur to accompany that of the spin–orbit splitting. In fact, B' and C' are quenched by factors of 2/5 when both modes are equally coupled, which is quite acceptable.

Hughes[89] has shown that this model provides a good quantitative understanding of the experimental results, and what is more, also explains the discrepancy in Huang–Rhys factors. With the occurrence of a complicated Jahn–Teller effect, the simple relations in Eq. (20) no longer apply. Defining Huang–Rhys factors for each type of mode by the relations $S_E = E_{\text{JT}}(E)/\hbar\omega$, etc., one finds[89] that in the equal-coupling case ($S_E = S_{T_2} = S'$, and ignoring coupling to A_{1g} breathing modes which would

Table IX. Stress and Strain Coupling Coefficients for the CaO F^+ Zero-Phonon Line[a]

Stress direction in measurement	Stress coupling coefficients, $cm^{-1}\,kg^{-1}\,mm^2$		
	A	B	C
[001]	0.40 ± 0.16	0.54 ± 0.09	—
[111]	0.42 ± 0.11	—	0.87 ± 0.19
[110]	0.54 ± 0.12	0.61 ± 0.14	1.00 ± 0.28
Average	0.45	0.58	0.94
	Strain coupling coefficients, $10^4\,cm^{-1}$		
	A'	B'	C'
	1.53	0.48	0.77

[a] Ref. 87.

be described by a Huang–Rhys factor S_A,

$$P_{00} \sim \exp(-S') \text{ for } S' \gg 1, \quad \delta \simeq S'\hbar\omega, \quad M_2 = 2.5S'\hbar^2\omega^2; \quad T = 0°\text{K} \tag{23}$$

The larger value of S derived from the band half-width (Table VIII) is therefore not surprising in view of the factor 2.5 in the expression for M_2.

The non-Gaussian band shape of the F^+ band (see Fig. 7) can also be understood in terms of this model. The larger ratio of M_2 to δ implies that the band centroid is closer to the zero-phonon line in units of the half-width than would be expected on the basis of Eq. (20), and explains the high-energy "tail." Semiclassical calculations[90] of the band shape when E_g and T_{2g} modes are coupled to the T_{1u} orbital level predict a triple-peaked band; the band in Fig. 7 does indeed show suggestions of a multipeak character. Quantum mechanically, the band shape is reflected in an evaluation of the moments. The ratio of the fourth moment M_4 to the square of M_2 is equal to 2.1 for equal coupling to E_g and T_{2g} modes,[89] compared with 3 for a Gaussian band (coupling to A_{1g} or E_g modes alone). This implies a "squarer than Gaussian" shape, and values of M_4/M_2^2 of about 2.4 have been measured for the CaO F^+ band.[89a,91] A recent quantum mechanical calculation of the band shape[89b] shows remarkable agreement with experiment. Finally, it is possible to estimate S_E, S_{T_2}, and S_A from the strain coupling coefficients B', C', and A' (Table IX). The result[89] is $S' = S_E = S_{T_2} \simeq 4$, $S_A \simeq 1.8$, which, aside from being in fortuitously good agreement with experiment, does indeed suggest that coupling to noncubic modes, E_g and T_{2g}, is more important than to cubic modes, A_{1g}.

Some recent results obtained by Merle d'Aubigné and Roussel[91a] have given a striking confirmation of these ideas. Their circular dichroism measurements on the broad band give $\lambda' = -31 \pm 6$ cm^{-1} (cf. Table VI) and by carefully measuring the first and third moments of the dichroism spectrum, they are able to show that the contribution of noncubic modes to the second moment M_2 is four times that of the cubic modes. Using $\omega = 280$ cm^{-1} for all modes and assuming equal coupling to E_g and T_{2g} modes, they deduce $S' = 3.15$ and $S_A = 2.05$. These values also explain the experimental values of P_{00} and δ, provided that Eq. (23) is extended to include the A_{1g} modes. Merle d'Aubigné and Roussel have also made measurements on the zero-phonon line with much higher resolution (0.4 cm^{-1}) than used by Kemp *et al.*,[49] and are able to detect both first- and higher-order spin–orbit coupling contributions to the circular dichroism. From the first-order effect, they estimate $\lambda'' = -0.58 \pm 0.06$ cm^{-1}, which gives $x = 4$. The deduced value of E_{JT} of about 750 cm^{-1} also fits the higher-order effects, i.e., the change in transition probabilities, and is clearly in line with the value of S'.

None of the other oxide F^+ bands have yet been found to show vibronic details, but they do provide some provocative comparisons with the situation in CaO. Both SrO and BaO have asymmetric bands, although, as we have explained, there may be an alternative explanation for BaO. As noted by Kappers *et al.*,[52] MgO may also be slightly asymmetric. Indeed, Henderson *et al.*[48] found that a single Gaussian was insufficient to represent absorption in the F^+-band region. Moreover, in explaining the Faraday rotation pattern for MgO in Fig. 4, Kemp *et al.*[50] point out that the rotation spectrum is rather narrower than one would expect on the rigid-shift model. Henry *et al.*[65] showed that this happens when noncubic modes are coupled, and a moments analysis by Kemp *et al.*[50] suggests that in MgO, the second moment M_2 may be almost entirely due to E_g and T_{2g} modes rather than A_{1g} modes.

One is therefore led to the rather surprising conclusion that in the oxides, coupling to noncubic modes appears to dominate. This is in contrast to the alkali halides, where the cubic modes usually give the largest contribution, and as a result, the F bands are closely Gaussian in most cases.

Even there, however, stress experiments by Schnatterly[92] indicate that E_g and T_{2g} modes are coupled to a similar extent, which may indicate that this particular feature is typical of F-type centers.

Some estimates of the Huang–Rhys factors for F^+ centers in the oxides have been made by Bennett.[58] He only includes the "breathing" A_{1g} mode and calculates $S = 3.56$ for MgO and CaO, and $S = 4.43$ for SrO. These may be compared with the experimental values in Tables VII and VIII. In view of what we have said about the importance of noncubic modes, Bennett's values must only be regarded as a first step toward completely understanding from a theoretical point of view the interesting vibronic properties of F^+ centers in the alkaline earth oxides. As yet, there have been no reported calculations of the coupling to noncubic modes, or lattice-dynamic calculations of the vibrational modes and frequencies of a lattice containing an F^+ center.

3.5. F^+ Centers in Emission

Considerably less information is available on emission from F^+ centers than on the absorption bands. Following the discovery of vibronic structure on the CaO F^+ band, the search for a "mirror image" emission band became imperative. Henderson and Tomlinson,[93] Evans *et al.*,[94] and, later, Henderson *et al.*[95] detected the fluorescence by exciting in the F^+ absorption band. More recently, Chen *et al.*[51] and Kappers *et al.*[52] have reported the F^+-center emission in MgO, while Evans and Kemp[96] have reported F^+ emission in SrO. The emission bands for MgO and CaO are shown in Fig. 11 and 12. The first point to notice is that the bands peak at higher energies than predicted on the basis of Huang–Rhys factors for the absorption bands. In Section 3.4, we predicted 2.4 or 1.65 eV for MgO and 2.5 eV for CaO, compared with the measured values of 3.13 and 3.3 eV, respectively. One explanation of the poor prediction is immediately suggested by our discussion of the importance of noncubic modes in contributing to S. The Stokes shift should be 2δ [Eq. (23)], which in terms of the values of S listed in Table VII, should be approximately $(2/2.5)S\hbar\omega$ instead of $2S\hbar\omega$ used before, because of the factor 2.5 in M_2 [Eq. (23)]. This revises the emission energies to 3.9 eV (MgO) and 3.25 eV (CaO). Despite the overcompensation in MgO (remember we have now neglected A_{1g} modes altogether), the Stokes shift clearly depends very strongly on the noncubic to cubic ratio.

The identification of the F^+ emission band in MgO is complicated by the overlap of the F and F^+ bands in absorption. Chen *et al.*[51] and later

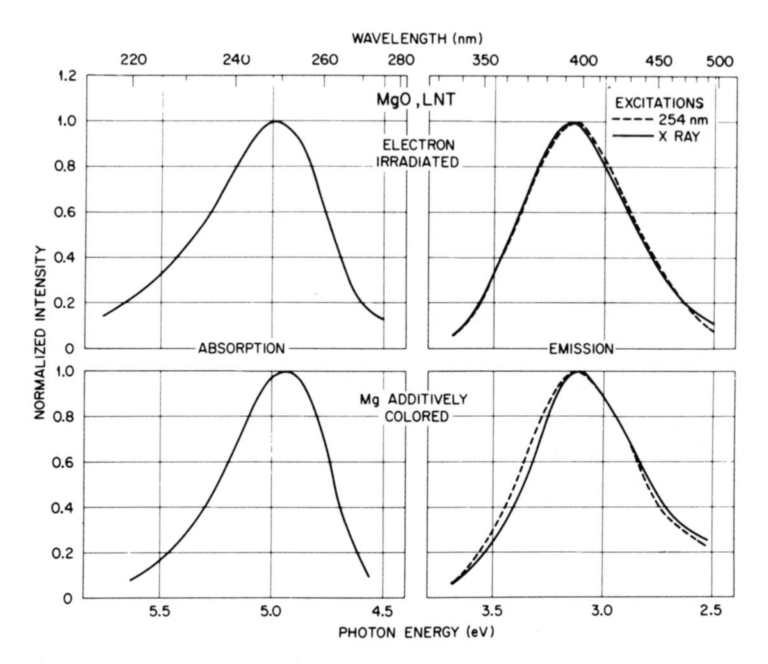

Fig. 11. Absorption and emission spectra of F^+ centers in MgO, taken at 77°K (after Chen *et al.*[51]).

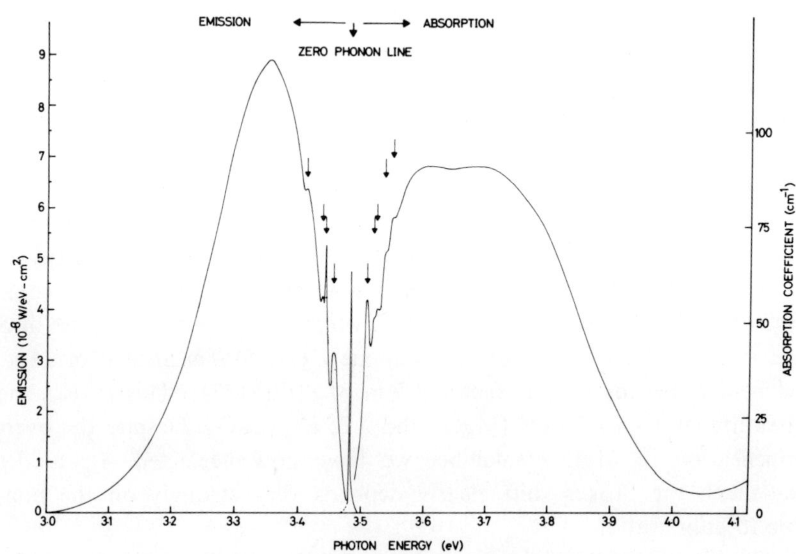

Fig. 12. Vibronic structure in the F^+ center emission and absorption in CaO, taken at 4°K (after Henderson *et al.*[95b]).

Kappers *et al.*[52] have nevertheless shown, by varying the F^+:F ratio, that the 3.13-eV emission is correlated with the F^+ centers. The half-width of the emission band is about 0.6 eV at 77°K, and it can be excited by X-rays as well as 5-eV ultraviolet radiation. The half-width may be compared with the value 0.5 eV measured for the absorption band at 77°K. Some variation in the F^+ emission band shape was observed in different samples, particularly those which had been neutron-irradiated. The quantum efficiency is stated to be 10 times higher at 77°K than at room temperature.[51]

In CaO, most attention has focused on the vibronic properties of the F^+ emission band. Some relevant data are collected in Table VIII. The zero-phonon line occurs at the same wavelength in emission and absorption, as indeed it should, but it can be seen from Fig. 12 that there is a lack of detailed "mirror symmetry" between absorption and emission. Lack of symmetry would not be too surprising in view of the complicated Jahn–Teller effect postulated to occur in the excited state. The absorption structure samples the vibronic structure of the excited state, whereas the emission samples the vibronic structure of the ground state. Since no Jahn–Teller effect occurs in the ground state, one would expect the emission band shape to be simpler than the absorption spectrum. The emission band half-width is less than that of the absorption band, as can be seen from the data in Table VIII. This is qualitatively what would be expected from the Jahn–Teller effect: The emission band should be closer to a Gaussian, and the factor 2.5 in M_2 in Eq. (23) should be replaced by something closer to unity.[89,89b] The emission band values of S should then be more self-consistent than those in absorption, which is seen from Table VIII to be so. The values of S are close to the values of S' [Eq. (23)] derived for the absorption band, which is satisfying.

The main sidebands in emission are $\sim$205 cm^{-1} and $\sim$302 cm^{-1} from the zero-phonon line.[95a,b] Similar peaks occur in absorption, but the form of the 302 cm^{-1} peak is somewhat different. In emission the peak is remarkably sharp, and Henderson *et al.*[95] suggested that it might be a second zero-phonon line or a vibronic side band involving a sharp local mode of the F^+ center. Recent studies show that the 205 cm^{-1} peak is associated with A_{1g} vibrational modes and the 302 cm^{-1} peak is a composite of E_g and T_{2g} modes.[96a,b]

Evans *et al.*[94] report a lifetime of <30 nsec for the F^+-center excited state in CaO at 77°K, but feel that in their neutron-irradiated samples, this does not represent an intrinsic decay. Henderson *et al.*[95] report a measurable lifetime of 60 nsec in accidentally colored crystals (Section 1). If this lifetime is intrinsic, it is much shorter than the radiative lifetimes of F

centers in the alkali halides, and may confirm the expectation that the excited state of the F^+ center in the oxides is fairly compact.

4. *F* CENTERS

4.1. Evidence for *F* Centers in Alkaline Earth Oxides

The F center in the alkaline earth oxides is the analog of the F^- center (formerly F') in the alkali halides: two electrons trapped at an anion vacancy. The principal difference that we expect between the alkali halides and the oxides is that in the former, the center is negatively charged, and as a result, it appears that there are no bound excited states.[97] Thus the F^- band is usually a broad, asymmetric absorption covering a good deal of the visible region, and corresponds to exciting the second electron into the conduction band. In the oxides, on the other hand, the F center is neutral, and owing to the apparently large trap depth of the F^+ center (Section 3.2), it is expected that both ground and excited states will be quite strongly bound. This conclusion may not apply to BaO, where some doubt exists about the absolute energies of the F^+ levels (Section 3.4).

Being diamagnetic, the F center has no EPR signal, and its detection must necessarily be optical. This is also the case for the F^- center in the alkali halides, and indeed, it is valid to point out that as a result, the F^- center is only identified by inference and not directly. In the alkali halides, the existence of the F^- center is implied by photochemical $F \leftrightarrow F^-$ conversion, which gives quantitative evidence of the F^- center, and by the qualitative nature of its properties (no bound excited state, low thermal stability, etc.). The same is largely true for the F center in the oxides, although the existence of bound excited states and greater stability allows a wider variety of identifying experiments, as we shall see.

We have already mentioned the F center in MgO during our discussion of F^+ centers. So far, detailed evidence for the structure of F centers is available only in MgO and CaO, there being rather more data available on the latter. Rather surprisingly, perhaps, neutron-irradiated crystals show mostly F^+ centers; the reasons for this are not too clear, but presumably impurities and other irradiation-induced defects favor the formation of positively charged anionic vacancy defects. The F centers in MgO and CaO were discovered at about the same time, although in different ways. Since we have already touched on the MgO situation, we discuss that first.

Following the correlation between the 4.95-eV optical band and the F^+-center EPR signal in neutron-irradiated MgO crystals,[46] Chen *et al.*[47] showed that this correlation did not exist in additively colored and electron-irradiated samples. In fact, there was only a small EPR signal, although the 4.95-eV optical band was strong. Since the F^+ EPR can be regenerated in additively colored crystals by ionizing radiation,[25] the ultimate conclusion is that the F center also absorbs near 4.95 eV, and the lack of correlation is due to the formation of F rather than F^+ centers. In fact, it is not unexpected that additive coloration produces mostly F centers, since these are the thermodynamically stable species. The close proximity of the F^+ and F bands in MgO is indicated by the theoretical calculations of Neeley and Kemp[98] and Wood and Öpik.[59] Neeley and Kemp calculate the F-band transition energy to be 5.4 eV, whereas Wood and Öpik find an energy of 5.05 eV. Chen *et al.*[51] have since shown that some interconversion of F and F^+ centers is possible in additively colored and electron-irradiated crystals. Bleaching with UV radiation at 77°K favors the formation of F^+ centers, whereas a similar bleach at room temperature favors F centers. Heat treatment of additively colored samples above 400°C converts F^+ centers to

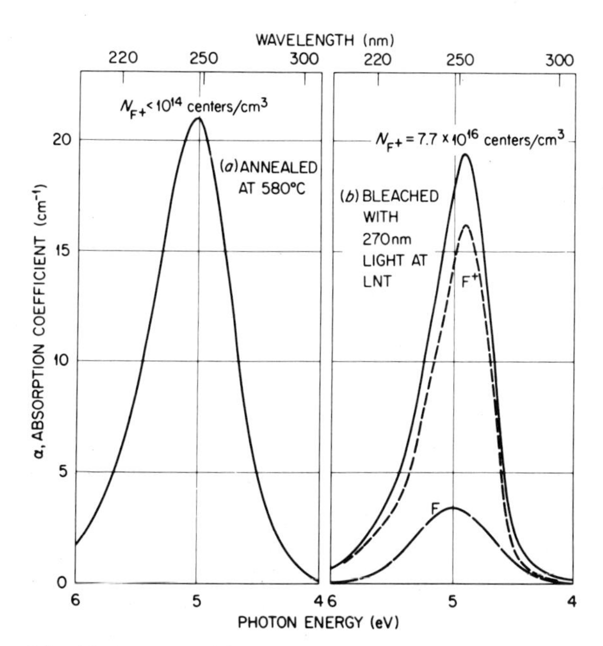

Fig. 13. Decomposition of the 4.95-eV band in MgO into F^+ and F bands. Data taken at room temperature on an additively colored sample. (After Chen *et al.*[51])

F centers. Using these techniques, Chen *et al.* are able to isolate the two bands, and conclude that the F band peaks at 5.01 eV at room temperature, with a half-width of 0.77 eV. They also determine the oscillator strength to be 1.25, by assuming a value 0.8 for the F^+ band, as found by Henderson and King.[46] The decomposition of the 4.95-eV band into F and F^+ is shown

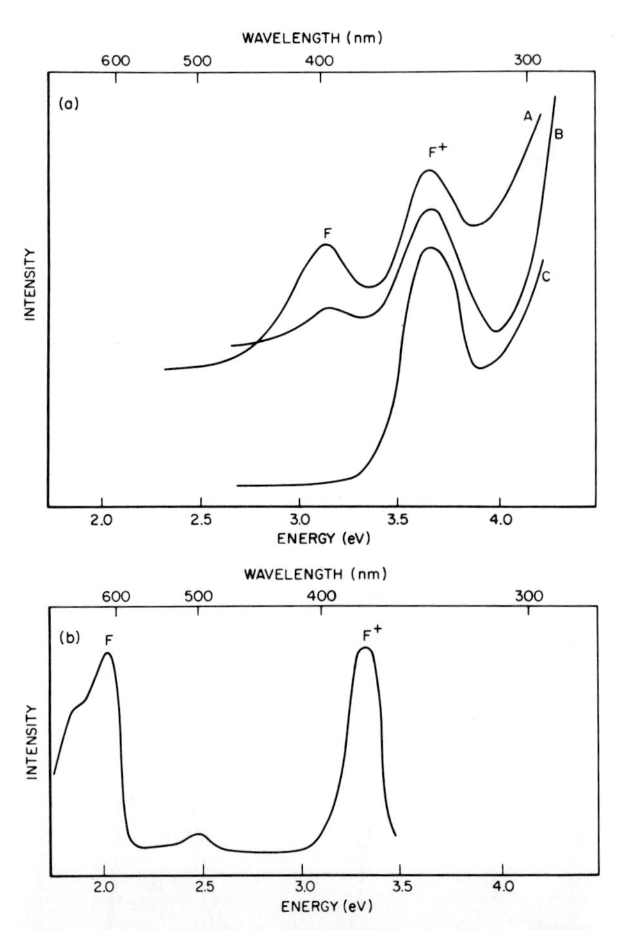

Fig. 14. (a) F^+ and F absorption bands in CaO at 300°K. Curves A and B refer to two different pieces of the same additively colored crystal, and curve C is for a neutron-irradiated sample. (b) Emission spectra from an additively colored crystal at 300°K. Excitation in the F absorption band only excites the emission bands at 2.5 and 2.0 eV. The F^+ emission band is excited only by absorption in the F^+ band. (After Henderson *et al.*[95])

in Fig. 13. The results of Kappers *et al.*[52] are not consistent with such a large value for the oscillator strength, and in fact require $f_F = 0.77$. On simple theoretical arguments, one expects $f_F/f_{F^+} \simeq 2$. This discrepancy remains to be resolved, as does the situation regarding the half-width of the F^+ band discussed in Section 3.4. Interconversion of F and F^+ centers in additively colored crystals of MgO is also implied by some of the Faraday rotation measurements of Kemp *et al.*,[50] who found that the powerful UV source used in their experiments converted F centers to F^+ centers. An emission band at 2.4 eV has recently been assigned to the F center in MgO,[52] although the nature of the excited states involved has not been established. Wood and Öpik[59] and Bennett[102] have predicted a rather small Stokes shift for the F band, although this appears inconsistent with the observed half-width of the absorption band (see also Section 4.2).

In CaO, the F and F^+ bands absorb at different energies, and as a result, it is rather easier to follow what is going on. The F band has been observed in additively colored crystals[99] and in crystals colored accidentally during growth.[24,95] Typical spectra are shown in Fig. 14. The band at 3.7 eV is the F^+ band, and the band at 3.1 eV has been identified as the F band. Evidence for the identification comes largely from photochemical interconversion, and from the interpretation of the emission spectrum, which we shall discuss in Section 4.2.

Kemp *et al.*[100] first showed that an interconversion process was possible in additively colored CaO crystals. By bleaching with "blue" light ($h\nu > 2.5$ eV) at 77°K, a metastable enhancement of the 3.7-eV (F^+) band occurs at the expense of the 3.1-eV (F) band. Bleaching with "red" light ($h\nu < 2.5$ eV) has the opposite effect. The processes are reversible, and can be cycled. The enhancement achieved under either bleaching treatment is 20–50%, and is stable for many minutes at 77°K. At 300°K, similar cycling is possible, but the results are stable for only about a minute, tending to revert to the situation after red-light bleaching (i.e., favoring the F band). At 4°K, no interconversion is possible. Although these photochemical processes are clearly consistent with the 3.1-eV band being due to the F center (as indeed the similar photoconversion effects described in MgO[51] are consistent with a similar interpretation), the mechanism is different from $F \leftrightarrow F^-$ conversion in additively colored alkali halides. There, bleaching at low temperatures in the F band releases the trapped electron, which traps at another F center to produce an F^- center, thus leaving behind an empty vacancy. Thus two F centers are destroyed for the creation of one F^- center. Bleaching with F^- light recreates the original situation, by trapping the released electron at an empty vacancy. In the oxides, this type of process is not possible because the F^+ centers cannot be bleached, and instead, it appears that the results of Kemp *et al.*[100] on CaO are to be interpreted in the following way. Blue light excites the F center and releases the second electron into the conduction band. It is then trapped at some impurity or other defect, so that the result of the "blue" process is the conversion of one F center to an F^+ center. "Red" light releases the electron from its (unidentified) trap, reforming the F center at the expense of an F^+ center. Slow thermal release of the electron from the trap has a similar effect. The kinetics of these reactions is by no means understood, and since it probably involves an impurity, it is likely that different investigators will produce a variety of $F^+ \leftrightarrow F$ recipies before the process is nailed down. For instance, Hall and Bessent[91] have effected $F^+ \rightarrow F$ conversion in neutron-irradiated CaO by low-temperature X-irradiation. Nevertheless, evidence so far is at least consistent with the 3.1-eV band in CaO being due to the F center, and highly suggestive that this is the case. The same conclusion applies for the F center in MgO. Neeley and Kemp[98] predict the F-band energy to be 4.4 eV in CaO, which is not as satisfactory as their prediction for MgO. Some recent work by Winterstein and Sieckmann[101] reports the observation of further products of photochemical reactions in CaO.

4.2. Electronic Structure of the F Center

In the sense that the F^+ center is analogous to a hydrogen atom (or He^+ ion in the oxides), the F center is analogous to the helium atom. The ground state is therefore $(1s)^2\,{}^1S$, and possible excited states are $(1s)(2s)\,{}^1S, {}^3S$ and $(1s)(2p)\,{}^1P, {}^3P$. The only allowed transition is ${}^1S \rightarrow {}^1P({}^1A_{1g} \rightarrow {}^1T_{1u}$ in group-theoretical notation; we shall tend to use the atomic nomenclature in this section to be consistent with the original work). The ${}^1S \rightarrow {}^1P$ transition is thus the F absorption band, and this is the transition to which we have referred in discussing calculated energies. Bennett[102] has recently performed more complete F-center energy level calculations than those mentioned in Section 4.1. He finds that for MgO and CaO, the lowest excited state is ${}^3P$, with ${}^1P$, ${}^3S$, and ${}^1S$ lying higher. Bartram[103] has also predicted this ordering. In the helium atom, the order of the excited states in increasing energy is ${}^3S$, ${}^1S$, ${}^3P$, and ${}^1P$. Bennett finds the ${}^1S \rightarrow {}^1P$ absorption energy to be 3.91 eV in MgO and 3.05 eV in CaO, compared with the experimental values of 5.01 and 3.10 eV. As far as emission transitions are concerned, the ${}^1P \rightarrow {}^1S$ transition is allowed, and we can expect that the ${}^3P \rightarrow {}^1S$ transition, although spin-forbidden, may be slightly allowed due to spin–orbit admixture of ${}^3P$ and ${}^1P$. The ${}^1S \rightarrow {}^1S$ and ${}^3S \rightarrow {}^1S$ transitions are parity-forbidden for electric dipole transitions. Bennett[102] predicts a negligibly small Stokes shift for the $^1P \leftrightarrow {}^1S$ transition, which is in sharp disagreement with experiment (see below) and inconsistent with the observed band half-widths [see Eq. (20) or (23)]. He predicts 2.53 eV in MgO and 1.93 eV in CaO for the $^3P \rightarrow {}^1S$ emission band. The band observed[52] at 2.4 eV in MgO may therefore possibly be this transition, a suggestion consistent with a measured lifetime of 7×10^{-4} sec at 1.6°K.[103a]

The emission spectrum obtained by exciting in the CaO 3.1-eV band has been observed by Henderson *et al.*[95] and is shown in Fig. 14. The emission bands are at 2.5 and 2.0 eV. The 2.0-eV band was found to have two components at 300°K (see Fig. 14), but only the higher-energy component remained at 4°K and has been studied in detail. At this temperature, the emission wavelengths are 500 and 601 nm. The lifetimes of these bands have been measured for optical excitation. At 4°K, the 500-nm band has a lifetime of 1 μsec, suggesting an allowed transition. The 601-nm band has short- and long-lived components. The lifetime of the short-lived component is 3×10^{-3} sec at 4°K, and at this temperature is the only observable component. At higher temperatures (77°K and above), the long-lived component appears and although the decay is found to be nonexponential, it has a persistence of several minutes. The temperature dependence suggests that a thermally activated process is responsible for the long-lived fluorescence, and that therefore the shorter lifetime should be associated with the lifetime of the excited electronic state itself.

The obvious interpretation of these results is that the 500-nm band (2.5 eV) corresponds to the $^1P \rightarrow {}^1S$ transition and that the 601-nm band (2.0 eV) is the $^3P \rightarrow {}^1S$ transition. The latter agrees well with Bennett's estimate.[102] This assignment is consistent with the values of the lifetimes measured for the emission bands, and also the separation of the 2.5-eV band and the F absorption band at 3.1 eV is about the expected value for the Stokes shift of the $^1S \leftrightarrow {}^1P$ transition based on the band half-widths. The oscillator strength of the $^3P \rightarrow {}^1S$ transition is found to be 5×10^{-7}, from the lifetime measurement. This is clearly reasonable for a spin-forbidden transition, and combined with the relative intensities of the two bands, suggests that the ^3P level is preferentially populated by excitation in the F band. This may occur either through a radiationless decay from the ^1P to the ^3P level, or by a process in which electrons are released into the conduction band from the ^1P state and then retrapped in the ^3P level. It is probable that the short-lived component of the 601-nm band is associated with the radiationless $^1P \rightarrow {}^3P$ decay, but that the second process explains the long-lived fluorescence through the intervention of a second electron trap, which traps the electrons released into the conduction band from the ^1P level and subsequently releases them slowly (cf. the mechanism of $F \leftrightarrow F^+$ conversion).

There is some evidence that Ti^{2+} impurities may be an important trap in this context, since Ti^+ formed by electron capture is reported to have a lifetime of 45 sec at 300°K and 200 sec at 77°K.[95,104] If this mechanism for the long-lived fluorescence is correct, then the fact that electrons can be thermally activated from the ^1P state into the conduction band at 77°K implies that the "ionization energy" of this state is less than a few tenths of an electron-volt. The $F \leftrightarrow F^+$ conversion experiments of Kemp *et al.* have similar implications. It is not clear which process quenches out first as the temperature is lowered: the thermal ionization of the ^1P level or the subsequent release of electrons from the (unknown) trap, although the slowness of the latter process at higher temperatures suggests the second possibility. Further experiments may help to clarify this point. Henderson *et al.*[95] found that rapid cooling to 77°K after excitation in the F band at 300°K causes quenching of the long-lived fluorescence. Thermoluminescence in the 601-nm band was then observed between 77°K and 300°K. This behavior is clearly consistent with the proposed model. More recently, Packwood and Hensley[105] have reported as many as five thermoluminescence peaks in this temperature range, indicating complex behavior. By pumping with F light to populate the ^3P level, Henderson *et al.*[95] found that light of wavelengths less than 1000 nm ($h\nu > 1$ eV) increased the F^+-center concentration, which was simultaneously being monitored with EPR. The experiment was performed at 300°K. Their interpretation is that the long-wavelength light excites electrons from the ^3P state into the conduction band, hence converting the F center into an F^+ center. The ^3P state is thus deduced to be about 1 eV below the bottom of the conduction band.

The energy level scheme for the F absorption and emission processes is shown in Fig. 15. The proximity of the excited states to the conduction band is reasonable for a neutral defect. The long-lived fluorescent component of the 601-nm band gives rise to a bright orange-red glow, which seems to be a feature of accidentally colored crystals from various sources. Although impurities may be involved indirectly as electron traps, the presence of F-band absorption at 3.1 eV is necessary for the appearance of the glow. The accidentally colored crystals are useful in this respect, since it has usually been found that some regions of the melt are colored and some are not, the variation often being over short distances of the order of a millimeter. Regions showing the 3.1-eV band give the fluorescence, neighboring regions with no 3.1-eV band (which are uncolored in appearance)

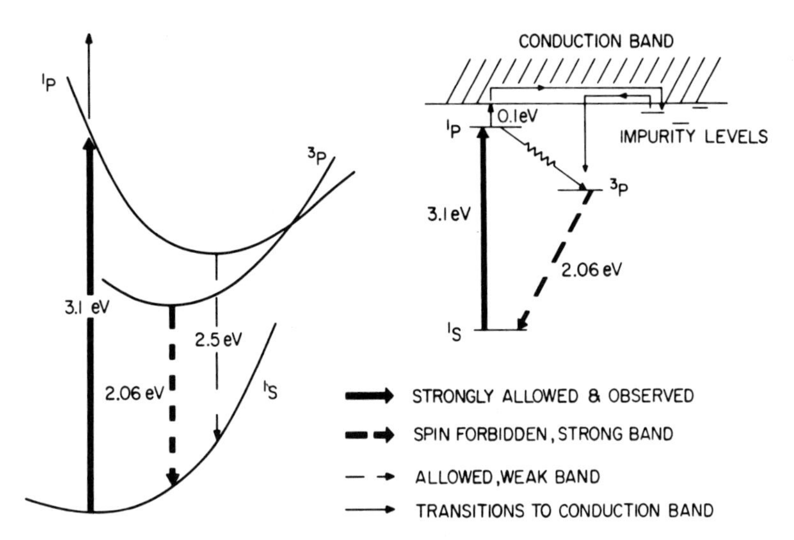

Fig. 15. Energy level scheme for absorption and emission processes associated with the F center in CaO. The left-hand diagram shows the configuration coordinate scheme for ${}^1P \leftrightarrow {}^1S$ and ${}^3P \rightarrow {}^1S$ transitions, and the right-hand diagram shows the location of F center and impurity levels within the band gap. (After Henderson *et al.*[95])

give no fluorescence. This seems to tie down the 601-nm band to the presence of F centers, assuming that the 3.1-eV band is the F band. The mechanism for the long-lived fluorescence we have described is clearly also in line with the $F \leftrightarrow F^+$ conversion experiments, where a second electron trap is postulated to play a similar role.

Measurements of the half-width of the 601-nm (${}^3P \rightarrow {}^1S$) band as a function of temperature[95b] show that [from Eq. (18)] $S \simeq 5.7$ and $\omega = 160$ cm^{-1}, suggesting the apperance of vibronic structure at low temperatures. This is indeed found; see Fig. 16. The zero-phonon line is at 574.0 nm and Henderson *et al.*[95] found that in their samples, the line was a partly resolved doublet. They suggested that this might be due to spin–orbit splitting of the ${}^3P$ level, although a consideration of selection rules for allowed transitions does not support this view. In fact, more careful measurements[106] show this splitting to be spurious, and due to strains in the original samples. Recently Edel *et al.*[106a] have observed EPR in the ${}^3P$ level. Their results are consistent with a tetragonal Jahn–Teller distortion of the F center when in the ${}^3P$ state.

The sidebands of the zero-phonon line indicated by arrows on Fig. 16 occur at approximately the same frequencies as those for the F^+ center.

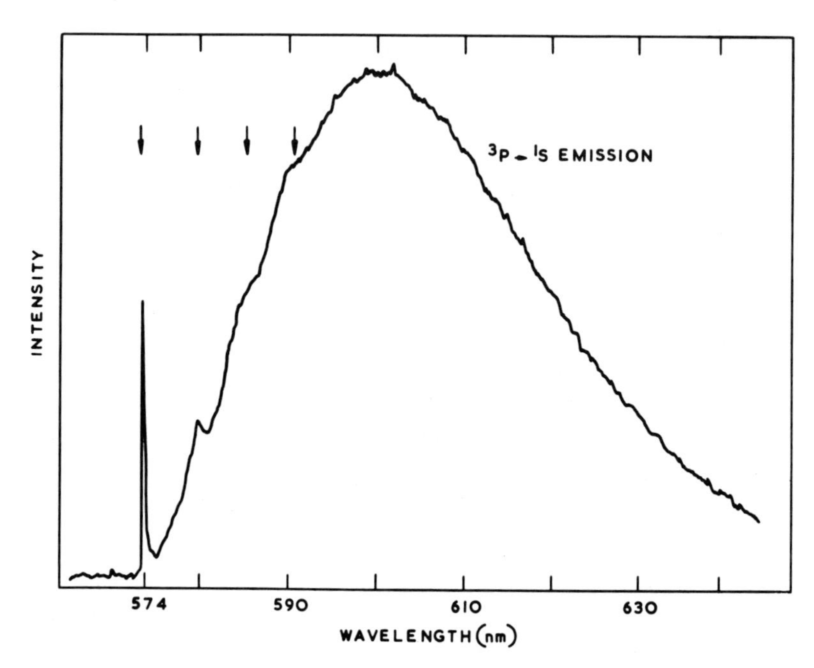

Fig. 16. The CaO F-center $^3P \rightarrow {}^1S$ emission band at 4°K (after Henderson *et al.*[95]).

No vibronic structure has been reported on the $^1P \leftrightarrow {}^1S$ emission or absorption bands, consistent with the relatively large half-width of the absorption band of about 0.3 eV, implying $S \approx 16$.

There are so far no experiments reported on MgO analogous to those just described on CaO. In SrO, the position of the F absorption band has been reported as 2.5 eV,[107] but no detailed experiments have been done. No F band has been identified in BaO as yet. Much of the work we have described relies on the known properties of F^+ centers to deduce the existence of F centers, but so far all the evidence points toward the correct identifications having been made. Further experiments, particularly on the more tractable CaO system, should prove very illuminating in providing more information on the structure and properties of the F center. The lack of detailed energy-level data on the less stable analogous alkali halide F^- center provides an area where the alkaline earth oxides can come into their own.

4.3. Annealing of *F*-Type Centers

Although this chapter is primarily concerned with the electronic properties of color centers in oxides, it is pertinent to briefly report the way in which vacancy centers anneal. In neutron- or electron-irradiated MgO, F and F^+ centers disappear near 500°C, whereas in additively colored crystals, annealing takes place at about 900–1000°C.[108] The kinetics of the annealing process have not been examined in detail, but it has been suggested[108] that in irradiated crystals, the anion interstitials become mobile at 500°C, and annihilate the vacancies. Alternatively, in neutron-irradiated crystals, the F^+ centers could be converted to P^- centers (see Section 5) as cation vacancies become mobile.[113] In additively colored samples, the annealing temperature must represent that of the diffusion of the vacancies themselves, although which charge state is involved is not known. No similar data appear to be available in CaO, SrO, or BaO. Various heat treatments can be used to produce aggregate centers in MgO, which are discussed in other sections.

4.4. *F*-Type Centers in Other Oxides

Very little conclusive evidence exists for the observation of F^+ and F optical bands in oxides other than the alkaline earth oxides. A large number of absorption bands are observed in irradiated ThO_2,[109] but none has been correlated with an anion vacancy defect. Perhaps the most promising example is the wurtzite structure ZnO,[110] where the so-called "*b*" band at 410 nm, close to the absorption edge, is reported to correlate to some extent with the EPR spectrum of the F^+ center[37] (Section 2.4). It is possible, however, that the band is due to interstitial zinc. Other oxides where EPR spectra possibly due to F^+ centers are observed include BeO, Al_2O_3, TiO_2, and $BaTiO_3$ (see Section 2.4). In no case does a convincing optical absorption band seem to have been found. However, Garrison *et al.*[111] have identified the F band at 188 nm in BeO electron-irradiated at 77°K. One should not be too pessimistic about the prospects in these materials, however, bearing in mind that only in the last ten years has appreciable progress been made on the alkaline earth oxides, which by and large are simpler in structure and now more readily available as well-categorized and nearly pure single crystals.

The problems of defect structure and nonstoichiometry of transition metal oxides form a field on their own. At present, there seems no clear point of contact between the detailed color center phenomena we have described and this other topic, but it is a goal worth aiming toward, perhaps where studies of color center aggregation and interactions will play their part.

5. PERTURBED F^+ CENTERS AND VACANCY PAIRS

5.1. EPR Studies

F^+ centers are introduced into alkaline earth oxides most efficiently by reactor irradiation. At low doses, the EPR transitions are usually narrow (width $\lesssim 0.6$–0.7 G) and easily resolved. At high doses, the lines broaden to such an extent that the hyperfine structure cannot be resolved. The magnitude of the effect for F^+ centers in MgO is such that the central transition is broadened by a factor of six as the F^+-center concentration increases from $\sim 10^{16}$ to $\sim 10^{19}$ cm^{-3}. The major sources of inhomogeneous broadening of EPR lines are (1) strain due to interstitial point defects and dislocation loops, (2) electric fields associated with the presence of charged defects, (3) dipolar broadening due to mutual magnetic interaction between close F^+ centers.* The progressive increase with neutron dose of the strain present in neutron-irradiated crystals is evidenced by the recent measurements of lattice parameter and crystal density.[113] If all the broadening is attributed to dipolar interaction, then for a linewidth of 4 G (observed when $N_F \approx 10^{19}$ cm^{-3}), it is required that on average, all F^+ centers must have other F^+ centers within a radius of 1.8 nm. This corresponds to an F^+-center concentration of at least 2×10^{19} cm^{-3}, rather larger than $N_F \approx 10^{19}$ cm^{-3} quoted above. Consequently, it appears that the effects of strain and electric fields are at least as important in perturbing the ground state of the F^+ center in neutron-irradiated MgO as are the dipolar interactions. In CaO, Kemp *et al.*[49,67] have observed a discrete splitting of the F^+-center resonance which they attribute to charged defects in the vicinity of the F^+ centers.

While the above effects are caused by remote perturbations of the F^+ center (remote in the sense that the first and second neighboring shells are complete), it is possible to produce more pronounced effects by association with other defects. Such an association between cation vacancies and F^+ centers can be promoted by annealing neutron-irradiated oxide crystals at temperatures above 300°C.[25] At these temperatures, cation vacancies are mobile and diffuse to the F^+ centers to form P^- centers.† Alternatively, oxide crystals may be γ-irradiated subsequent to plastic deformation (Section 5.2). Figure 17 shows the change in symmetry which accompanies

* A review of the inhomogeneous broadening of spectroscopic lines is given by Stoneham.[112]

† As noted in Section 1, we adopt the nomenclature suggested by Sonder and Sibley. These centers were originally designated F_2 centers[25] and later F_c^+ centers.[1]

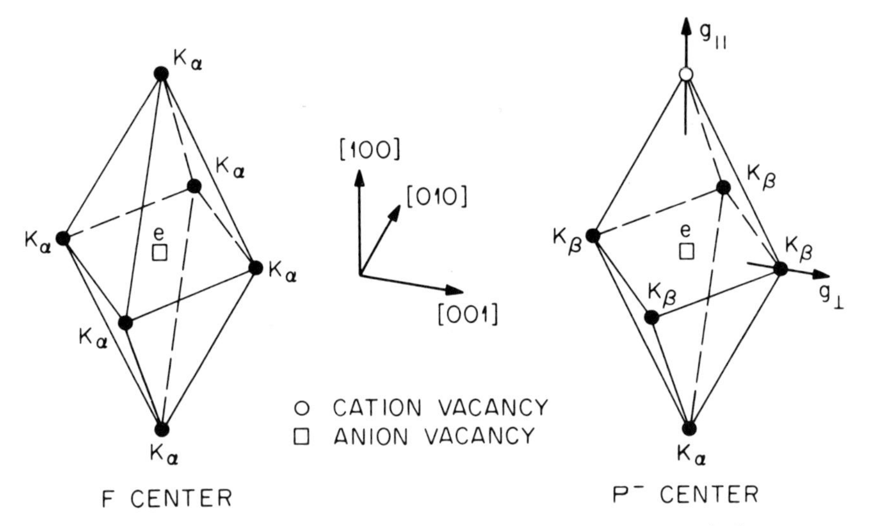

Fig. 17. Comparison of the F-center and P^--center structures. Nonequivalent nearest-neighbor nuclei of the P^- center are denoted by K_α and K_β.

the F^+-center $\rightarrow P^-$-center reaction. The absence of one nearest-neighbor cation results in a more complex hyperfine interaction than for the F^+ center. According to the structure shown in Fig. 17, there should be two groups of hyperfine lines because the four cation neighbors (K_β) in the plane perpendicular to the [100] axis are no longer equivalent to the single cation (K_α) situated on the defect axis. The hyperfine structure has been observed in MgO for nucleus K_α.[114] The observed spectrum can be fitted to Eq. (1) with $g_\parallel = 2.0004$, $g_\perp = 2.0012$, $a = -48.6$ MHz, and $b = -1.9$ MHz. The intensity of this spectrum is approximately 16% of that in the central line, in agreement with the predictions of Eq. (8). Note that this hyperfine interaction is roughly four times that for the F^+ center (Table III), showing the extent to which the cation vacancy affects the charge distribution. The hyperfine interaction with the four equivalent nuclei K_β is much weaker, possibly no more than 5 MHz. The hyperfine structure observed for P^- centers in MgO gives reasonable confidence that the model proposed in Fig. 17 is correct. In the other alkaline earth oxides, no such hyperfine interaction has been reported, although the appropriate g-tensor anisotropy has been observed for EPR signals in irradiated and annealed CaO and SrO (Table X). In each case, the g shift for P^- centers is greater than in the corresponding F^+ center. This is only to be expected since the cation vacancy results in considerable reduction in the binding energy for the electron at the divacancy. Thus the electron will overlap more onto the

Table X. Optical and EPR Constants for P^- Centers in Alkaline Earth Oxides

	MgO	CaO	SrO	BaO
$g_{\parallel}$	2.0004	1.9995	1.9839	—
$g_{\perp}$	2.0012	1.9980	1.9804	—
a, MHz	−48.6 (Expt.)	—	—	—
	−50.3 (Theory)	—	—	—
b, MHz	−1.9	—	—	—
E_b (Theory), eV	2.42	1.72	1.70	1.24
Bleaching peak, eV	3.60	—	—	—

ion-core orbitals, so admixing additional orbital momentum into the ground state.

The P^- centers have similar properties to S centers, F^+ centers situated near the surface of powder samples.[115] Both the g tensor and hyperfine structure testify to this similarity.

No analog of the P^- center has been observed in the alkali halides, although Pincherle[116] has investigated their possible properties theoretically. The P^- center does bear some resemblance to the F_A centers in the alkali halides[117] (F centers in which one of the six neighboring cation sites is occupied by an alkali ion impurity). In this case, it is notable that the hyperfine structure is greater for nuclei K_α than for K_β, although the effect is less pronounced than in the P^- centers.[19] It will be seen in Section 5.2 that no such similarity exists with the optical properties of F_A centers, where the lowering of the symmetry consequent upon the presence of the impurity ion splits the bound excited $2p$ level so that two characteristic transitions are observed.[117] In fact, the optical properties of P^- centers in the alkaline earth oxides have rather more in common with the F_A^- centers in the alkali halides.[117]

5.2. Optical Studies of Vacancy Pair Centers

The first indications of the nature of optical transitions expected from P^- centers came from Stoneham's unpublished calculations (see also Ref. 114). Using a semicontinuum model, in which the electron was assumed to be bound to the field of an electric dipole, it was shown that only the ground state was further than 0.01 eV from the conduction band. The calculated binding energies are given in Table X, where it can be seen that the depth of the trapping potential decreases with increasing lattice constant. These theoretical studies have now been extended to the magnitude of the hyperfine interaction,[114] within the original framework of Stoneham's calculation. The result of $a_\alpha = -50.3$ MHz is fortuitously good (see Table X). However, the theoretical ratio a_α/a_β is approximately equal to two, whereas experiment indicates $a_\alpha/a_\beta \approx 9$. More sophisticated treatments which include the discrete lattice as well as effects due to polarization, distortion, and finite ion size are underway. These improve the ratio of a_α/a_β, but increase the magnitude of a_α by a factor of two.

Since no bound states other than the ground $|1s\rangle$-like state are predicted, the P^- center should be bleachable with optical photons. This has been observed for P^- centers in MgO.[118] The concentration of P^- centers was optimized by annealing MgO crystals irradiated to a dose of 6×10^{19} nvt first at 720°C for 4 hr, followed by 1/2 hr at 900°C, and then X-irradiation with 33-kV X-rays. No F^+ centers survived this treatment, nor were significant numbers of V^- centers (Section 6) produced by the irradiation. The P^--center concentration was then measured continuously in the resonant cavity of an EPR spectrometer while bleaching with monochromatic radiation in the wavelength range 250–900 nm. The results in Fig. 18 show that the efficiency of bleaching is greatest with photons of energy ~3.6 eV.

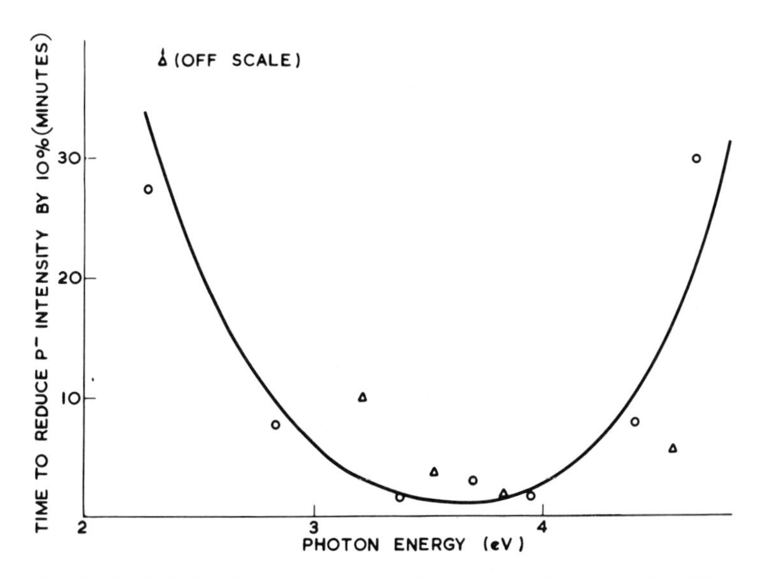

Fig. 18. Optical bleaching efficiency for the P^- center in MgO (after King and Henderson[118]).

Although this is larger than the predicted binding energy, it does demonstrate the essential correctness of the model. It should be remembered that the bleaching peak anyway only gives an upper limit to the ionization energy.

Recently, Sibley *et al.*[119] have investigated the formation of various color centers in MgO after plastic deformation of single-crystal samples. They observed no changes in EPR spectra of impurity ions even after deforming up to 14% in compression. However, such treatments do produce a new optical absorption band at 216.5 nm (Fig. 19a), the intensity of which increases linearly with compression up to 3% plastic strain. Turner *et al.*[120]

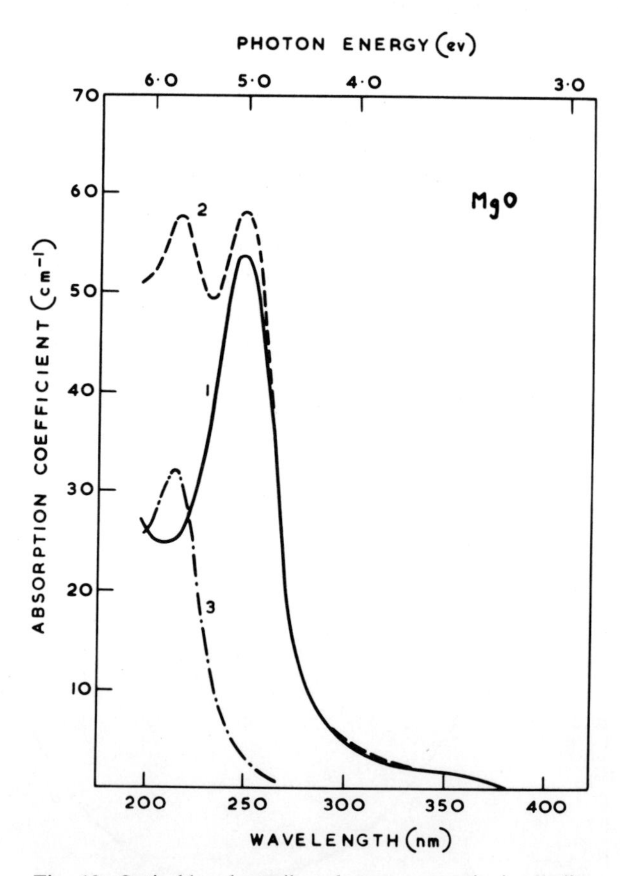

Fig. 19. Optical bands attributed to vacancy pairs in alkaline earth oxides. (a) MgO. Curve 1 is the F, F^+ composite band in a neutron-irradiated sample. Curve 2 results from a 1% deformation and curve 3 is the difference between curves 1 and 2. (After Turner *et al.*[120])

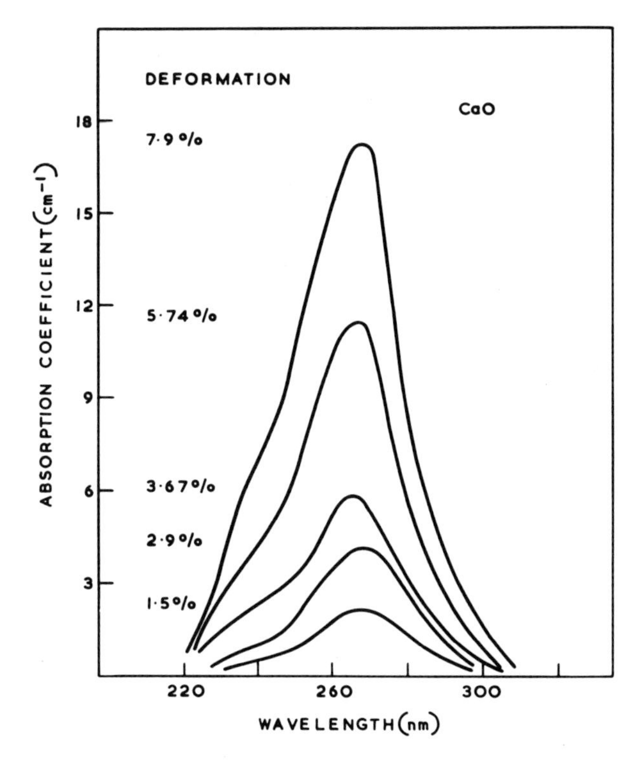

Fig. 19 (b). CaO. The deformation-induced band for various amounts of deformation. (After Turner *et al.*[120])

also find deformation-induced absorption bands in CaO and SrO at 268.0 and 304.0 nm, respectively (Figs. 19b and 19c). They attribute these deformation-induced bands in MgO, CaO, and SrO to the same center: a bound exciton formed in the neighborhood of a cation–anion vacancy pair. Arguments in favor of this proposal are the similar modes of production in the three oxides, the linear dependence of the absorption intensity on deformation, and the dichroic absorption of the bands (at least in CaO and SrO). The effect of the dichroism on the absorption spectra taken with polarized light is shown for SrO in Fig. 19(c); the results are consistent with electric dipole transitions occurring along a $\langle 100 \rangle$ axis of the crystal. Chen[121] has demonstrated that excitation in the CaO 268.0-nm band results in an intense blue luminescence peaking at 347.0 nm.

When the deformed crystals are γ- or X-irradiated, both F^+ and P^- centers are formed, as evidenced by their EPR spectra, indicating that single anion vacancies, in addition to the vacancy pairs, result from the inter-

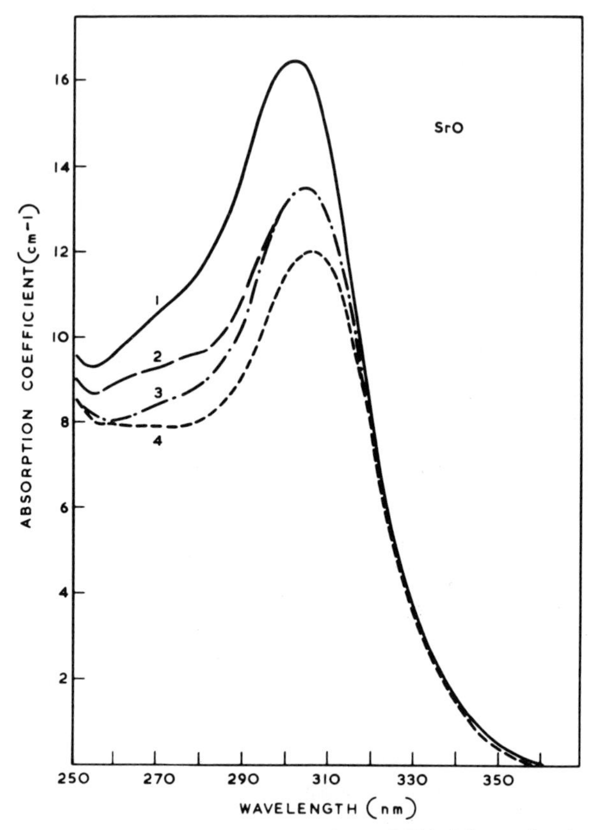

Fig. 19 (c). SrO. Spectra following a 2.5% deformation in compression along [100]. Curves 1 and 4 are with the electric vector **E** of the measuring light parallel and perpendicular to the [100] axis of the crystal. Curves 2 and 3 are with **E** at $\pm 45°$ to the [100] axis. (After Turner *et al.*[120])

section of dislocations. The concomitant decrease in the empty vacancy pair concentration which must result from the reaction $P + e \rightarrow P^-$ has not been unambiguously observed. Sibley *et al.*[119] are careful to note that in their samples, background changes near the 216.5-nm band make observation of the intensity changes on γ-irradiation difficult to follow quantitatively. Turner *et al.*,[120] however, report that X-irradiation does result in an initial small decrease in the intensity of the perturbed exciton band. The band then slowly returns to its original intensity, due to the thermal release of the X-ray-induced positive holes from their trapping sites, favoring recombination according to the reaction $P^- + h \rightarrow P$.

6. TRAPPED-HOLE CENTERS

6.1. Properties of Trapped-Hole Centers in the Alkaline Earth Oxides

The numerous EPR, optical, and magnetooptical studies of F^+ centers have given a very detailed description of their electronic structure (Sections 2 and 3). Less is known about an equally fundamental defect, the antimorph of the F^+ center. This defect is referred to here as the V^- center: it consists of a hole trapped on an oxygen ion neighboring a cation vacancy. Seitz proposed that an analogous defect was responsible for the V_1 band in the alkali halides,[122] but subsequent work has shown this speculation to be incorrect.[123] V^- centers may however, be produced in high concentrations in the alkaline earth oxides.[124] In the alkali halides, EPR studies demonstrate that positive holes trapped at cation vacancies achieve a lower-energy configuration by being localized between two adjacent halide ions. The resulting defect is a molecular ion X_2^- adjacent to the cation vacancy,[125] originally called the V_F center. The new notation would be V. Note the striking dissimilarity between the structure of the F^+ center, its antimorph the V^- center, and the V center in the alkali halides. In the F^+ center, the electron resides in the "1*s*" ground state of a center with cubic symmetry. The V^- center has tetragonal symmetry, with the hole trapped in a *p* state of the O^{2-} ion, whereas the alkali halide V center has orthorhombic symmetry due to the location of the trapped hole in a *p*-like state of the molecular ion. The only evidence for molecular ion formation in the alkaline earth oxides was reported by Rius and Cox,[126] who identified the mixed molecular ion $(OF)^{2-}$ in MgO crystals irradiated at 27°K. Several other examples of trapped-hole centers associated with impurity ions have been identified by EPR. Their structures are portrayed in Fig. 20.

In both MgO and CaO, large concentrations of V^- centers are produced by irradiation with ultraviolet light, X-rays, and low-energy electrons. The required trapping sites, the M^{2+} vacancies, are incorporated in the crystals in excess of the thermal equilibrium concentrations as charge compensators for impurities such as OH^-, F^-, Al^{3+}, Si^{4+}, etc. There are also present in all commercially available oxide crystals cation vacancies which compensate for the positive charge excess of the transition metal ions Fe^{3+}, Cr^{3+}, V^{3+}, etc. When present in high concentration, these transition metal ions do not noticeably enhance the concentration of V^- centers produced by irradiation. Rather, they compete with the cation vacancies as trapping sites for the holes released by the irradiation.

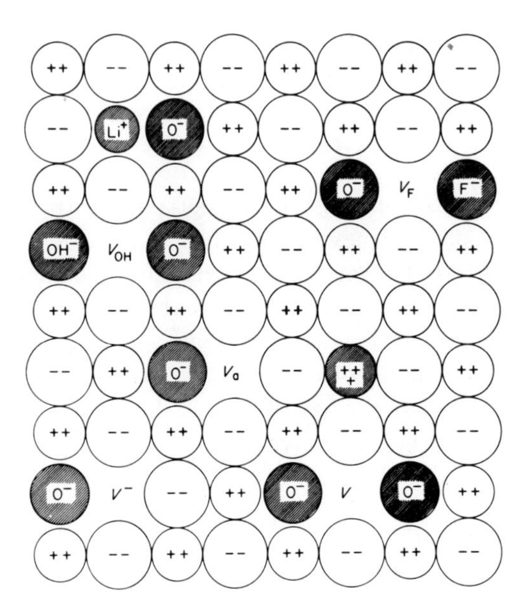

Fig. 20. Models for V centers in alkaline earth oxides.

The presence of V^- centers in X-irradiated MgO was first reported by Wertz *et al.*,[124] who observed the EPR spectrum shown in Fig. 21. Low temperature ($\sim$77°K) is required for best observation of this spectrum, but not for the production of the defects. At arbitrary orientation of magnetic field relative to the crystalline axes, the spectrum consists of three anisotropic lines. With the magnetic field directed along a $\langle 111 \rangle$ axis, a single line is observed. In the (100) plane, a maximum and minimum g value are observed 90° apart corresponding to the magnetic field being either perpendicular to or parallel to a $\langle 100 \rangle$ axis of the crystal. With $\mathbf{B} \parallel [100]$, the $g_{\parallel}$ line is half the intensity of the $g_{\perp}$ line, since both the [010] and [001] axes are perpendicular to [100] and are equally populated with defects. Thus, the EPR spectrum demonstrates the tetragonal symmetry of the defect, eliminating the O_2^{3-} molecular ion as a possible candidate for the spectrum. The EPR transitions are well described by the following spin Hamiltonian:

$$\mathscr{H} = g_{\parallel}\beta B_z S_z + g_{\perp}\beta(B_x S_x + B_y S_y) \tag{24}$$

with $g_{\parallel} = 2.0032$ and $g_{\perp} = 2.0385$. It is evident in Fig. 21 that at 77°K, the EPR linewidths are narrow; they are broader by a factor of 30 at

290°K. There is no temporal averaging of the spectrum due to hole hopping between the six equivalent anion sites and which would be evidenced by the three-line spectrum coalescing into a single line. Thus the lifetime of the hole on a particular site must be long compared with the inverse of the linewidth, i.e., $\tau > 10^{-7}$ sec. Similar EPR spectra have been associated with trapped-hole centers in CaO,[127] SrO,[26] and Al_2O_3,[128] and the appropriate g values are given in Table XI. Speculative identification of V^- centers in BeO[128a] and ZnO[128b] has now been reported.

Magnesium oxide crystals which contain V^- centers have a characteristic violet coloration due to a broad optical absorption band with peak intensity at 2.3 eV. Although the band peak energy is insensitive to temperature, the half-width decreases from 1.07 eV at 295°K to 0.96 eV at 77°K.[129] Parallel optical absorption and EPR measurements identify this band as arising from V^- centers. Estimates of the oscillator strength vary from[1] 0.07–0.08 to[129] 0.1–0.2. The V^- band can be bleached either optically

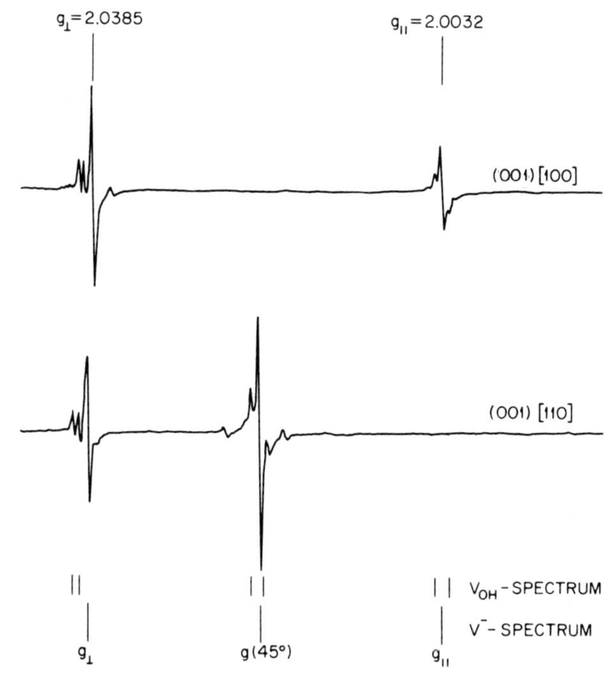

Fig. 21. EPR spectrum of the V^- center in MgO at 77°K. Spectra are shown for **B** in the (001) plane parallel to [100] and [110] directions, and the positions of weaker lines due to the V_{OH} center are indicated.

Table XI. Comparison of EPR Parameters for O^- Ions in Oxides and Some Alkali Halides

Crystal	Ref.	$g_\parallel$	$g_\perp$	Δ, eV		$-l_x^2\lambda/\Delta$	
				Obs.	Theory[a]	Obs.	Theory[b]
MgO	124	2.0032	2.0385	2.30	1.47	0.016	0.012
CaO	127, 144a	2.0011	2.0710	1.91	1.00	0.034	0.017
SrO	26	2.0010	2.0703	—	0.79	0.034	0.021
BaO	—	—	—	—	0.66	—	0.025
Al_2O_3	128	2.006	2.020	3.08	2.93	0.0073	0.0057
NaI	132	1.9769	2.2931	—	—	0.174	—
KI	132	1.9733	2.3023	—	—	0.182	—
RbI	132	1.9733	2.2888	—	—	0.174	—

[a] Ref. 130.
[b] Calculated assuming $\lambda = -135\ \text{cm}^{-1}$,[130] with $l_x = 1$ and the *theoretical* value of Δ.

or thermally. Optical bleaching occurs over the entire visible spectrum, but is most efficient with photons of energy $h\nu = 2.3$ eV. Although light polarized along $\langle 100 \rangle$ or $\langle 110 \rangle$ bleaches the V^- band, no dichroism is observed, even at 4°K. Consequently, the lifetime of the hole must be short compared with the time required for bleaching and measuring.[129] It must be concluded that the lifetime of the hole on a particular oxygen site is between 10^{-7} sec and several seconds.

6.2. The Electronic Structure of the V^- Center

The EPR spectrum in Fig. 21 demonstrates that the paramagnetic species occupies three equivalent sites in the crystal, and has tetragonal symmetry about the $\langle 100 \rangle$ crystal axes. The mode of formation strongly suggests that hole centers are involved. The ground configuration for a free O^- ion is $1s^2 2s^2 2p^5$, corresponding to a single ^2P term. If the O^- ion were in a cubic crystalline environment, then the orbital degeneracy of the ^2P state would be unaffected. However, in tetragonal symmetry, the ^2P term is split into two terms: an orbital singlet $^2P(A)$ and an orbital doublet $^2P(E)$. The splitting can be calculated using crystal field theory;[130] accordingly, the operator equivalent for the axially symmetric crystal field potential is

$$V_c = -\tfrac{3}{5} e C_{20} \langle r^2 \rangle [L_z^2 - \tfrac{1}{3} L(L+1)] \tag{25}$$

where e is the electronic charge and $\langle r^2 \rangle$ denotes the integral for the radial part of the O^- $2p$ function

$$\langle r^2 \rangle = \int_0^\infty [\psi_{2p}(r)]^2 r^4 \, dr$$

We are interested only in that part of the crystal field potential which results from the cation vacancy; this enters the crystal field as an effective charge $Z = -2$ on the z axis. Thus the crystal field term C_{20} is

$$C_{20} = -2e/R^3 \tag{26}$$

where R is the anion–cation separation, and the operator equivalent in Eq. (25) becomes

$$V_c = \tfrac{6}{5}(e^2/R^3)\langle r^2 \rangle [L_z^2 - \tfrac{1}{3}L(L+1)] \tag{27}$$

It is readily shown that this crystal field operator splits the ^2P term into the two terms at energies

$$\begin{aligned} W(E) &= \tfrac{2}{5}(e^2/R^3)\langle r^2 \rangle \\ W(A) &= -\,\tfrac{4}{5}(e^2/R^3)\langle r^2 \rangle \end{aligned} \tag{28}$$

As intuitively expected, $^2P(A)$ is the ground state. This corresponds to a hole in the $2p_z$ orbital, which points into the cation vacancy. Optical absorption corresponding to the transition $^2P(A) \to {}^2P(E)$ should thus occur at a photon energy Δ where

$$\Delta = \tfrac{6}{5}(e^2/R^3)\langle r^2 \rangle \tag{29}$$

In deriving Eqs. (26)–(29), it has been assumed that the vacancy-induced crystal field is the dominant factor in determining Δ. Configurational mixing and changes in the crystal field due to rearrangements of the ion positions around the vacancy are ignored. Neglect of such considerations is not particularly well justified. Transitions between the states $^2P(A)$ and $^2P(E)$ can only be induced by light polarized perpendicular to the z axis. Since the hole apparently migrates among the six equivalent positions at rates which are rapid compared with the measuring time, it is evident that the $^2P(A) \to {}^2P(E)$ transition is observed with essentially unpolarized light.

The transition energy [Eq. (29)] can now be predicted only if a reasonable value is known for the radial integral $\langle r^2 \rangle$. Bartram *et al.*[130] have calculated $\langle r^2 \rangle$ for the free O^- ion using a $2p$ wave function due to Hartree *et al.*[131] They find $\langle r^2 \rangle = 3.04$ atomic units. Estimated positions of these transitions for $MgO \rightarrow BaO$ as well as for Al_2O_3 are given in Table XI. It should be pointed out that these transitions involve a redistribution of electrons within the same type of orbital, in a single quantum shell. As such, they are strictly forbidden by the Laporte selection rule. That the oscillator strength for the 2.3-eV band is as high as 0.1–0.2 in MgO may be due to the transition being partially allowed by mixing of the $2p$ wave functions with s and d orbitals. This mixing is allowed by the lack of inversion symmetry at the center. Evidently, the transition should occur at increasingly long wavelengths as the lattice parameter of the crystals increases. It is clear from the result for MgO that the method underestimates the strength of the crystal field. Nevertheless, bearing in mind the simplifications of the model, the agreement is regarded as encouraging. It also serves to confirm that the trapping site is a cation vacancy rather than an impurity atom.

We now proceed to analyze the g values determined in the EPR experiments. To do so, we write an effective Hamiltonian

$$\mathscr{H} = \Delta[L_z^2 - \tfrac{1}{3}L(L+1)] + \lambda \mathbf{L} \cdot \mathbf{S} + \beta \mathbf{B} \cdot [\mathbf{L} + 2\mathbf{S}] \tag{30}$$

The first term is the familiar crystal field operator described in Eqs. (25) and (27). The second and third terms represent spin–orbit and Zeeman interactions, respectively. The eigenstates of the spin–orbit coupling operator,

$$\lambda \mathbf{L} \cdot \mathbf{S} = \lambda[L_z S_z + \tfrac{1}{2}(L_+ S_- + L_- S_+)] \tag{31}$$

are determined by factorizing the appropriate 6×6 matrix. Referring to "spin up" and "spin down" states by $(+)$ and $(-)$, respectively, it is easy to show that the spin–orbit interaction couples $p_z(+)$ to $p_x(-)$ and $p_y(-)$ (i.e., to the p_x and p_y functions with electron spins in the opposite sense). Similarly, the state $p_z(-)$ couples only to $p_x(+)$ and $p_y(+)$. The perturbed wave function $\phi(\pm)$ is written as[132,133]

$$\phi(\pm) = \mp A p_x(\mp) - iB p_y(\mp) + C p_z(\pm) \tag{32}$$

where

$$\begin{aligned} C &= \{1 + [2l_x^2 \lambda^2/(2\Delta - l_z\lambda)^2]\}^{-1/2} \\ A &= B = -l_x \lambda C/(2\Delta - l_z\lambda) \end{aligned} \tag{33}$$

Here, we have written $l_z = -i\langle p_y \mid L_z \mid p_x \rangle$ etc. For true atomic p functions, $l_x = l_z = 1$, but this need not be the case when admixtures from both the $3p$ functions of oxygen and from suitable combinations of neighboring core wave functions are taken into account.[132] Our Eq. (33) differs from that given by Brailsford *et al.*[132] in that we allow $l_x \neq l_z \neq 1$ in the spin–orbit interaction as well as the Zeeman interaction ($\mathbf{L} + 2\mathbf{S}$). The g values are found from the expectation value of the latter over $\phi(\pm)$ in Eq. (32). Thus, to order $(\lambda/\Delta)^2$,

$$\begin{aligned} g_{\parallel} &= 2\langle\phi(+) \mid L_z + 2S_z \mid \phi(+)\rangle \\ &= 2\{1 - [2l_x{}^2\lambda^2(2 - l_z)/(2\Delta - l_z\lambda)^2]\} \\ g_{\perp} &= 2\langle\phi(+) \mid L_x + 2S_x \mid \phi(-)\rangle \\ &= 2\{1 - [2l_x{}^2\lambda/(2\Delta - l_z\lambda)] - [2l_x{}^2\lambda^2/(2\Delta - l_z\lambda)^2]\} \end{aligned} \tag{34}$$

Evidently, since the shift in $g_{\parallel}$ involves only quadratic terms in $\lambda/(2\Delta - l_z\lambda)$ and since $\lambda \ll \Delta$, $g_{\parallel} \simeq 2.00$. Similarly, the quadratic term contributes very little to the value of $g_{\perp}$. The significance of l_x is that it scales λ, the spin–orbit coupling constant in the free O^- ion.

The important parameters appropriate to the EPR spectra of V^- centers in the alkaline earth oxides and in Al_2O_3 are given in Table XI. Evidently, for both MgO and Al_2O_3, $g_{\parallel}$ is greater than the free-electron g value, in contrast to the predictions of Eq. (34). This seems to imply that in these two cases, covalent bonding to neighboring ions is significant. Furthermore, the underestimated theoretical values of Δ for Al_2O_3 and MgO lead to fortuitously good agreement between experimental and computed values of $l_x{}^2\lambda/\Delta$. When the experimental value for Δ is used, the magnitude of $l_x{}^2\lambda/\Delta$ is much smaller than required by the experimental g values. Since $l_x \approx 1$ for O^- ions in numerous alkali halides, it is evident that λ must be greatly modified in the crystal, relative to the free ion.[32a] To explain the experimental value of $g_{\perp}$ using the observed Δ thus requires $|\lambda| \approx 250\ \text{cm}^{-1}$. No convincing explanation of this anomaly has yet been put forward. The parameters quoted in Table XI for O^- ions in alkali halides are taken from the work of Brailsford *et al.*[132] In these cases, it is found that $g_{\parallel} < 2.00$. This is expected in the alkali halides, at least in comparison with MgO and Al_2O_3, since the bonding is almost entirely ionic.

6.3. The Stability of the Trapped Hole

It was noted in Section 6.1 that the V^--center concentration is impurity-dependent, the transition group elements cobalt, chromium, iron, and titanium being particularly effective in suppressing the V^--center formation. However, in crystals which are relatively free of transition metal ions, the production of V^- centers depends upon thermal history. For example, the maximum defect concentration attained in a particular crystal of MgO is suppressed by slow cooling from about 1000°C.[129] This process results in the precipitation of impurities from solid solution, so eliminating the need to charge compensate for various aliovalent ions. Conversely, rapid quenching from 1000°C *enhances* the peak intensity in the V^- band. As Fig. 22 shows, the most pronounced increase is observed in the range 600–900°C, which is just the temperature range over which precipitation is most efficient. Quenching from above ~1000°C does not further increase the concentration of V^- centers.

At 4°K, the V^--center concentration is stable almost indefinitely unless exposed to light, whereas at 77°K, a small decay rate is observed. At room temperature, the V^- centers decay with a half-life of 2–7 hr, depending upon crystal purity.[129] Searle and Glass[134] have examined in detail the thermal decay of the V^- centers in MgO over the temperature

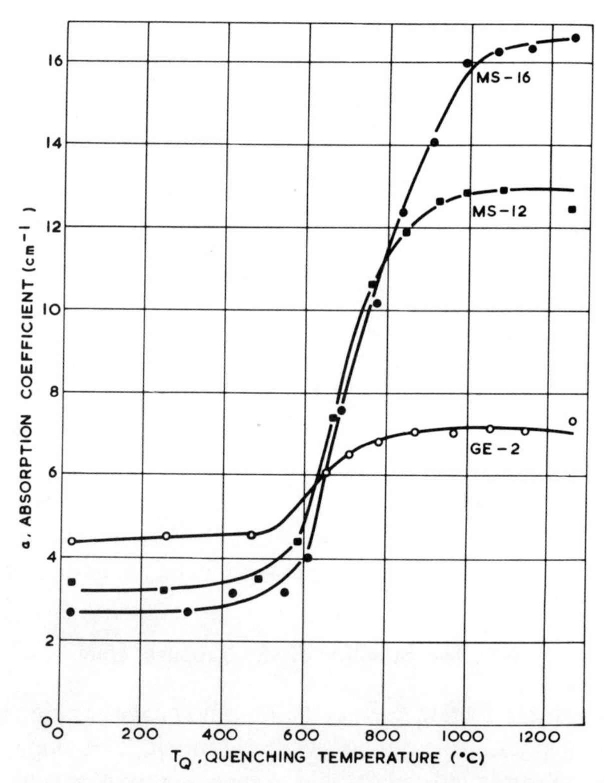

Fig. 22. Intensity of the V^- optical band as a function of quenching temperature for three different MgO samples (after Chen and Sibley[129]).

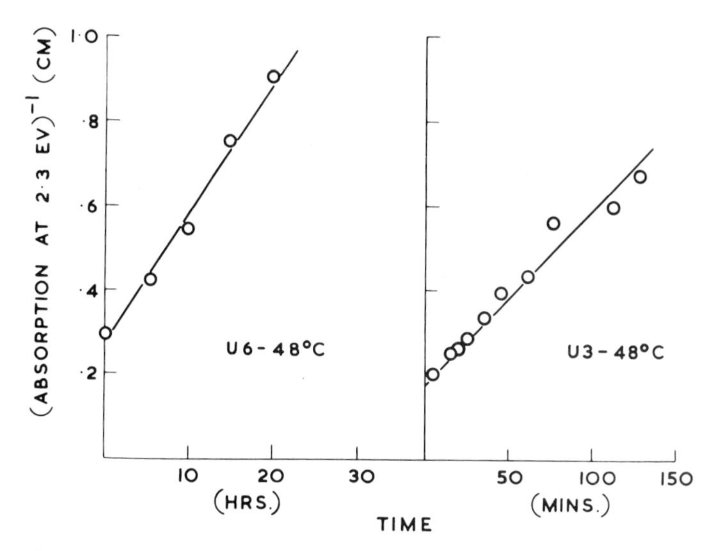

Fig. 23. Annealing plots for the V^- center in two different samples of MgO, as measured by Searle and Glass.[134] The temperature was 48°C and the V^- centers were monitored through the optical band at 2.3 eV.

range 18–50°C. Their results for two different crystals annealed at 48°C are shown in Fig. 23. They deduce that the decay process is bimolecular, involving the V^- center and only one electron trapping level. The hole is released from the V^- center with an activation energy of 1.13 ± 0.05 eV, whence it recombines with the trapped electron or is retrapped at the cation vacancy. Searle and Glass attribute the electron trap to Fe^{2+}, since they observe a bright red luminescence to accompany the decay. This luminescence had been assigned by Hansler and Segelken[135] to the reaction

$$Fe^{2+} + (\text{free hole}) \rightarrow Fe^{3+} + h\nu$$

More recent work shows that the $Fe^{2+} \rightarrow Fe^{3+}$ conversion is accompanied by an intense blue luminescence,[119] whereas the red luminescence corresponds to the reaction[136]

$$Cr^{2+} + \text{hole} \rightarrow Cr^{3+*} \rightarrow Cr^{3+} + h\nu$$

Careful measurements of the temperature dependence of optical bleaching of the V^- center in the 2.3-eV band can be used to estimate the position of the excited levels of the V^- center. Searle and Bowler[137] show that there is a bound excited level 17 ± 2 meV above the valence band edge. Using an effective mass model to describe such an excited state, they determine

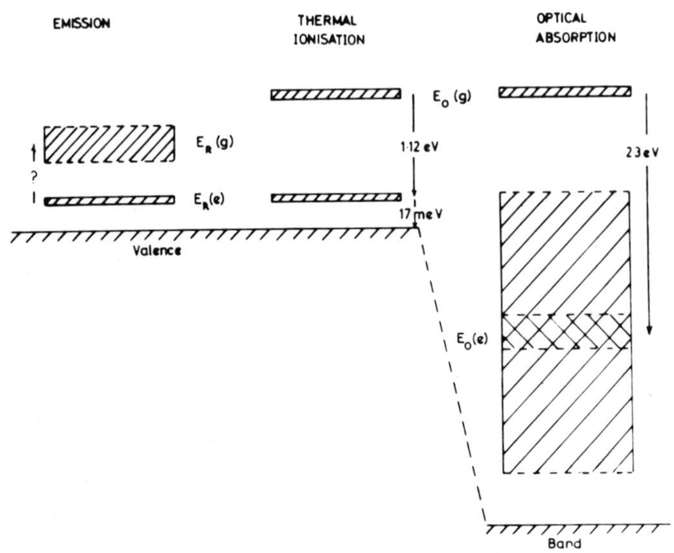

Fig. 24. Energy level diagram for the V^- center in MgO, after Searle and Bowler.[137] The differences between emission, thermal ionization, and optical absorption diagrams represent the relaxation of the center.

a Bohr radius of ~4 nm, consistent with a diffuse state. The average effective mass is $m^* = 0.12m$. The radiative lifetime of the excited state is determined as $\tau_R = (11 \pm 7) \times 10^{-7}$ sec. A simplified energy level diagram is given in Fig. 24. In including the thermal ionization it is assumed that relaxation in the excited state occurs prior to ionization.

6.4. Impurity Effects

The properties of the V^- center may be modified by interaction with neighboring impurities. Such effects have been recognized by EPR, ENDOR, and infrared spectroscopy, and some of the results are summarized in Table XII. The modified V^- centers shown in Fig. 20 are referred to as V_{OH}, V_{F}, and V_a according as the cation vacancy has a neighboring OH^-,[138,138a] F^-,[139,139a] or M^{3+} ion.[1] In the latter case, the defect is charge-neutral overall: The O^- ion is expected to occupy a site on the opposite side of the vacancy to the trivalent cation on account of their mutual repulsion. This center has an EPR spectrum[1] described in MgO by Eq. (24) with $g_{\parallel} = 2.0032$ and $g_{\perp} = 2.0390$. In the V_{OH} center, the OH bond is directed into the vacancy. The presence of the nearby proton is evidenced by the resolved hyperfine structure in the EPR spectrum.[138] The EPR and ENDOR transi-

Table XII. Spin-Hamiltonian Parameters for Trapped-Hole Centers

Crystal	Center	$g_{\parallel}$	$g_{\perp}$	a,[a] MHz	b,[a] MHz	P, MHz	Ref.
MgO	V^-	2.0032	2.0385	—	—	—	124
	V_{OH}	2.0033	2.0396	± 0.044	± 2.3755	—	138
	V_{OD}	2.0033	2.0396	± 0.0073	± 0.363	0.173	143a
	V_F	2.0031	2.0388	0.234	0.923	—	139, 139a
	$[Li]^0$	2.0043	2.0542	∓ 4.80	± 2.12	—	141
	V_a	2.0032	2.0390	—	—	—	1
CaO	V^-	2.0011	2.0710	—	—	—	127
	V_{OH}	2.0011	2.0702	∓ 0.0110	± 1.3650	—	138a
	V_F	2.0015	2.0714	0.24	0.61	—	141
	$[Li]^0$	2.0017	2.0885	∓ 2.49	± 1.34	0.0027	141, 144a
	$[Na]^0$	2.0002	2.1234	∓ 9.14	± 1.84	1.2	144a
SrO	V^-	2.0010	2.0703	—	—	—	26
	$[Li]^0$	1.9999	2.0931	∓ 1.8	± 0.9	—	141
BeO	$[Li]^0$	2.0023	2.0207	∓ 5.85	± 3.61	—	144
ZnO	$[Li]^0$	2.0028	2.0253	∓ 3.20	± 1.90	—	144

[a] The absolute signs of a and b cannot be determined experimentally, and relative signs are given here. The negative value of a for the $[Li]^0$ and $[Na]^0$ centers (see text) is deduced by assuming that the anisotropic dipolar interaction b has the same sign as the nuclear g value ($+ve$ for Li and Na).[144]

tions may be used to obtain the following spin-Hamiltonian parameters: $S = I = 1/2$, $g_{\parallel} = 2.0033$, $g_{\perp} = 2.0396$, $a = 0.044$ MHz, $b = 2.3755$ MHz. In their ENDOR investigation of this spectrum, Kirklin *et al.*[138] ascribed a shift in the mean position of the two proton ENDOR transitions relative to the resonance frequency of a free proton in the same magnetic field to a chemical shift of unknown origin. Davies[140] has shown that if the nuclear spin of the proton is quantized along the direction of the total effective magnetic field rather than along the applied field, then the discrepancy is removed. The anisotropy of the hyperfine structure is consistent with the proton being situated along the tetragonal axis of the defect, and separated by 0.322 nm from the trapped hole. Assuming that the oxygen ion in the OH^- complex is separated from the O^- ion on the opposite side of the cation vacancy by the normal cell length 0.42 nm, one obtains 0.098 nm for the OH^- bond length, in close agreement with the bond length in $Mg(OH)_2$. The decay of the V_{OH} center at 20°C has been monitored with

EPR by Kappers and Wertz,[140a] who have shown that the center decays by release of the trapped hole. The optical band of the V_{OH} center in MgO has been isolated, and peaks at 2.2 eV.[140b,144a]

In CaO, the V_{OH} center has been identified by EPR and ENDOR studies.[141,138a] The hyperfine parameters are given in Table XII. The V_{OH} center has also been identified in SrO.[138b]

The presence of OH^- in the crystal is also detected by a series of infrared absorption lines in the spectral region 3200–3700 cm^{-1}.[138,142] A prominent line at 3296 cm^{-1} has been shown to be due to the OH^- stretching frequency modified by the presence of an "empty" cation vacancy in a nearest-neighbor site (i.e., the V_{OH}^- center). The trapping of a hole by the V_{OH}^- center to form a V_{OH} center is accompanied by a decrease in the 3296-cm^{-1} absorption and the growth of another line at 3314 cm^{-1} due to the V_{OH} center. Searle[143] has used these lines to study the kinetics of dissociation of the V_{OH}^- centers in the temperature range 470–620°C. He finds that the V_{OH}^- centers dissociate according to the reaction

$$V_{OH}^- \rightleftharpoons [OH^-]^+ + V^{2-}$$

with an activation energy of 1.43 eV (charge differences from the normal lattice are shown). An unidentified cation impurity was also assumed to participate in the reaction and the binding energy of the vacancy to the OH^- ion was deduced to be 0.23 eV.

The OD^- ion has also been studied in MgO.[143a] When located next to a cation vacancy, the OD^- stretching vibration absorbs at 2444.7 cm^{-1}. This frequency is shifted to 2464.5 cm^{-1} when a hole is trapped in forming the V_{OD} center. Note that the frequency ratio $\bar{\nu}_{OH}/\bar{\nu}_{OD}$ is 1.347 for both the cation vacancy–OH^- (or -OD^-) associate and the V_{OH} (or V_{OD}) configuration. The theoretical ratio for the stretching mode is 1.37. The EPR and ENDOR measurements confirm the structure of the V_{OD} center; the spin-Hamiltonian parameters are given in Table XII.

The V_F center is also charge neutral. The MgO spin-Hamiltonian parameters are $S = I = 1/2$, $g_{\parallel} = 2.0031$, $g_{\perp} = 2.0388$, $a = 0.234$ MHz, and $b = 0.923$ MHz.[139,139a] Similar results have been obtained for the V_F center in CaO (see Table XII). Hyperfine structure from ^{17}O has been observed for the V_F center in MgO crystals enriched with ^{17}O. The splitting is described by $a = 67.3$ MHz, $b = 124.0$ MHz.[139b] These results demonstrate conclusively that the V_F center consists essentially of an O^- ion.

Recently, Schirmer has observed an interesting variation on the V^--center structure, not only in MgO, CaO, and SrO,[141] but also in BeO and

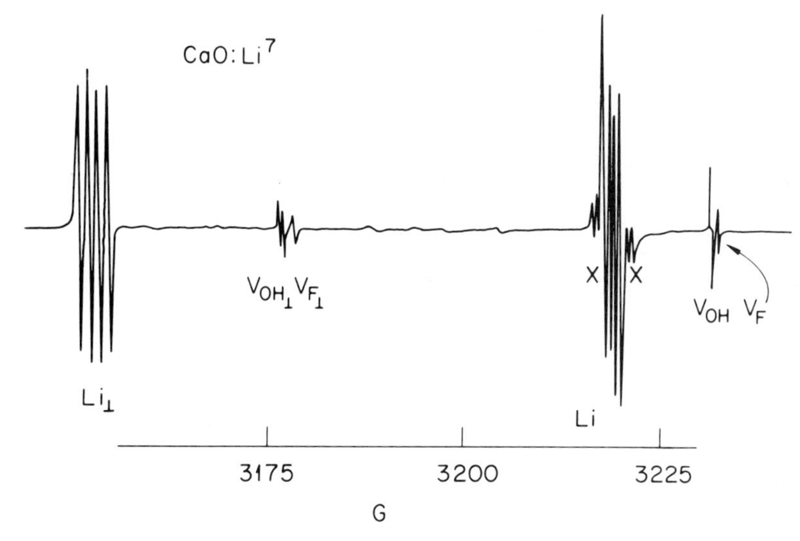

Fig. 25. EPR spectrum of the $[Li]^0$ center in CaO (after Schirmer[141]).

ZnO.[144] In this case, the positive hole is trapped on an oxygen ion neighboring a Li^+ impurity ion. A spectrum showing resolved ^7Li hyperfine structure from this center in CaO is shown in Fig. 25. According to the present notation, the center should be called $[Li]^0$. The hyperfine structure clearly identifies the structure of the defect. Other authors have also made EPR and optical measurements on $[Li]^0$ and $[Na]^0$ in MgO, CaO, and SrO.[141a,144a] The spin-Hamiltonian parameters for the $[Li]^0$ centers in MgO, CaO, and SrO and the $[Na]^0$ center in CaO are given in Table XII. An obvious point of interest is the very significant positive g shift for the $[Li]^0$ and $[Na]^0$ centers relative to the V^- centers. This can be understood by reference to Eq. (26), where it can be seen that the crystal field term C_{20} for the $[Li]^0$ or $[Na]^0$ center is half that appropriate to the V^- center. Thus the value of Δ in Eq. (29) is reduced, so increasing the value of $g_\perp$ given by Eq. (34). Consequently the experimental result $g_\perp[Li]^0 > g_\perp(V^-)$ is anticipated from the simple theory outlined in Section 6.2. Optical experiments show that although Δ is reduced in the $[Li]^0$ and $[Na]^0$ centers relative to the V^- center, the change is not as large as predicted by Eq. (34).[144a] A further interesting feature of these results is the apparently negative value of the contact hyperfine interaction for the $[Li]^0$ and $[Na]^0$ centers (see Table XII). Schirmer originally explained this result for the $[Li]^0$ centers in terms of exchange polarization of the Li $1s$ electrons,[144] but his later work shows that this is not an adequate interpretation.[141]

7. AGGREGATE CENTERS

7.1. Zero-Phonon Lines in MgO

Some types of aggregate centers are described in Sections 5 and 8, principally those studied using EPR. In this section, we discuss centers whose properties have been elucidated mainly through optical studies exploiting zero-phonon lines. It will be seen that these apparently probe rather different defects than do the magnetic measurements.

Wertz *et al.*[145] first reported an abundance of zero-phonon line structure in heavily neutron-irradiated MgO (dose $>10^{19}$ nvt). Subsequently, these lines have received considerable attention in view of the information on defect symmetry types available from uniaxial stress splitting data. Faraday rotation measurements have also contributed to this area of research. Many of the lines studied in this way have been attributed to defects of symmetry lower than cubic, and models proposed for the centers have drawn heavily on the types of *F*-aggregate centers identified in the alkali halides. Zero-phonon-line studies in the alkaline earth oxides have proceeded more or less in parallel with similar work in the alkali halides,[81] but the photochemical and kinetic evidence for particular types of aggregate center in the latter materials[146] has been largely lacking in the oxides. As a result, some of the detailed models are more speculative and rely to some extent on the alkali halide analogy. There is some evidence, however, that thermal annealing studies in the oxides may help to put the models on a firm basis.[108,147] The divalent nature of the oxide lattice again produces some interesting comparisons and contrasts with the alkali halides.

We have already mentioned in Section 3.4 the type of information about the electronic structure of a color center which can be obtained from experiments on zero-phonon lines. The detailed theory of uniaxial stress effects has been comprehensively reviewed by Kaplyanskii[148] and Runciman,[149] The essential concept is the removal of various types of degeneracy of the center upon the application of the perturbation, resulting in a splitting of the spectral line. If the line is narrow, as for a zero-phonon line, this splitting can be directly resolved and the polarization of the components studied in detail. The types of degeneracy possessed by a color center may be of the following types[148,149]:

1. Orientational degeneracy. If the defect symmetry is lower than cubic, then there is degeneracy associated with equivalent orientations in the lattice.
2. Electronic degeneracy. Centers possessing relatively high symmetry (with at least a threefold or fourfold axis of symmetry) may have energy

levels which are degenerate (other than Kramers degeneracy, which cannot be removed by a stress).

3. Orientational plus electronic degeneracy may occur for centers with tetragonal or trigonal symmetry.

In general, the information available from a uniaxial stress experiment is restricted to the symmetry of the center and of the electronic states responsible for the optical transition. Further information, particularly about the spin of the center and whether or not it has inversion symmetry may be obtained from magnetic[150] and electric field[151] measurements, respectively. Relatively little work on these two techniques has been reported on the oxides, although we shall mention Faraday rotation work in Section 7.2.2.

Usually, the half-widths of zero-phonon lines in MgO at low temperatures tend to be rather larger than the widths of 5–20 cm^{-1} typically observed

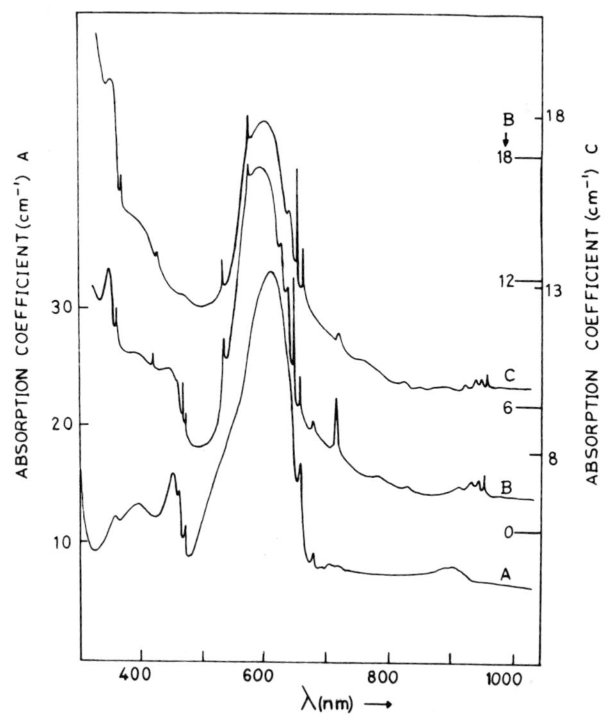

Fig. 26. Zero-phonon line structure observed at 77°K in neutron-irradiated MgO. (A) 6.8×10^{19} nvt at 150°C followed by 400°C anneal, (B) 6.8×10^{19} nvt at 600°C, (C) 7×10^{18} nvt at 45°C and annealed at 600°C. (After Henderson and Wertz.[1])

in the alkali halides. This is undoubtedly due to the strains produced in the lattice by the large doses of neutrons necessary to form the defects. It is not unusual to find some lines with half-widths at 4°K as large as 100 cm^{-1} [152] The widths may be reduced to workable proportions for stress-splitting experiments by annealing treatments, or by performing the neutron irradiation at elevated temperatures (450°C). Details of the types of treatment resulting in narrow lines are given by King and Henderson.[152] Annealing at 600°C also produces some changes in the intensities of the lines. Some typical spectra are shown in Fig. 26, and similar results can be found in Fig. 24 of the chapter by Sonder and Sibley.

A wealth of vibrational structure accompanies some of the zero-phonon lines, and various attempts have been made to assign the phonon-assisted sidebands to particular critical points in the phonon dispersion curves of MgO. Experience in the alkali halides suggests that this approach can sometimes be confused by serious modifications of the density of states by the defect, but nevertheless, a surprising number of cases are known where host lattice phonons appear to play an important role.[81] Table XIII gives the wavelengths of some of the zero-phonon lines observed in MgO and the tentative assignments of phonon-assisted peaks. In general, it can be seen that the defects couple to several different modes of vibration, as indeed do most *F*-aggregate centers in the alkali halides. This is expected for complex aggregate centers, where the low symmetry allows many different modes to couple to the electronic states. One prominent zero-phonon line at 524.8 nm (see Section 7.2.2) appears to have no detectable vibrational sidebands.

The intensities and half-widths of the 361.8- and 1045.0-nm zero-phonon lines have been followed as a function of temperature by Kazumata *et al.*[156] The fractional intensity can be fitted to the expression $P_{00} = \exp\{-S[1 + 6.6(T/\theta_D)^2]\}$ which is equivalent to the expression given in Section 3.4 when a Debye spectrum of coupled modes is assumed.[74] The required value of θ_D was 625°K, rather lower than the thermodynamic value. This trend has also been found for zero-phonon lines in the alkali halides. Kazumata *et al.* were able to fit the half-widths W to the relation $W = A[\coth(\hbar\omega/2kT)]^2$ with $\hbar\omega = 160$ cm^{-1} for the 361.8-nm line and 217 cm^{-1} for the 1045.0-nm line. This expression strictly is not appropriate to zero-phonon lines, where the temperature dependence of the half-width would be expected to vary according to the related function $[\sinh(\hbar\omega/2kT)]^{-2}$ (see Ref. 74). In a more recent study, Nakagawa and Ozawa have interpreted the temperature dependence of the halfwidths using a Debye spectrum of modes.[156a]

Table XIII. Phonon-Assisted Transitions and Zero-Phonon Lines in MgO

Zero-phonon line		Phonon-assisted line, cm^{-1}	Energy difference, cm^{-1}	Phonon assignment	Ref.
nm	cm^{-1}				
361.8	27651	27839	188	Local mode A	153
		27943	292	TA [100], TA [110] or TA [111] (B)	
		28014	363	2×A	
		28052	401	TO [110] or TO [000]	
		28114	463	LA [110]	
		28227	576	3×A or 2×B	
		28285	634	TO [111]+(TA [100] or TA [110])	
468.5	21345	21594	249	Local mode	152
		21758	413	LO [110], TO [100]	
		22183	838	2×LO [110] or 2×TO [100]	154
470.0	21276	21701	425	LO [110], TO [100]	152
					154
642.0	15579	15853	274	TA [111]	154
		15936	357	TO [111]	
		16000	421	LA or TO [100], LO [110]	
		16041	462	LA [110]	
		16279	700	LO [000]	
		16447	868	2×LA [100]	
649.0	15406	15691	285	TA [111]	154
		15753	347	TO [111]	
670.6	14912	15204	292	TA [110]	152
705.8	14168	14384	216	Local mode	154
		14453	285	TA [111]	
		14522	354	TO [111]	
		14556	383	TO [110]	
993.1	10070	10450	380	TO [110]	155
		10840	770	2×TO [110]	
		11050	980	2×(LO [100] or LA [111])	
		11210	1140	3×TO [110]	
		11550	1480	3×LO [100]	
		12030	1960	4×LO [100]	
1045.0	9570	9850	280	TA [111] or TA [110]	155
		9950	380	TO [110]	
		10210	640	TA [111] + TO [110]	
		10330	760	2×TO [110]	
		10690	1120	3×TO [110]	

7.2. Uniaxial Stress Experiments

The results of uniaxial stress experiments on zero-phonon lines can be conveniently divided into groups according to the deduced symmetry of the color centers involved. So far, three classes have been observed in MgO: orthorhombic centers with a ⟨110⟩ symmetry axis, trigonal centers with electronic degeneracy in one of the participating states, and monoclinic centers. In Kaplyanskii's notation,[148] the orthorhombic and monoclinic centers have orthorhombic I and monoclinic I symmetry, respectively.

7.2.1. *Orthorhombic Centers*

Orthorhombic I centers have been associated with the zero-phonon lines at 361.8 and 413.2 nm studied by Ludlow and Runicman,[153] the 468.5-, 470.0-, and 670.6-nm lines studied by King and Henderson,[152] and the 1045.0-nm lines studied by Ludlow[155] and Stettler *et al.*[157] The

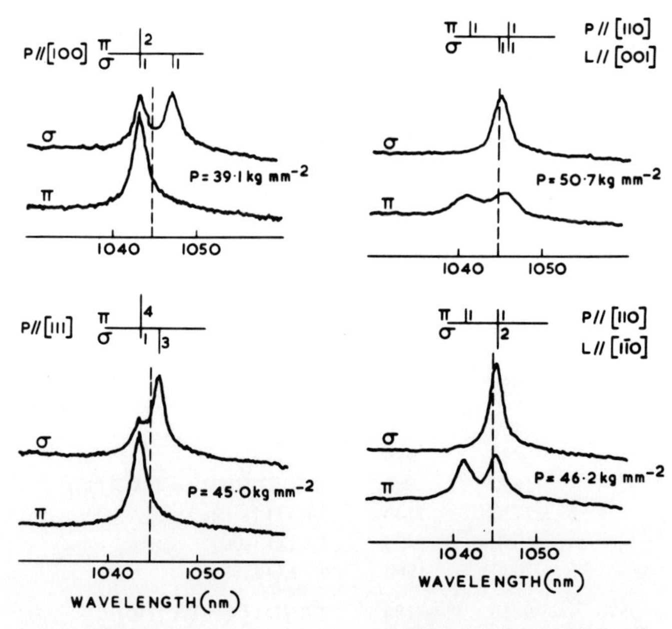

Fig. 27. Stress-splitting patterns for the 1045.0-nm line in MgO. The schematic patterns show the lines expected for a center of orthorhombic I symmetry with a ⟨110⟩ transition moment. π and σ refer to light polarized with the electric vector parallel and perpendicular to the stress, respectively. (After Ludlow.[155])

Table XIV. Stress and Strain Coupling Coefficients for Orthorhombic Centers in MgO

Line position, nm	Stress coefficients, $cm^{-1}\ kg^{-1}\ mm^2$			Strain coefficients, $10^4\ cm^{-1}$ per unit strain			Ref.
	A_1	A_2	A_3	B_1	B_2	B_3	
361.8	0.98	0.70	0.38	4.21	2.97	4.20	153
413.2	0.62	0.55	0.55	3.81	2.04	0.89	153
468.5	−0.17	0.54	0.40	2.53	1.24	0.64	152
470.0	−0.15	0.52	0.49	2.62	1.04	0.82	152
670.6	0.11	0.05	0.45	2.21	0.45	0.41	152
1045.0	−0.60	0.38	0.33	1.50	0.44	−1.15	155

splitting patterns for the 1045.0-nm line observed by Ludlow[155] are shown in Fig. 27. There is excellent agreement with the theoretical patterns for an orthorhombic center with a transition electric dipole moment along a $\langle 110 \rangle$ axis. The results of Stettler *et al.*[157] are similar as regards the line intensities, although the magnitude of their line shifts are not entirely in agreement with Ludlow's results. One of the authors (BH) has checked the line shifts, and found close agreement with Ludlow's values.

The splitting of the lines is found to be linear with applied stress. The stress perturbation can therefore be represented by a perturbing potential[148]

$$V = \sum_{ij} A_{ij}\sigma_{ij} = \sum_{kl} B_{kl}e_{kl} \tag{35}$$

where σ_{ij} and e_{kl} are the stress and strain tensor components, respectively. The number of independent parameters A_{ij} and B_{kl} is determined by the symmetry of the center, there being three stress parameters A_1, A_2, and A_3 (coupling coefficients) for an orthorhombic I center. Strain coupling coefficients B_1, B_2, and B_3 can also be defined which are related to A_1, A_2, and A_3 by the elastic constants of the crystal. Strictly speaking, these should be the local elastic constants, but these are generally unavailable and the bulk constants are used.

Values of the A's and B's for the orthorhombic centers are shown in Table XIV. The B's have the order of magnitude of $10^4\ cm^{-1}$ per unit strain, which is typical for strain coupling coefficients of color centers (cf. the strain coupling parameters for the CaO F^+ center, Section 3.4). Although all the centers have B coefficients of the same order of magnitude, only the pair of lines at 468.5 and 470.0 nm have closely similar values for B_1, B_2, and B_3.

The assignment of detailed atomic models on the basis of these results must be somewhat speculative, but a number of clear features have emerged. Two pairs of lines appear to be correlated, but in different ways. We have already mentioned the similarity of the coupling coefficients for the 468.5- and 470.0-nm lines. Despite this similarity, the relative intensities of the lines do not change in parallel fashion though various heat and bleaching treatments.[152] Thus the lines are not due to the same center, but may be due to related centers. It has been suggested that one may be a slightly perturbed version of the other, for instance, by an impurity ion along one of the twofold rotation axes $\langle 110 \rangle$ of the defect.[152]

Ludlow[155] has suggested that the 361.8- and 1045.0-nm lines may be due to different transitions within the same center, since the lines change in parallel in samples irradiated to different neutron doses. Their intensities increase for doses up to 10^{19} nvt and decrease for doses between 10^{19} and 10^{20} nvt. This has been confirmed by Chen *et al.*,[108,147] who have also shown

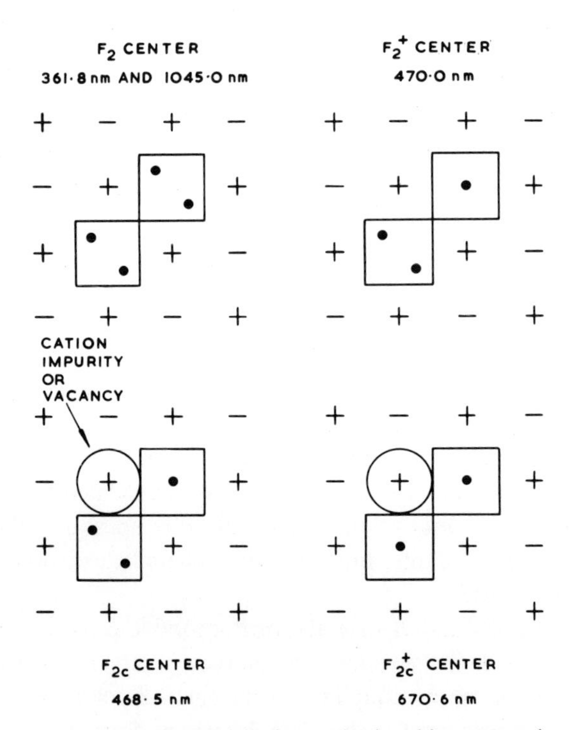

Fig. 28. Some models for orthorhombic centers in MgO (after Henderson and Wertz[1]).

that parallel behavior of the lines occurs during annealing. The different stress coupling coefficients and the different vibrational frequencies derived by Kazumata *et al.*[156] for the two lines do not, of course, conflict with the lines being due to different transitions within the same center.

The growth of the zero-phonon lines during irradiation shows that the the color centers involve intrinsic defects, possibly in association with impurities. Ludlow[155] found no effect of transition metal impurities on the intensity of the 361.8- and 1045.0-nm lines. The most obvious models for orthorhombic centers are analogs of the F_2 (M) center in the alkali halides. Some possible models are shown in Fig. 28, there being various possibilities due to different charge states and impurity association. It is also likely that cation vacancies could be included, since these are also produced by neutron irradiation.

Some doubts about the identification of these lines with F-aggregate centers have emerged from the results of Chen *et al.*[108,147] They were unable to produce the lines by annealing additively colored and electron-irradiated (2 MeV) samples above 900°C, the temperature at which F and F^+ centers diffuse (see Section 4.3). Instead, new zero-phonon lines at 1081.6 and 1324.2 nm were observed. However, by irradiating with fast electrons with energies up to 29 MeV, which produce similar secondary cascades to neutrons, the orthorhombic I centers were created.[158] The conclusion seems to be that whereas the centers may well involve small clusters of anion and cation vacancies (or possibly interstitials), the simple F_2 models may be incorrect. Unfortunately, it appears that there are no orthorhombic centers observed by EPR[25,159] which can be correlated with any of the zero-phonon lines.

7.2.2. *Trigonal Centers*

Some rather more detailed and provocative results have evolved from uniaxial stress experiments on the lines at 524.8, 642.0, and 649.0 nm. These have been shown to be due to $A \leftrightarrow E$ type transitions within centers of trigonal symmetry.[152,160] Furthermore, the stress measurements are able to show which state is the ground state through the observation of stress-induced dichroism.[161,162] Removal of the twofold degeneracy associated with an E ground state results in two levels split by amounts comparable to kT at low temperatures. The transitions from these levels to the A excited state have intensities weighted by a Boltzmann factor, and temperature-dependent dichroism arises in the spectrum. No such effect occurs with an A ground state, since there is then no electronic degeneracy to be removed by the stress.

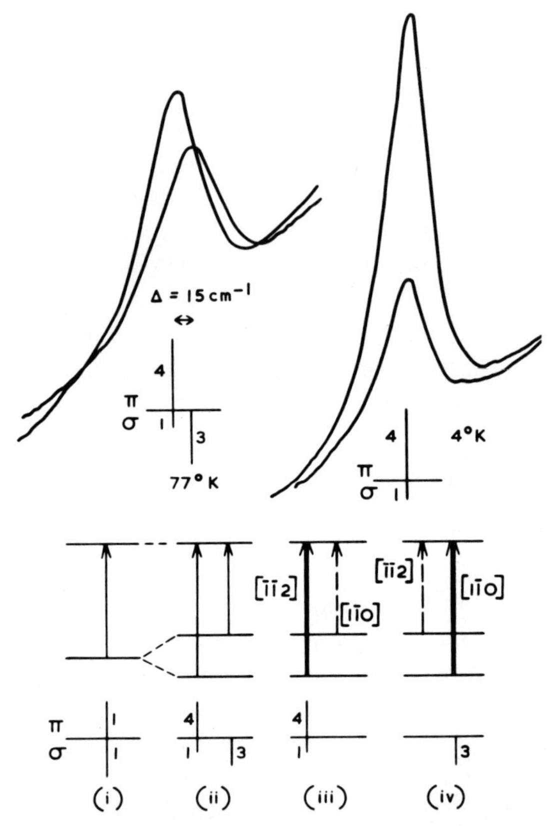

Fig. 29. Stress-induced dichroism of the 524.8-nm line in MgO under a [001] stress of 52 kg mm^{-2}. The lower diagram shows the expected lines for an $E \rightarrow A$ transition at 77°K under (i) zero stress; (ii) high stress. At 4°K, (iii) applies if the lowest level gives rise to a [$\bar{1}\bar{1}2$] transition moment and (iv) if it has a [$1\bar{1}0$] transition moment. (After King.[154])

Stress-induced dichroism for the 524.8-nm line[154] is shown in Fig. 29 for a stress applied along a ⟨100⟩ direction. The sign of the dichroism implies that the lower-lying component of the E state under a [001] stress gives rise to a transition moment in the [$\bar{1}\bar{1}2$] direction for a [111]-oriented center.

The 642.0- and 649.0-nm lines give no dichroism and are attributed to $A \rightarrow E$ transitions.[154] Ludlow[160] has found that in high resolution, the 649.0-nm line is in fact a doublet with a separation of 9 cm^{-1} between the components. Since there is no temperature dependence of the intensities of the two components, the splitting is in the excited state. It is tempting to

Table XV. Stress and Strain Coupling Coefficients for Trigonal Centers in MgO

Coupling coefficients		524.8 nm	642.0 nm	649.0 nm
Stress coefficients, $cm^{-1}\ kg^{-1}\ mm^2$	A_1	0.18	0.077	0.055
	A_2	−0.18	−0.190	−0.165
	B	−0.08	−0.130	0.041
	C	−0.13	−0.190	−0.106
Strain coefficients, $10^4\ cm^{-1}$ per unit strain	$(c_{11} + 2c_{12})A_1$	0.86	0.37	0.26
	$c_{44}A_2$	−0.29	−0.30	−0.26
	$(c_{11} - c_{12})B$	−0.17	−0.28	0.09
	$c_{44}C$	−0.21	−0.30	−0.17

interpret this splitting as being due to spin–orbit coupling in a ^2E state, and a detailed analysis of the uniaxial stress effects by Ludlow[160] shows this to be a consistent interpretation. However, a later study[160a] has cast doubt on this scheme. Ludlow has also confirmed King and Henderson's assignment of the 642.0-nm line to an $A \rightarrow E$ transition in a trigonal center.

The energy shifts of the components in a stress splitting pattern for a trigonal $A \leftrightarrow E$ transition may be expressed in terms of four stress coupling coefficients A_1, A_2, B, and C,[163] where A_1 and A_2 describe the removal of orientational degeneracy and B and C the removal of electronic degeneracy. Strain coupling coefficients may again be derived using the crystal elastic constants. The values of the parameters derived for the three zero-phonon lines are given in Table XV. There is no clear similarity between the values for the different centers, except perhaps for A_1 and A_2 between the 642.0- and 649.0-nm lines. This may be compared with the situation in the alkali halides, where in LiF, three trigonal centers are found, the F_3 (R), F_3^+, and F_3^- centers.[81,164] The coupling coefficients for their zero-phonon lines are all rather similar.[164] This does not rule out the possibility that the MgO lines are due to related centers, but is an interesting contrast. All three lines in MgO appear to be due to different centers, since their intensities do not vary in the same way during heat treatment. King[154] has found that some reversible interconversion of the 642.0- and 649.0-nm lines is possible by optical bleaching, which suggests different charge states of the same defect.

In searching for suitable models for these three trigonal centers, we have to bear in mind that satisfactory models should be able to explain the occurrence of A or E ground states, and the possibility that at least the 642.0- and 649.0-nm lines are related. By analogy with the alkali halides, the obvious models involve F_3-type centers. The assignments of the lines on this basis have been discussed by Henderson and King[159] and Ludlow.[160] These assignments are based on a molecular orbital description of the electronic states of the F_3 centers, an approach which is extraordinarily successful in predicting the lowest states of F_3-type centers in the alkali halides.[164,165] Two different defect aggregates have been proposed: the simple F_3 center consisting of three anion vacancies lying in an equilateral triangle on the (111) plane, and F_{3c} centers in which a cation vacancy is added along the [111] axis. The symmetry is C_{3v}. These models are shown in Fig. 30, along with the simple molecular orbital prediction of the energy levels. The basic idea is that from the 1*s* orbitals of the constituent F^+ centers, one can form molecular orbitals of a_1 and *e* symmetry (lower case letters are used for one-electron orbitals). Intuitively, one expects that the a_1 orbital lies lower because of its higher electron density in the center of the equilateral triangle. Ludlow[160] has suggested that in the F_{3c} configuration, the *e* orbitals could be lower, which produces a third variation on the scheme. This is not shown in Fig. 30, but is included in the discussion below.

Consider first the F_3 models. The most reasonable charge states $F_3{}^+$, F_3, and $F_3{}^-$ involve five, six, and seven electrons, respectively. For the

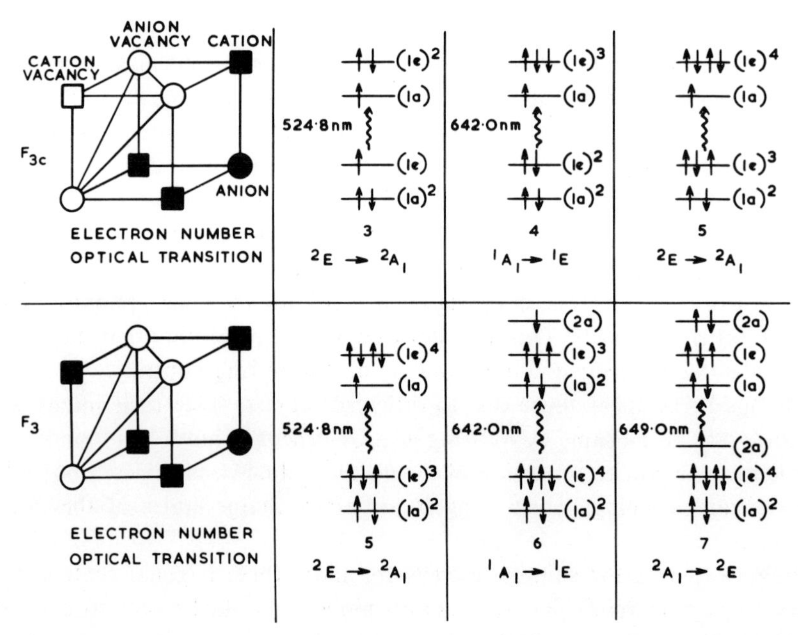

Fig. 30. Models and molecular orbital schemes for trigonal centers in MgO. The possible assignment to zero-phonon lines is shown. (Adapted from Henderson and Wertz.[1])

seven-electron center, it is necessary to include another a_1 molecular orbital, derived from the F^+-center $2p$ states. The predicted ground states are ^2E, $^1A_1$, and $^2A_1$, so that the assignment to the 524.8-, 642.0-, and 649.0-nm lines follows naturally. The excited states can also be reasonably accounted for with the same molecular orbitals. The interconversion of the 642.0- and 649.0-nm lines by bleaching[154] is also reasonable since the neutral and electron-excess F_3 and F_3^- centers are involved.

The F_{3c} models are not quite so satisfactory all round, since the simple molecular orbital prediction for F_{3c}^+, F_{3c}, and F_{3c}^- centers (three, four, and five electrons) gives ^2E, $^1A_1$, and ^2E ground states, respectively, as in Fig. 30. In the case of F_{3c}, the configuration $(1a_1)^2(1e)^2$ also gives rise to $^3A_2$ and ^1E levels, and the choice of $^1A_1$ as the ground state is somewhat arbitrary. In the analogous four-electron F_3^- center in LiF, the $^3A_2$ is lowest.[166] Thus the only really satisfactory assignment involving the F_{3c} models is that of the F_{3c}^+ or F_{3c}^- centers to the 524.8-nm line.

Ludlow[160] has suggested that if the $1e$ orbitals are placed below the $1a_1$, then the F_{3c} models predict the observed ground states for all three lines. Although it seems unlikely that the cation vacancy could actually reverse the order of the one-electron orbitals, the configurations used by Ludlow may be important when electron–electron interactions are included in the calculation, allowing configuration interactions. This configuration mixing may be very important where large numbers of trapped electrons are involved, as for these trigonal centers. No attempt at a proper molecular calculation of the electronic states of aggregate centers in MgO has yet been made.

There are some other experimental results which have an important bearing on the structure of the trigonal centers. Glaze and Kemp[167] have observed a Faraday rotation signal from the 524.8-nm line, which they quote as being at 523.3 nm. The wavelength discrepancy is probably not significant, since Glaze and Kemp report a half-width of 73 cm^{-1} for the line (cf. Fig. 29). The rotation signal is shown in Fig. 31, and can be seen to have a dispersion-like shape characteristic of orbital singlet-to-singlet transitions (cf. the CaO F^+ center, Section 3.4). The rotation is paramagnetic, following a $\tanh(g\beta B/2kT)$ temperature and magnetic field variation characteristic of a spin of $\frac{1}{2}$. This is consistent with the spin doublet ground states predicted for the F_3^+, F_{3c}^+, and F_{3c}^- centers by the molecular orbital approach. At first sight, the dispersion-like rotation shape is inconsistent with the twofold orbital degeneracy, but in an exhaustive analysis, Glaze and Kemp show that large, random strain splitting of the orbital E state is responsible for the observed shape. The essential argument is that if the

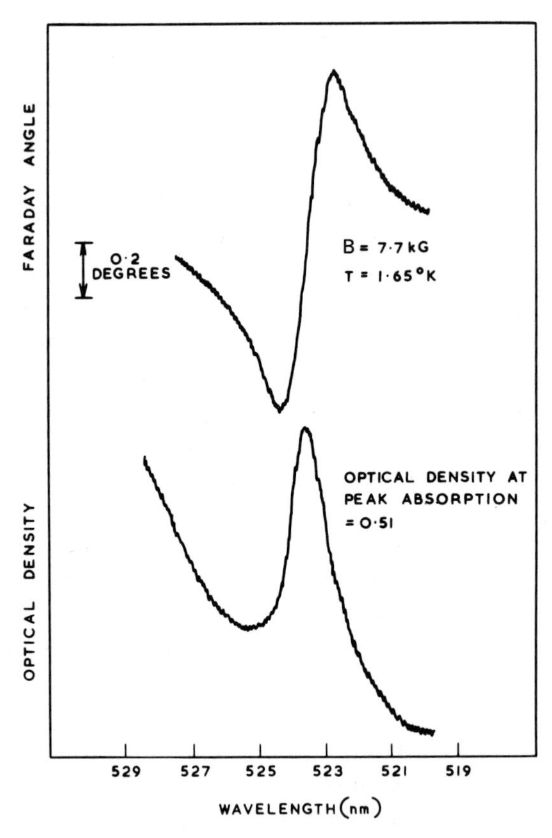

Fig. 31. Faraday rotation and optical density for the 524.8-nm line in MgO (after Glaze and Kemp[167]).

strain splitting is much larger than kT, the transition is from an orbitally nondegenerate state. The analysis has much in common with the work on circular dichroism of the R_2 zero-phonon line in the alkali halides.[168] From their results, Glaze and Kemp are able to derive the spin–orbit coupling constant for the ^2E ground state, and make some estimates of the orbital g value. In common with work on the E states of F_3 centers in the alkali halides,[166,168,169] the orbital g value is very small, suggesting orbital reduction by a dynamic Jahn–Teller effect.[170] Glaze and Kemp find $g_{\text{orb}} \lesssim 0.07$ and $1\ \text{cm}^{-1} \lesssim \lambda g_{\text{orb}} \lesssim 2\ \text{cm}^{-1}$. Here, g_{orb} refers to the reduced orbital g value, sometimes written $g_{\text{orb}} = a\Lambda$, where a is the Jahn–Teller reduction factor[170] and Λ the orbital angular momentum of the *electronic* state. The most important result is the *positive* sign of λ, opposite to the negative sign observed for the F_3 center in alkali halides.[168] This suggests that a "hole" in the $1e$ "shell" may be responsible for the positive spin–orbit coupling in the case of the 524.8-nm line. (The alkali halide F_3 center has one *electron* in the $1e$ shell.) Going back to our discussion of the relative merits of the F_3^+, F_{3c}^+, and F_{3c}^- centers as candidates for the 524.8-nm line, the "holelike" situation favours the five-electron F_3^+ or F_{3c}^- centers as opposed to the three-electron F_{3c}^+ center.

Glaze and Kemp[167] were unable to find any Faraday rotation associated with the 642.0-nm line, consistent with its assignment to centers with spin singlet ground states, F_3 or F_{3c}. Unfortunately, they were unable to observe the 649.0-nm line with sufficient intensity to perform magnetooptical experiments and hence explore further the model of a ^2E excited state.

Spin-resonance data on trigonal centers are rather meagre. Wertz *et al.*[25] reported a trigonal system referred to as the V_5 center, which they proposed was either a local body-centred structure or an interstitial O^- ion with an adjacent cation vacancy. However, King and Henderson[152] were unable to detect this center in the samples used for their zero-phonon-line work. More recently, Henderson[171] has reported a trigonal spectrum observed in MgO neutron-irradiated at 600°C. Under these conditions, the concentration of centers giving rise to the trigonal centers is comparable with the F^+-center concentration, and hence the latter does not necessarily swamp any other paramagnetic centers. Measurements at Q band showed that the spectrum could be described by a spin Hamiltonian

$$\mathscr{H} = \beta[g_{\parallel} B_z S_z + g_{\perp}(B_x S_x + B_y S_y)] + \mathbf{I} \cdot \mathsf{A} \cdot \mathbf{S}$$

with $g_{\parallel} = 2.0026$ and $g_{\perp} = 2.0039$. The hyperfine structure with ^{25}Mg was consistent with the structure of an F_3 aggregate rather than an F_{3c} aggregate, since all four cation near-neighbors are required. No correlation of this system with any one of the "trigonal" zero-phonon lines has been specifically reported, but the $^2A_1$ ground state of the F_3^- center and hence the 649.0-nm line seems the most likely candidate. Spin resonance in the ^2E state of the center associated with the 524.8-nm line might be difficult to observe due to residual strain problems similar to those found for the ^2E ground state of the F_3 center in alkali halides.[172]

The assignment of the 524.8-, 642.0-, and 649.0-nm lines to the various trigonal aggregate centers is summarized in Table XVI. There is possibly some reason to favor the F_{3c}^- center for the 524.8-nm line, since it does not participate in the interconversion by bleaching[154] with the 642.0- and 649.0-nm lines, which are attributed to F_3 aggregates. At present, there is no detailed confirmation of any of the models, but as in the case of the ortho-

Table XVI. Summary of Trigonal Centers Assigned to Zero-phonon Lines in MgO

Line position, nm	Ground state	Model	Comments
524.8	^2E	F_3^+, F_{3c}^-	Positive λ from Faraday rotation favors "holelike" state
642.0	$^1A_1$	F_3	No Faraday rotation suggests diamagnetic center; bleaching favors same defect configuration as for 649.0-nm line
649.0	$^2A_1$	F_3^-	Possible EPR spectrum favors F_3 rather than F_{3c} defect configuration

rhombic centers discussed in Section 7.2.1, it seems clear that intrinsic defects must be involved. Whether *F*-aggregate centers are really involved is subject to the same criticism as for the orthorhombic centers, in that Chen *et al.*[108,147] were unable to produce the lines by annealing additively colored and 2-MeV electron-irradiated samples. Only more detailed studies of the structure of the centers and of their kinetics of formation will answer these questions.

King[154] has observed a red luminescence from crystals containing high concentrations of trigonal centers. The zero-phonon lines at 642.0 and 649.0 nm appear in the emission spectrum, which peaks at 720.0 nm. Further studies of this emission should result in more information about the electronic states and possible Jahn–Teller distortions of the centers.

7.2.3. *Monoclinic I Centers*

The one remaining zero-phonon line for which observable splittings have been obtained under uniaxial stress is the 705.8-nm line.[152] The splitting pattern conforms to that of a monoclinic I center with a C_2 axis along [110] and transition moment along $[1\bar{1}2]$. This is similar to the symmetry assignments for some lines in the N_1 band region in alkali halides,[81,173] and evokes the speculation that the same kinds of model are appropriate to MgO. These involve F_2, F_3, and F_4 configurations, with two of the constituent anion vacancies joined by a $\langle 112 \rangle$ direction.

Bessent[174] has reported an EPR spectrum associated with a monoclinic I center in neutron-irradiated CaO. The principal axes of the g tensor for $S = \frac{1}{2}$ were found to be $x = [01\bar{1}]$, $y \sim [\bar{1}11]$, and $z \sim [211]$, the actual z direction being tilted about 6.5° from [211] toward [111]. There is good evidence that an intrinsic defect is involved, and Bessent proposes an F_4 configuration, in one of its possible charge states. Optical bleaching at room temperature destroyed up to 75% of the centers, which could be restored by X-irradiation. The center was found to anneal rapidly at about 100°C.

7.3. Other Studies

One of the unresolved questions regarding the optical spectrum of irradiated MgO is the origin of a prominent broad band at about 570.0 nm (2.2 eV) (see Fig. 26). Although many of the zero-phonon lines we have described occur close to the band, none are directly associated with the major portion of it. The results of Chen *et al.*[108,147] indicate that its production rate and annealing in neutron-irradiated samples are similar to those for the F^+ band and some of the orthorhombic zero-phonon lines. This suggests that the 570.0-nm band is due to a small cluster-type defect, but little more can be said. Like the orthorhombic and trigonal center zero-phonon lines, it is not observed in annealed additively colored and 2-MeV-electron-irradiated samples.

Copious numbers of zero-phonon lines have been produced by annealing neutron-irradiated MgO above 900°C, followed by 77°K electron irradiation.[147] This general behavior is similar to the alkali halides, where vast numbers of zero-phonon lines may be produced by heat-treating or bleaching irradiated samples.[81,162] The overall conclusion is that the nature and variety of the centers formed by annealing can be extremely complex.

Little work has been done on "aggregate" zero-phonon lines in oxides other than MgO. Neeley[175] first reported fine structure on color center bands in CaO, and Henderson and King[176] have shown that a line at 572.0 nm is due to an orthorhombic I center. They also analyzed the phonon-assisted spectrum of this line and one at 500.0 nm. Nishi *et al.*[176a] have reported an emission system corresponding to the 572.0 nm line. Apart from this work, the only other zero-phonon lines studied in CaO are the F^+- and F-center lines described in Sections 3 and 4. There appear to have been no color center zero-phonon lines reported in irradiated SrO or BaO. Bessent[174a] has reported an orthorhombic I EPR spectrum in CaO with $S = \frac{1}{2}$.

To summarize, there are a large number of optical bands and zero-phonon lines which appear to be associated with aggregate or cluster centers in the alkaline earth oxides. As yet, none of these centers has been conclusively identified, although a great deal of evidence has accumulated from optical studies which suggests that some of the defects may be analogs of the F-aggregate centers in the alkali halides. Further work, particularly detailed correlation of EPR and optical spectra, should be of great help in establishing more definitely the structure of these centers.

8. EXCHANGE-COUPLED DEFECTS

8.1. Spin Hamiltonian for an Exchange-Coupled Pair

When defect concentrations are very high, it is probable that some defects occupy near-neighbor sites. In this case, there may be overlap of the defect wave functions resulting in the presence of exchange interactions, which are manifested by changes in the EPR and optical spectra. The broadening of F^+-center EPR lines due to interactions between closely spaced defects has been mentioned in Section 5. In this section, we discuss only the consequences of exchange interaction between pairs of defects in nearest-neighbor and next-nearest-neighbor sites.

The effect of the exchange interaction is to couple together the individual spins $\mathbf{S}_1$ and $\mathbf{S}_2$ of the defect pair to form a new total spin angular momentum vector $\mathbf{S}$ such that

$$\mathbf{S} = \mathbf{S}_1 + \mathbf{S}_2 \tag{36}$$

where the magnitude of $\mathbf{S}$ is $\{S(S+1)\}^{1/2}$, etc. From Eq. (36), we find

$$\mathbf{S}_1 \cdot \mathbf{S}_2 = \tfrac{1}{2}[S(S+1) - S_1(S_1+1) - S_2(S_2+1)] \tag{37}$$

For example, in the case that $S_1 = S_2 = 3/2$, the isotropic exchange interaction is*

$$J\mathbf{S}_1 \cdot \mathbf{S}_2 = \tfrac{1}{2}J[S(S+1) - (15/2)] \tag{38}$$

in which S may take values 0, 1, 2, and 3 according to the different ways of taking the vector sum in Eq. (36). For $J > 0$ (antiferromagnetic interactions), the singlet state lies lowest with energy $-15J/4$. The energies of the other states, shown in Fig. 32, follow a Landé interval pattern according to Eq. (38). Figure 32 also shows the modified level pattern when a higher-order term $-j(\mathbf{S}_1 \cdot \mathbf{S}_2)^2$, with $j = J/10$, is included in the isotropic

* Note that this definition of the exchange energy is different from that conventionally used in magnetism, where it is usually written $-2J\mathbf{S}_1 \cdot \mathbf{S}_2$.

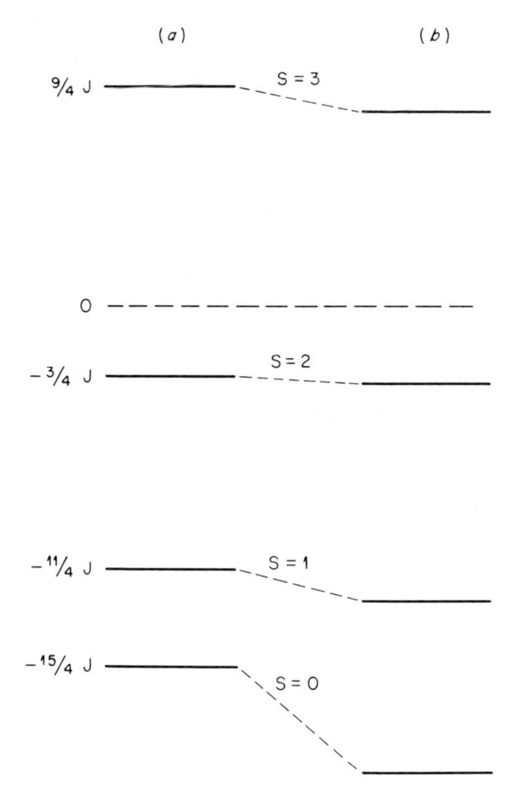

Fig. 32. The energy level scheme for a pair of ions with $S_1 = S_2 = \frac{3}{2}$ for: (a) isotropic exchange with $j/J = 0$ (Landé pattern) and (b) for $j/J = 0.1$.

exchange energy, Eq. (38). This biquadratic contribution to the exchange energy changes the levels by amounts $-(j/4)[S(S+1) - (15/2)]^2$.

For the purposes of EPR observation, the particular spin state is only observed if it is thermally populated. The exchange energy J can be measured by observing the temperature dependence of the transitions within the energy levels $S > 0$.

When exchange effects are dominant, EPR transitions within the spin state S of the coupled system may be described by a spin Hamiltonian

$$\mathscr{H} = g\beta\mathbf{B} \cdot \mathbf{S} + J\mathbf{S}_1 \cdot \mathbf{S}_2 + D_S[S_z^2 - \tfrac{1}{3}S(S+1)] + E_S(S_x^2 - S_y^2) + \tfrac{1}{2}A\mathbf{S} \cdot (\mathbf{I}_1 + \mathbf{I}_2) \tag{39}$$

where $D_S = 3\alpha D_e + \beta D_c$ and $E_S = \alpha E_e + \beta E_c$ represent the zero-field splittings for $S > \frac{1}{2}$ in terms of D_c and E_c, the zero-field splittings of the uncoupled defects, and D_e and E_e, which are due to the interactions between the spins. D_e is the sum of the magnetic dipolar term D_d and an anisotropic exchange term D_E which is of order $(\Delta g)^2 J$, where Δg is the g shift for the individual centers comprising the pair. The parameters α and β are defined by

$$\begin{aligned} \alpha &= \tfrac{1}{2}[S(S+1) + 4S_1(S_1+1)]/(2S-1)(2S+3) \\ \beta &= [3S(S+1) - 4S_1(S_1+1) - 3]/(2S-1)(2S+3) \end{aligned} \tag{40}$$

when $S_1 = S_2$.[177] Equation (39) shows that the hyperfine structure gives the clearest indication in the EPR spectrum of the presence of exchange interactions between defects. For $\Delta M_S = \pm 1$ transitions, each energy level is shifted by $\frac{1}{2}A(m_{I_1} + m_{I_2})$. Thus there are $[2(I_1 + I_2) + 1]$ hyperfine lines for each fine-structure transition, since $(m_{I_1} + m_{I_2}) = M_I$ takes values from $+(I_1 + I_2)$ to $-(I_1 + I_2)$. The intensity of each hyperfine line then depends upon the number of ways in which m_{I_1} and m_{I_2} can be added to give a resultant value of M_I. Thus the intensities of the hyperfine lines follow a stairstep pattern, $1{:}2{:}3 \cdots (I_1 + I_2 + 1) \cdots 3{:}2{:}1$. An example of such a pattern is given in Fig. 33, which portrays hyperfine lines from V^{2+} ion pairs in next-nearest-neighbor sites in MgO (see Section 8.3).

In the following, we consider a number of examples of pair spectra associated with both intrinsic lattice defects and impurity ions. In the simplest case (F^+ center pairs, V^- center pairs, and Co^{2+} ions), $S_1 = S_2 = \frac{1}{2}$.

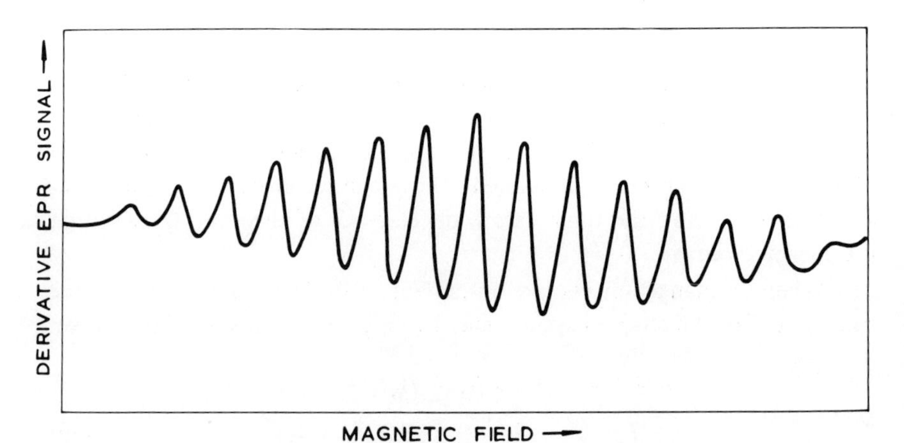

Fig. 33. EPR spectrum and hyperfine structure of next-nearest-neighbor V^{2+} pairs in MgO (after Codling and Henderson[187]).

Table XVII. Spin-Hamiltonian Parameters for Defect Pairs

Crystal	Pair system	g Value	$\mid D_S \mid$ for $S = 1$, cm^{-1}	$\mid E_S \mid$ for $S = 1$, cm^{-1}	J, cm^{-1}	j	Ref.
MgO	F^+ centers	2.003	0.0286	0	56 ± 7	—	179
	V^- centers	2.0032 (∥)	0.02125	0	—	—	124
		2.0401 (⊥)					
	Co^{2+} ions	4.276 (g_x)					
	(nn)	4.494 (g_y)	0.046	0.030	6.3 ± 1.0	—	183
		4.092 (g_z)					
	V^{2+} ions						
	nn	—	—	—	-6.5 ± 2	—	187
	nnn	1.980	0.0718 ($S=2$)	0	54 ± 4	$0.07J$	187
	Mn^{2+} ions						
	nn	2.00	0.776	0.149	10.2	$0.05J$	193, 194
	nnn	2.00	0.261	0	19.6	—	192
CaO	F^+ centers	2.001	0.0158	0	35 ± 14	—	178
	V^- centers	2.0023 (∥)	0.0106	0	—	—	24
		2.0683 (⊥)					

Some parameters obtained from EPR measurements are given in Table XVII.

8.2. Pair Spectra with $S_1 = S_2 = \frac{1}{2}$

In both MgO and CaO, exchange-coupled pairs of F^+ centers[178,179] and V^- centers[24,124] have been reported. In these cases, $S_1 = S_2 = \frac{1}{2}$, so in Eq. (39), terms in D_c are not involved. Furthermore, since the g values are isotropic, there is no anisotropic exchange coupling to be considered (i.e., $D_E = 0$). Thus the fine-structure splitting will be due entirely to the weak dipolar coupling energy

$$\mathscr{H}_d = g^2\beta^2\{(\mathbf{S}_1 \cdot \mathbf{S}_2/R^3) - [3(\mathbf{S}_1 \cdot \mathbf{R})(\mathbf{S}_2 \cdot \mathbf{R})/R^5]\} \tag{41}$$

R being the separation between the constituent defects. The use of Eq. (41) gives $D_d = -g^2\beta^2/R^3$ and hence $D_S = -\frac{3}{2}g^2\beta^2/R^3$, since $\alpha = \frac{1}{2}$, $\beta = 0$ for $S = 1$, $S_1 = S_2 = \frac{1}{2}$.

Tanimoto *et al.*[178] reported observations of a thermally excited triplet ($S = 1$) state resonance in neutron-irradiated calcium oxide. The spectrum

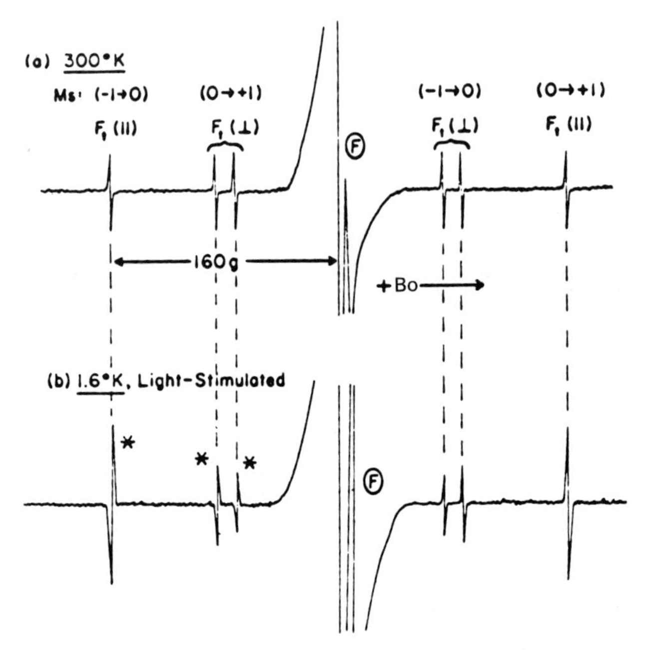

Fig. 34. EPR spectra of exchange-coupled F^+ centers in CaO. The large central line marked F is due to isolated F^+ centers. (a) $T = 300$°K, showing the pair lines (marked F_t) for **B** slightly off a ⟨100⟩ axis. (b) $T = 1.6$°K, showing the pair lines stimulated by light in the region of the F^+ absorption band. Emissive lines are indicated with an asterisk. Intensity scales in (a) and (b) are unrelated. (After Tanimoto *et al.*[178])

is shown in Fig. 34 to consist of six equally intense lines in an arbitrary field orientation. The maximum splitting occurs when the magnetic field is parallel to a crystal ⟨100⟩ direction. In this orientation, only four lines are observed, the inner pair of which is doubly intense. The lines have been assigned to the various transitions assuming D_S to be negative, as expected for dipolar coupling. Detailed orientation dependence studies show that the associated defects have $S = 1$ and are oriented along the ⟨100⟩ crystal directions. The spin-Hamiltonian parameters for this spectrum are given in Table XVII. In the absence of hyperfine structure, Tanimoto *et al.*[178] proposed that F^+-center pairs were involved from the variation with temperature of the intensity of the lines, which decreased to zero as the measuring temperature was reduced to 4°K. Thus the singlet ($S = 0$) state lies lowest and J is positive. According to Eq. (38), the two states are separated by an energy of J and the intensity is determined by the relative occupation of $S = 0$ and $S = 1$ states. The relative population of a given S level is given by Boltzmann statistics as

$$R(S) = (2S + 1)e^{-E(S)/kT} \Big/ \sum_{0}^{S} (2S + 1)e^{-E(S)/kT} \tag{42}$$

Measuring intensities over the temperature range 4–300°K, it was inferred from Eq. (42) that $J = 35 \pm 14\ \mathrm{cm}^{-1}$.

From the measured value of the fine-structure splitting D_S quoted in Table XVII and using Eq. (41), Tanimoto *et al.* deduce that the separation between the two neighboring defects is 0.55 nm, approximately 14% larger than the unit cell dimension. It was concluded that two F^+ centers were involved. In order to conserve charge neutrality, they suggested that a linear trivacancy, in which a cation vacancy is interposed between the two F^+ centers, is the most probable defect structure.

Evidence for F^+-center pairs in MgO is somewhat more direct since it is possible to discern a weak hyperfine structure associated with the EPR lines.[179] Since the ^{25}Mg nuclides are only 10% abundant, the probability that one of the neighboring sites is occupied by ^{25}Mg is 40%. The intensity of the observed hyperfine pattern corresponds with this probability. The separation between the adjacent components in the hyperfine pattern is 5.55 MHz, precisely one half the hyperfine structure constant for single F^+ centers. The value of the fine structure splitting D_S quoted in Table XVII corresponds to a separation between the F^+-center pairs of 0.45 nm.* This is to be compared with a lattice constant of 0.42 nm. Henderson[190] also observed the $\Delta M_S = \pm 2$ transitions which become allowed with **B** off-axis by the second-order mixing of the $| \pm 1\rangle$ spin states by the D term in the Hamiltonian. As expected, these transitions are observed at half-field with approximate intensity D_S^2/B.

Recent computations of the exchange energy J give support to the proposal that the cation site between the two F^+ centers is vacant.[180–182] It is found that the values of J for the linear trivacancy are very close to those measured experimentally. When the cation sites are not vacant, much larger exchange energies are predicted.[181]

In both CaO and MgO, the EPR spectra of the F^+-center pairs at 4°K are too weak to be observed since the triplet state is virtually empty. The

* In Refs. 24 and 179, the zero-field splitting was incorrectly written as $|D_S| = 111g^2\beta^2/10R^3$, which applies only for $S = 1$, $S_1 = S_2 = \frac{5}{2}$.[177,192] However, in calculating R from the measured splitting, the simple form of the dipolar splitting $|D_d| = g^2\beta^2/R^3$ was used, underestimating R, according to Eq. (41), by $(\frac{3}{2})^{1/3}$. This latter error occurs in other publications.[124,139a]

triplet state may be optically populated for both crystals by pumping with F^+-band light below 4°K.[178,190] As shown in Fig. 34, for CaO, the low-field lines in the EPR spectra are emissive rather than absorptive. These striking examples of spin polarization may be effected with unpolarized, broadband light. Tanimoto *et al.*[178] explained the effect as due to level inversion resulting from selective generation of either the $M_S = 0$ or $M_S = \pm 1$ levels over the other. In particular, the fact that the low-field lines are emissive and the high-field lines absorptive in Fig. 34 means that preferential population of the $M_S = 0$ level occurs when the magnetic field is parallel to the pair axis and the $M_S = \pm 1$ levels when the magnetic field **B** is perpendicular to the pair axis. For observation of the emission transitions at all it is required that the direct spin–lattice relaxation time in the triplet state is long compared with the triplet lifetime.[178] The population mechanism results from anisotropic spin–orbit coupling in the excited states, and its role in the selection rules for the singlet–triplet transition. The explanation of the spin polarization as given by Tanimoto *et al.*[178] is as follows.

We can consider the F^+-center pair to have a ground state $^1A_1$ composed of 1s-like wave functions, with electron spins aligned antiparallel. Optical excitation promotes one of the electrons into a 2p-like state: the possible Heitler–London configurations of the (sp) representation are shown as $^1A_{1g}$, $^1A_{2u}$, ^1E_g, ^1E_u, $^3A_{1g}$, $^3A_{2u}$, ^3E_g, and ^3E_u in Fig. 35. The tetragonal crystal field parallel to the defect z axis quenches orbital angular momentum in excited states perpendicular to this axis, while allowing orbital angular momentum to be generated parallel to the z direction. Spin–orbit interaction mixes the excited $^1E_{u,g}$ and $^3E_{u,g}$ states, so allowing direct excitation into the excited triplet levels from the ground singlet level. Alternatively, optical excitation into singlet (s, p) states may be followed by lattice-modulated spin–orbit interaction inducing singlet to triplet transitions. Thus direct excitation into the triplet excited levels by near-F^+-band light allows direct electric dipole transitions into the lower triplet state [i.e., in the (s, s) configuration] such that spin is conserved (i.e., $\Delta M_S = 0$). Consequently, for **B** parallel to the pair axis (z), only the $M_S = 0$ level is populated. When spin is quantized normal to the z axis, the low spin–orbit coupling perpendicular to z causes $\Delta M_S = \pm 1$ transitions to dominate. There is then preferential population of the $M_S = \pm 1$ levels for **B** perpendicular to z. This is portrayed in Fig. 35, where the resulting emissive low-field transitions are indicated by an asterisk.

It is easy to show that the selection rules for return to the singlet ground state are just the reverse of those for generation of the triplet levels. These transitions are also shown in Fig. 35.

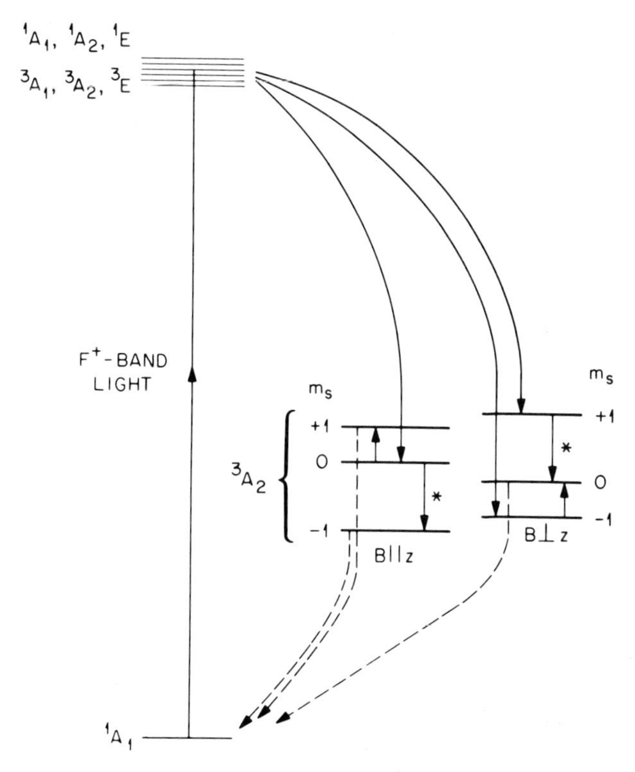

Fig. 35. Optical pumping scheme for exchange-coupled F^+ centers. Emissive transitions within the triplet state $^3A_2$ are indicated by an asterisk. There are actually two E excited states of each spin (see text). (Adapted from Tanimoto *et al.*[178])

EPR studies of pairs of trapped-hole centers have also been reported in MgO[124] and in CaO.[24] The two positive holes occupy a collinear orientation on opposite sides of a cation vacancy. This defect is charge neutral and is labeled as the V center in Fig. 20. In both MgO and CaO, the defects require low temperature (77°K) for production and observation of the EPR spectrum. The spectrum consists of pairs of lines, due to $\Delta M_S = -1 \rightarrow 0$ and $\Delta M_S = 0 \rightarrow +1$ transitions, approximately centered upon the lines of the V^- center. Since there are three equivalent $\langle 100 \rangle$ directions, each equally populated with defects, six equally intense lines are observed for an arbitrary field orientation. The values of $g_\parallel$, $g_\perp$, and D_S are given in Table XVII. The mean separation of the holes estimated from Eq. (41) is 0.521 nm for MgO and 0.68 nm for CaO. These values are larger than originally quoted (see footnote on p. 477). Since extensive lattice relaxation is expected, these rather large separations compared with the normal lattice constants ($a_{MgO} = 0.420$ nm and $a_{CaO} = 0.480$ nm) do not really contradict the model of the defect. Trapped-hole pairs with $S = 1$ have also been identified in Al_2O_3,[39,171,182a] BeO,[182b] and ZnO.[128b] An interesting feature of all these $S = 1$ hole centers is that the temperature dependence of the EPR spectra shows that the exchange coupling is *ferromagnetic*, i.e., the $S = 1$ state lies lowest.[190,190a]

Recently, Smith[183] has observed EPR spectra from pairs of Co^{2+} ions in nearest-neighbor sites in MgO. The defect pair has orthorhombic symmetry with the z axis being parallel to a $\langle 110 \rangle$ direction of the crystal. The Co^{2+} ion has the configuration $3d^7$, which in octahedral symmetry has a $^4T_1$ ground term split by spin–orbit coupling ($\lambda \approx -180$ cm^{-1}) into three levels described in lowest order by $J = \frac{1}{2}, \frac{3}{2}$, and $\frac{5}{2}$. The lowest state, $J = \frac{1}{2}$, thus has an effective spin $S' = \frac{1}{2}$. Exchange coupling between two such ions can be represented by an exchange Hamiltonian

$$\mathscr{H}' = J'\mathbf{S}_1' \cdot \mathbf{S}_2', \qquad \text{where} \quad J' = (25/9)J \tag{43}$$

J is the isotropic interaction between the real spins of the ions. The effective spins couple to form a spin state with total effective spin $S' = 0$ or 1. Unfortunately, EPR lines due to Co^{2+} ions broaden above 30°K owing to spin–lattice relaxation. However, the pair lines are observed in the temperature range 13.8–20.4°K by the distinctive hyperfine structure predicted by Eq. (39) for pair spectra of defects with $I_1 = I_2 = \frac{7}{2}$. The orientation dependence of these lines demonstrates that the pair of Co^{2+} ions undergoes a distortion, in which the pair system is twisted away from the [110] direction. This twist of the pair axis is through a small angle θ away from [110] in the (001) plane plus a displacement through an angle ϕ out of the plane. The spin-Hamiltonian parameters (Table XVII) were calculated from the experimental results by neglecting the distortion. It should be noted that the distortion is probably related to the tetragonal distortion of CoO below the Néel temperature, since both originate in magnetoelastic coupling between the orbital components of the $^4T_1$ state and lattice distortions.[184] This coupling is, of course, closely related to the Jahn–Teller effect.[185] CoO is isomorphous with MgO. In the antiferromagnetic state, the spins along a [001] direction are aligned antiparallel, parallel spins being arranged in sheets in the (111) plane. It should be possible to relate the exchange constant for the Co^{2+} pair in MgO to those for CoO. Molecular field theory predicts for the Néel temperature[186]

$$T_N = z_2 J_2 S(S + 1)/3k \tag{44}$$

and for the Curie–Weiss temperature

$$\theta = -S(S+1)(z_1 J_1 + z_2 J_2)/3k \tag{45}$$

where z_1 and z_2 are the numbers of nearest and next-nearest neighbors for each ion, and J_1 and J_2 are the corresponding exchange constants. The experimental values $\theta = -330°K$ and $T_N = 292°K$ give the exchange energies as $J_1 = 9.5\ cm^{-1}$ and $J_2 = 30\ cm^{-1}$,[186] provided Eqs. (44) and (45) are modified to take account of the role played by the unquenched orbital angular momentum.[184] Smith[183] finds $J = 6.3 \pm 1.0\ cm^{-1}$, in good agreement with the above value for J_1, bearing in mind that the pair site is distorted and that J varies rapidly with ion separation.

8.3. Impurity Pair Spectra with $S_1 = S_2 > \frac{1}{2}$

There have been no investigations of intrinsic lattice defects with $S > \frac{1}{2}$ interacting by exchange mechanisms in the way described in Section 8.1. Impurity ion systems, e.g., V^{2+} and Mn^{2+}, have been investigated quite thoroughly. We showed in Fig. 33 the hyperfine structure from next-nearest-neighbor V^{2+} ion pairs in MgO. For V^{2+} ions, $S = \frac{3}{2}$ and $I = \frac{7}{2}$. Thus the hyperfine structure in Fig. 33 clearly identifies the pair system, since the 15 components correspond to the allowed values of the total quantum number $M_I = 7, 6, \ldots, 0, \ldots, -6, -7$. The relative intensities are close to the expected ratios 1:2:3: $\cdots$:8: $\cdots$ 3:2:1, corresponding to the different ways of compounding each M_I value. Due to complex broadening mechanisms, which leave the $S = 2$ level unaffected, only transitions within this level are observed.[187] From the orientation dependence of the spectrum, the spin-Hamiltonian parameters given in Table XVII are deduced. The sign of D_S was deduced to be negative; thus

$$D_S = 3\alpha D_e + \beta D_c = -0.0718\ cm^{-1}$$

and for $S = 2$, $S_1 = S_2 = \frac{3}{2}$, $\alpha = \frac{1}{2}$, and $\beta = 0$, we find $D_e = -0.0479\ cm^{-1}$. The dipolar contribution to D_e computed from Eq. (41) is $-0.0246\ cm^{-1}$ assuming that $R = 0.413$ nm, i.e., half-way between the lattice parameters of MgO (0.420 nm) and VO (0.406 nm). Thus the dipolar term underestimates D_e by as much as $-0.0233\ cm^{-1}$. Symmetry arguments lead one to reject the Dzyaloshinsky–Moriya[188] antisymmetric interaction as contributing to D_E, since in the presence of a center of symmetry, this vanishes. However, anisotropic exchange of the type discussed by Van Vleck[189] might account for some of the anisotropy. The magnitude of this part of the anisotropy is given by[183,187] $D_E \approx +(\Delta g)^2 J/96$. As we discuss below, $J \simeq 54\ \text{cm}^{-1}$, and $\Delta g = 0.02$,[187] so that $D_E = +0.0002\ \text{cm}^{-1}$ and is clearly too small to be of importance. The anisotropy thus appears to involve some interactions which remain unidentified. The magnetoelastic effect discussed below for Mn^{2+}–Mn^{2+} pairs will only contribute ~6% to the value of D_e.

The value of J for these V^{2+} pairs was estimated from the temperature dependence of the intensity of the pair lines. Although only the $S = 2$ state was observed, a very careful study in the temperature range 40–300°K showed that it was necessary to include a biquadratic term in the exchange energy. The result was $J = 54 \pm 2\ \text{cm}^{-1}$, $j = (0.07 \pm 0.03)J$. A preliminary investigation of nearest-neighbor V^{2+} pairs suggests that J is small ($-6.5 \pm 2\ \text{cm}^{-1}$) and negative (i.e., ferromagnetic).[187]

The value of $J = 54\ \text{cm}^{-1}$ for next-nearest-neighbor pairs in MgO is an order of magnitude larger than that determined by Smith for V^{2+} pairs in $KMgF_3$.[183,183a] The difference is presumably due to the nature of the superexchange, which in the one lattice is via O^{2-} and in the other is carried by F^-. For Cr^{3+} pairs in Al_2O_3, however, J may be larger or smaller than that determined for V^{2+} pairs in MgO.[191] For first-nearest-neighbor pairs, $J = 240\ \text{cm}^{-1}$; while the second-nearest-neighbor pairs have $J = 83.6\ \text{cm}^{-1}$ and $j = -9.7\ \text{cm}^{-1}$. Third-nearest-neighbor pairs have exchange energies somewhat smaller, with $J = 11.59\ \text{cm}^{-1}$ and $j = 0.06\ \text{cm}^{-1}$. The fourth-nearest-neighbor pairs have been extensively investigated both by optical and EPR techniques. This interaction is ferromagnetic ($J = -6.99\ \text{cm}^{-1}$ and $j = 0.14\ \text{cm}^{-1}$).

Nearest-neighbor and next-nearest-neighbor Mn^{2+} ($S = \frac{5}{2}$, $I = \frac{5}{2}$) pairs have been investigated in considerable detail.[192–194] For the nearest-neighbor pairs, measurements on the states $S = 1, 2, 3$, and 4 were reported, whereas for next-nearest-neighbor pairs, only $S = 1$ resonances were observed. The EPR parameters of these systems are also recorded in Table XVII. Note that for the nearest-neighbor pairs, a significant biquadratic interaction ($j = 0.05J$) must be included. This changes quite significantly the original estimate of $J = 19.6\ \text{cm}^{-1}$ for nearest-neighbor pairs by Coles *et al.*[192] To explain the values of D_S for the various spin states, it is necessary to assume that D_c and D_e depend on S.[194] For the $S = 1$ state $D_c = 0.024\ \text{cm}^{-1}$ and $D_e = -0.056\ \text{cm}^{-1}$. This value of D_e should be compared with that deduced from the classical dipolar interaction, which gives $D_d = -0.060\ \text{cm}^{-1}$ for a lattice spacing midway between MgO and MnO (the nearest-neighbor cation–cation distance is 0.294 nm in MgO). Anisotropic exchange should only make a contribution of order $(g - 2)^2 J$

$\approx 0.001\ \text{cm}^{-1}$ to the magnitude of D_e. The dependence of D_c and D_e on S may result in part from a balance between the exchange and elastic forces. The isotropic exchange interaction is a rapidly varying function of ion separation, and thus creates a force between the ions which is balanced by the elastic forces. If we assume $J = J_0(R/R_0)^{-n}$, then it is easily shown that the ion separation R changes in the spin state S by an amount

$$\Delta R/R_0 = (nJ_0/2KR_0)[S(S+1) - (35/2)] \tag{46}$$

where K is an appropriate elastic constant. For $R_0 = 0.294$ nm and assuming $n = 15$, the change in $\Delta R/R_0$ between the $S = 1$ and $S = 4$ states is of order 6×10^{-3}. Since D_d varies as R^{-3}, such a change in ion separation should lead to a 2% change in D_d. The sign of ΔR is such as to increase $|D_d|$ for the lower spin states, as is observed.

Note that this magnetoelastic effect would also give rise to a biquadratic term in the exchange energy. Writing the total energy of an ion pair as the sum of the exchange energy and the elastic energy,[194] we have

$$E = J\mathbf{S}_1 \cdot \mathbf{S}_2 + (K/2R_0)(\Delta R)^2 \tag{47}$$

Substituting for $(\Delta R)^2$ from Eq. (46), we find using $J \approx J_0 + \Delta R(\partial J/\partial R)_{R_0}$

$$E = J_0\mathbf{S}_1 \cdot \mathbf{S}_2 - (n^2 J_0^2/2KR_0)(\mathbf{S}_1 \cdot \mathbf{S}_2)^2 \tag{48}$$

Harris[194] finds that a magnetoelastic effect of this kind can explain the apparent biquadratic exchange term required for the nearest neighbor pairs.

The values of J_1 and J_2 from EPR results on dilute solid solutions of MnO in MgO given in Table XVII are rather larger than those from bulk magnetic measurements on MnO[186] which give $J_1 = 10\ \text{cm}^{-1}$ and $J_2 = 4.9\ \text{cm}^{-1}$. However, the inclusion of an effective biquadratic term in the exchange energy, as required for the pair results, does account for the anomalous behavior of both the sublattice magnetization and transverse susceptibility of MnO.[195] Similar anomalies in the magnetic behavior of NiO can also be explained on the basis of a biquadratic exchange term being added to the exchange energy. For Mn^{2+} pairs in CaO, the measured value of J is smaller than in MgO; too small for a reliable estimate of j.[194] This is consistent with the increased ionic separation in the CaO lattice.

ACKNOWLEDGMENTS

We should like to thank Drs. M. M. Abraham, T. P. P. Hall, W. A. Runciman, W. A. Sibley, and A. M. Stoneham for many useful discussions and comments on the manuscript. One of us (BH) gratefully acknowledges the support by the Science Research Council of the experimental program relating to this work.

REFERENCES

1. B. Henderson and J. E. Wertz, *Advan. Phys.* **17**, 749 (1968).
2. A. R. Hutson, in *Semiconductors*, Ed. by N. B. Hannay (Reinhold, New York, 1959), p. 569.
3. G. H. Reiling and E. B. Hensley, *Phys. Rev.* **112**, 1106 (1958).
4. *American Institute of Physics Handbook* (McGraw-Hill, New York, 1957).
4a. R. Bever and R. L. Sproull, *Phys. Rev.* **83**, 801 (1951).
5. D. M. Roessler and W. C. Walker, *Phys. Rev.* **159**, 733 (1967).
6. E. Burstein, in *Phonons and Phonon Interactions*, Ed. by T. A. Bak (Benjamin, New York, 1964), p. 276.
7. G. Peckham, *Proc. Phys. Soc.* **90**, 657 (1967).
8. M. J. L. Sangster, G. Peckham, and D. Saunderson, *J. Phys. C.* (*Solid State Phys.*) **3**, 1026 (1970).
9. J. L. Jacobson and E. R. Nixon, *J. Phys. Chem. Solids* **29**, 967 (1968).
9a. G. E. Pynchon and E. F. Sieckmann, Phys. Rev., **143**, 595 (1966).
9b. M. Parodi, *Compt. Rend.* **204**, 1636 (1937).
10. M. Haase, *Z. Krist.* **65**, 510 (1927).
11. R. C. Whited and W. C. Walker, *Phys. Rev.* **188**, 1380 (1969).
12. H. H. Glascock and E. B. Hensley, *Phys. Rev.* **131**, 649 (1963).
13. F. W. Webster and E. A. D. White, *J. Cryst. Growth* **5**, 167 (1969).
14. R. J. Gambino, *J. Appl. Phys.* **36**, 656 (1965); F. Galtier and R. Collongues, *Compt. Rend.* **264**, C87 (1967).
15. C. T. Butler, B. Sturm, and B. Quincey, *J. Cryst. Growth* **8**, 197 (1971).
16. W. A. Sibley and Y. Chen, *Phys. Rev.* **160**, 712 (1967).
17. R. G. Bessent, B. C. Cavenett, and I. C. Hunter, *J. Phys. Chem. Solids* **29**, 1523 (1968).
18. E. B. Hensley, W. C. Ward, B. P. Johnson, and R. L. Kroes, *Phys. Rev.* **175**, 1227 (1968).
19. H. Seidel and H. C. Wolf, in *Physics of Color Centers*, Ed. by W. Beall Fowler (Academic Press, New York, 1968), p. 538.
20. B. S. Gourary and F. J. Adrian, *Phys. Rev.* **105**, 1180 (1957); also in *Solid State Physics*, Vol. 10, Ed. by F. Seitz and D. Turnbull (Academic Press, New York), p. 127.
21. W. C. Holton and H. Blum, *Phys. Rev.* **125**, 89 (1962).
22. T. E. Feuchtwang, *Phys. Rev.* **126**, 1616, 1622 (1962).
23. G. Feher, *Phys. Rev.* **114**, 1219, 1249 (1959).
24. B. Henderson and A. C. Tomlinson, *J. Phys. Chem. Solids* **30**, 1801 (1969).
25. J. E. Wertz, P. Auzins, R. A. Weeks, and R. H. Silsbee, *Phys. Rev.* **107**, 1535 (1957); J. E. Wertz, J. W. Orton, and P. Auzins, *Disc. Faraday Soc.* **31**, 140 (1961).
26. J. W. Culvahouse, L. V. Holroyd, and J. L. Kolopus, *Phys. Rev.* **140**, 1181 (1965).
27. W. P. Unruh, private communication.

28. W. P. Unruh and J. W. Culvahouse, *Phys. Rev.* **154**, 861 (1967).
29. J. C. Kemp and V. I. Neeley, *Phys. Rev.* **132**, 215 (1963); *Bull. Am. Phys. Soc.* **8**, 484 (1963).
30. L. Halliburton, D. L. Cowan, and L. V. Holroyd, *Bull. Amer. Phys. Soc.* **15**, 760 (1970) and private communication.
31. K. E. Mann, L. V. Holroyd, and D. L. Cowan, *Phys. Stat. Sol.* **33**, 391 (1969).
32. R. H. Bartram, C. E. Swenberg, and Y. La., *Phys. Rev.* **162**, 759 (1967).
32a. W. E. Hagston, *J. Phys. C* (*Solid State Phys.*) **3**, 1233 (1970).
33. J. L. Kolopus and G. J. Lapeyre, *Phys. Rev.* **176**, 1025 (1968); W. P. Unruh, private communication.
34. A. J. Tench and R. L. Nelson, *Proc. Phys. Soc.* **92**, 1055 (1967).
35. H. Seidel, *Z. Phys.* **165**, 218 (1961).
36. R. C. DuVarney, A. K. Garrison, and R. H. Thorland, *Phys. Rev.* **188**, 657 (1969).
36a. R. C. DuVarney and A. K. Garrison, Int. Conf. on Colour Centres in Ionic Crystals, Reading, 1971, Abstract A7.
37. J. M. Smith and W. E. Vehse, *Phys. Letters* **31A**, 147 (1970).
38. V. I. Neeley, J. B. Gruber, and W. J. Gray, *Phys. Rev.* **158**, 809 (1967); J. L. Kolopus, C. B. Finch and M. M. Abraham, *Phys. Rev.* **B2**, 2040 (1970).
39. F. T. Gamble, R. H. Bartram, C. G. Young, O. R. Gilliam, and P. W. Levy, *Phys. Rev.* **138**, A577 (1965).
40. T. Purcell and R. A. Weeks, *Phys. Chem. Glasses* **10**, 198 (1969).
41. T. Purcell and R. A. Weeks, *J. Chem. Phys.* **54**, 2800 (1971).
42. T. Takeda and A. Wanatabe, *J. Phys. Soc. Japan* **21**, 267 (1966).
43. G. D. Watkins, in *Radiation Effects in Semiconductors* (Plenum Press, New York, 1968), p. 67.
44. R. A. Weeks, in *Interaction of Radiation with Solids*, Ed. by A. Bishay (Plenum Press, New York, 1967), p. 55.
45. J. E. Wertz, G. S. Saville, L. Hall, and P. Auzins, *Proc. Brit. Ceram. Soc.* **1**, 59 (1964).
46. B. Henderson and R. D. King, *Phil. Mag.* **13**, 1149 (1966).
47. Y. Chen, W. A. Sibley, F. D. Srygley, R. A. Weeks, E. B. Hensley, and R. L. Kroes, *J. Phys. Chem. Solids* **29**, 863 (1968).
48. B. Henderson, R. D. King, and A. M. Stoneham, *J. Phys. C* (*Proc. Phys. Soc.*) *Ser.* 2, **1**, 586 (1968).
49. J. C. Kemp, W. M. Ziniker, J. A. Glaze, and J. C. Cheng, *Phys. Rev.* **171**, 1024 (1968).
50. J. C. Kemp, J. C. Cheng, E. H. Izen, and F. A. Modine, *Phys. Rev.* **179**, 818 (1969).
51. Y. Chen, J. L. Kolopus, and W. A. Sibley, *Phys. Rev.* **186**, 865 (1969).
52. L. A. Kappers, R. L. Kroes, and E. B. Hensley, *Phys. Rev.* **B1**, 4151 (1970).
53. T. J. Turner, *Solid State Comm.* **7**, 635 (1969).
54. W. Hayes and A. M. Stoneham, *Phys. Letters* **29A**, 519 (1969).
55. R. G. Bessent and P. Feltham, *Phys. Stat. Sol.* **25**, K107 (1968).
56. K. C. To, Ph. D. thesis, University of Keele (1969, unpublished); K. C. To and B. Henderson, *J. Phys. C.* (*Solid State Phys.*) **4**, L216 (1971).
57. A. M. Stoneham, W. Hayes, P. H. S. Smith, and J. P. Stott, *Proc. Roy. Soc.* **A306**, 369 (1968).
58. H. S. Bennett, *Phys. Rev.* **169**, 729 (1968); **184**, 918 (1969).
58a. H. S. Bennett, *Phys. Rev.* **B3**, 2763 (1971).

59. R. F. Wood and U. Öpik, to be published.
60. G. Chiarotti, U. M. Grassano, G., Margaritondo, and R. Rosei, *Nuovo Cimento* **64B**, 159 (1969).
60a. A. Harman and W. Hayes, private communication.
60b. A. L. Harmer, W. Hayes, and M. C. M. O'Brien, *J. Phys. C.* (*Solid State Phys.*) **4**, L108 (1971).
61. D. Y. Smith, Int. Symp. on Color Centers in Alkali Halides, Rome, 1968, Abstract 171, p. 252.
62. D. L. Dexter, *Phys. Rev.* **111**, 119 (1958).
63. F. Lüty and J. Mort, *Phys. Rev. Letters* **12**, 45 (1964); J. Mort, F. Lüty, and F. C. Brown, *Phys. Rev.* **137**, A566 (1965).
64. N. V. Karlov, J. Margerie, and Y. Merle d'Aubigné, *J. Phys.* (*Paris*) **24**, 717 (1963).
65. C. H. Henry, S. E. Schnatterly, and C. P. Slichter, *Phys. Rev.* **137**, A583 (1965).
66. C. H. Henry and C. P. Slichter, in *Physics of Color Centers*, Ed. by W. Beall Fowler (Academic Press, New York, 1968), p. 352.
67. J. C. Kemp, W. M. Ziniker, and J. A. Glaze, *Phys. Letters* **22**, 37 (1966); *Proc. Brit. Ceram. Soc.* **9**, 109 (1967).
68. J. Margerie, in *Electronic Magnetic Resonance and Solid Dielectrics*, Proc. XIIth Colloque Ampère (North-Holland, Amsterdam, 1964), p. 69.
69. D. Y. Smith, *Phys. Rev.* **137**, A574 (1965).
70. R. H. Bartram, private communication to R. G. Bessent, B. C. Cavenett, and I. C. Hunter quoted in Ref. 17. See also Ref. 32.
71. H. Rabin and J. H. Schulman, *Phys. Rev.* **125**, 1584 (1962); T. A. Fulton and D. B. Fitchen, *Phys. Rev.* **179**, 846 (1969).
72. W. C. Dash, *Phys. Rev.* **92**, 68 (1953).
73. T. H. Keil, *Phys. Rev.* **140**, A601 (1965).
74. J. J. Markham, *Rev. Mod. Phys.* **31**, 956 (1959); M. H. L. Pryce, in *Phonons*, ed. by R. W. H. Stevenson (Oliver and Boyd, Edinburgh, 1966), p. 403; A. A. Maradudin, in *Solid State Physics*, Vol. 18, Ed. by F. Seitz and D. Turnbull (Academic Press, New York, 1966), p. 274; R. H. Silsbee, in *Optical Properties of Solids*, Ed. by S. Nudelman and S. S. Mitra (Plenum Press, New York, 1969).
75. M. Lax, *J. Chem. Phys.* **20**, 1752 (1952).
76. R. G. Bessent, private communication.
77. G. M. Borgonovi and G. W. Carriveau, *Phys. Rev.* **174**, 953 (1968).
77a. D. Saunderson and G. Peckham, *J. Phys. C.* (*Solid State Phys.*) **4**, 2009 (1971).
78. J. P. Mon, *Phys. Stat. Sol.* **33**, 641 (1969).
79. R. K. Dawson and D. Pooley, *Phys. Stat. Sol.* **35**, 95 (1969).
80. W. Gebhardt and H. Kühnert, *Phys. Letters* **11**, 15 (1964).
81. C. B. Pierce, *Phys. Rev.* **135**, A83 (1964); A. E. Hughes, *J. Phys.* (*Paris*) **28**, C4-55 (1967); D. B. Fitchen, in *Physics of Color Centers*, Ed. by W. Beall Fowler (Academic Press, New York, 1968), p. 294; W. von der Osten, *Z. Angew. Phys.* **24**, 365 (1968).
82. Y. R. Shen, *Phys. Rev.* **133**, A511 (1964).
83. A. H. Jahn and E. Teller, *Proc. Roy. Soc.* **A161**, 220 (1937); M. D. Sturge, in *Solid State Physics*, Vol. 20, Ed. by F. Seitz, D. Turnbull and H. Ehrenreich (Academic Press, New York, 1967), p. 92.
84. U. Öpik and M. H. L. Pryce, *Proc. Roy. Soc.* **A238**, 425 (1957).
85. F. S. Ham, *Phys. Rev.* **138**, A1727 (1965).

86. M. Caner and R. Englman, *J. Chem. Phys.* **44**, 4054 (1966).
87. A. E. Hughes and W. A. Runciman, *J. Phys. C* (*Solid State Phys.*) **2**, 37 (1969).
88. M. C. M. O'Brien, *Phys. Rev.* **187**, 407 (1969).
89. A. E. Hughes, *J. Phys. C* (*Solid State Phys.*) **3**, 627 (1970).
89a. C. Escribe and A. E. Hughes, *J. Phys. C.* (*Solid State Phys.*) **4**, 2537 (1971).
89b. M. C. M. O'Brien, *J. Phys. C.* (*Solid State Phys.*) **4**, 2524 (1971).
90. K. Cho, *J. Phys. Soc. Japan* **25**, 1372 (1968).
91. T. P. P. Hall and R. G. Bessent, *Solid State Comm.* **8**, 1151 (1971).
91a. Y. Merle d'Aubigné and A. Roussel, *Phys. Rev.* **B3**, 1421 (1971).
92. S. E. Schnatterly, *Phys. Rev.* **140**, A1364 (1965).
93. B. Henderson and A. C. Tomlinson, private communication, cited in Ref. 94.
94. B. D. Evans, J. C. Cheng, and J. C. Kemp, *Phys. Letters* **27A**, 506 (1968).
95. B. Henderson, S. E. Stokowski, and T. C. Ensign, *Phys. Rev.* **183**, 826 (1969).
95a. B. D. Evans and J. C. Kemp, *Phys. Rev.* **B2**, 4179 (1970).
95b. B. Henderson, Y. Chen, and W. A. Sibley, to be published.
96. B. D. Evans and J. C. Kemp, private communication.
96a. A. E. Hughes, G. P. Pells, and E. Sonder, *J. Phys. C.* (*Solid State Phys.*) **5**, 709 (1972).
96b. J. Duran, Y. Merle d'Aubigné, and R. Romestain, *J. Phys. C.* (*Solid State Phys.*) to be published.
97. J. H. Schulman and W. D. Compton, *Color Centers in Solids* (Pergamon Press, Oxford, 1963).
98. V. I. Neeley and J. C. Kemp, *Bull. Am. Phys. Soc.* **8**, 484 (1963).
99. W. C. Ward and E. B. Hensley, *Phys. Rev.* **175**, 1230 (1968).
100. J. C. Kemp, W. M. Ziniker, and E. B. Hensley, *Phys. Letters* **A25**, 43 (1967).
101. D. F. Winterstein and E. F. Sieckmann, *J. Phys. Chem. Solids* **30**, 2887 (1969).
102. H. S. Bennett, *Phys. Rev.* **B1**, 1709 (1970).
103. R. H. Bartram, private communication.
103a. B. Henderson, unpublished.
104. A. C. Tomlinson and B. Henderson, *J. Phys. Chem. Solids* **30**, 1793 (1969).
105. D. L. Packwood and E. B. Hensley, *Bull. Am. Phys. Soc., Ser. II*, **15**, 370 (1970).
106. B. Henderson, unpublished.
106a. P. Edel, C. Hennies, Y. Merle d'Aubigné, R. Romestain, and Y. Twarowski, to be published.
107. B. P. Johnson and E. B. Hensley, *Phys. Rev.* **180**, 931 (1969).
108. Y. Chen, R. T. Williams, and W. A. Sibley, *Phys. Rev.* **182**, 960 (1969).
109. B. G. Childs, P. J. Harvey, and J. B. Hallett, *J. Am. Ceram. Soc.* **53**, 431 (1970).
110. W. E. Vehse, W. A. Sibley, F. J. Keller, and Y. Chen, *Phys. Rev.* **167**, 828 (1968).
111. A. K. Garrison, J. C. Pigg, and O. E. Schow, *Bull. Am. Phys. Soc., Ser. II*, **15**, 1369 (1970).
112. A. M. Stoneham, *Rev. Mod. Phys.* **41**, 82 (1969).
113. B. Henderson and D. H. Bowen, *J. Phys. C* (*Solid State Phys.*) **4**, 1487 (1971).
114. K. C. To, A. M. Stoneham, and B. Henderson, *Phys. Rev.* **181**, 1237 (1969).
115. R. L. Nelson and A. J. Tench, *J. Chem. Phys.* **40**, 2736 (1964); R. L. Nelson, A. J. Tench, and B. J. Harmsworth, *Trans. Faraday Soc.* **63**, 1427 (1967).
116. L. Pincherle, *Proc. Phys. Soc.* **A64**, 648 (1951).
117. F. Lüty, in *Physics of Color Centers*, Ed. by W. Beall Fowler (Academic Press, New York, 1968), p. 182.

118. R. D. King and B. Henderson, *Proc. Brit. Ceram. Soc.* **9**, 63 (1967).
119. W. A. Sibley, J. L. Kolopus, and W. C. Mallard, *Phys. Stat. Sol.* **31**, 223 (1969).
120. T. J. Turner, N. N. Isenhower, and P. K. Tse, *Solid State Comm.* **7**, 1661 (1969).
121. Y. Chen, private communication.
122. F. Seitz, *Rev. Mod. Phys.* **18**, 384 (1946); **26**, 7 (1954); *Phys. Rev.* **79**, 529 (1950).
123. C. J. Delbecq, E. Hutchinson, D. Schoemaker, E. L. Yasaitis, and P. H. Yuster, *Phys. Rev.* **187**, 1103 (1969); F. W. Patten and F. J. Keller, *Phys. Rev.* **187**, 1120 (1969).
124. J. E. Wertz, P. Auzins, J. H. E. Griffiths, and J. W. Orton, *Disc. Faraday Soc.* **28**, 136 (1959).
125. W. Känzig, *J. Phys. Chem. Solids* **17**, 80 (1960); D. Schoemaker, Int. Symp. on Color Centers in Alkali Halides, Urbana, 1965, Abstract 167, p. 129.
126. G. Rius and R. Cox, *Phys. Letters* **27A**, 76 (1968).
127. A. J. Shuskus, *J. Chem. Phys.* **39**, 849 (1963).
128. F. T. Gamble, R. H. Bartram, C. G. Young, O. R. Gilliam, and P. W. Levy, *Phys. Rev.* **134**, A589 (1964).
128a. A. Herve and B. Maffeo, *Phys. Letters* **32A**, 247 (1970); B. Maffeo, A. Herve, G. Rius, C. Santier, and R. Picard, *Solid State Comm.*, to be published.
128b. D. Galland and A. Herve, *Phys. Letters* **33A**, 1 (1970); A. L. Taylor, G. Filipovitch, and G. K. Lindeberg, *Sol. State Comm.* **8**, 1359 (1970).
129. Y. Chen and W. A. Sibley, *Phys. Rev.* **154**, 842 (1967).
130. R. H. Bartram, C. E. Swenberg, and C. T. Fournier, *Phys. Rev.* **139**, A941 (1965).
131. D. R. Hartree, W. Hartree, and B. Swirles, *Phil. Trans. Roy. Soc.* **A238**, 229 (1939).
132. J. R. Brailsford, J. R. Morton, and L. E. Vannotti, *J. Chem. Phys.* **49**, 2237 (1968).
133. L. E. Vannotti and J. R. Morton, *Phys. Rev.* **174**, 448 (1968).
134. T. M. Searle and A. M. Glass, *J. Phys. Chem. Solids* **29**, 609 (1968).
135. R. L. Hansler and W. G. Segelken, *J. Phys. Chem. Solids* **13**, 124 (1960).
136. J. E. Wertz, L. C. Hall, J. Hegelson, C. C. Chao, and W. S. Dykoski, in *Interaction of Radiation with Solids*, Ed. by A. Bishay (Plenum Press, New York, 1967), p. 617; C. C. Chao, *J. Phys. Chem. Solids* **32**, 2517 (1971).
137. T. M. Searle and B. Bowler, *J. Phys. Chem. Solids* **32**, 591 (1971).
138. P. W. Kirklin, P. Auzins, and J. E. Wertz, *J. Phys. Chem. Solids* **26**, 1067 (1965).
138a. W. C. O'Mara and J. E. Wertz, *Solid State Comm.* **8**, 807 (1970).
138b. W. B. J. Blake, H. A. Gitelson, and J. E. Wertz, *J. Phys. C.* (*Solid State Phys.*) **4**, L261 (1971).
139. J. E. Wertz and P. Auzins, *Phys. Rev.* **139**, A1645 (1965).
139a. W. C. O'Mara, J. J. Davies, and J. E. Wertz, *Phys. Rev.* **179**, 816 (1968).
139b. A. Schoenberg, J. T. Suss, S. Szapiro, and Z. Luz, *Phys. Rev. Lett.* **27**, 1641 (1971).
140. J. J. Davies, *Phys. Lett.* **28A**, 9 (1968).
140a. L. A. Kappers and J. E. Wertz, *Solid State Comm.* **9**, 1755 (1971).
140b. L. A. Kappers, F. Dravnieks, and J. E. Wertz, *Solid State Comm.*, to be published.
141. O. F. Schirmer, *J. Phys. Chem. Solids* **32**, 499 (1971).
141a. G. Rius, R. Cox, P. Picard, and C. Santier, *Compt. Rend.* **271**, 824 (1970).
142. A. M. Glass and T. M. Searle, *J. Chem. Phys.* **46**, 2092 (1967).
143. T. M. Searle, *J. Phys. Chem. Solids* **30**, 2143 (1969).
143a. B. Henderson, J. L. Kolopus, and W. P. Unruh, *J. Chem. Phys.* **55**, 3519 (1971).
144. O. F. Schirmer, *J. Phys. Chem. Solids* **29**, 1407 (1968).
144a. H. T. Tohver, B. Henderson, M. M. Abraham, and Y. Chen, *Phys. Rev.*, to be

published; H. T. Tohver, M. M. Abraham, Y. Chen, and J. Kolopus, *Bull.Amer. Phys. Soc.* **16**, 817 (1971); B. Henderson, to be published.
145. J. E. Wertz, G. Saville, P. Auzins, and J. W. Orton, *J. Phys. Soc. Japan* (*Suppl. II*), **18**, 305 (1963).
146. W. D. Compton and H. Rabin, in *Solid State Physics*, Vol. 16, Ed. by F. Seitz and D. Turnbull (Academic Press, New York 1964), p. 121.
147. Y. Chen and W. A. Sibley, *Phil. Mag.* **20**, 217 (1969).
148. A. A. Kaplyanskii, *Opt. i Spektroskopiya* **16**, 602; **16**, 1031 (1964) [English transl.: *Opt. Spectry.* **16**, 329; **16**, 557 (1964)].
149. W. A. Runciman, *Proc. Phys. Soc.* **86**, 629 (1965).
150. W. A. Runciman, in *Physics of Solids in Intense Magnetic Fields*, Ed. by E. D. Haidemenakis (Plenum Press, New York, 1969), p. 344.
151. A. A. Kaplyanskii and V. N. Medvedev, *Opt. i Spectroskopiya* **23**, 743 (1967) [English transl.: *Opt. Spectry.* **23**, 404 (1967)].
152. R. D. King and B. Henderson, *Proc. Phys. Soc.* **89**, 153 (1966).
153. I. K. Ludlow and W. A. Runciman, *Proc. Phys. Soc.* **86**, 1081 (1965).
154. R. D. King, Ph. D. thesis, University of Reading (1967, unpüblished).
155. I. K. Ludlow, *Proc. Phys. Soc.* **88**, 763 (1966).
156. Y. Kazumata, K. Ozawa, and M. Nakagawa, *Phys. Letters* **19**, 529 (1965).
156a. M. Nakagawa and K. Ozawa, *J. Phys. Soc. Japan* **24**, 96 (1968).
157. J. D. Stettler, R. A. Shatas, and G. A. Tanton, *Phys. Letters* **23**, 70 (1966).
158. Y. Chen, J. L. Trueblood, O. E. Schow, and H. T. Tohver, *J. Phys. C.* (*Solid State Phys.*) **3**, 2501 (1970).
159. B. Henderson and R. D. King, *J. Phys.* (*Paris*), **28**, C4-75 (1967).
160. I. K. Ludlow, *J. Phys. C.* (*Proc. Phys. Soc.*), Ser. 2, **1**, 1194 (1968).
160a. I. K. Ludlow, *Ph.D. thesis*, University of London (1970, unpublished).
161. R. H. Silsbee, *Phys. Rev.* **138**, A180 (1965).
162. A. E. Hughes and W. A. Runciman, *Proc. Phys. Soc.* **86**, 615 (1965).
163. A. E. Hughes and W. A. Runciman, *Proc. Phys. Soc.* **90**, 827 (1967).
164. A. E. Hughes, D. Phil. thesis, University of Oxford (1966, unpublished).
165. C. Z. Van Doorn, *Philips Res. Repts.* (*Suppl.* No. 4) (1962); Shao-Fu Wang, *Phys. Rev.* **147**, 521 (1966); Shao-Fu Wang and Chin Chu, *Phys. Rev.* **147**, 527 (1966).
166. J. A. Davis and D. B. Fitchen, *Solid State Comm.* **6**, 505 (1968).
167. J. A. Glaze and J. C. Kemp, *Phys. Rev.* **178**, 1502, 1507 (1969).
168. I. W. Shepherd, *Phys. Rev.* **165**, 985 (1968); W. Burke, *Phys. Rev.* **172**, 886 (1968); Y. Merle d'Aubigné and P. Duval, *J. Phys.* (*Paris*) **29**, 896 (1968).
169. H. R. Fetterman and D. B. Fitchen, *Solid State Comm.* **6**, 501 (1968).
170. F. S. Ham, *Phys. Rev.* **166**, 307 (1966).
171. B. Henderson, Symp. on Mass Transport in Oxides, NBS publication 296, p. 41 (1968).
172. D. C. Krupka and R. H. Silsbee, *Phys. Rev.* **152**, 816 (1966).
173. D. B. Fitchen, R. H. Silsbee, T. A. Fulton, and E. L. Wolf, *Phys. Rev. Letters* **11**, 275 (1963); G. Johannson, F. Lanzl, W. von der Osten, and W. Waidelich, *Phys. Letters* **15**, 110 (1965) and *Z. Phys.* **201**, 430 (1967); A. E. Hughes, *Proc. Phys. Soc.* **87**, 535 (1966).
174. R. G. Bessent, *J. Phys. C* (*Solid State Phys.*) **2**, 1101 (1969).
174a. R. G. Bessent, *Solid State Comm.* **9**, 1155 (1971).
175. V. I. Neeley, Ph. D. thesis, University of Oregon (1964, unpublished).

176. B. Henderson and R. D. King, *Phys. Stat. Sol.* **26**, K147 (1968).
176a. M. Nishi, T. Fujita, A. Fujii, and S. Kato, *J. Phys. Soc. Japan* **31**, 612 (1971).
177. J. Owen, *J. Appl. Phys.* (*Suppl.*) **32**, 213 (1961).
178. D. H. Tanimoto, W. M. Ziniker, and J. C. Kemp, *Phys. Rev. Letters* **14**, 645 (1965).
179. B. Henderson, *Brit. J. Appl. Phys.* **17**, 851 (1966).
180. A. M. Stoneham, unpublished.
181. M. J. Norgett, *J. Phys. C.* (*Solid State Phys.*) **4**, 1289 (1971).
182. A. A. Berezin, *Fiz. Tverd. Tela* **10**, 2882 (1968) [English trans.: *Soviet Phys.—Solid State* **10**, 2880 (1969)]; *Phys. Stat. Sol.* **49**, 51 (1972).
182a. R. Cox, in *Magnetic Resonance and Relaxation* (Proc. XIVth Colloque Ampère) (North-Holland, Amsterdam, 1967), p. 279.
182b. B. Maffeo, A. Herve, and R. Cox, *Solid State Comm.* **8**, 2169 (1970); A. K. Garrison, private communication.
183. S. R. P. Smith, D. Phil. thesis, University of Oxford (1966, unpublished).
183a. S. R. P. Smith and J. Owen, *J. Phys. C.* (*Solid State Phys.*) **4**, 1399 (1971).
184. J. Kanamori, *Prog. Theor. Phys.* **17**, 177, 197 (1957).
185. A. E. Hughes, *Phys. Rev.* **B3**, 877 (1971).
186. J. S. Smart, in *Magnetism*, Vol. III, Ed. by G. T. Rado and H. Suhl (Academic Press, New York, 1963), p. 63.
187. A. J. B. Codling and B. Henderson, *J. Phys. C.* (*Solid State Phys.*) **4**, 1409 (1971).
188. P. W. Anderson, in *Solid State Physics*, Vol. 14, Ed. by F. Seitz and D. Turnbull (Academic Press, New York, 1963), p. 99; T. Moriya, *Phys. Rev.* **120**, 91 (1960).
189. J. H. Van Vleck, *Phys. Rev.* **52**, 1178 (1937); *J. Phys.* (*Paris*) **12**, 262 (1951); *Conf. on Magnetism and Magnetic Materials*, Boston (1957), p. 6.
190. B. Henderson, unpublished.
190a. R. Cox, private communication.
191. M. J. Bergren, G. F. Imbush, and P. L. Scott, *Phys. Rev.* **188**, 675 (1969).
192. B. A. Coles, J. W. Orton, and J. Owen, *Phys. Rev. Letters* **4**, 116 (1960).
193. E. A. Harris and J. Owen, *Phys. Rev. Letters* **11**, 9 (1963); D. S. Rodbell, I. S. Jacobs, J. Owen, and E. A. Harris, *Phys. Rev. Letters* **11**, 10 (1963).
194. E. A. Harris, D. Phil. thesis, University of Oxford (1963, unpublished); *J. Phys. C.* (*Solid State Phys.*) **5**, 338 (1972).
195. M. E. Lines, *Phys. Rev.* **139**, A1304 (1965); M. E. Lines and E. D. Jones, *Phys. Rev.* **139**, A1313 (1965). See also ref. 193.

Chapter 8

CONDUCTION BY POLARONS IN IONIC CRYSTALS

Frederick C. Brown
Department of Physics
University of Illinois
Urbana, Illinois

1. INTRODUCTION—HISTORY AND DEVELOPMENT OF POLARON CONCEPTS

Strongly ionic crystals such as the alkali or silver halides have filled valence bands which are separated from empty conduction bands by forbidden regions as wide as several electron-volts. Consequently these materials are insulators at low temperature. On the other hand, small photocurrents can be made to flow when the crystals are illuminated with light capable of producing electrons and holes. Photoexcitation takes place in highly perfect crystals either by a one-photon process across the band gap or, at very high light intensity, by a two-photon process. In the latter case, the sum of the photon energies must be equal to or greater than the band gap. Defects which become electron donors can be introduced into ionic crystals just as in semiconductors. In the alkali halides, the most common donor is the F center, or electron trapped at a negative-ion vacancy, and this defect can be formed either by radiation, for example, X-rays, or by heating in alkali metal vapor, i.e., additive coloration. The F-center electron in its ground state is bound with too great an energy to be thermally ionized at room temperature. Therefore, colored alkali halide crystals can be made conducting most conveniently by illuminating with light whose quantum energy is equal to or greater than the ionization energy of the F center, usually 3 or 4 eV.

Electrons are the principal carriers of electricity in the alkali halides at low temperatures because holes in the valence band of these crystals rapidly become self-trapped to form molecular-ion or V_K-center configurations. The self-trapped hole has only a very limited mobility and, in fact, has not been observed to contribute to photoconduction. Holes apparently do not self-trap in AgBr, although they are very rapidly captured by impurities such as iodine, which is difficult to eliminate totally from the silver halides. In the thallous halides TlCl and TlBr, both holes and electrons are mobile, although one type of carrier usually predominates in a given temperature range.

Conduction processes in ionic crystals always indirectly involve polaron phenomena, at least as corrections to the band masses and scattering processes predicted by the usual semiconductor theory. Until recently, very few experiments have been performed whose outcomes depend in a unique way upon the existence of polarons. This situation has changed, however, with the measurement of optical response and conductivity at high magnetic fields and in the far-infrared. There is now also a general realization of the practical importance of polaron corrections in a wide variety of insulators, including the more complicated crystals of ionic nature such as the titanates, the spinels, and many others. An extensive literature on polarons, both experimental and theoretical, has developed out of academic interest in the problem of an electron coupled to a phonon field. The polaron is a particle–field problem closely analogous to, but perhaps simpler to solve than, the nucleon–meson field problem. A number of reviews of the subject have appeared, for example, the early paper by Fröhlich[1] and the small book by Pekar.[2] A conference on polarons and excitons was held in St. Andrews, Scotland, in the summer of 1962 and the Proceedings were published in a book edited by Kuper and Whitfield.[3]* More recently, Appel has written a chapter entitled "Polarons" which appeared in Volume 21 of the Seitz–Turnbull series, *Solid State Physics*.[4]

The polaron concept was first introduced by Landau in 1933.[5] Landau had the idea that an electron near the bottom of the conduction band of an ionic crystal might self-trap, or become localized due to the displacement of the surrounding lattice ions. The situation is as shown in Fig. 1, where the extra electron is imagined to be localized in the vicinity of a positive ion at the face center of a sodium chloride lattice. Notice that the four nearest-neighbor negative ions in the two dimensional diagram of Fig. 1 are displaced outward from their equilibrium positions, whereas the four positive

* For a review of polaron experiments, see Brown.[3a]

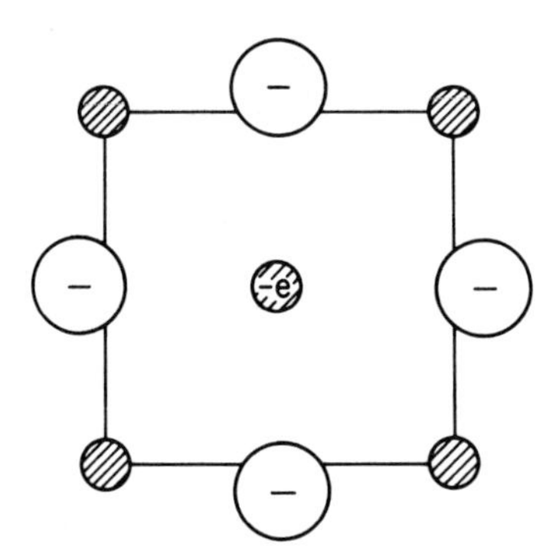

Fig. 1. An electron (*-e*) localized on a positive ion. The positive and negative ions are displaced from their equilibrium positions due to Coulomb interaction with the point electron.

ions at the corners of the square are displaced slightly inward. This distortion and the fact that the ions are charged result in a lowering of the energy of the extra electron and the formation of a potential well, causing localization. In Landau's model, the highly localized electron has a ground state of the order of a few electron-volts deep as well as some excited states. Optical excitation from the ground to the excited states was expected to account for the *F* absorption band. Self-trapped electrons were thus identified with *F* centers, about which an extensive literature was beginning to accumulate in the early thirties.

Polarons and *F* centers have always been closely associated in the literature. It is interesting that in open debate, a decision was firmly made against Landau's polaron explanation of *F* centers at a conference which was held in Bristol in 1937. A detailed report of this Conference on Conduction of Electricity in Solids, together with a summary of the informal discussions, appears in an extra part of the *Proceedings of the Physical Society* in August 1937.[6] Extensive experimental work on *F* centers was also reviewed by R. W. Pohl in these *Proceedings* without offering a hypothesis as to the nature of *F* centers. The self-trapping idea introduced by Landau, and developed further by Frenkel and von Hippel, was discussed. It was then pointed out by J. H. de Boer that an attractive alternative model for the *F* center is an electron trapped at a negative-ion vacancy. This model is in agreement with the fact that when an alkali halide crystal is heated in alkali metal vapor, the number of *F* centers formed is proportional to the pressure of the vapor. If, on the other hand, an alkali atom dissociated into a self-trapped electron plus a positive ion (interstitial or absorbed at an internal crack), the number of F centers would be proportional to the square root of pressure, contrary to what is observed. De Boer's model was thus generally favored at this early Bristol conference and has turned out to be correct. There are, of course, many other properties of F centers which substantiate the vacancy electron model, for example, the fact that an F center can trap a second electron to become an F' center, etc.

The polaron concept was pursued in later years, especially by Fröhlich and co-workers. The application of quantum field concepts enabled these workers to treat the scattering, as well as the self-energy, of an electron in an ionic crystal[7] and to realistically take into account the strong interaction between an electron and the lattice polarization. It is now known that an electron in an alkali halide crystal is not as deeply trapped nor as highly localized as implied by Landau's early model and Fig. 1. The motion of the electron within the potential well associated with the lattice distortion must be taken into account; therefore, the polaron essentially involves a dynamic interaction between the electron and the lattice. The potential which localizes the electron is Coulombic at large distances, and dynamic considerations dictate that the potential be truncated close to the electron. The radius at which the potential is terminated determines the degree of localization. Thus a characteristic length enters into the problem.

In continuum polaron theory for slow electrons, as usually applied to alkali halide crystals, a single characteristic lattice frequency ω is involved. This ω is the longitudinal optical (LO) mode, the lattice frequency to which the electrons are most strongly coupled because of the polarization wave associated with the longitudinal motion of the ions (see Fig. 2). Dispersion of the LO mode is usually neglected, mainly because of the overwhelming importance of the long-wavelength $k = 0$ phonons and the cutoff in momentum. The highly localized or small polaron is a separate case and is

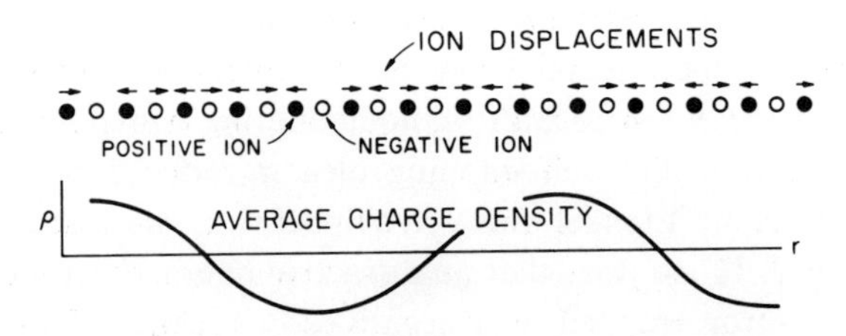

Fig. 2. A long-wave longitudinal optical mode of lattice vibration in an ionic crystal (linear chain). The displacements of positive and negative ions are shown above and the average charge density of the polarization wave ϱ is shown below.

responsible for a different kind of conduction mechanism, the hopping mechanism.[4] In the present chapter, we will be less concerned with the hopping mechanism than with band transport characteristics of large polarons in the alkali halides at low temperature.

In the next section, we discuss some of the central features of the polaron, including the Fröhlich Hamiltonian. The effective mass and self-energy of the polaron is derived by a perturbation treatment. A list of continuum properties and the polaron coupling constants for a number of materials are given in Section 3. Here, new, more reliable values of effective mass obtained by cyclotron resonance are listed. Experimental methods are discussed in Section 4. In Section 5, a comparison is made between polaron mobility theory and experiment. This comparison points up apparent difficulties with solutions of the Boltzmann transport equation as obtained so far. It suggests that results of the recent dc mobility theory of Thornber and Feynman may be more nearly correct; however, this theory also has approximations which are open to question. Finally, interesting non-Ohmic effects observable at low temperature and high electric field are surveyed in Section 6. Experimental evidence is cited for effects due to the nonparabolic nature of polaron bands and for quantum-limit phenomena and spin-dependent interactions in high magnetic fields.

2. ESSENTIAL FEATURES OF THE POLARON

The essential features of the polaron can be illustrated by an artificial model of the electron-plus-lattice polarization. This model, which is due to Fröhlich,[8] was arrived at in hindsight after a more complete solution of the problem. As discussed qualitatively above, the electron induces a polarization in the surrounding lattice equivalent to a potential well in which the electron tends to be localized. At large distances, the potential is Coulombic and the lattice ions distort about a point electron at the center. Because of the electron's finite extension, the potential must be cut off, and in the simplest case, is taken to be constant inside some radius r_1 as shown in the upper part of Fig. 3. In the limit of a highly localized electron moving very fast in the potential well, the lattice is unable to adjust to this motion, and we have a static charge distribution and potential. This is Landau's picture of a localized polaron. In the other extreme, the electron is less localized, moves slower, and we are forced to consider the dynamic problem of an electron of velocity v interacting with a lattice whose characteristic frequency is ω. The lattice will tend to follow the motion of the electron and this

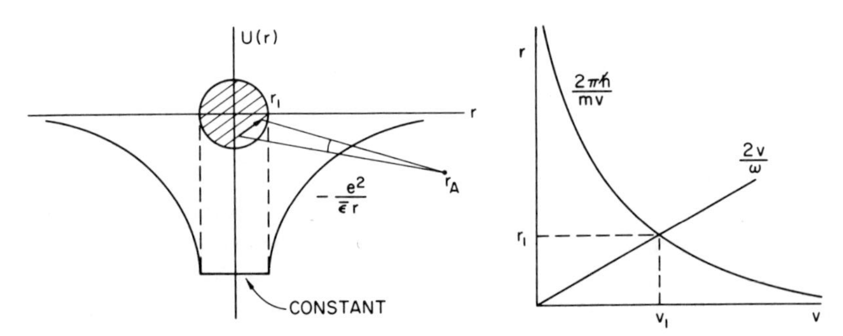

Fig. 3. Model of a polaron is shown in left part of the figure. The potential energy $U(r)$ is Coulombic outside of the region r_1 within which the electron moves. The angle subtended by the moving electron at a distant point in the lattice, r_A, is also shown. A graph to determine r_1 is shown in the right-part of the figure.

adjustment will be important, especially in close. We will estimate the cutoff radius r_1 below and then, neglecting kinetic energy, estimate the electron's energy from

$$U_1 = -eV = -e^2/\bar{\varepsilon}r_1 \tag{1}$$

where $\bar{\varepsilon}$ is an effective dielectric constant characterizing the lattice polarization.

The response of the lattice to applied fields is described in terms of the polarization **P**, which is the sum of ionic and electronic contributions

$$\mathbf{P} = \mathbf{P}_i + \mathbf{P}_e \tag{2}$$

Polarization **P**, electric field **E**, and displacement **D** are related by

$$\mathbf{D} = \mathbf{E} + 4\pi\mathbf{P} \tag{3}$$

By using the equation $\mathbf{D} = \varepsilon_s\mathbf{E}$, where ε_s is the static dielectric constant, we can eliminate the electric field **E** in Eq. (3). The polarization is thus related to the displacement **D**, which we know originates on true charges in the crystal. The resulting relation is

$$4\pi\mathbf{P} = [1 - (1/\varepsilon_s)]\mathbf{D} \tag{4}$$

At high frequencies, greater than the *Reststrahlen* frequency, the electronic part of the polarization $\mathbf{P}_e = \mathbf{P} - \mathbf{P}_i$ is characterized by an optical or high-frequency dielectric constant ε_∞ as follows:

$$4\pi(\mathbf{P} - \mathbf{P}_i) = [1 - (1/\varepsilon_\infty)]\mathbf{D} \tag{5}$$

In the lattice polaron problem, we are interested in the ionic part of the polarization, which can be found by subtracting Eq. (5) from Eq. (4), resulting in

$$4\pi \mathbf{P}_i = [(1/\varepsilon_\infty) - (1/\varepsilon_s)]\mathbf{D} = (1/\bar{\varepsilon})\mathbf{D} \quad (6)$$

Thus the reciprocal of the effective dielectric constant $1/\bar{\varepsilon}$ can be obtained from the difference in $1/\varepsilon_\infty$ and $1/\varepsilon_s$. In a continuum model, the constant $\bar{\varepsilon}$ used in Eq. (1) allows for the screening of charge due to the displacement of lattice ions.

Let us now estimate the value r_1 so that the energy U_1 or self-energy in the polaron problem can be approximately evaluated. We do this by considering the ability of the lattice to follow the motion of the electron in close, and we also apply the uncertainty principle. It is crucial from the point of view of the lattice that beyond r_1, the electron look like a point charge. Thus at a distance r_A (Fig. 3), the velocity of the electron v in the well must be such that the electron appears to cover only a small angular distance in a time of the order of a lattice vibration $2\pi/\omega$. The condition is therefore $v(2\pi/\omega)/r_A \ll \pi$. The points in the lattice r_A for which the electron appears to be a point charge all lie outside of a limiting value given by

$$r = 2v/\omega \quad (7)$$

This relation is indicated by a line of constant slope in the lower part of Fig. 3. At the same time that the lattice follows the electron's motion down to an r and v given by Eq. (7), the electron cannot have an infinite de Broglie wavelength. That is, the velocity of the electron v cannot approach zero. From the uncertainty principle, since the electron is localized to within the region r_1, one must have the relation

$$r = 2\pi\hbar/mv \quad (8)$$

as also shown in Fig. 3. The intersection of these two curves gives the smallest $r = r_1$ and therefore the potential U_1 of lowest energy. By equating Eqs. (7) and (8), we find v_1 and therefore

$$r_1 = (2h/m\omega)^{1/2} \quad (9)$$

Neglecting kinetic energy, an approximate value of the electron's self-energy is found from Eq. (1) as follows:

$$U_1 = (-e^2/\bar{\varepsilon})(m\omega/2h)^{1/2} \quad (10)$$

The actual values of r_1 and U_1 obtained from a more complete analysis (perturbation theory) are the same as Eqs. (9) and (10) except that h is replaced by $\hbar$.

Longitudinal optical frequencies ω are of the order of $2\text{–}3 \times 10^{13}$ sec^{-1} (see Table II). From Eq. (9), we see that r_1 is of the order of $3\text{–}4 \times 10^{-7}$ cm for a free electron mass. Fortunately, r_1 comes out greater than the lattice parameter, so that using the continuum approximation (use of ε_s and ε_∞ and a constant ω) to describe the lattice is not such a bad approximation. It should be further remarked that the self-energy U_1 is frequently written

$$U_1 = -\alpha\hbar\omega \tag{11}$$

where the dimensionless coupling constant α is given by

$$\alpha = (e^2/\hbar)(m/2\hbar\omega)^{1/2}[(1/\varepsilon_\infty) - (1/\varepsilon_s)] \tag{12}$$

Equations (11) and (12) follow from Eq. (10) by writing out $1/\bar{\varepsilon}$ and replacing h by $\hbar$ as noted above.

This simple model due to Fröhlich is in agreement with the perturbation treatment of the electron–lattice problem, but a number of approximations are made. Among these is the effective mass approximation. The free electron's momentum exchange with the lattice is incorporated by introducing a fixed lattice or bare mass m obtainable from band theory. This bare mass m can be any value depending upon the results of the band calculation. In the case of the alkali halides, m usually turns out to be less than the free-electron mass m_e, of the order of 0.5 m_e. Thus for these materials, the polaron radius is even larger than the value computed from Eq. (9) using m_e. The corresponding values of polaron self-energy U are of the order of a few times $\hbar\omega$, for example, 0.02–0.07 eV. These values of self-energy are very much smaller than the 2–5 eV suggested by the early localized model devised to explain the F center. It is furthermore fundamental that the electron–polaron is translationally invariant and can move through the crystal. Although the lattice polarization is expected to follow the slow electron's motion, it can be shown to provide additional inertia and to affect the response in applied fields. A polaron mass m_p thus results which is always larger than the fixed lattice mass m.

Most detailed theoretical work on polarons carried out to date begins with the electron–lattice Hamiltonian first derived by Fröhlich. A careful discussion and derivation of this Hamiltonian is given in the first chapter of *Polarons and Excitons*.[3] The problem is set up using classical mechanics to take proper account of lattice coupling. A constant lattice frequency ω is assumed. By means of a transformation, the Hamiltonian can be separated into easily identifiable terms closely analogous to the normal mode vibrational problem of the free lattice. Introducing annihilation and creation operators a_k and a_k^+, respectively, the Hamiltonian takes the form

$$H = H_e + H_l + H_{\text{int}} \tag{13}$$

where

$$H_e = p^2/2m \tag{14}$$

the kinetic energy of the band electron, and the term

$$H_l = \sum_k \hbar\omega a_k^+ a_k \tag{15}$$

is simply that part of the quantized Hamiltonian corresponding to the phonon field. A constant self-energy term, $\sum_k \hbar\omega/2$, has been neglected since it does not contribute to the dynamics of the polaron problem. Finally, the electron–lattice interaction term is given by

$$H_{\text{int}} = \sum_k (V_k a_k e^{ik\cdot r} + V_k^+ a_k^+ e^{-ik\cdot r}) \tag{16}$$

and

$$V_k = -(i\hbar\omega/k)(\hbar/2m\omega)^{1/4}(4\pi\alpha/V)^{1/2} \tag{17}$$

where V is the volume of the crystal. Again the dimensionless constant α enters the equation and is defined by Eq. (12).

A number of approximations lie behind the Fröhlich Hamiltonian, as follows:

1. A free electron interacts mainly with the longitudinal optical mode of the lattice, and other modes arc disregarded. Arguments are given by various authors[3] which show that this is a reasonable approximation; nevertheless, the polaron concept can be worked out for acoustic phonons as well.[9] In this case, a deformation potential interaction between the electron and the acoustic mode of lattice vibration is involved. The acoustic polaron effects are very much weaker than optical mode polaron phenomena, and they have not been incorporated into a general polaron theory, at least for crystals such as the alkali halides. Other types of electron–lattice interaction may also be more or less important. For example, many crystals lack a center of symmetry and the piezoacoustic interaction is quite strong. In other, centrosymmetric ionic crystals, especially those

with different ion masses, a polarization wave is associated with long-wavelength acoustic rather than optical vibrations of the lattice. Consequently, in this case, one has strong scattering due to acoustic vibrations and, in principle, a polaron can be associated with this interaction.

2. Interactions between the conduction electron and other electrons in the crystal are neglected except for those interactions that contribute to the band effective mass. Screening by other conduction electrons can be taken into account, but is neglected for the crystals with which we will be dealing. Electron–plasmon effects are neglected. It is assumed that the core electrons and valence band electrons adjust adiabatically to the motion of the conduction electron. A kind of polaron effect is involved here, but one which results in a constant energy shift of the conduction band with respect to the valence band.[10]

3. Neglect of dispersion in the lattice frequency ω. The frequency ω is taken to be a constant because only a small region of k space corresponding to long-wavelength phonons is primarily involved in the interaction. This is reasonable since a cutoff in k vector is involved in the polaron problem [refer to the qualitative arguments behind Eq. (9)]. Since the electron's speed $v = p/m$ must be less than a certain value, $\hbar k = p$ has an approximate upper limit. Moreover, the electron will not interact very strongly with lattice waves of higher k value. Insertion of numbers into Eqs. (8) and (9) shows that k is always only a small fraction of the Brillouin zone boundary value π/a. The polaron problem is characterized by a cutoff in momentum as well as by a length r_1.

4. The continuum approximation associated with the use of ε_s, ε_∞, and ω is an important assumption. This was commented on above in connection with the size of the polaron, Eq. (9). It should be mentioned that here we are considering large polarons, which can be treated within this approximation. Small polarons have been discussed and may be important in uranium oxide, the titanates, and other materials.[4] Small polarons, rather than drifting through the crystal, transport by means of a hopping mechanism. Their mobility should increase exponentially with increasing temperature, a dependence just opposite to that expected for lattice scattering of free carriers.

5. Neglect of electron-magnon and other magnetic interactions.

6. The use of a single band mass parameter corresponding to standard band shape, that is, an isotropic energy band is assumed. A treatment which goes beyond this approximation has recently been given by Kahn.[11] His results show the nature of polaron effects for spheroidal energy surfaces.

It appears that anisotropy between the longitudinal and transverse masses tends to be smoothed-out or reduced by the inclusion of polaron effects.

7. Although not specifically limited to binary compounds such as the alkali halides, most of the work so far would have to be extended to apply to ternaries or more complex crystals. Several frequencies ω may be involved. This is certainly an important direction for further work on polarons as understanding of the conduction and optical properties of more complex crystals is advanced.

8. The Hamiltonian given above does not include other fields, such as might be introduced by a charged defect or impurity. For example, the bound polaron problem has been treated by including an additional $-e^2/\varepsilon r$ term in the Hamiltonian.[12–14] The bound or trapped polaron is of special importance in certain infrared optical properties of ionic crystals.

We will now briefly discuss the simple perturbation theory of the polaron self-energy and effective mass, starting with the Fröhlich Hamiltonian and using Eq. (13). Actually, it can be shown that a perturbation theory treatment is not very useful for materials such as the alkali halides. This is because for these materials, the coupling constant α is of the order of 3 and the perturbation result, which is essentially a series expansion in α, will not necessarily converge rapidly. On the other hand, some materials, such an InSb, have small α and perturbation theory is thought to be quite useful. It has been the starting point for more complicated problems, such as the polaron in a magnetic field.[15,16]

Without the interaction term in the Hamiltonian, a solution of the problem could be written as the product of plane waves and the phonon states of the lattice, i.e.,

$$\psi = (e^{ik\cdot r}/N) \mid n_1, n_2, \ldots, n_j, \ldots\rangle \tag{18}$$

We use the ket notation to describe the phonon field, where the indices n_j give the number of phonons with wave vector k_j. The vacuum state where no phonons are present is described by $\mid 0\rangle$. The states in Eq. (18) are not proper solutions to the polaron problem, but we expect from perturbation theory that a weak H_{int} will mix appropriate states. It is clear that first-order perturbation theory cannot modify the ground-state energies of the system. The interaction term $H_{\text{int}} \mid 0\rangle$ is orthogonal to $\langle 0 \mid$ since the interaction term always creates or annihilates a phonon. On the other hand, H_{int} will mix the ground-state energy with those terms for which $\mid n_j\rangle$ corresponds to a state with one phonon of any k value present in the field. All the other combinations of two states are orthogonal. The change in energy from second-order perturbation theory is given by

$$\Delta E = -\sum_n \frac{|\langle n | H_{\rm int} | 0\rangle|^2}{E_n - E_0}$$
$$= -\sum_k \frac{V_k^2}{[(\mathbf{p} - \hbar\mathbf{k})^2/2m] + \hbar\omega - (p^2/2m)} \tag{19}$$

where n is a state with one phonon in any k value. For low momentum $\mathbf{p}$, the energy change ΔE can be developed as a power series in $\mathbf{k} \cdot \mathbf{p}$:

$$\Delta E = -\sum_k \frac{V_k^2}{\hbar\omega + (\hbar^2 k^2/2m_e)}$$
$$\times \left[1 + \frac{(\hbar/m)\mathbf{k} \cdot \mathbf{p}}{\hbar\omega + (\hbar^2 k^2/2m)} + \frac{(\hbar^2/m^2)(\mathbf{k} \cdot \mathbf{p})^2}{[\hbar\omega + (\hbar^2 k^2/2m)]^2} + \cdots\right] \tag{20}$$

By substituting for V_k [Eq. (17)], choosing fixed direction $\mathbf{p}$, and replacing the $\sum_k$ by integrals (remember the appropriate factors $V/8\pi^3$), one obtains

$$\Delta E = \frac{4\pi V}{8\pi^3} \frac{2\pi e^2 \hbar\omega}{V} \frac{1}{\bar{\varepsilon}} \int_0^{k_{\max}} \frac{dk}{\hbar\omega + (\hbar^2 k^2/2m)} \tag{21}$$

Little error is made by replacing the upper limit $k_{\max}$ by infinity. The result becomes

$$\Delta E = -\alpha\hbar\omega - (\alpha/6)(p^2/2m) \tag{22}$$

Therefore the total energy of the ground state of the polaron is

$$(p^2/2m) + \Delta E = -\alpha\hbar\omega + (p^2/2m)[1 - (\alpha/6)] + O(p^4) \tag{23}$$

where $O(p^4)$ designates higher-order terms in an expansion in terms of $|\mathbf{p}|$ which are assumed to be small. The coefficient of $p^2/2m$ indicates that the effective mass in the perturbation result is given approximately by

$$m_p = m/[1 - (\alpha/6)] \tag{24}$$

The perturbation result therefore contains the essential results for small α, namely a self-energy of the order of $-\alpha\hbar\omega$ and an increase in the bare mass to give the polaron mass m_p.

Because the coupling constant α is greater than one for real crystals such as the alkali halides, the simple perturbation treatment must be replaced by some other approach. The so-called intermediate coupling theory was introduced by Lee *et al.*[17] This approach has been reexamined by Larson,[18] who proposes that the effective mass for $\alpha \lesssim 4.5$ is quite accurately given by

$$m_p = m(1 + \tfrac{1}{6}\alpha + 0.02363\alpha^2 + 0.0014\alpha^3) \tag{25}$$

Devreese *et al.*[19] have also given a derivation of the Lee *et al.* result and have discussed the physical meaning of the transformations used. These investigators explore the idea that when α is sufficiently large, the lattice tends to follow the motion of the electron point by point. They have also obtained the internal excited states of the polaron and suggest that these might be observable in free carrier absorption when a sufficiently high density of polarons can be generated.

There is also a school, largely due to Pekar and co-workers, which has worked out the so-called strong coupling theory of the polaron.[2] The results of this theory are supposed to be correct in the limit of very large α, in which case, the polaron mass is given by

$$m_p = m(1 + 0.02\alpha^4 + \cdots) \tag{26}$$

which allows the mass to increase nearly as α^4 for large α.

Finally, our brief survey would be very incomplete if we did not mention the Feynman model of the polaron[20] and also recent Greens function formulations of the polaron. It is generally accepted that the Feynman theory of the polaron gives the lowest (or nearly the lowest) self-energies over a wide range of α including the perturbation, intermediate, and strong coupling ranges. This does not necessarily mean that it gives the best effective mass at all α, although it probably yields reasonable values of m_p. In the Feynman theory of the polaron, the path integral formulation of quantum mechanics is used. The problem is set up in terms of an approximate Lagrangian which describes the way in which the system moves in space and time. The lattice phonons are removed from the problem and it can be shown that the model is equivalent to an electron of bare mass m coupled by means of a spring of constant $\varkappa$ to a mass M. The electron and lattice are therefore treated as one in a kind of harmonic approximation. The problem is solved by minimizing the total energy, essentially determining the spring constant and mass M variationally. Because the variational parameters have to be determined at each value of α, considerable numerical work is involved in obtaining the relationship between polaron mass and band mass. An interpolation formula which closely approximates the Feynman mass has been given by Langreth[23] and is as follows:

$$m_p = m(1 - 0.0008\alpha^2)/(1 - \tfrac{1}{6}\alpha + 0.0034\alpha^2) \qquad (27)$$

Very recently, Porsch[21] and Matz[22] have given unified theories of the polaron using retarded Greens functions. Their methods are in principle applicable for arbitrary electron–phonon coupling constant.

3. SURVEY OF CONTINUUM PROPERTIES AND COUPLING CONSTANTS FOR IONIC CRYSTALS

It is interesting to compare the size of the polaron effect for a number of real materials. In Table I, we list values of m_p ranging from 0.1 to 10 for three different bromides together with the corresponding band masses and α's consistent with the Feynman theory. We have used the Langreth interpolation formula Eq. (27) in preparing this table. Remember that α depends upon the continuum properties of the lattice and the band effective mass as given by Eq. (12). The values of lattice frequency and dielectric constants used are given in the table for AgBr, TlBr, and KBr. These are all materials

Table I. The Band Masses m and Coupling Constants α Corresponding to Various Polaron Masses m_p

AgBr[a]											
m_p	0.1	0.2	0.4	0.6	0.8	1.0	2.0	4.0	6.0	8.0	10.0
m	0.084	0.156	0.283	0.394	0.494	0.586	0.959	1.47	1.82	2.08	2.28
α	1.00	1.36	1.84	2.17	2.43	2.65	3.40	4.19	4.66	4.90	5.21
TlBr[b]											
m_p	0.1	0.2	0.4	0.6	0.8	1.0	2.0	4.0	6.0	8.0	10.0
m	0.078	0.142	0.248	0.337	0.414	0.482	0.739	1.05	1.23	1.36	1.45
α	1.36	1.83	2.42	2.82	3.12	3.37	4.17	4.96	5.39	5.51	5.87
KBr[c]											
m_p	0.1	0.2	0.4	0.6	0.8	1.0	2.0	4.0	6.0	8.0	10.0
m	0.078	0.141	0.246	0.333	0.409	0.476	0.727	1.03	1.21	1.33	1.42
α	1.38	1.86	2.45	2.85	3.15	3.41	4.21	5.00	5.43	5.69	5.89

[a] $1/\lambda = 139\ \mathrm{cm}^{-1}$, $\varepsilon_\infty = 4.62$, $\varepsilon_s = 10.60$, $\bar{\varepsilon} = 8.19$, observed electron $m_p/m_e = 0.289 \pm 0.01$ ($m/m_e = 0.215$, $\alpha = 1.6$).

[b] $1/\lambda = 116\ \mathrm{cm}^{-1}$, $\varepsilon_\infty = 5.41$, $\varepsilon_s = 35.4$, $\bar{\varepsilon} = 6.39$ observed electron mass, $m_p/m_e = 0.52 \pm 0.03$ ($m/m_e = 0.31$, $\alpha = 2.7$).

[c] $1/\lambda = 168\ \mathrm{cm}^{-1}$, $\varepsilon_\infty = 2.42$, $\varepsilon_s = 4.52$, $\bar{\varepsilon} = 5.21$, observed electron $m_p/m_e = 0.700 \pm 0.03$ ($m/m_e = 0.37$, $\alpha = 3.0$).

where the actual polaron mass has been determined from cyclotron resonance experiments. The values of m_p and α are therefore given in each case. This emphasizes the point that it is m_p which is determined experimentally and the bare mass m is to be regarded as a parameter either to be computed from band theory or from the observed m_p using polaron theory. The materials shown in Table I have increasingly strong polaron effects in the order listed, AgBr, TlBr, and KBr. Actually, the highly polarizable substance TlBr would have as strong a polaron effect as KBr if it had as large a band mass.

A list of dielectric constants, lattice frequencies, and coupling constants α is given for a larger number of ionic crystals in Table II. Many of the static dielectric constants and lattice frequencies have now been accurately measured by Lowndes and Martin at low temperature. The static constants decrease 5% or more upon cooling to low temperature, so that this must be taken into account for careful work on polaron properties that are temperature-dependent, i.e., mobility. On the other hand, the decreases in ε_s are partially offset by increases in ω_t, so we list only the low-temperature values when available. The optical dielectric constants vary only slightly, less than 1%, with temperature. Inelastic neutron scattering data, when available, compare favorably with the optical data. The cyclotron resonance masses (cold polaron) of Hodby and of others are given in the third from the last column. Values of m/m_e and α are then computed according to the Feynman model using Eq. (27). A comparison with results obtained by the Green's function formulation of Matz[22] is given in Table IV on page 526.

4. TRANSPORT MEASUREMENTS—EXPERIMENTAL TECHNIQUES

The transport properties of electrons and holes in ionic crystals can be measured in a number of ways. These include the observation of photocurrent versus electric field, the spectral response of photoconductivity, photo-Hall effect, magnetoresistance, and resonance phenomena such as cyclotron resonance. Special problems arise in measuring conductivity since all of the materials we will be concerned with are highly insulating, especially at low temperature, even in the presence of illumination. In general, the methods of measurement fall into two classes: steady-state methods (including direct current and alternating current) and transient methods. Factors such as contact resistance, the available light intensity, and, especially, the resistivity under illumination, determine whether steady

Table II. Dielectric Constants, Long-Wavelength Transverse ($1/\lambda_t$) and Longitudinal ($1/\lambda$) Optical Frequencies with LO Debye Θ, and Effective Masses when Known[a]

	ε_s	Ref.	ε_∞	Ref.	$1/\lambda_t$, cm^{-1}	Ref.	$1/\lambda$, cm^{-1}	Ref.	Θ, °K	m_p/m_e	Ref.	m/m_e	α
LiH	12.9	121	3.6	123	590	121	1120	121	1612	—		—	—
LiF	8.50	122	1.93	124	318	124	668	122, 124	960	—		—	—
LiCl	10.92	122	2.79	124	221	124	438	122, 124	630	—		—	—
LiBr	11.95	122	3.22	124	187	124	361	122, 124	519	—		—	—
LiI	—		3.89	124	152	124	—		—	—		—	—
NaF	4.73	122	1.75	124	262	124	429	122	617	—		—	—
NaCl	5.45	122	2.25	124	178	124	270	122	389	—		—	—
NaBr	5.78	122	2.64	124	146	124	211	122	304	—		—	—
NaI	6.62	122	3.08	124	124	124	184	122	265	—		—	—
KF	5.11	122	1.86	124	202	124	335	122, 124	482	—		—	—
KCl	4.49	122	2.20	124	151	124	216	122, 124	310	0.922±0.04	105	0.432	3.46
KBr	4.52	122	2.39	124	123	124	169	122, 124	243	0.700±0.03	105	0.367	3.07
KI	4.68	122	2.68	124	110	124	143	122	206	0.536±0.03	129	0.325	2.51

RbF	5.99	122	1.94	124	163	124	287	122, 124	413	—		—	—
RbCl	4.53	122	2.20	124	126	124	180	122	259	1.038±0.10	105	0.430	3.81
RbBr	4.51	122	2.36	124	94.5	124	130	122	187	—		—	—
RbI	4.55	122	2.61	124	81.5	124	108	122	155	0.72 ±0.07	105	0.365	3.18
CsF	7.27	122	2.17	124	134	124	245	122, 124	353	—		—	—
CsCl	6.75	122	2.67	124	106	124	165	122, 124	237	—		—	—
CsBr	6.39	122	2.83	124	78.5	124	105	122	151	—		—	—
CsI	6.29	122	3.09	124	65.5	124	93	122, 124	134	—		—	—
AgF	10.8	132	2.99	132	176	132	322	132	465	—		—	—
AgCl	9.50	122	3.97	124	120	124	197	125	283	0.431±0.04	105	0.303	1.86
AgBr	10.60	122	4.68	124	91.5	124	139	125	200	0.289±0.01	105, 130	0.215	1.60
AgI	7.0	133	4.91	134	110	133	131	133	189	—		—	—
TlCl	37.6	122	5.00	124	60.4	124	174	122	250	0.56 ±0.02	131	0.34	2.5
TlBr	35.1	122	5.64	124	47.2	124	116	126	167	0.52 ±0.03	131	0.31	2.7
TlI	21	127	~6		54	128	102	127, 128	147	—		—	—

[a] Low-temperature values are given except those under Ref. 121. In general, the LO frequencies are computed from $(1/\lambda) = (\varepsilon_s/\varepsilon_\infty)^{1/2}(1/\lambda_t)$.

or transient techniques are used. In general, the sample in the form of a thin dielectric slab is placed between plane parallel electrodes. In the absence of polarization effects, the electric field in the sample is given by the ratio of applied voltage to electrode separation, $E = V/d$. The conductivity is defined as the ratio of the current component in the direction of applied field to the field strength. For accurate measurements, attention must be given to edge or fringing field effects. Sometimes, the electrodes are applied directly to the crystal by evaporation or with the use of colloidal graphite or silver conducting paint. Photocharge can be released by illumination parallel to the electrodes from the side or by illumination through a transparent semiconducting glass (NESA) electrode. Transparent electrolytes and fine wire mesh have also been used as electrodes. In the latter case, care must be taken to account for field inhomogeneity and also for a gap if the electrode is located separately from the sample surface.

In some electrical measurements on insulators, technical considerations dictate the electrode configuration. This is the case for studies of dielectric breakdown, a subject in which the alkali halides have figured prominently.[24] These materials have breakdown strengths in the range 10^5–10^6 V/cm. Fields of this magnitude can be produced in very thin sections between electrodes of planoconcave geometry.[25] We will not discuss dielectric breakdown in detail here except to comment that the electron–LO phonon interaction is fundamental in determining the velocity of electrons in ionic crystals. It seems likely that the new response theory of Thornber and Feynman, which is discussed toward the end of Section 5, will be very useful in understanding how carriers acquire sufficient velocity to multiply and to initiate breakdown.

A large amount of work has been done or is now underway on device-like structures where the insulating material is in the form of a thin film overlying a semiconductor or a metal, for example, SiO_2 on Si or Al_2O_3 on Al.[26] An obvious application is to thin-film capacitors, which have been greatly improved and reduced in size. On the other hand, much recent interest in the properties of thin-film insulators centers around the technological importance of insulating layers in metal–insulator–semiconductor (MIS or MOS) charge-control devices. Photoemission into the insulator from the metal or semiconductor is an important means for investigating thin dielectric films.

The techniques for studying internal photoemission, with special emphasis on SiO_2, have been reviewed by Goodman.[27] The current situation regarding the mobility and trapping of electrons injected into SiO_2 is interesting. Thin layers of this substance on Si are vitreous and noncrystalline, so that trapping is significant, especially at low electric fields. On the other hand, at high electric fields, 10^5–10^6 V/cm, it was discovered earlier by R. W. Williams that photoinjected electrons drift through several microns of oxide layer without appreciable trapping. He also estimated values of electron mobility which could be either approximately 17 or 34 cm^2/V-sec, depending upon assumptions. This mobility is perhaps comparable to that produced by optical mode lattice scattering at room temperature as discussed in Section 5. Apparently, the disorder is less effective than optical scattering.

The above values of estimated mobility in SiO_2 at room temperature have been verified by a Hall experiment employing an unusual Hall geometry for steady-state measurements. A sketch of the three-terminal method employed by Goodman[27] is shown in Fig. 4. The opaque metal electrodes on the face of the SiO_2 have a gap or slit which defines the illuminated portion of the photoemitting silicon surface. In the absence of a transverse magnetic field, the current flow lines are symmetric about the midplane as shown in the figure. When a transverse magnetic field is applied, an unbalance occurs which can be detected by a differential electrometer connected across the two metal counterelectrodes. An approximate analysis of the current distribution shows that the method is capable of yielding reasonable estimates of Hall mobility.

A four-terminal method has been used by Dressner[28] to measure the photo-Hall effect in anthracene and also in vitreous selenium. The four electrodes in these dc measurements are on one surface of the sample and incident light excites carriers in a very narrow region close to the surface. The various difficulties connected with drift, space charge, and spurious

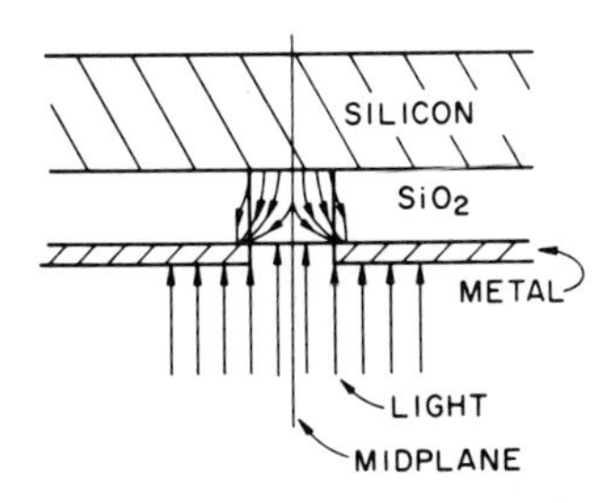

Fig. 4. Three-electrode photo-Hall geometry used by Goodman[27] for SiO_2. Note that a kind of "sandwich-cell" geometry is used and that the gap width is about equal to the sample thickness. Lines of current flow are shown within the SiO_2.

potentials inherent to this technique have been discussed. They are extremely difficult to overcome.

Steady-state ac Hall techniques were employed quite early by Macdonald and Robinson[29] in order to determine the mobility of electrons released by light from *F* centers in KBr at room temperature. Intense illumination was used, but even so, the signal impedance at the Hall electrodes was of the order of 100 megohms, necessitating the use of a special type of preamplifier. Difficulties associated with space charge and blocking or partially blocking electrodes were analyzed, and the results indicated that meaningful Hall mobilities could be obtained for frequencies of applied voltage greater than 50–500 Hz. Very similar electronic techniques were employed by Hanson in order to determine the Hall mobility of holes injected by an excess of bromine in AgBr.[30] In this case also, the Hall signal impedance was high, of the order of megohms, but with care, mobility values were obtained which have stood the test of time. Steady-state ac measurements were also carried out on KBr containing *F* centers by Onuki,[31] who succeeded in showing the increase in mobility as the lattice temperature is decreased toward liquid nitrogen temperature. Extreme difficulties arose because of the very high impedance encountered at low temperature.

The problem of conduction, especially mixed electronic and ionic conduction in high-impedance materials, has long been a subject for investigation. One of the earliest definitive papers on the subject appeared in 1933.[32] The subject is of importance in connection with the interpretation of steady-state Hall measurements under normal illumination intensities with blocking or partially blocking electrodes on crystals such as the alkali halides. It has also become of interest recently in connection with analysis of the metal-oxide thin-film studies. When experiments are carried out, ideal conditions are not always achieved in practice. The problem is complicated by the injection of charge at one or more electrodes which may be partially blocking. The necessity of solving the continuity equation including both drift and diffusion as well as the nonlinearities introduced by Poisson's equation are such complications. A critique of some of the approaches used for solving this problem is given by Macdonald,[33] who also provides a table which serves as a guide to the various works covering a wide variety of possible cases (one or more blocking electrodes, one or more carriers mobile, etc.).

There is a way around the contact and space-charge problems which complicate transport measurements on very-high-impedance photoconductors: this is to use transient techniques, especially high-speed pulse techniques, with blocking electrodes. Such methods allow one to avoid the problems associated with making Ohmic contact to wide-band-gap insulating crystals. Pulse methods permit the observation of Hall effect, transit-time mobility, magnetoconduction, and other processes. They are especially suited to conduction at low excitation levels in bulk crystals, but may also apply to the thin-film device configurations. In these measurements, the sample is usually in the form of a rectangular parallelepiped sandwiched between plane-parallel electrodes. Photocarriers can be injected at one surface by illuminating through a transparent electrode with light which is strongly absorbed in the crystal, or by photoinjection from the metal electrode as mentioned above.

The transit time of electrons or holes across an insulating photoconductor can sometimes be directly observed so that the drift mobility can be calculated from the drift velocity and applied field. This can be done only when the lifetime of carriers before trapping is sufficiently long. In such cases, the carriers might be produced as a sheet of charge by short pulse of light which is strongly absorbed at one surface or by photoemission at one electrode. The direct observation of drift mobility in this way is analogous to the early experiments on the drift of minority carriers injected into filamentary transistor material. However, in the experiments on insulators, space-charge neutrality does not hold in the volume of the solid. In order to make a meaningful mobility determination, the dark conductivity must be low enough to avoid the smoothing out of the sheet of drifting charge during the transit time. The electrodes must be blocked, either by proper choice of electrode material and work function, or by the deliberate insertion of insulating barriers. Some of the early transit-time measurements on AgCl carried out by Haynes and Schockley[34–35] actually predate the pioneer work on drift mobility in germanium and silicon. The use of AgCl as a crystal counter and the direct observation of drift mobility by means of oscillographic techniques was reported shortly after the early work of Haynes.[36–39] The silver halides have been improved over the years so that now highly accurate measurements of electron drift mobility in both AgCl[40] and in AgBr[41] are available. Similar timing measurements have yielded the mobility of charge carriers in organic solids such as anthracene.[42] Very recently, using nanosecond light flashes, the drift mobility of electrons photoinjected into KCl has been observed.[43]

In order to avoid polarization and space-charge effects with the transient method, a very small amount of charge must be released by the light or ionization pulses during the experiment. These released carriers then drift in a known electric field which can be simply calculated from the voltage and spacing of the electrodes. Frequently, the electrodes are blocking; nevertheless, charge is induced on the electrodes and current flows in the external circuit, as can be proven by conservation of energy. The small 10^{-14}-C induced charge is detected by means of a vibrating reed electrometer or, in the fast pulse method, by a preamplifier having an integrating time constant. The equations used for relating induced charge to absorbed photons and quantum efficiency are simple and were first derived by Gudden and Pohl (primary photocurrents).[44] This type of experiment can also be carried out on high-resistance semiconductors; however, it is necessary that the RC time constant of the sample be large compared to the transit time of carriers $\tau_d = d/\mu_d$.[45] That is,

$$\tau_d \ll RC \qquad \text{or} \qquad \tau_d \ll \varrho\varepsilon_s/4\pi \tag{28}$$

where ϱ is the resistivity of the material and ε_s the dielectric constant.

There is another set of conditions frequently encountered known as space-charge-limited current (SCLC). These conditions are approached at high light intensities on the illuminated surface, or when sufficient charge is photoemitted from an electrode into the insulator. If the pulse light intensity is sufficiently high, the initial boundary condition at the illuminated electrode will be one of zero field. Obviously, this condition is favored at low applied voltages as well as high injected charge density. Under SCLC conditions, the transit time of charge carriers is about one-third less than under low carrier concentrations. Moreover, the photocurrent waveform has a characteristic shape, increasing with time, which allows one to recognize the conditions. Transient space-charge-limited currents in the absence of trapping have been analyzed by Mark and Helfrich[46] and by Many and Rakavy.[47] A unified treatment which includes the effect of traps has recently been given.[48] Although the SCLC analysis is important, it is possible to achieve conditions in many cases (low light intensity) where the effect of space charge is minimal and can be neglected.

The low-intensity transient techniques can be employed to determine Hall mobility, thus avoiding the necessity of applying Ohmic contacts. A method for producing the longitudinal electric field required by the Hall geometry was invented in 1954 by Redfield[49] while working on his thesis at the University of Illinois. In the course of his investigations, Redfield obtained values of the Hall mobility for carriers in diamond and also in a few selected alkali halides. Initially, a vibrating reed electrometer was used to observe the Hall effect and, as a null detector, to establish a balance in the presence of crossed E and H fields. In the summer of 1954, an attempt was made by the present author to apply this method to crystals of AgCl.

Great difficulties due to drift and space charge were encountered, however. Consequently, a modification was made which replaced the vibrating reed electrometer by a fast pulse amplifier and the steady light source by a short-duration spark or flash lamp. This enabled the determination of Hall mobility in AgCl at liquid nitrogen temperatures for comparison with available drift mobility results. As expected, the AgCl Hall mobility was approximately equal to or slightly larger than the drift mobility, $\mu_H = 370 \pm 70$ cm^2/V-sec at 89°K.[50] Since that time, the fast-pulse techniques and Redfield geometry have been used at the University of Illinois to determine carrier mobility in AgCl,[51,52] AgBr,[53] KCl,[54] and other alkali halides such as KBr, KI, and NaCl containing *F* centers at low temperature.[55] Kobayashi and co-workers at the Institute for Solid State Physics, Tokyo have applied the method to the thallous halides,[56] cadmium sulfide at low temperature,[57] and again to KCl in order to study mobility at high fields and low temperature.[58] The method yielded estimates of the electron mobility in the cesium halides from work carried out in Belgium by Jacobs and co-workers.[59] An extensive investigation of polaron mobility in CsBr, CsI, and other alkali halides has recently been carried out by Seager and Emin[60] at the Sandia Laboratories. Modifications of the fast transient Hall technique using modern pulse averaging methods have been described by Borders and Hodby[61] and Smith.[62] Such methods are currently being employed by J. W. Hodby and co-workers at the Clarendon Laboratory, Oxford, for extensive investigations on insulating ionic crystals over a wide range of temperatures and fields.

A variety of transport experiments can be carried out using transient methods besides the photo-Hall effect. These include transverse and longitudinal magnetoconductivity[63] and cyclotron resonance. In the vicinity of room temperature, electrons in ionic crystals are strongly scattered by optical-mode lattice vibrations, with the consequence that their mobility is quite low, usually less than 50 cm^2/V-sec. On the other hand, below about 30°K, the optical modes of vibration are frozen out, with the consequence that Hall mobilities in many crystals exceed 10,000 cm^2/V-sec. The mobility in such crystals at helium temperatures seems to be determined by defect concentrations. The highest values found to date are in extremely high-purity samples of AgBr, where electron mobility in excess of 10^5 cm^2/V-sec have been obtained below 20°K. Because of these high mobilities at helium temperatures, hot-electron phenomena are frequently important and can be readily studied. In fact, in the cyclotron resonance, hot-electron and other phenomena can be detected through the changes in mobility associated with electron heating.

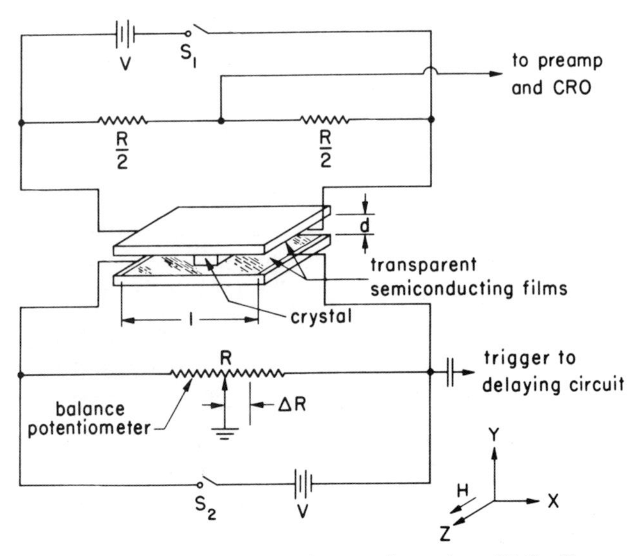

Fig. 5. Electrode geometry for transient photo-Hall effect.

The transient methods discussed above have been compared with more conventional semiconductor techniques elsewhere.[3] Only a brief discussion of the photo-Hall geometry will be given here. Figure 5 shows the transient photo-Hall geometry. The resistance plates indicated schematically in the figure are frequently glass or quartz with a transparent tin oxide layer of the order of 2000–5000 ohms/square.[64] Conduction in semiconducting glass films of this low resistivity is degenerate and usually remains high even when cooled to liquid helium temperatures. Higher-resistivity coatings become extremely resistive at very low temperatures. Identical voltage gradients are applied to the two resistant films, thus establishing the longitudinal electric field in the x direction. Electrons are released by pulses of illumination through the transparent electrodes, or, sometimes, from the side. A fast pulse amplifier similar to that used with silicon particle detectors is connected to the midpoint of the upper resistance as shown. With the balance potentiometer adjusted symmetrically, no transient photosignal is detected (charge flow in the y direction) for a pulse of illumination in the absence of a magnetic field. Under these conditions, the carriers drift only in the x direction. When a magnetic field is applied in the z direction, one observes an induced transient charge Q,

$$Q = \iiint_{V_c} Nev_y(\tau_t/d)\, d\tau_r \tag{29}$$

Here, V_c is the crystal volume, N is the number of carriers released per unit volume by the light, v_y is the y component of drift velocity produced by the Lorentz force (proportional to $v_x H$), and τ_t is the mean time before trapping at deep traps. The thickness of the sample is d and charge of the electron e. This is an approximate relation useful in the absence of carrier collection or saturation effects. The total charge released during each pulse must be kept small so that a Hall potential does not develop. As actually employed, the transient Hall method is a null technique. Instead of measuring the magnitude of Q in Eq. (29), the balance point of the lower-resistance potentiometer in Fig. 5 is adjusted by a small amount so as to rotate the applied electric field in the crystal. In this way, the transient Hall signal can be reduced to zero. In principle, one then has a way of directly observing the Hall angle θ_{H},

$$\theta_{\mathrm{H}} \approx \tan \theta_{\mathrm{H}} = J_y/J_x \tag{30}$$

and the Hall mobility,

$$\mu_{\mathrm{H}} = c\theta_{\mathrm{H}}/H \tag{31}$$

Analysis shows that the following equation holds between mobility and the observed quantities:

$$\mu_{\mathrm{H}} H/c = \tfrac{1}{2}(l/d)\, \Delta R/R \tag{32}$$

where l is the length of resistance films, d the electrode spacing, and ΔR the change in balance potentiometer setting in going from positive to negative polarities of magnetic field H.

A modification of the above Hall geometry has recently been made by Delacote and Schott.[65] In this variation, two counterelectrodes separated by a small gap, much as in Fig. 4, are connected to differential pulse amplifiers. The applied electric field is perpendicular to these plane-counter electrodes. Under balanced conditions, no photosignal will be detected in the absence of a transverse or perpendicular magnetic field. In the presence of a magnetic field, the Hall effect produces an unbalance which can be detected and related to mobility. This modified electrode arrangement shows promise being applicable to thin samples and should permit the extension of the transient methods to lower-mobility materials and perhaps to device geometries. Work is currently under way along these lines at the University of Illinois.[66]

Figure 6 shows an outline of the pulse averaging system used by Hodby and Borders[61] to observe cyclotron resonance at 2-mm wave frequencies and fields in excess of 100 kOe at the National Magnet Laboratory. Repeated pulses of light ($\sim$1 μsec duration) and collecting voltage (100 μsec

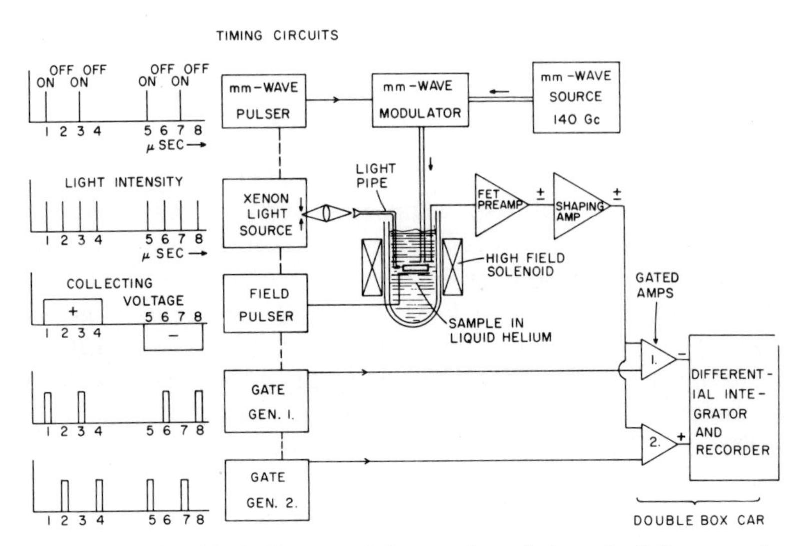

Fig. 6. Simplified block diagram of the transient photoconductivity apparatus used to observe cyclotron resonance by electron heating. The sequence of light, field, and gating pulses is shown on the left, after Borders and Hodby.[61]

duration) were applied to the insulating photoconductor (blocking electrodes) in such a way as to minimize space-charge and polarization effects. From the timing sequence shown in the figure, it can be seen that exactly equal amounts of charge (four pulses) are detected with the collecting voltage pulse positive as with the voltage negative. Of course, the signal pulses reverse sign as the applied field is reversed and this must be taken into account during detection. The time-averaged response of either channel 1 or channel 2 of the "boxcar" integrating circuit would be close to zero were it not for the fact that rf power is applied during every other pulse. As will be explained in Section 6, the millimeter-wave power causes carrier heating at a longitudinal magnetic field H_0 corresponding to cyclotron resonance with a resulting decrease in mobility and observed photocharge. This tends to drive channel 1 negative and channel 2 positive. The integrated output of the boxcar is thus a differential output. By averaging over many pulses, a difference in collected photocharge as small as six electrons per pulse could be detected as the rf power was switched on and off. The high sensitivity of this differential system permitted the observation of cyclotron resonance as the magnetic field was scanned in a number of insulating photoconductors.[67] A very high 2-mm wave frequency (150 GHz) was employed, together with very high magnetic fields, in order to satisfy the scattering or relaxation time condition for cyclotron resonance, $\omega\tau > 1$.

The 140-Hz cyclotron resonance experiments yield polaron mass, since the lattice frequencies are even higher. Larsen,[18] for example, has worked out the theory of the polaron in a magnetic field within the framework of a perturbation treatment. For cyclotron frequencies small compared to the longitudinal optical frequency, he finds

$$\hbar\omega_{\mathrm{cR}} = (\hbar eH/mc)(1 - \tfrac{1}{6}\alpha) \simeq \hbar eH/m_p c \qquad (33)$$

At higher frequencies and magnetic fields ($\omega_{\mathrm{cR}} \simeq \omega_{\mathrm{LO}}$), "pinning" effects occur, and these have been observed even in weak polaron materials such as CdTe.[68]

The feasibility of carrying out cyclotron resonance by this cross-modulation technique was first demonstrated by Mikkor *et al.*[69] The results at 4 mm were definitive, but not highly accurate. Moreover, they had to use values of collecting voltage which were sufficiently high that electron heating effects were almost certainly present at low temperature. In fact, this was also the case, as it has turned out, in the work by Hodby *et al.*[67] A careful investigation of the field dependence of the observed resonance signals has revealed that the masses reported for the various ionic crystals at helium or pumped helium temperatures were all somewhat high, of the order of 10 or 20%. In fact, the numbers given in Ref. 67 represent the effective mass of hot polarons rather than of carriers in equilibrium with the lattice at low temperature. Recent estimates of the cold polaron masses have been given by Hodby[70] and these are still being refined. The best available values to date are listed, with uncertainties, in Table II. These newer values of cold polaron mass are in good agreement with the cyclotron resonance values determined by the more conventional cyclotron resonance power absorption experiment in those few cases where this comparison can be made. For example, the value given by Hodby for AgBr is $m_p/m_e = 0.289 \pm 0.01$, in good agreement with the early cyclotron resonance absorption estimates of Ascarelli and Brown.[71]

5. MOBILITY DETERMINED BY OPTICAL MODE SCATTERING—COMPARISON OF EXPERIMENT WITH THEORY

A great deal of progress has been made on the mobility problem since the 1962 Polaron Summer School cited in Ref. 3. There, the mobility formulas of Howarth and Sondheimer,[72]* Low and Pines,[74] Schultz,[75]

* See also Petritz and Scanlon.[73]

and Feynman, Hellwerth, Iddings, and Platzman (FHIP)[76] were displayed and compared with each other. Quite different values of scattering rate and therefore mobility were obtained in the limit of small coupling constant, even though one would expect these different theoretical results to agree with each other and with the exact perturbation result as $\alpha \ll 1$. The weak coupling perturbation result for temperatures much less than the Debye temperature $T \ll \Theta$ is just[77]

$$\mu = (e/2m\omega\alpha) \exp \beta \tag{34}$$

In this equation, e/m is the charge-to-mass ratio for the bare electron in cgs units. α is the coupling constant, ω is the longitudinal optical mode lattice frequency in $\sec^{-1}$, and $\beta = \hbar\omega/kT = \Theta/T$. One must divide Eq. (34) by 300 in order to obtain the mobility μ in cm^2/V-sec. A formula such as this is the exact perturbation result at low temperature and very small α in the usual relaxation time approach. It is not appropriate in the intermediate coupling region, where α is of the order of 3–6. The intermediate coupling mobility has been evaluated by Low and Pines. In its corrected form, the Low–Pines mobility now agrees with the perturbation result, Eq. (34), in the limit of small α. The Low–Pines formula as modified by Langreth[78] is

$$\mu = (e/2m_p\omega\alpha)f(\alpha) \exp \beta \tag{35}$$

where m_p is the polaron mass and $f(\alpha)$ the slowly varying function given in Ref. 74. Langreth has also modified the mobility formula given by Schultz in order to restore agreement with the perturbation results at small α.

Kadanoff[79] has used the Feynman model of the polaron to write down a Boltzmann equation which can then be solved, at least to first order, to find the microscopic mobility. A low-temperature mobility formula emerges which agrees with an equation previously given by Osaka.[80] It furthermore coincides with Eqs. (34) and (35) in the limit of small coupling constants. Thus a rather satisfactory agreement now exists between the various mobility results in the limit of very small α and low temperature. The FHIP formula[76] is an exception to this statement and stands apart.

Of the above-mentioned mobility theories, only the Howarth–Sondheimer formula is meant to apply at higher temperature T approaching Θ. It is, in fact, not a polaron mobility theory but simply a solution of the Boltzmann equation for band electrons with phonon absorption and emission so that the highly inelastic nature of the scattering process for optical phonons is taken into account. More recently, Langreth has given an expression for polaron mobility which is suitable for low but finite temperatures. This expression, which agrees with the above-mentioned results in the extremes of weak coupling and low temperature, is suitable for comparison with experimental results, but not at temperatures approaching the Debye temperature. The Langreth finite-temperature result is as follows[81]:

$$\mu = (e/2m_p\omega\alpha)[1 + (1.53/\beta)][(\exp\beta) - 1] \tag{36}$$

Except for the FHIP result, the solutions obtained so far for the polaron mobility make use of a Boltzmann transport equation or its equivalent. It is implicit in these approaches that the particle can be thought of as separate from the lattice field even though the particle–field interaction is very strong. The particle in this case is the polaron, which is assumed to be scattered by the lattice phonons. For convenience, a relaxation time is frequently assumed and eventually the average scattering probability obtained. Schultz[82] has shown how the conventional scattering approach can be justified under certain circumstances for the polaron at low temperature. At low temperature, the scattering processes are fairly easy to handle; for example, the rate of phonon absorption is found to be much slower than the rate of phonon emission. This means that for cold polarons at low temperature, the scattering, and therefore mobility, is controlled by $\bar{N} = 1/[(\exp\beta) - 1]$, the number of phonons available for absorption. Figure 7 shows the various processes. The predominant scattering process is a second-order one in which an optical phonon is absorbed and then very rapidly emitted so that overall, the collisions are elastic. Such a scattering process has been called resonance scattering. The exponential β which appears in Eqs. (34)–(36) simply comes from the $\bar{N}$ terms.

Further study of the polaron mobility problem shows that the Boltzmann equation approach is much more difficult to justify at intermediate or high temperatures. Because of the strong interaction and high probability for scattering as T approaches Θ, the situation is one approaching continuous interaction. It is not clear that a quasiparticle approach with a meaningful relaxation time can be justified. In order to surmount such difficulties, Thornber and Feynman[83] have calculated the expectation value of the steady-state velocity of the electron in such a way that the electron and lattice system are considered as a whole. The rate of momentum loss of an electron drifting through the crystal is expressed in a form in which the lattice coordinates have been eliminated. Path integral methods are used to obtain the response of an electron in an ionic crystal without the usual limitations of low field and low temperature. Numerical results for the applied electric field versus electron velocity are given in Ref. 83. At

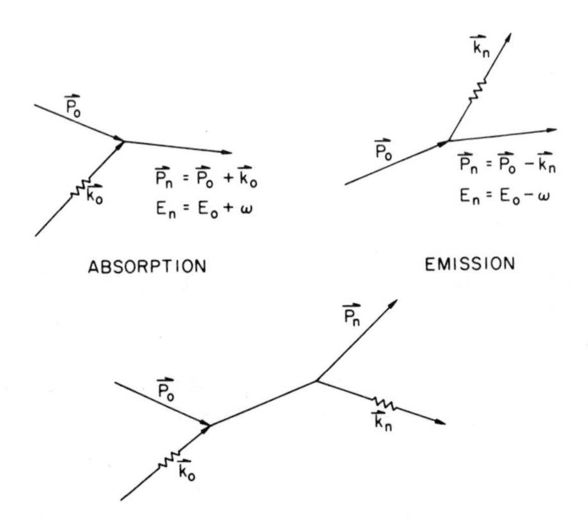

Fig. 7. First-order polaron–LO phonon scattering processes important at finite temperatures are shown above. At very low temperature, only the second-order process (resonant scattering) shown below is important.

room temperature, the steady-state energy loss per unit distance in Al_2O_3 turns out to be very high, of the order of 0.01–0.06 eV/Å, similar to that found from the photoemission experiments mentioned in Section 4. The Thornber–Feynman dc mobility formula in cgs units is given as follows:

$$\mu = \frac{3}{2}\,\frac{e}{m_e\omega\alpha}\left(\frac{m_e}{m}\right)\frac{\sinh(\beta/2)}{\beta}\,A^{3/2}\left(\frac{C\pi}{\beta}\right)^{1/2}\frac{1}{K_1(\frac{1}{2}\beta\sqrt{C})} \tag{37}$$

where

$$A = 1 + \left(1 - \frac{w_0{}^2}{v_0{}^2}\right)\left(\frac{v_0\beta}{2\sinh(\frac{1}{2}v_0\beta)} - 1\right)$$

$$B = \beta\,\frac{w_0{}^2}{v_0{}^2}\left[\frac{(v_0{}^2/w_0{}^2) - 1}{v_0}\left(\tanh\frac{\beta v_0}{4}\right) + \frac{\beta}{4}\right]$$

$$C = 4B/A\beta^2$$

(a typographical error occurs in Ref. 83 in the denominator of the expression for C). Here, m_e is the free electron mass, m the band or fixed lattice mass, $\beta = \Theta/T$ as before, and K_1 is the modified Bessel function of order one tabulated, for example, in *NBS Handbook* No. *AMS55*, p. 417. The quantities v_0 and w_0 are the variational parameters of the Feynman polaron

Table III. The Variational Parameters, Self-Energies, Effective Mass, and Radius of the Feynman Polaron Theory As a Function of Coupling Constant α

α^a	1	3	5	7	9	11
v_0	3.22	3.44	4.02	5.81	9.85	15.5
w_0	3.0	2.55	2.13	1.60	1.28	1.15
E_f	−1.01	−3.13	−5.44	−8.11	−11.5	−15.7
m_f/m_e	1.19	1.89	3.89	14.4	62.8	1.85
r_f	2.61	1.42	1.00	0.748	0.557	0.443

[a] v_0 and w_0 are given in units of ω, E_f in units of $\hbar\omega$, and r_f in units of $(\hbar/2m\omega)^{1/2}$. From Schultz.[75]

theory. They have been computed numerically as a function of α by Schultz[75] and are given in Table III. Since they are smooth functions of α, it is easy to interpolate between the values given. Again, in order to obtain the mobility in cm^2/V-sec, divide Eq. (37) by 300.

The situation is now more favorable than ever before for comparison of polaron mobility theory with experiment. Hall mobility measurements were discussed at the end of Section 4, where the cyclotron resonance experiments were also introduced. Reliable cyclotron resonance mass values are now available for a number of simple ionic crystals and these are listed in Table II. The values of m_p/m listed in Table II are more reliable now because polaron heating effects have been taken into account; also, errors in the experiment have been reduced by use of the higher 2-mm wave frequencies. Given the experimental polaron masses, the band mass can be evaluated from polaron effective mass theory using either Eq. (25) or Eq. (27) and the expression for α, Eq. (12). It is most useful to set up a simple computer program which gives the values of α and m_p for a range of values of band mass m. The proper continuum parameters ε_s, ε_∞, and ω must be inserted for each material.

The dielectric constants and the transverse optical frequency have been determined for a large number of crystals as a function of temperature down to liquid helium temperatures.* The Lyddane–Sachs–Teller relation

$$\omega = (\varepsilon_s/\varepsilon_\infty)^{1/2}\omega_t \tag{38}$$

appears to hold quite well in cases where it can be accurately tested. There-

* See especially Refs. 121–124.

fore one expects that ω can be obtained from observed *Reststrahlen* wavelengths. It is also possible to determine ω by the technique of Berreman[84] and sometimes from phonon periodicities or from Raman scattering experiments. In a few cases, the detailed lattice vibrational spectrum has been obtained from inelastic neutron scattering—in particular, new results are available for KCl[85] and for AgCl.[86]

The comparison between mobility theory and experiment must be made in a temperature range where optical mode scattering predominates and yet not so close to the Debye temperature that the theory is incomplete. In alkali halides containing F centers,[54,55] the mobility has been shown to rise to values of about 10,000 cm^2/V-sec and to level off at about 20°K. In the silver halides, mobilities can rise to much higher values, of the order of 10^5, which in principle should permit a wider range of pure lattice scattering for comparison with theory. In some cases, both drift mobility and Hall mobility have been measured and they appear to be the same to within 20%. Very recently, Thornber has completed calculations of the polaron response in a magnetic field and he finds the Hall mobility equal to the microscopic mobility for optical mode interaction.

The points in Fig. 8 show the measured values of Hall mobility determined by Brown and Inchauspe,[54] Ahrenkiel,[55] and by Seager[60] for KCl, a crystal with an effective mass $m_p = 0.92 m_e$.[70] Inelastic neutron scattering data[85] are now available confirming that the LO Θ is close to 300°K. The coupling constant for KCl is 3.62, so that moderately large polaron effects are present. The solid curves in Fig. 8 show the mobility computed from the theory of Thornber and Feynman, Eq. (37), and also by the finite-temperature formula of Langreth, Eq. (36). The latter closely follows the intermediate coupling results of Low and Pines, Eq. (35). It can be seen that the Thornber–Feynman formula seems to be in closest agreement with the observed mobilities. On the other hand, the Boltzmann equation mobilities are of the order of 2–3 times higher at a given temperature.

Figure 9 shows a similar comparison between experiment and theory for KBr. In this case, the observed polaron mass is $0.7 m_e$ and from Table I, it can be seen that the band mass is close to 0.37, and the coupling constant is about 3.0. The Debye temperature in potassium bromide is somewhat lower than in KCl, the value being 242°K. Again, it can be seen that the Thornber–Feynman mobility most closely approximates the observed values in the intermediate temperature range.

Figure 10 shows the comparison between experiment and theoretical mobilities for AgCl using the new values of lattice frequency[86] and the cold polaron masses given in Table II.[105] The Hall mobility values of Burnham[53]

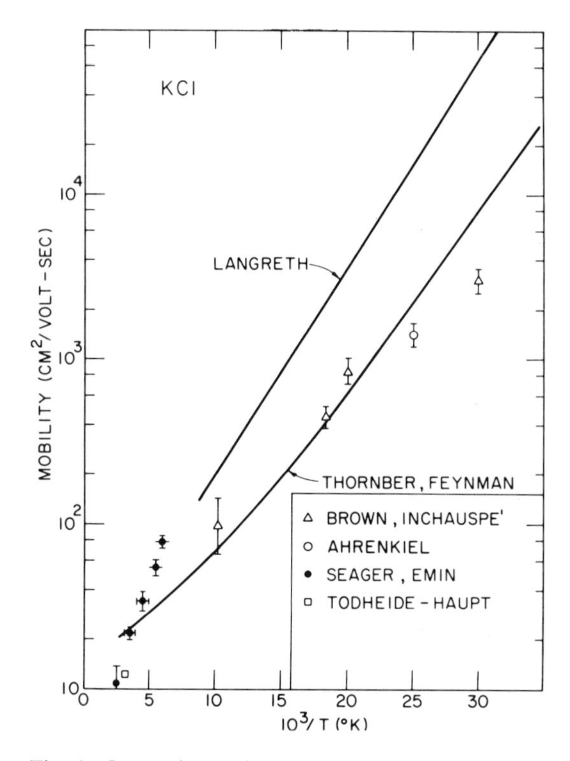

Fig. 8. Comparison of experimental and theoretical mobilities for KCl. The points shown by triangles, open, and closed circles are all photo-Hall data. The squares are drift mobility data. The two different curves were drawn for $m/m_e = 0.412$ and $\Theta = 300°K$ (see Table II).

are used at low temperatures and the accurate drift mobility values of Van Heyningen[40] yield the dotted curve at high temperature. Silver chloride is a substance with a longitudinal Debye temperature of 195°K. The other continuum parameters, including the effective mass, are such, however, that the coupling constant α is more nearly 1.9 (see Table II). Here again, the Thornber–Feynman mobilities appear to be in best agreement with experiment. It should be stressed that no arbitrary parameters are employed in these comparisons.

At high temperature, in the vicinity of the Debye temperatures, the recent results of Seager and Emin[60] indicate that the observed mobility decreases much more rapidly with increasing temperature than predicted by the continuum mobility theories. This trend can be seen in Fig. 8, where

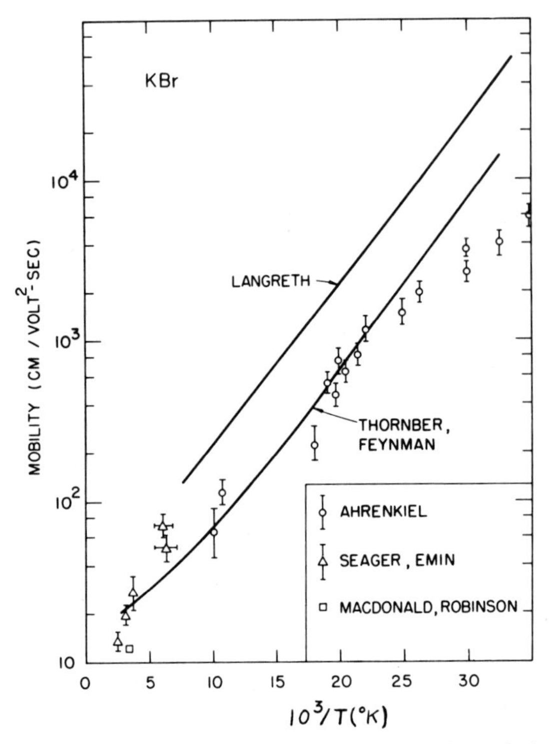

Fig. 9. Comparison of experimental and theoretical mobilities for KBr. All of the points shown are Hall data. The theoretical curves were drawn as solid lines for $m/m_e = 0.370$ and $\Theta = 242°K$ (see Table II).

the points of Seager and Emin show a much steeper slope than at low temperatures. These authors argue that a new phenomenon is taking place at and above the Debye temperature, namely polaron band narrowing. This is not taken into account in the mobility theories discussed above. Further investigations are called for in order to understand this extremely interesting possibility, which might serve as a bridge between the small- and large-polaron phenomena. It might also be that the fall off in μ at high temperatures is simply due to polaron band effects as discussed below.

The difference between the Thornber–Feynman mobility and the Boltzmann-equation results of other authors amounts numerically to approximately the factor $3/2\beta$ which enters into Eq. (37). It is not really possible to confirm the slope or temperature dependence introduced by this factor at low temperatures, because of the overwhelming effect of the exponential term. The same discrepancy $3/2\beta$ occurred in the earlier FHIP formula,[75] but at that time, there was some question about the validity of the approximation which these authors used in arriving at the dc low-temperature result. The correctness of their formula within the framework of the Feynman approach has been verified in the recent work of Thornber and Feynman. It should be emphasized that this last work is appropriate to intermediate and high temperatures and is not limited to weak coupling. The favorable comparison with experiment given above indicates that this theory may be more appropriate for high temperature in the range of coupling constants found for real materials. On the other hand, this new mobility theory is still approximate in the sense that it predicts that the electron's velocity is characterized by a drifted Maxwellian or isotropic distribution in velocities. The approach does not specifically take into account streaming such as might occur if the distribution becomes anisotropic at high fields (low carrier densities). Finally, the difficulties with

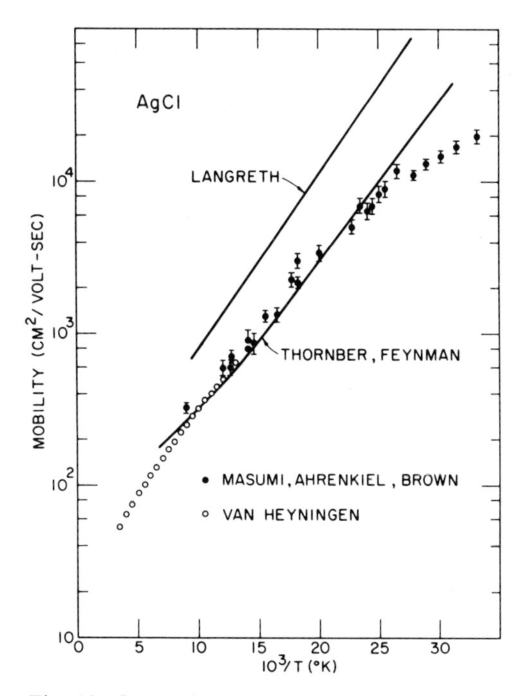

Fig. 10. Comparison of experimental and theoretical mobilities for AgCl. The points shown by solid circles are photo-Hall data, whereas the open circles are drift mobility data. The theoretical curves shown by solid lines were computed for $m/m_e = 0.303$ and $\Theta = 284°K$ (see Table II).

solutions of the Boltzmann equation as presently obtained are not understood.

Upper and lower bounds for the ground-state energy of the polaron in the intermediate coupling regime have been given by Larsen.[87] These indicate that the Feynman self–energies are fairly accurate near $\alpha = 2$–3. It should be emphasized, however, that the Feynman polaron as usually treated is a single harmonic oscillator approximation. Thornber has recently been successful in finding the appropriate distribution in oscillators for the problem.[88] His results indicate that only small changes probably occur in the low-field linear region of the dc response of the polaron.

Other polaron theories may be more accurate in the last analysis. For example, the unified theory of Matz[22] includes the important electron–lattice processes and correctly takes into account translational invariance. According to this new theory, the ratio of polaron mass to band mass tends toward a saturation value of 1.9 at very high coupling constants; thus for real materials, band masses and coupling constants turn out somewhat larger than in the Feynman model. For example, as shown in Table IV, the Matz mass for KI is 0.40 instead of the Feynman mass of 0.325. Incidently, this larger mass is in much better agreement with a value calculated from the band theory of KI by Kunz.[99] Also, Eq. (35) indicates that larger coupling constants yield lower mobilities, although not enough lower to remove the discrepancy between mobility theory and experiment. This comparison may not be very meaningful, however, since a cyclotron

Table IV. Comparison of Band Masses and Coupling Constants Computed from the Feynman Model Using the Langreth Interpolation formula, Eq. (37); and Those from the Recent Greens Function Formulation of Matz[a]

	m_p/m_e	$(m/m_e)_L$	α_L	$(m/m_e)_M$	α_M
KCl	0.922 ± 0.04	0.432	3.46	0.61	4.2
KBr	0.700 ± 0.03	0.367	3.07	0.50	3.5
KI	0.536 ± 0.03	0.325	2.51	0.40	2.8
RbCl	1.038 ± 0.10	0.430	3.81	0.70	5.0
RbI	0.72 ± 0.07	0.365	3.18	0.50	3.7
AgCl	0.431 ± 0.04	0.303	1.86	0.36	2.0
AgBr	0.289 ± 0.01	0.215	1.60	0.26	1.8
TlCl	0.56 ± 0.02	0.34	2.5	0.42	2.8
TlBr	0.52 ± 0.03	0.31	2.7	0.39	2.9

[a] The Feynman–Langreth results are subscripted L, and the Matz[22] results M. The polaron masses (m_p/m_e) are cyclotron resonance values as listed in Table II.

resonance mass and mobility have not yet been obtained within the framework of the Matz theory.

In the next section, we discuss the high-field behavior of electrons in ionic crystals as a first step toward understanding the low-temperature results; an energy-balance approach which does not include polaron effects will be presented. In this theory, acoustic phonons play a dominant role. Later in Section 6, we will return to the Thornber–Feynman polaron treatment for high fields and optical phonon interaction.

6. HIGH FIELDS AND LOW TEMPERATURE

The experimental results of the preceding section show that the mobility of polarons in ionic crystals can be extremely high, in excess of 10^4 cm^2/V-sec at very low temperatures. The possibility therefore arises of observing hot polaron effects at high electric fields. Such phenomena were first observed by Masumi[89] in highly purified AgCl crystals at liquid helium temperatures. Masumi's experiments were carried out with the transient photocharge technique using thin ~1–2 mm, slabs of AgCl placed between plane parallel electrodes with blocking contacts. He also measured the Hall effect by the Redfield technique. When n electrons are released in the volume of a crystal by a short pulse of light, these electrons drift in a uniform electric field E with a drift mobility μ and an average time before deep trapping τ_t. Equation (29) for the induced photocharge in the case of uniform field and n electrons released simplifies as follows:

$$Q = ne\mu E\tau_t/d \tag{39}$$

where d is the electrode separation. This is the result appropriate to small *Schubweg*, $w = \mu E\tau_t$, in the absence of saturation or collection effects related to trapping of the electrons at the surfaces. This equation states that the photocharge Q should depend linearly on E, corresponding to Ohm's law, if both μ and τ_t are independent of E at high electric field. Any deviations from Ohm's law might be attributed to the change in either the mobility or the trapping time. What was specifically observed at 6.4°K but not at 77°K was a photoresponse which increased linearly with increasing electric field up to about 20 V/cm and thereafter increased very approximately as $E^{1/2}$ up to and in excess of 1000 V/cm. The non-Ohmic behavior was attributed to decreasing mobility, since above about 20 V/cm, the Hall mobility was observed to decrease with the increasing average component of electric field. Likewise, careful magnetoresistance measurements on the same material at liquid helium temperatures indicated a decrease in mobility with increasing electric field.[63]

Hall mobility versus temperature results on the samples of AgCl used by Masumi showed that the scattering of slow electrons at temperatures near 10°K is structure-dependent, varying from sample to sample and probably associated with residual impurities in the crystals. However, such scattering processes were thought to be less important for the scattering of fast electrons, and it was also realized that they could not, in general, provide the energy transfer mechanisms whereby the energy gained from the electric field is dissipated. Some type of inelastic scattering such as due to lattice vibrations was thought to become increasingly important at high electric fields. At high fields, the observed $E^{1/2}$ dependence brought to mind an energy-balance theory first given for semiconductors by Schockley[91] which assumed energy dissipation by acoustic phonon emission. The importance of acoustic phonons in ionic crystals was noted by Seitz in 1949.[92]

In Schockley's energy-balance theory, it is assumed that an electron temperature T_e can be defined characteristic of the electron distribution and higher than the lattice temperature T_l. If, for example, the conduction electron distribution in the presence of the electric field were simply a displaced Maxwellian distribution, the electron temperature T_e would be related to the mean velocity of the electron v_1 in the following manner:

$$\varepsilon = \tfrac{1}{2}mv_1{}^2 = \tfrac{3}{2}kT_e \tag{40}$$

The mean speed v_1, and therefore T_e, is obtained from an energy-balance equation:

$$(d\varepsilon/dt)_{\text{field}} + (d\varepsilon/dt)_{\text{lattice}} = 0 \tag{41}$$

which relates the rate of energy gained from the electric field to the rate of energy lost to the lattice, or

$$v_d eE = \mu_{\text{ac}} eE^2 = \langle \delta\varepsilon \rangle (1/\tau_{\text{ac}}) \tag{42}$$

In this last expression, v_d is the drift velocity, also assumed to be controlled by scattering with acoustic vibrations. The quantity $\langle \delta\varepsilon \rangle$ is the average energy lost per collision, and $1/\tau_{\text{ac}}$ is the probability for acoustic phonon scattering. Assuming equipartition of energy for the lattice modes (appropriate at intermediate or high temperatures), Schockley showed that for a simple hard-sphere model, the last equation becomes approximately

$$e^2 lE^2/mv_1 = (m^2 s^2 v_1{}^2/kT_l) v_1/l \tag{43}$$

where s is the sound velocity in the crystal and l the mean free path for acoustic phonon scattering. This equation can be solved for v_1 and inserted into the expression for the drift velocity,

$$v_d = (e/m)(l/v_1)E \tag{44}$$

One then obtains an $E^{1/2}$ dependence at high field and also a transition from Ohmic to non-Ohmic behavior near a critical field E_c which is given by

$$E_c \approx 1.51 s/\mu_{\text{ac}} \tag{45}$$

When the analysis is carried out assuming a Maxwellian distribution, accurately evaluating the energy loss and taking appropriate averages, the coefficient 1.51 in Eq. (45) becomes 1.84. The result is not strongly dependent on the nature of the distribution function assumed. What is important is that the conduction electrons interact with each other, and with the lattice, in such a way that their effective temperature T_e can be defined.

The simple theory can be easily modified to include the possibility that the mobility on the left side of Eq. (43) (momentum exchange) is controlled by some scattering mechanism other than acoustic scattering. A similar critical field expression arises except that a ratio of mobilities occurs as follows:

$$E_c \approx 0.41(\mu_{\text{ac}}^0/\mu_i)s/\mu_i \tag{46}$$

Here, μ_{ac}^0 is the acoustic scattering mobility that would prevail if the electrons were in equilibrium with the lattice at the temperature of the measurements; μ_i is the much lower mobility due to elastic scattering by the impurities present. Because the sound velocity in silver chloride is quite small ($s = 3.29 \times 10^5$ cm/sec) and the mobilities $\mu_i \approx 10^4$–10^5 cm^2/V-sec quite high, Eq. (46) gives critical fields in the range 10–100 V/cm at 4.2°K.

This type of energy-balance theory (basically a semiconductor theory for nondegenerate carriers) has been extended to ionic crystals by Matz and Garcia-Moliner.[93] In particular, they write the energy loss term for lattice scattering as follows:

$$(\partial\varepsilon/\partial t)_{\text{coll}} = (32/3\pi)(m^* s^2/\tau_{\text{ac}}^0)(1 - \gamma) \tag{47}$$

where τ_{ac}^0 is the low-field acoustic scattering time and γ is the ratio of electron to lattice temperatures, $\gamma = T_e/T_l$. The low-field acoustic mobility

μ_{ac}^0 and scattering time would be related as follows for a Maxwellian distribution:

$$\begin{aligned}\mu_{ac}^0 &= (e/m)\langle\tau(\varepsilon)\rangle \\ &= (e/m)(4/3\sqrt{\pi})\tau_{ac}^0 \int_0^\infty x^{-1/2}x^{3/2}(\exp -x)\,dx \\ &= (e/m)(4/3\sqrt{\pi})\tau_{ac}^0 \end{aligned} \tag{48}$$

where $x = \varepsilon/kT_e$. They furthermore modify the rate at which electrons gain energy from the electric field to include the effect of an applied magnetic field. If the magnetic field is transverse to the applied electric field, the rate at which carriers acquire energy from the lattice is decreased. A transverse magnetoresistance effect occurs, but at high-E fields, the hot electrons are cooled down. The energy gain term in Eq. (41) can simply be written as

$$(d\varepsilon/dt)_{\text{field}} = ev \cdot \mathbf{E} = (e/n)\sigma(H)E^2 \tag{49}$$

where $\sigma(H)$ is one of the conductivity coefficients relating the current $\mathbf{j}$ and field $\mathbf{E}$ in the presence of an applied H. These coefficients occur in the usual solutions of the Boltzmann transport equation. The various transport integrals which enter can be evaluated numerically for mixed scattering processes. Such calculations are relatively easy in the relaxation time approximation, where τ takes the form

$$\tau = \tau_0(\varepsilon/kT_e)^p \tag{50}$$

where $p = -\frac{1}{2}$, 0, $\frac{1}{2}$, and $\frac{3}{2}$ for acoustic, neutral, dipole, and charged impurity scattering, respectively. The results have to be evaluated numerically and extensive computations of this type have been carried out by Mikkor and others.[94] It is remarkable how well the various magnetic experiments can be explained by an analysis of this type, which must be only approximate at best.

The points in Fig. 11 show the data taken by Mikkor on crystals of KBr at 4.2°K. The solid lines were computed from the energy-balance theory, all three with the same parameters, i.e., $\mu_{ac}^0 = 40{,}000$ cm^2/V-sec and neutral scattering mobility $\mu_n = (e/m)\tau_n{}^0 = 15{,}200$ cm^2/V-sec. Three conditions of field were used, $H = 0$, 18, and 48.7 kOe. The upper curve in the absence of magnetic field shows that the transition from Ohmic to non-Ohmic behavior occurs in the vicinity of 100 V/cm. When a transverse magnetic field is applied, the conductivity decreases due to the expected magnetoresistive effect.[90] As a function of E, however, a linear relation

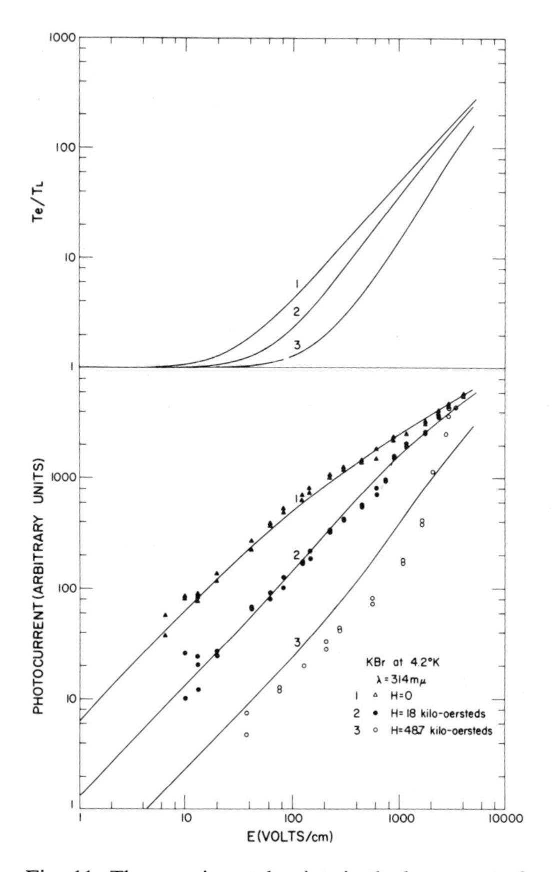

Fig. 11. The experimental points in the lower part of the figure show the photoconductivity versus electric field in KBr at 4.2°K for $H = 0$ and for transverse magnetic fields of 18 and 48.7 kOe. The curves below and electron temperatures above were obtained from an energy-balance theory with $\mu_{\text{ac}}^{0} = 40{,}000$ cm²/V-sec and $\mu_n{}^0 = 15{,}200$ cm²/V-sec. From Mikkor *et al.*[94]

holds to higher electric field. At very high electric field, all of the curves approach one another in the limit of extreme high-field behavior. In the presence of a transverse magnetic field, the transition from the low- to high-field region can be superlinear. Curves in the upper part of Fig. 11 indicate the electron temperatures which would be deduced from the energy balance theory in the vicinity of several hundred to 1000 V/cm. These electron temperatures can be very high, and it would seem that the distributions do not actually reach such effective temperatures. Various omissions of the theory, including the exclusion of optical mode scattering, invalidate the approach, at least at extremely high fields. An approximate method for taking into account the optical modes and also for getting around the equipartition of energy approximation for acoustic interaction is described by Matz and Garcia-Moliner.[93]

Figure 12 shows the decrease in Hall mobility associated with the high-field effects in KBr as obtained in further study of the problem by Borders.[95] The high-field effects have also been observed in KCl by Nakazawa and Kanzaki[96] and in a variety of other ionic crystals by Borders and Hodby. The transition from ohmic to non-ohmic behavior in KI at 4.2°K is shown in Fig. 13. Two different crystals and types of photocharge measurement are shown. In the upper curve, the conduction electrons were released by ultraviolet light from *F* centers.[95] The samples were addively colored to an *F*-center density of about 10^{16} cm^{-3}. In the lower curve, an uncolored crystal of KI was used and the photocarriers were released by a two-quantum effect across the band gap, using frequency-doubled radiation from a *Q*-switched ruby laser.[97] Notice that the critical field region separating Ohmic from non-Ohmic behavior occurs at much lower values, of the order of 20 V/cm rather than 100 V/cm obtained in the colored crystals. This is undoubtedly related to the residual mobilities and scattering by imperfections (*F* centers).

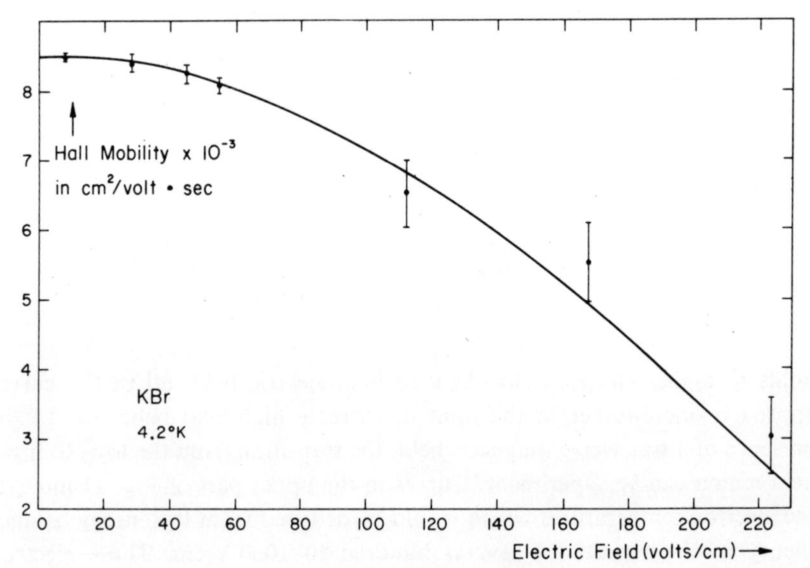

Fig. 12. Observed Hall mobility versus electric field for an additively colored (3×10^{16} cm^{-3}) KBr crystal at 4.2°K. From Borders.[95]

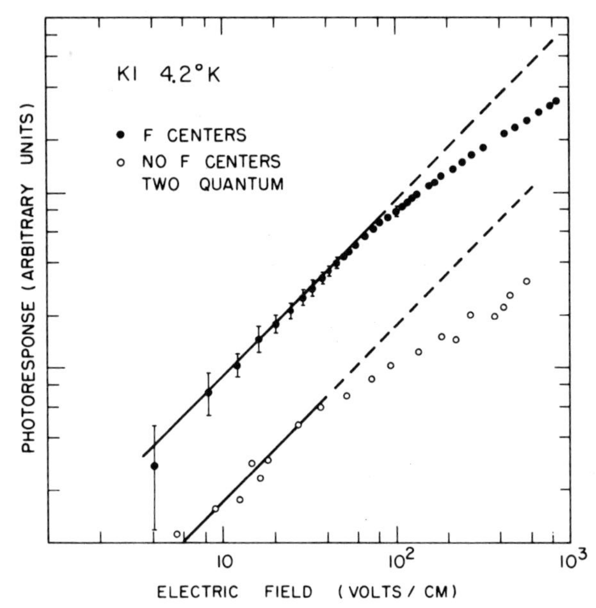

Fig. 13. The photoresponse *versus* electric field for a KI crystal containing $10^{16}F$ centers/cm³ (above) and for an uncolored KI crystal (below) using two-photon absorption. After Bolte.[97]

In the hot electron experiments discussed so far, energy is fed into the electron system by an applied electric field which also produces a drift of carriers. Assuming spherical energy bands and isotropic scattering, the current and electric field are in the same direction ($H = 0$). The lower conduction bands of the alkali halides indeed seem to be of this type, as proven by a recent study of transverse magnetoconductivity.[98] In these experiments, carriers photoinjected from F centers thermalize very rapidly to states near the bottom of the conduction bands (near Γ_1 in the Brillouin zone), where they drift in the applied fields. That the lower bands are isotropic agrees with the present extensive knowledge of the band structure of these materials.[99]

The idea that energy can be fed into the electron system from the electric field, resulting in an increase in effective electron temperature, and therefore in a shift of the electron distribution, has been tested experimentally in the first cyclotron resonance–electron heating experiments.[100] In these experiments of Mikkor on KBr, the carrier temperature was increased at cyclotron resonance by the absorption of rf power at 70 GHz, resulting in increased scattering and an observed decrease in electron drift mobility. The samples were in the form of cylindrical disks which served as a dielectric post resonator. The rf electric field was in the plane of the disk, whereas the drift of photocarriers was detected on plane parallel electrodes in an orthogonal direction. An approximate analysis of the observed resonance line shape was given by Mikkor, based upon an energy-balance theory outlined by Hanamura *et al.*[101] Some difficulties were present because dc voltages around 100 V were used; however, the general idea that at resonance, energy can be fed in by an orthogonal (rf) field was well demonstrated. More recent experiments of this type will be discussed below.

The principal omissions or approximations of the energy balance theory can be listed as follows:

1. The theory is not a polaron theory. The carriers are treated as band electrons with fixed mass. A Boltzmann transport approximation is assumed valid as in semiconductors.

2. The parameter $\gamma = T_e/T_l$ characterizes the distribution in the presence of the high field in an unknown fashion. The actual distribution is almost certainly skewed somewhat and is not Maxwellian. In semiconductors, when only acoustic phonons act as an energy loss mechanisms, it has been estimated[102] that a minimum concentration of 5×10^{10} cm^{-3} would be required to justify the concept of an electron temperature for a lattice temperature of 2°K. This is very much larger than the maximum carrier concentrations employed in the transient photocharge experiments.

3. Optical phonons have been omitted from most of the numerical calculations. The only possible justification for their omission[93] was because the average rate of LO phonon emission is not quite competitive with acoustic scattering in the very-low-temperature region and possibly in the vicinity of, or below, the critical field. No sharp onset of optical phonon emission is observed in the current versus electric field curves. On the other hand, the estimated temperatures achieved at very high electric fields are thought to be unrealistic. It would appear that the acoustic modes are important at low or intermediate fields, but an additional mechanism for energy loss (optical modes?) is active at high field. Finally, streaming or anisotropy in the distribution function is neglected, which would further limit the validity of the theory to very low electric fields. One way around these difficulties would be to solve the Boltzmann transport equation for the actual distribution function.

The way to handle the Boltzmann equation in high electric fields has been pointed out Budd[103] for both spherical and ellipsoidal constant-energy surfaces. We outline his method for setting up the problem in the isotropic case. Assume that a distribution function $f(k)$ can be defined so that $f(k)$ is the number of electrons in the sample having momentum vector ending within an infinitesimal volume element of k space centered at the point $\mathbf{k}$. The time-independent Boltzmann equation in the presence of applied electric and magnetic fields can be written as

$$(e/\hbar)[\mathbf{E} + (1/e)(\mathbf{v} \times \mathbf{H})] \cdot \nabla_k f = \hat{C} f \tag{51}$$

Here, the net collision term $(\partial f/\partial t)_c$ is written as a collision operator $\hat{C}$ acting on f. More precisely, the rate of scattering into, and out of, the elemental volume in k space can be written as

$$(\partial f/\partial t)_c = (V/4\pi^3) \iiint P(\mathbf{k}', \mathbf{k}) f(\mathbf{k}')[1 - f(\mathbf{k})] - P(\mathbf{k}, \mathbf{k}') f(\mathbf{k})[1 - f(\mathbf{k}')]\, dk_x\, dk_y\, dk_z \tag{52}$$

where $f(k)[1 - f(k)]$ is equal to the probability that the initial state $\mathbf{k}$ is occupied and the final state $\mathbf{k}'$ unoccupied. The scattering probabilities $P(\mathbf{k}', \mathbf{k})$ and $P(\mathbf{k}', \mathbf{k})$ could in principle be properly evaluated from perturbation theory.[78] More frequently, a simple relaxation-time approximation is employed.

In the case of energy bands of standard shape, the carrier energy is given by

$$\varepsilon = \hbar^2 k^2/2m \tag{53}$$

and the carrier velocity by

$$\mathbf{v} = (1/\hbar)\, \nabla_k \varepsilon = \hbar \mathbf{k}/m \tag{54}$$

In order to solve the problem, one now separates f into two parts, an isotropic part $S(\varepsilon)$ depending only upon energy and an anisotropic term A, as follows:

$$f = S(\varepsilon) + A \tag{55}$$

The isotropic part S is not necessarily the equilibrium Maxwellian $f_0 = \varepsilon^{3/2} \times \exp(-\varepsilon/kT)$, but rather a new distribution appropriate to the high electric field. Equation (55) is really the first of two terms in an expansion of the distribution in Legendre polynominals as follows:

$$f = S_0(\varepsilon) + S_1 P_1(\cos\theta) + S_2 P_2(\theta) \tag{56}$$

where θ is the angle between the vector **k** and electric field **E**. Substituting Eq. (55) into (51), the following results are obtained:

$$(e\mathbf{E}/\hbar) \cdot \nabla_k S + \text{Aniso}\{e\mathbf{E} \cdot \nabla_k A\} + (e/mc)\mathbf{k} \times \mathbf{H} \cdot \nabla_k A = \hat{C}A \tag{57}$$

$$\text{Iso}\{e\mathbf{E} \cdot \nabla_k A\} = \hat{C}S \tag{58}$$

Here,

$$(\mathbf{v} \times \mathbf{H}) \cdot \nabla_k S = \hbar(dS/d\varepsilon)(\mathbf{v} \times \mathbf{H}) \cdot \mathbf{v} = 0$$

since it follows from Eq. (54) that $\nabla_k S = (\hbar\mathbf{k}/m)\, dS/d\varepsilon$. We designate the isotropic and anisotropic parts by "Iso" and "Aniso," respectively. We now assume, after Budd, that Aniso$\{e\mathbf{E} \cdot \nabla_k A\}$ is small and

$$\hat{C}A = -A/t(\varepsilon) \tag{59}$$

the latter being the relaxation-time approximation. If this approximation is not made, the collision integrals must be written out as in Eq. (52). Within the limits of a relaxation-time approximation, the solution for the anisotropic part of the distribution function is as follows:

$$A = -\tau \frac{dS}{d\varepsilon} \mathbf{k} \cdot \frac{(e\hbar/m)[\mathbf{E} + (e\tau/mc)\mathbf{E} \times \mathbf{H} + (e\tau^2/mc)(\mathbf{H} \cdot \mathbf{E})\mathbf{H}]}{1 + (e\tau H/mc)^2} \tag{60}$$

This can be recognized as similar to the usual isotropic solution. See, for example, Eq. (21), p. 330 of Ref. 3. Using Eq. (54) and the current density $\mathbf{j} = ne\mathbf{v}$, the current or corresponding conductivity coefficient can be obtained by integrating over **k** space as follows:

$$\mathbf{j} = (\hbar e/4\pi^3 m) \iiint \mathbf{k} f(\mathbf{k})\, dk_x\, dk_y\, dk_z \tag{61}$$

In the theory of Budd,[103] the isotropic part of the distribution function S can be found by substituting back into Eq. (58). In the Ohmic or low-electric-field region, S approximates the equilibrium distribution f_0 and the current density is given by

$$\mathbf{j} = \sigma(H)\mathbf{E} + \delta(H)\mathbf{E} \times \mathbf{H} + \gamma(H)(\mathbf{H} \cdot \mathbf{E})\mathbf{H} \tag{62}$$

Expressions for the conductivity coefficients σ, δ, and γ as appropriate averages are given in Ref. 3. An approach such as mentioned above, in which the actual S and A are found with proper and realistic account of the various scattering processes, would be an improvement over the simple

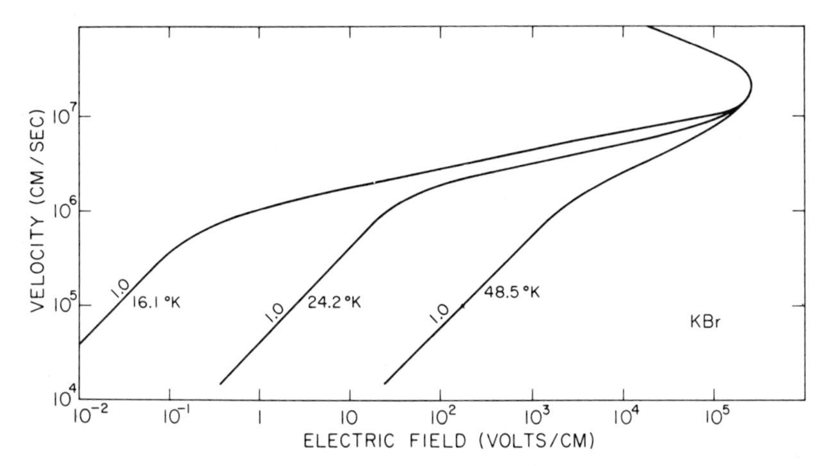

Fig. 14. The theoretical electron drift velocity versus electric field KBr at three different temperatures assuming only optical mode interaction. From the theory of Thornber and Feynman,[83] $\alpha = 3$, $\Theta = 242°K$, $m/m_e = 0.37$.

energy-balance theory. It is not clear, however, that the results will be drastically different than the simpler calculations used so far.

It was pointed out in Section 5 that the Boltzmann approach to transport may not be appropriate for polarons in the intermediate temperature range. The theory of Thornber and Feynman[83] gives the response of the polaron to arbitrary electric fields without using the assumptions inherent in the Boltzmann equation. On the other hand, certain approximations are involved, the principal ones being the inclusion of optical modes only (acoustic modes are ignored) and an approach which is equivalent to a spherical distribution function shifted in k space. It is interesting to see what this new theory predicts for current versus electric field in a real crystal. Figure 14 shows the results for KBr at three different lattice temperatures. It can be seen that an Ohmic region exists at very low field followed by a transition to high-field behavior rather similar to experiment as shown in Fig. 11. Notice that the transition field E_c depends upon the lattice temperature, as was actually observed by Masumi in AgCl.[89] On the other hand, for very low lattice temperatures, this theory predicts a transition from Ohmic to non-Ohmic behavior (10^{-2} V/cm at 11.2°K in KBr) which is very much lower than actually observed ($E_c = 20$–100 V/cm). Notice also that the electron velocities predicted at low lattice temperature are greater than the velocity of sound in KBr (3.8×10^5 cm/sec). This indicates that additional scattering mechanisms exist and that acoustic as well as optical modes must be taken into account, at least at very low temperature (10°K).

At extremely high electric field strength, a critical region (independent of temperature) appears above which the velocities can be maintained with even lower applied fields. That is, above this region, there is a decreasing rate of energy loss with velocity for very fast electrons. This behavior has not yet been seen in the transient photocharge experiments, although it may play a role in dielectric breakdown. Avalanche or pair-production processes are not included in the Thornber–Feynman work, although they must certainly become important at very high velocities in the region of instability. On the other hand, this new theory appears to be an important step forward, in that it is able to quantitatively account for the large rates of energy transfer to the lattice by optical phonon interaction.

Finally, let us return to the recent experimental results of J. W. Hodby, which indicate that both quantum-limit and spin-dependent scattering effects are present for electron transport in ionic crystals at low temperatures. These new discoveries have come about, at least in part, because of significant improvements in pulse averaging and detection.[61,104] It has become increasingly apparent that phenomena observed in the alkali halides in intense magnetic fields are very different from those observed in ordinary semiconductors. For one thing, no thermally generated carriers are present in these ionic crystals at liquid helium temperatures, so that carrier freezeout effects are absent. Impurity band conduction does not occur; furthermore, the number of carriers present per unit volume at any one time is always quite small, so that carrier degeneracy effects are entirely absent.

The first cross-modulation cyclotron resonance experiments on KBr[69] were interpreted in terms of carrier heating effects at resonance resulting in an increased rate of acoustic phonon emission and therefore a decrease in conductivity. A lineshape theory based upon the energy-balance concept with constant carrier mass was developed. Further experiments using the improved apparatus shown in Fig. 6 were subsequently carrier out on a variety of crystals[67] and it was discovered that the line shapes could not be entirely understood in terms of the simple theory. Subsequent work[70,105] indicated the importance of carrying out the measurements at as low electric fields as possible. It was found, for example, that at small dc fields, the maximum resonance effects corresponded to considerably lower effective polaron masses than those obtained at high electric fields. Apparently, both cold and hot polaron masses could be measured, depending upon the magnitude of the collecting voltage. At present, this is interpreted as due to the sweeping of carriers through a polaron band of the type predicted by Whitfield and Puff[106] and Larsen.[107,108] Figure 15 shows how the polaron

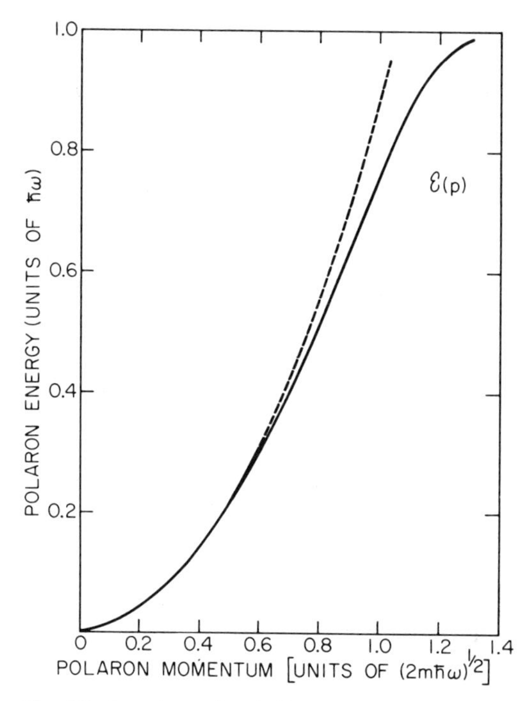

Fig. 15. A polaron band diagram (E versus k) for $\alpha = 1$ after the theory of Larsen.[107]

energy depends upon polaron momentum for $\alpha = 1$ according to Larsen.[107] Notice that considerable departure from a parabolic dependence occurs away from the minimum of the band but well before energies comparable to that of longitudinal optical phonons are reached. The departures are expected to be larger for greater α. It is not clear what happens when the carrier reaches the optical phonon energy $\hbar\omega$. The states above $\hbar\omega$ correspond to a polaron plus one or more real phonons and these one-phonon states must be taken into account in some experiments. On the other hand, in the cyclotron resonance experiments at 2 mm, it is likely that only states well below $\hbar\omega$ are involved. The cold polaron masses observed by Hodby[70,105] and listed in Table II are of the order of 10–20% less than the hot polaron masses previously reported.[67]

Quantum-limit effects were also observed in the cyclotron resonance electron heating experiments at very high magnetic fields and low temperature. The evidence for these interesting phenomena occurred under moderate electric fields, where the cyclotron resonance effect was observed to reverse sign.[70] Under certain conditions, application of rf power caused the apparent mobility of carriers excited from *F* centers in the alkali halides to increase rather than decrease as expected. Careful measurements of the current versus electric field curves at high longitudinal magnetic field suggested that these curves might be slightly concave upward rather than linear, even below the transition region. The effect is small and difficult to see by direct observation in data such as shown in Fig. 11, for example. Hodby[109] has been able to study this effect using an apparatus sensitive to the curvature of current versus electric field curve rather than directly to the observed photocharge. At 1.7°K, in the presence of fields 30 kG or above, the curvatures are positive. This is probably due to the separation of the conduction bands into Landau subbands. Electrons photoexcited into the conduction band are scattered mainly by *F* centers which resemble neutral donors. In very high magnetic fields at low temperature, sharp peaks occur in the conduction-band density of states at the minima of each Landau band. Thus, if at very low electric field the carriers occupy a distribution whose average energy is much less than the separation of a Landau levels, there will be many states into which an electron can be scattered by the *F* centers. The scattering will be very strong and the mobility will be limited and determined by the number of *F* centers present. As the electric field is increased, the average energy of the electrons will increase; the distribution may in fact encompass several Landau levels. The density of states into which the carriers can be scattered will be less than in the bottom of the subbands; consequently a higher mobility will result. This mechanism could conceivably explain the superlinear current–voltage characteristic observed in the presence of an *H* field well below the critical region (~100 V/cm).

As first discussed by Tippins[63] for ionic photoconductors, one expects to find no longitudinal magnetoconductivity (electric and magnetic fields parallel) for the case of spherical energy surfaces and isotropic scattering. Band studies[99] indicate that the alkali halides should fall in this category. Transverse magnetoconductivity experiments[98] in which the angle between magnetic field and crystal axis was continuously varied indicated that this spherical, isotropic band assumption applied. Yet, in Borders's experiments at high magnetic field,[95] a large longitudinal magnetoconductive increase appeared. This is now understood in terms of spin-dependent trapping of electrons by *F* centers. We thus have an experimental method for investigating the quantum mechanics of interaction between spin-polarized polarons and spin-polarized *F* centers. A similar problem occurs in semiconductors such as germanium or silicon at low temperature when carriers are scattered by neutral hydrogenlike impurities, for example, phosphorus in silicon. The effect of a magnetic field on such neutral impurity scattering in semiconductors has been described by Honig[110] and Maxwell and Honig.[111]

The experimental evidence for spin effects in the case of alkali halides containing F centers seems to be even clearer than in the semiconductor case. A simple model apparently applies in which the electrons are trapped by hydrogenlike atoms (F centers). During the interaction, the conduction electron and trapping (or scattering) electron are indistinguishable, so that antisymmetric wave functions must be formed from products of the spatial wave functions and appropriate spin functions. Scattering amplitudes corresponding to no interchange of the incident and scattering electrons as well as those arising from interchange appear. Furthermore, the ratio of singlet (antiparallel) to triplet (parallel) cross sections can be very much greater than one. A calculation of the total cross section proceeds as given by Schiff,[112] but includes the spin polarization of electrons and interacting centers. The result is[110]

$$\sigma_{\text{tot}} = \tfrac{1}{4}(1 - P^2)\sigma_s + \tfrac{1}{4}(3 + P^2)\sigma_t \qquad (63)$$

where the polarizations of conduction electrons and F-center electrons are assumed equal and are given by

$$P = (n_\uparrow - n_\downarrow)/(n_\uparrow + n_\downarrow) = \tanh(g\beta H/2kT) \qquad (64)$$

where β is the Bohr magniton. The F-center electrons are distinguishable one from another since they are on separate point defect sites. Therefore Boltzmann statistics apply and Eq. (64) is appropriate. In transient experiments, the conduction electrons are photoexcited from F centers, and it appears that they remember the spin of the center from which they are excited for times longer than their trapping or conduction times (nanoseconds). It is thus reasonable to assume equal polarizations $P_e = P_F = P$ given as a function of H/T by Eq. (64.).

In the alkali halide studies of Hodby,[98] about 10^{16} F centers per cm^3 were present and these were the most important scattering and trapping defects. A lower concentration N_0 of nonmagnetic centers also occurred. In order to take the nonmagnetic defects into account, Eq. (63) was multiplied by the number of F centers per unit volume N_F, and a term $N_0\sigma_0$, independent of polarization, was added.

The photocharge observed in transient experiments is proportional to reciprocal total cross section as follows:

$$Q(H) \propto (N_F{}^2\sigma_{\text{scat}}\sigma_{\text{trap}})^{-1} \qquad (65)$$

In this equation, σ_{scat} is a cross section for scattering of conduction electrons, such as might enter into the mobility relation, and σ_{trap} is the cross section for deep trapping of electrons. We know that F centers primarily determine the mobility in these crystals at low temperatures[55] and they also determine the ultimate fate of a photoexcited electron. The photoexcited electrons are trapped at F centers to form F' centers, that is, an F to F' conversion takes place. The conversion is incomplete and usually only a small concentration of F' centers builds up in the experiments at liquid helium temperature. Figure 16 and the arguments given below show that the spin polarization dependence cannot enter into both σ_{scat} and σ_{trap} of Eq. (65). In Ref. 98, the question of whether the magnetic effects are trap-controlled or due to scattering was left open, although preliminary evidence for trapping was cited. Further transport work by Hodby is in favor of trapping. Very recently, a study by Porret and Lüty[113] of F to F' conversion and luminescence under extreme values of H/T has appeared. In KCl at 1.2°K and 80 kG, the ratio $Q(H)/Q(0)$ was as high as 1.75. The increase can be understood from Eqs. (63) and (65) since for large H/T, and therefore polarization approaching one, σ_{tot} becomes approximately equal to σ_t, the triplet cross section, which is very much smaller than σ_s. The results for KCl as a function of polarization P are shown in Fig. 16. Here, the reciprocal $1/Q(H)$ is plotted as a function of P^2 with the crystal at three different temperatures. The results lie very close to the P^2 dependence.

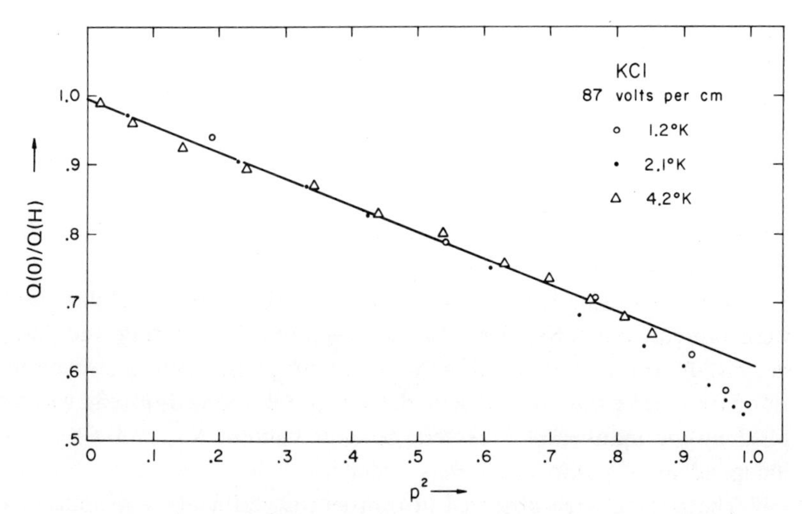

Fig. 16. Effect of a longitudinal magnetic field on $1/Q$ for a KCl crystal containing 10^{16} F centers/cm³. The solid line is drawn according to Eq. (66). Circles, 1.21°K; dots, 2.16°K; triangles, 4.2°K. From Hodby *et al.*[98]

If both σ_{scat} and σ_{trap} were polarization-dependent, a plot more nearly P^4 would be expected. Substituting Eq. (63) for σ_{trap} in Eq. (65), the ratio $Q(0)/Q(H)$ should be of the form

$$Q(0)/Q(H) = 1 - C \tanh^2(gBH/2kT) \tag{66}$$

where C is a constant independent of magnetic field and Eq. (64) has been used for the polarization.

A lower limit for the ratio of singlet to triplet F-center trapping cross section was found to be 4.4, whereas the upper limit is of the order of $\sigma_s/\sigma_t \approx 38$. No detailed theories of the trapping (or scattering) of polarons by F centers have been given. On the other hand, the present experiments as well as similar studies carried out with two-quantum excitation[97] yield detailed and surprisingly well-defined information on such processes at low temperature. Experiments also indicate that even at these low temperatures, transport occurs by a more or less conventional process involving diffusive motion of conduction electrons between trapping events. The mean trapping time $\langle\tau_{\text{tr}}\rangle$ appears to be greater than the mean scattering time $\langle\tau_{\text{sc}}\rangle$. That is, the spin-dependent results are not consistent with an electron mobility limited or determined by the trapping time.[104]

That spin-polarized F centers are responsible for the observed longitudinal magnetoconductivity can be seen by the results of an experiment carried out in the presence of a small amount of microwave power at 138 GHz. This frequency and power were such as to fully saturate the spin resonance of F centers at a value of H very close to 50 kG. A plot of the observed photoresponse (boxcar-averaged) in the presence of a magnetic field is shown as a function of H in Fig. 17.[104] The increasing response above about 30 kG is the spin-controlled magnetoconductive effect which we have been talking about. The effect of reducing the F-center polarization to nearly zero can be clearly seen superimposed upon this graph at just the right place, $g = 2.00$. This experiment confirms the above explanation of a polarization-dependent conductivity and incidentally serves as another independent means for observing electron spin resonance.

A number of interesting auxiliary experiments involving these extreme low-T, high-H phenomena have been carried out. For example, relaxation times can be estimated by changing the magnetic field rapidly to a new value. In some cases, the contact hyperfine interactions between F centers and the nearby nuclei can be investigated. Such effects have recently been revealed both in transport studies and also through the observation of the Faraday rotation pattern for F centers at extreme H/T.[114] When the magnetic

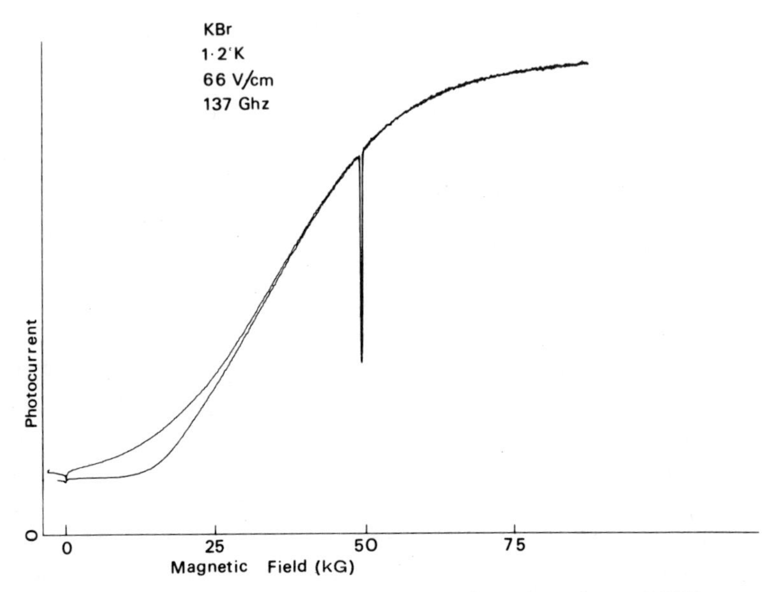

Fig. 17. A recorder trace [proportional to transient photocharge $Q(H)$] versus a longitudinal magnetic field for a KBr crystal containing $5 \times 10^{15} F$ centers/cm^3 at 4.2°K. A small amount of 138-GHz microwave power was present and the effect of a saturated F-center resonance can be seen very close to 50 kG.

field is decreased rapidly from a high value, the ordering of F centers present at high field decays very slowly in times, of the order of minutes, characteristic of a long spin–lattice relaxation time. Because of the hyperfine coupling to the nuclei, however, the polarization of the F-center electrons can be reversed many times by weak magnetic fields of the order of ± 100 G. In other words, in a sufficiently high magnetic field, electron and nuclear states are effectively uncoupled from each other and can be specified by their individual spin quantum numbers m_s and m_i, i.e., by $\langle m_s, m_i |$. On the other hand, at low or zero magnetic fields, the electron and nuclear spin states are coupled by the hyperfine interactions. Consequently states such as $\langle 1/2, -1/2 |$ are mixed with states $\langle -1/2, +1/2 |$. The electron spin projection of such a mixed state depends strongly upon the magnitude of applied fields. These methods for studying hyperfine interaction, although not universally applicable, can sometimes be used where electron spin resonance or ENDOR experiments are difficult to perform. Inhomogeneous broadening and materials with loss are two examples.

The low-temperature mobilities observed in alkali halide crystals containing F centers suggest that an F center is a much more effective scattering center than a simple hydrogenlike impurity.[55] Furthermore, as mentioned above, the scattering probabilities are independent of spin polarization. Perhaps this can be explained by noting that exchange effects are short-range, and important in trapping ($F \rightarrow F'$ conversion), whereas the scattering interaction is of longer range. Preliminary calculations by S. Bolte and A. B. Kunz indicate that lattice polarization must be taken into account when electrons are scattered by F centers. A polaron theory of scattering is called for.

Finally, in this review, we have not discussed the accumulating evidence for polaron phenomena in the infrared optical response of solids. Free polaron effects have been invoked by Baer[115] to explain the observed infrared absorption in barium titanate, and significant results of similar type have been obtained on a number of related materials. Bound polarons have also recently been studied in the silver halides in both zero and applied magnetic field.[116] They may also enter into the optical and electrical properties of cadmium fluoride.[117–119] Recently, polaron effects have turned out to be essential for understanding the high-field cyclotron resonance of carriers in materials such as cadmium telluride.[120] Perhaps it is best to leave these matters for a later discussion. It would appear that we are on the verge of discovery of a variety of specifically polaron effects important for a wide range of ionic solids.

ACKNOWLEDGMENTS

The author is grateful to Drs. K. K. Thornber, J. W. Hodby, D. Matz, and J. Devreese for numerous discussions and for conveying recent results prior to publication. He also appreciates the assistance of many other persons, especially J. Wei, S. Bolte, A. B. Kunz, A. M. Goodman, C. H. Seager, and J. A. Borders. Finally, the work carried out at the University of Illinois has had the support of the U. S. Army Research Office (Durham) and the Advanced Research Projects Agency. An appointment to the University of Illinois Center for Advanced Study is also gratefully acknowledged.

REFERENCES

1. H. Fröhlich, *Advan. Phys.* **3**, 325 (1954).
2. S. Pekar, *Untersuchungen über die Electronentheorie der Kristalle* (Akademie-Verlag, Berlin, 1954). L. D. Landau and S. I. Pekar, *Zh. Eksperim. i Teor. Fiz.* **18**, 419 (1948).
3. C. G. Kuper and G. D. Whitfield (Eds.), *Polarons and Excitons* (Oliver and Boyd, Edinburgh, 1963).

3a. F. C. Brown, *Ibid.*, p. 323.
4. J. Appel, in *Solid State Physics*, Vol. 21, Ed. by F. Seitz, D. Turnbull, and H. Ehrenreich (Academic Press, New York, 1968).
5. L. D. Landau, *Phys. Z. Sowjet.* **3**, 664 (1933).
6. N. F. Mott, *Proc. Phys. Soc. London* (extra part) **49**, 3 (1937).
7. H. Fröhlich, H. Pelzer, and S. Zienau, *Phil. Mag.* **41**, 221 (1950).
8. H. Fröhlich, in Ref. 3, Chapter 1.
9. C. Kittel, *Quantum Theory of Solids* (Wiley, New York, 1963).
10. W. B. Fowler, *Phys. Rev.* **135**, A1725 (1967).
11. A. H. Kahn, *Phys. Rev.* **172**, 813 (1968).
12. P. M. Platzman, *Phys. Rev.* **125**, 1961 (1962).
13. D. M. Larsen, *Phys. Rev.* **187**, 1147 (1969).
14. A. M. Stoneham, *J. Phys. C, Solid State Physics* **3**, L131 (1970).
15. D. M. Larsen, *Phys. Rev.* **180**, 919 (1969); **135**, A419 (1964).
16. K. K. Bajaj, *Phys. Rev.* **170**, 694 (1968).
17. T. Lee, F. Low, and D. Pines, *Phys. Rev.* **90**, 297 (1953).
18. D. M. Larsen, *Phys. Rev.* **174**, 1046 (1968).
19. J. Devreese and R. Evrard, *Phys. Stat. Sol.* **3**, 2133 (1963); *Physics Letters* **11**, 278 (1966); E. Kartheuser, R. Evrard, and J. Devreese, *Phys. Rev. Letters* **22**, 94 (1969).
20. R. P. Feynman, *Phys. Rev.* **97**, 660 (1955).
21. M. Porsch, *Phys. Stat. Sol.* **39**, 477 (1970).
22. D. Matz and B. C. Burkey, *Phys. Rev.* **B3**, 3487 (1971); D. Matz, to be published.
23. D. C. Langreth, *Phys. Rev.* **159**, 717 (1967).
24. J. J. O'Dwyer, *Theory of Dielectric Breakdown of Solids* (Oxford University Press, 1964); J. J. O'Dwyer, *J. Appl. Phys.* **37**, 599 (1966); *J. Phys. Chem. Solids* **28**, 1137 (1967).
25. J. R. Hanscomb, *J. Appl. Phys.* **41**, 3597 (1970).
26. R. M. Handy, *J. Appl. Phys.* **37**, 4620 (1966); E. P. Savoy and D. E. Anderson, *J. Appl. Phys.* **38**, 3245 (1967).
27. A. M. Goodman, *J. Electrochem. Soc.* **115**, 276C (1968); *Phys. Rev.* **164**, 1145 (1967).
28. J. Dressner, *Phys. Rev.* **143**, 558 (1966); *J. Phys. Chem. Solids* **25**, 505 (1964).
29. J. R. Macdonald and J. E. Robinson, *Phys. Rev.* **95**, 44 (1954).
30. R. C. Hanson, *J. Phys. Chem.* **66**, 2376 (1962).
31. M. Onuki, *J. Phys. Soc. Japan* **16**, 981 (1961).
32. G. Jaffe, *Ann. Physik* **16**, 217 (1933).
33. J. R. Macdonald, *J. Chem. Phys.* **40**, 1 (1964).
34. J. R. Haynes and W. Schockley, Physical Society Bristol Conference Report, p. 151 (1948); see also *Solid Luminescent Materials* (*Cornell Conference*) (Wiley, New York, 1948), p. 430.
35. J. R. Haynes and W. Schockley, *Phys. Rev.* **82**, 935 (1951).
36. R. Hofstadter, *Nucleonics* **4** (4), 2 (1949); **4** (5), 29 (1949).
37. K. A. Yamakawa, *Phys. Rev.* **82**, 522 (1951).
38. F. C. Brown and J. C. Street, *Phys. Rev.* **84**, 1183 (1951); F. C. Brown, *Phys. Rev.* **97**, 355 (1955).
39. F. C. Brown and F. E. Dart, *Phys. Rev.* **108**, 281 (1957); F. C. Brown, *J. Phys. Chem. Solids* **4**, 206 (1958).

40. R. S. Van Heyningen, *Phys. Rev.* **128**, 2112 (1962).
41. R. K. Ahrenkiel, *Phys. Rev.* **180**, 859 (1969).
42. R. G. Kepler, *Phys. Rev.* **119**, 1226 (1960).
43. H. Hirth and U. Todheide-Haupt, *Phys. Stat. Sol.* **31**, 425 (1969).
44. N. F. Mott and R. W. Gurney, *Electronic Processes in Ionic Crystals* (Oxford University Press, 1948), p. 117–124; see also R. S. Van Heyningen and F. C. Brown, *Phys. Rev.* **111**, 462 (1958).
45. P. J. Van Heerden, *Phys. Rev.* **106**, 468 (1957); **108**, 230 (1957); S. M. Ryvikin, *Soviet Phys.—Tech. Phys.* **1**, 2580 (1956).
46. P. Mark and W. Helfrich, *J. Appl. Phys.* **33**, 205 (1961).
47. A. Many and G. Rakavy, *Phys. Rev.* **126**, 1980 (1962).
48. I. P. Batra and H. Seki, *Phys. Rev.* **B1**, 3409 (1970).
49. A. G. Redfield, *Phys. Rev.* **94**, 526 (1954); **94**, 537 (1954).
50. F. C. Brown, *Phys. Rev.* **92**, 502 (1953).
51. K. Kobayashi and F. C. Brown, *Phys. Rev.* **113**, 507 (1959); *J. Phys. Chem. Solids* **8**, 300 (1959).
52. T. Masumi, R. K. Ahrenkiel, and F. C. Brown, *Phys. Stat. Sol.* **11**, 163 (1965).
53. D. C. Burnham, F. C. Brown, and R. S. Knox, *Phys. Rev.* **119**, 1560 (1960).
54. F. C. Brown and N. Inchauspe, *Phys. Rev.* **121**, 1303 (1961).
55. R. K. Ahrenkiel and F. C. Brown, *Phys. Rev.* **136**, A223 (1964).
56. T. Kawai, K. Kobayashi, and H. Fujita, *J. Phys. Soc. Japan* **21**, 453 (1966).
57. H. Fujita, K. Kobayashi, and T. Kawai, and K. Shiga, *J. Phys. Soc. Japan* **20**, 109 (1965).
58. F. Nakazawa and H. Kanzaki, *J. Phys. Soc. Japan* **27**, 1185 (1969).
59. G. Jacobs, private communications.
60. C. H. Seager and D. Emin, *Phys. Rev.* **B2**, 3421 (1970); C. H. Seager, *Phys. Rev.* **B3**, 3479 (1971).
61. J. A. Borders and J. W. Hodby, *Rev. Sci. Instr.* **39**, 722 (1968).
62. G. C. Smith, *Rev. Sci. Instr.* **40**, 1454 (1969).
63. H. H. Tippins and F. C. Brown, *Phys. Rev.* **129**, 2554 (1963).
64. D. N. Lyon and T. H. Geballe, *Rev. Sci. Instr.* **21**, 769 (1950).
65. G. Delacote and M. Schott, *Solid State Comm.* **4**, 177 (1966).
66. J. Wei, private communication.
67. J. W. Hodby and J. A. Borders, F. C. Brown, and S. Foner, *Phys. Rev. Letters* **19**, 952 (1967).
68. J. Waldman, D. M. Larsen, P. E. Tannenwald, C. C. Bradley, D. R. Cohn, and B. Lax, *Phys. Rev. Letters* **23**, 1033 (1969).
69. M. Mikkor, K. Kanazawa, and F. C. Brown, *Phys. Rev.* **162**, 848 (1967).
70. J. W. Hodby, *Solid State Comm.* **7**, 811 (1969).
71. G. Ascarelli and F. C. Brown, *Phys. Rev. Letters* **9**, 209 (1962).
72. D. J. Howarth and E. H. Sondheimer, *Proc. Roy. Soc.* **A219**, 53 (1953).
73. R. L. Petritz and W. W. Scanlon, *Phys. Rev.* **97**, 1620 (1958).
74. F. Low and D. Pines, *Phys. Rev.* **98**, 414 (1958).
75. T. Schultz, *Phys. Rev.* **116**, 529 (1959).
76. R. P. Feynman, R. W. Hellwerth, C. K. Iddings, and P. M. Platzman, *Phys. Rev.* **127**, 1004 (1962).
77. D. C. Langreth and L. P. Kadanoff, *Phys. Rev.* **133**, A1070 (1964).
78. D. C. Langreth, *Phys. Rev.* **137**, A760 (1965).

79. L. P. Kadanoff, *Phys. Rev.* **130**, 1364 (1963).
80. Y. Osaka, *Progr. Theoret. Phys.* (*Kyoto*) **22**, 437 (1959).
81. D. C. Langreth, *Phys. Rev.* **159**, 717 (1967).
82. T. Schultz, Thesis, Massachusetts Institute of Technology, 1956 (unpublished).
83. K. K. Thornber and R. P. Feynman, *Phys. Rev.* **B1**, 4099 (1970).
84. D. W. Berreman, *Phys. Rev.* **130**, 2193 (1963).
85. J. R. D. Copley, R. W. Macpherson, and T. Timusk, *Phys. Rev.* **182**, 965 (1969).
86. P. R. Vijayaraghavan, R. M. Nicklow, H. G. Smith, and M. K. Wilkinson, *Phys. Rev.* **B1**, 4819 (1970).
87. D. M. Larsen, *Phys. Rev.* **172**, 967 (1968).
88. K. K. Thornber, *Phys. Rev.* **B3**, 1929 (1971).
89. T. Masumi, *Phys. Rev.* **129**, 2564 (1963); *Phys. Rev.* **159**, 761 (1967).
90. H. H. Tippins and F. C. Brown, *Phys. Rev.* **129**, 2554 (1963).
91. W. Schockley, *Bell System Tech. J.* **30**, 991 (1951).
92. F. Seitz, *Phys. Rev.* **76**, 1376 (1949).
93. D. Matz and F. Garcia-Moliner, *Phys. Stat. Sol.* **5**, 495 (1964); *J. Phys. Chem. Solids* **26**, 551 (1965).
94. M. Mikkor and F. C. Brown, *Phys. Rev.* **162**, 841 (1967).
95. J. Borders, Thesis, University of Illinois, 1968 (unpublished).
96. F. Nakazawa and H. Kanzaki, *J. Phys. Soc. Japan* **27**, 1184 (1969); **22**, 844 (1967); **20**, 468 (1965).
97. S. Bolte and F. C. Brown, in *Proc. Third Int. Conf. on Photoconductivity* (*J. Phys. Chem. Solids Suppl.*), Ed. by E. M. Pell (Pergamon Press, New York, 1970).
98. J. W. Hodby, J. A. Borders, and F. C. Brown, *J. Phys. C, Solid State Phys.* **3**, 335 (1970).
99. A. B. Kunz, *Phys. Stat. Sol.* **29**, 115 (1968); *J. Phys. Chem. Solids* **31**, 265 (1970); *J. Phys. C, Solid State Phys.* **3**, 1542 (1970).
100. M. Mikkor, K. Kanazawa, and F. C. Brown, *Phys. Rev.* **162**, 848 (1967).
101. E. Hanamura, T. Inui, and Y. Toyozawa, *J. Phys. Soc. Japan* **17**, 666 (1963).
102. R. Stratton, *Proc. Roy. Soc.* **A242**, 355 (1957); **A246**, 406 (1958).
103. H. F. Budd, *Phys. Rev.* **131**, 1520 (1964); *Phys. Rev.* **140**, A2170 (1965).
104. J. W. Hodby, *J. Phys. E, Sci. Instr.* **2**, 796 (1969); *J. Phys. E. Sci. Instr.* **3**, 229 (1970).
105. J. H. Hodby, *J. Phys. C., Solid State Phys.* **4**, L9 (1971).
106. G. Whitfield and R. Puff, *Phys. Rev.* **139**, A338 (1965).
107. D. M. Larsen, *Phys. Rev.* **144**, 697 (1966).
108. B. Velicky, in *International School of Physics, Enrico Fermi*, Vol. XXXIV, Ed. by J. Tauc (Academic Press, New York, 1966), p. 379; E. Kartheuser, Ph.D. thesis, University of Liege, 1968 (unpublished).
109. J. W. Hodby, *Phys. Rev. Letters* **23**, 1235 (1969).
110. A. Honig, *Phys. Rev. Letters* **17**, 186 (1966).
111. R. Maxwell and A. Honig, *Phys. Rev. Letters* **17**, 188 (1966).
112. L. I. Schiff, *Quantum Mechanics*, 3rd Ed. (McGraw-Hill, New York, 1968), p. 393.
113. F. Porret and F. Lüty, *Phys. Rev. Letters* **26**, 843 (1971).
114. J. W. Hodby, *J. Phys. C* **3**, 592 (1970).
115. W. S. Baer, *Phys. Rev.* **144**, 734 (1966); *Phys. Rev.* **154**, 785 (1967).
116. R. C. Brandt and F. C. Brown, *Phys. Rev.* **181**, 1241 (1969); R. C. Brandt, D. M. Larsen, P. P. Crooker, and G. B. Wright, *Phys. Rev. Letters* **23**, 240 (1969).
117. R. P. Khosla, *Phys. Rev.* **183**, 827 (1970).

118. F. Moser, D. Matz, and S. Lyu, *Phys. Rev.* **182**, 808 (1968).
119. K. K. Bajaj and T. D. Clark, *Solid State Comm.* **8**, 825 (1970).
120. J. Waldman, D. M. Larsen, P. E. Tannenwald, C. C. Bradley, D. R. Cohn, and B. Lax, *Phys. Rev. Letters*, **23**, 1033 (1969).
121. E. Burstein, in *Lattice Dynamics, Proc. Int. Conf., Copenhagen, August 1963* (Pergamon Press, New York, 1964).
122. R. P. Lowndes, *Physics Letters* **21**, 26 (1966); R. P. Lowndes and D. H. Martin, *Proc. Roy. Soc.* (*London*) **A316**, 351 (1970).
123. K. Højendahl, *K. dansk vidensk. Selsk.* **16**, No. 2 (1938).
124. R. P. Lowndes and D. H. Martin, *Proc. Roy. Soc.* **A308**, 473 (1969).
125. R. C. Brandt and F. C. Brown, *Phys. Rev.* **181**, 1241 (1969).
126. E. R. Cowley and A. Okazaki, *Proc. Roy. Soc.* (*London*) **A300**, 45 (1967).
127. R. Z. Bachrach, *Solid State Comm.* **7**, 1023 (1969).
128. G. O. Jones, D. H. Martin, P. A. Mawer, and C. H. Perry, *Proc. Roy. Soc.* **A261**, 10 (1961).
129. J. T. Jenkin, J. W. Hodly, and U. Gross, *J. Phys. C.* (*Solid State Phys.*) **4**, L89 (1971).
130. J. E. Baxter, G. Ascarelli, and S. Rodriguez, *Phys. Rev. Letters* **27**, 100 (1971).
131. H. Tamura, T. Masumi, and K. Kobayashi, in *Proc. Third Int. Conf. on Photoconductivity* (*J. Phys. Chem. Solids Suppl.*), Ed. by E. M. Pell (Pergamon Press, New York, 1970).
132. A. P. Marchetti and G. L. Bottger, *Phys. Rev.* **B3**, 2604 (1971); also G. L. Bottger, private communication.
133. G. L. Bottger and A. L. Geddes, *J. Chem. Phys.* **56** (1972).
134. *American Institute of Physics Handbook*, 2nd ed., McGraw-Hill, New York (1963), Vol. 6, p. 97.

INDEX